TEMPERATE FRUITS
Production, Processing, and Marketing

Innovations in Horticultural Science

TEMPERATE FRUITS
Production, Processing, and Marketing

Edited by
Debashis Mandal, PhD
Ursula Wermund, PhD
Lop Phavaphutanon, PhD
Regina Cronje, MSc

First edition published 2021

Apple Academic Press Inc.
1265 Goldenrod Circle, NE,
Palm Bay, FL 32905 USA

4164 Lakeshore Road, Burlington,
ON, L7L 1A4 Canada

CRC Press
6000 Broken Sound Parkway NW,
Suite 300, Boca Raton, FL 33487-2742 USA

2 Park Square, Milton Park,
Abingdon, Oxon, OX14 4RN UK

© 2021 Apple Academic Press, Inc.

Apple Academic Press exclusively co-publishes with CRC Press, an imprint of Taylor & Francis Group, LLC

Reasonable efforts have been made to publish reliable data and information, but the authors, editors, and publisher cannot assume responsibility for the validity of all materials or the consequences of their use. The authors, editors, and publishers have attempted to trace the copyright holders of all material reproduced in this publication and apologize to copyright holders if permission to publish in this form has not been obtained. If any copyright material has not been acknowledged, please write and let us know so we may rectify in any future reprint.

Except as permitted under U.S. Copyright Law, no part of this book may be reprinted, reproduced, transmitted, or utilized in any form by any electronic, mechanical, or other means, now known or hereafter invented, including photocopying, microfilming, and recording, or in any information storage or retrieval system, without written permission from the publishers.

For permission to photocopy or use material electronically from this work, access www.copyright.com or contact the Copyright Clearance Center, Inc. (CCC), 222 Rosewood Drive, Danvers, MA 01923, 978-750-8400. For works that are not available on CCC please contact mpkbookspermissions@tandf.co.uk

Trademark notice: Product or corporate names may be trademarks or registered trademarks and are used only for identification and explanation without intent to infringe.

Library and Archives Canada Cataloguing in Publication

Title: Temperate fruits : production, processing, and marketing / edited by Debashis Mandal, PhD, Ursula Wermund, PhD, Lop Phavaphutanon, PhD, Regina Cronje, MSc.

Other titles: Temperate fruits (Palm Bay, Fla.)

Names: Mandal, Debashis, editor. | Wermund, Ursula, editor. | Phavaphutanon, Lop, editor. | Cronje, R. (Regina), editor.

Series: Innovations in horticultural science.

Description: Series statement: Innovations in horticultural science | Includes bibliographical references and index.

Identifiers: Canadiana (print) 20200329391 | Canadiana (ebook) 20200329472 | ISBN 9781771889193 (hardcover) | ISBN 9781003045861 (ebook)

Subjects: LCSH: Fruit-culture. | LCSH: Fruit.

Classification: LCC SB359 .T46 2021 | DDC 634/.04—dc23

Library of Congress Cataloging-in-Publication Data

Names: Mandal, Debashis, editor. | Wermund, Ursula, editor. | Phavaphutanon, Lop, editor. | Cronje, R. (Regina), editor.

Title: Temperate fruits : production, processing, and marketing / edited by Debashis Mandal, Ursula Wermund, Lop Phavaphutanon, Regina Cronje.

Other titles: Innovations in horticultural science.

Description: 1st edition. | Palm Bay, FL, USA : Apple Academic Press, 2021. | Series: Innovations in horticultural science | Includes bibliographical references and index. | Summary: "This volume, Temperate Fruits: Production, Processing, and Marketing, presents the latest pomological research on the production, postharvest handling, processing and storage, and information on marketing for a selection of temperate fruits. These include apple, pear, quince, peach, plum, sweet cherry, kiwifruit, strawberry, mulberry, and chestnut. With chapters from fruit experts from different countries of the world, the book provides the latest information on the effect of climate change on fruit production, organic fruit growing and advanced fruit breeding, the nutraceutical value and bioactive compounds in fruits and their role in human health, and new and advanced methods of fruit production. Topics include microirrigation, sustainable nutrient management, crop protection and plant heath management, and farm mechanization. The volume considers crop diversity, species variability and conservation strategies, production technology, plant architecture management, plant propagation and nutrition management, organic farming, dynamics in breeding and marketing of fruit crops, postharvest management and processed food production of fruit crops, and crop protection and plant health management. The book looks at the advancements in agro-techniques, timely harvests, and proper postharvest handling and care that have paved the way for enhanced market share of fruit crops. It also considers the extreme challenges of climate vagaries, erratic rainfall, drought, rapid urbanization, particularly in South Asia, that have affected the major fruit-producing belt, as well as other challenges, such as land degradation, long-term use of inorganic inputs, land fertility depletion, vulnerabilities of pest and diseases and more. This volume provides a wealth of diversified and contemporary information on temperate fruit and will be valuable for those involved in research and industry in temperate fruit production, processing, and marketing"-- Provided by publisher.

Identifiers: LCCN 2020040581 (print) | LCCN 2020040582 (ebook) | ISBN 9781771889193 (hardcover) | ISBN 9781003045861 (ebook)

Subjects: LCSH: Fruit-culture. | Fruit.

Classification: LCC SB359 .T392 2021 (print) | LCC SB359 (ebook) | DDC 634--dc23

LC record available at https://lccn.loc.gov/2020040581

LC ebook record available at https://lccn.loc.gov/2020040582

ISBN: 978-1-77188-919-3 (hbk)
ISBN: 978-1-00304-586-1 (ebk)

INNOVATIONS IN HORTICULTURAL SCIENCE

Editor-in-Chief:

Dr. Mohammed Wasim Siddiqui Assistant Professor-cum- Scientist
Bihar Agricultural University | www.bausabour.ac.in
Department of Food Science and Post-Harvest Technology
Sabour | Bhagalpur | Bihar | P. O. Box 813210 | INDIA
Contacts: (91) 9835502897
Email: wasim_serene@yahoo.com | wasim@appleacademicpress.com

The horticulture sector is considered as the most dynamic and sustainable segment of agriculture all over the world. It covers pre- and postharvest management of a wide spectrum of crops, including fruits and nuts, vegetables (including potatoes), flowering and aromatic plants, tuber crops, mushrooms, spices, plantation crops, edible bamboos etc. Shifting food pattern in wake of increasing income and health awareness of the populace has transformed horticulture into a vibrant commercial venture for the farming community all over the world.

It is a well-established fact that horticulture is one of the best options for improving the productivity of land, ensuring nutritional security for mankind and for sustaining the livelihood of the farming community worldwide. The world's populace is projected to be 9 billion by the year 2030, and the largest increase will be confined to the developing countries, where chronic food shortages and malnutrition already persist. This projected increase of population will certainly reduce the per capita availability of natural resources and may hinder the equilibrium and sustainability of agricultural systems due to overexploitation of natural resources, which will ultimately lead to more poverty, starvation, malnutrition, and higher food prices. The judicious utilization of natural resources is thus needed and must be addressed immediately.

Climate change is emerging as a major threat to the agriculture throughout the world as well. Surface temperatures of the earth have risen significantly over the past century, and the impact is most significant on agriculture. The rise in temperature enhances the rate of respiration, reduces cropping periods, advances ripening, and hastens crop maturity, which adversely affects crop productivity. Several climatic extremes such as droughts, floods, tropical cyclones, heavy precipitation events, hot extremes, and heat waves cause a negative impact on agriculture and are mainly caused and triggered by climate change.

In order to optimize the use of resources, hi-tech interventions like precision farming, which comprises temporal and spatial management of resources in horticulture, is essentially required. Infusion of technology for an efficient utilization of resources is intended for deriving higher crop productivity per unit of inputs. This would be possible only through deployment of modern hi-tech applications and precision farming methods. For improvement in crop production and returns to farmers, these technologies have to be widely spread and adopted. Considering the above-mentioned challenges of horticulturist and their expected role in ensuring food and nutritional security to mankind, a compilation of hi-tech cultivation techniques and postharvest management of horticultural crops is needed.

This book series, Innovations in Horticultural Science, is designed to address the need for advance knowledge for horticulture researchers and students. Moreover, the major advancements and developments in this subject area to be covered in this series would be beneficial to mankind.

Topics of interest include:

1. Importance of horticultural crops for livelihood
2. Dynamics in sustainable horticulture production
3. Precision horticulture for sustainability
4. Protected horticulture for sustainability
5. Classification of fruit, vegetables, flowers, and other horticultural crops
6. Nursery and orchard management
7. Propagation of horticultural crops
8. Rootstocks in fruit and vegetable production
9. Growth and development of horticultural crops
10. Horticultural plant physiology
11. Role of plant growth regulator in horticultural production
12. Nutrient and irrigation management
13. Fertigation in fruit and vegetables crops
14. High-density planting of fruit crops
15. Training and pruning of plants
16. Pollination management in horticultural crops
17. Organic crop production
18. Pest management dynamics for sustainable horticulture
19. Physiological disorders and their management
20. Biotic and abiotic stress management of fruit crops
21. Postharvest management of horticultural crops
22. Marketing strategies for horticultural crops
23. Climate change and sustainable horticulture
24. Molecular markers in horticultural science
25. Conventional and modern breeding approaches for quality improvement
26. Mushroom, bamboo, spices, medicinal, and plantation crop production

BOOKS IN THE SERIES

- **Spices: Agrotechniques for Quality Produce**
  Amit Baran Sharangi, PhD, S. Datta, PhD, and Prahlad Deb, PhD

- **Sustainable Horticulture, Volume 1: Diversity, Production, and Crop Improvement**
  Editors: Debashis Mandal, PhD, Amritesh C. Shukla, PhD, and Mohammed Wasim Siddiqui, PhD

- **Sustainable Horticulture, Volume 2: Food, Health, and Nutrition**
  Editors: Debashis Mandal, PhD, Amritesh C. Shukla, PhD, and Mohammed Wasim Siddiqui, PhD

- **Underexploited Spice Crops: Present Status, Agrotechnology, and Future Research Directions**
  Amit Baran Sharangi, PhD, Pemba H. Bhutia, Akkabathula Chandini Raj, and Majjiga Sreenivas

- **The Vegetable Pathosystem: Ecology, Disease Mechanism, and Management**
  Editors: Mohammad Ansar, PhD, and Abhijeet Ghatak, PhD

- **Advances in Pest Management in Commercial Flowers**
  Editors: Suprakash Pal, PhD, and Akshay Kumar Chakravarthy, PhD

- **Diseases of Fruits and Vegetable Crops: Recent Management Approaches**
  Editors: Gireesh Chand, PhD, Md. Nadeem Akhtar, and Santosh Kumar

- **Management of Insect Pests in Vegetable Crops: Concepts and Approaches**
  Editors: Ramanuj Vishwakarma, PhD, and Ranjeet Kumar, PhD

- **Temperate Fruits: Production, Processing, and Marketing**
  Editors: Debashis Mandal, PhD, Ursula Wermund, PhD, Lop Phavaphutanon, PhD, and Regina Cronje

- **Diseases of Horticultural Crops: Diagnosis and Management, Volume 1: Fruit Crops**
  Editors: J. N. Srivastava, PhD, and A. K. Singh, PhD

- **Diseases of Horticultural Crops: Diagnosis and Management, Volume 2: Vegetable Crops**
  Editors: J. N. Srivastava, PhD, and A. K. Singh, PhD

- **Diseases of of Horticultural Crops: Diagnosis and Management, Volume 3: Ornamental Plants and Spice Crops**

  Editors: J. N. Srivastava, PhD, and A. K. Singh, PhD

- **Diseases of Horticultural Crops: Diagnosis and Management, Volume 4: Important Plantation Crops, Medicinal Crops, and Mushrooms**

  Editors: J. N. Srivastava, PhD, and A. K. Singh, PhD

ABOUT THE EDITORS

Debashis Mandal, PhD

Assistant Professor, Department of Horticulture, Aromatic & Medicinal Plants, Mizoram University, Aizawl, Mizoram, India.

Dr. Mandal is a young academician and research fellow working in sustainable hill farming for past 10 years. He did his PhD from BCKV, India, and was a postdoctoral project scientist in IIT, Kharagpur. He previously worked as an Assistant Professor at Sikkim University, India, and has published over 40 research papers and book chapters in reputed journals and books. He has also published six books from Apple Academic Press and American Academic Press, USA.

In addition, he is a member in a working group on Lychee and other Sapindaceae Crops, ISHS, Belgium, and is also a member in the ISHS section of tropical–subtropical fruits, organic horticulture, and commission on quality and postharvest horticulture. Currently, he is working as an Editor-in-Lead (Horticulture) for the *International Journal of Bio Resources & Stress Management (IJBSM)*. Dr. Mandal is also an Editorial Advisor in Horticulture Science to Cambridge Scholar Publishing, United Kingdom, and regular reviewer of journals such as *Fruits, HortScience, Acta Physiologica Plantarum, African Journal of Agricultural Research*.

Further, he is a Consultant Horticulturist in the Department of Horticulture & Agriculture (Research & Extension), Government of Mizoram, India, and Himadri Specialty Chemicals Ltd., and is also handling externally funded research projects. He was Convener for the International Symposium on Sustainable Horticulture, 2016, India, and Co-convener for the International Conference of Bio-Resource and Stress Management, 2017, Jaipur, India.

He was also a Session Moderator and Keynote Speaker at the ISHS Symposium on Litchi, India, in 2016, on postharvest technology in Vietnam, 2014, and at South Korea, 2017, and AFSA Conference in 2018, Cambodia.

He has visited countries including Thailand, China, Nepal, Bhutan, Vietnam, South Korea, South Africa, and Cambodia for professional meetings, seminars, and symposia. His thrust areas of research are organic horticulture, pomology, postharvest technology, plant nutrition, and microirrigation.

Ursula Wermund, PhD

Project Manager and R&D Coordinator Greenyard Fresh—Greenyard Fresh Trade International (UNIVEG) GmbH, Bremen, Germany

Dr. Wermund is an active and dynamic woman and has marked experience in professional-corporate management, particularly in line with postharvest handling and marketing of fresh fruits and vegetables apart from teaching assignments. She received her doctoral degree in Agricultural Science from Cranfield University, UK, and started her career as a Research Assistant at Writtle College, University of Essex, Chelmsford, UK. Later, she joined the prestigious Imperial College, Wye, UK, and became the Head of Post-Harvest Group. During that period, she was actively associated with teaching and research related to temperate fruits production and post-harvest management.

Subsequently, she started her corporate assignment as Head, Quality Management in Petter Vetter Group, GmbH, Kehl, Germany. Currently, she is a Project Manager and R&D Coordinator for UNIVEG Group (presently known as GREENYARD), Bremen, Germany. Her past 13 years of corporate affairs led her to deal with quality assurance and management at Surinamese Banana, Madagascar Litchi, Italian & Turkish Grapes, and Kenyan French Bean etc., in coordination with the German and European Fruit Trading Associate and Food Safety Working Group.

Thermal pest control and pesticide residue analysis has been added experience for her at working with UNIVEG. She has published 14 research papers in reputed international journals in addition to participating in international meetings, conferences, and symposiums in different foreign countries. Her key areas of work in horticulture are postharvest technology and packaging and marketing of fruits and vegetables.

Lop Phavaphutanon, PhD

Deputy Head and Chief of the Tropical Fruit Research and Development Center, Department of Horticulture, Faculty of Agriculture at Kamphaeng Sean, Kasetsart University, Thailand

Dr. Phavaphutanon is an academician with 30 years of teaching experience at Kasetsart University, Thailand, where he joined just after his MS. He received his doctoral degree from Texas A&M University, USA, and has worked on many tropical and subtropical fruits of Thailand. Currently, he is a Deputy Head, Department of Horticulture, Kasetsart University and is also working as a Chief of the Tropical Fruit Research and Development Centre at the University.

He has published more than 30 research papers, two book chapters, and 24 seminar papers during his active research career. Currently, he is handling three research projects related to aromatic coconut and pummelo fruit. His key areas of work in horticulture are physiology and nutrition of fruits and other horticultural crops.

Regina Cronje, MSc
Horticulturist, Agricultural Research Council—Institute for Tropical and Subtropical Crops, Nelspruit, South Africa

Mr. Cronje is an enthusiastic and extremely energetic researcher working at the Agricultural Research Council—Institute for Tropical and Subtropical Crops, Nelspruit, South Africa for the past 13 years after her MSc in Crop Science from the University of Hohenheim, Stuttgart, Germany. She is actively associated with the research of crop production technology of subtropical fruits and is currently focused on litchi and mandarin. She is serving on the Board of Directors of the South African Litchi Growers' Associate and an active member of the South African Society for Horticultural Science and International Society for Horticultural Science.

In addition, she is a reviewer of reputed journals *HortScience* and *Agricultural Science Journal*. She has published 49 research papers, 15 book chapters, and was a chief editor for book volume published as *Acta Hort.* (*Proceedings of 4th International Symposium on Lychee, Longan and Other Sapindaceae Fruits*). Besides, she was the recipient of the Lindsey Milne Industry Award for outstanding contribution to the South African Litchi Industry.

CONTENTS

Contributors .... *xv*

Abbreviations .... *xix*

Preface .... *xxiii*

1. **Apple** 1
   Graciela María Colavita, Mariela Curetti, Dolores Raffo, María Cristina Sosa, and Laura I. Vita

2. **Pear** 107
   Graciela María Colavita, Mariela Curetti, María Cristina Sosa, and Laura I. Vita

3. **Quince** 183
   Hamid Abdollahi

4. **Peach** 247
   Monika Gupta, Rachna Arora, and Debashis Mandal

5. **Plum** 297
   Lobsang Wangchu, Thejangulie Angami, and Debashis Mandal

6. **Sweet Cherry** 333
   Berta Gonçalves, Alfredo Aires, Ivo Oliveira, Sílvia Afonso, Maria Cristina Morais, Sofia Correia, Sandra Martins, and Ana Paula Silva

7. **Kiwifruit** 417
   Vishal S. Rana and Gitesh Kumar

8. **Strawberry** 449
   G. Quintero-Arias, J. Vargas, J. F. Acuña-Caita, and J. L. Valenzuela

9. **Mulberry** 491
   Jer-Chia Chang and Yi-Hsuan Hsu

10. **Chestnut** 537
    Gabriele L. Beccaro, Dario Donno, Michele Warmund, Feng Zou, Chiara Ferracini, Paolo Gonthier, and Maria Gabriella Mellano

Index *559*

CONTRIBUTORS

Hamid Abdollahi
Temperate Fruits Research Centre, Horticultural Sciences Research Institute, Agricultural Research, Education and Extension Organization (AREEO), Karaj, Iran

J. F. Acuña-Caita
Departamento de Ingeniería Civil y Agrícola Universidad Nacional de Colombia Bogotá, República de Colombia

Sílvia Afonso
Centre for the Research and Technology for Agro-Environmental and Biological Sciences, CITAB, Universidade de Trás-os-Montes e Alto Douro, UTAD, Quinta de Prados 5000-801, Vila Real, Portugal

Alfredo Aires
Centre for the Research and Technology for Agro-Environmental and Biological Sciences, CITAB, Universidade de Trás-os-Montes e Alto Douro, UTAD, Quinta de Prados 5000-801, Vila Real, Portugal

Thejangulie Angami
Department of Fruit Science, College of Horticulture and Forestry, CAU, Pasighat 791102, Arunachal Pradesh, India

G Quintero-Arias
Departamento de Ingeniería Civil y Agrícola Universidad Nacional de Colombia Bogotá, República de Colombia

Rachna Arora
Department of Fruit Science, Punjab Agricultural University, Ludhiana, India

Gabriele L Beccaro
Department of Agricultural, Forest and Food Sciences, University of Turin, Grugliasco, Turin, Italy
Chestnut R&D Centre, Cuneo, Italy

Jer-Chia Chang
Department of Horticulture, National Chung Hsing University 145, Taichung 40227, Taiwan, Republic of China

Graciela María Colavita
Plant Physiology, Comahue Institute of Agricultural Biotechnology, Faculty of Agricultural Sciences, Comahue National University, National Council for Science and Technology (UNCo-CONICET), Km 11, 5 Ruta 151, Cinco Saltos, Río Negro, Patagonia, Argentina

Sofia Correia
Centre for the Research and Technology for Agro-Environmental and Biological Sciences, CITAB, Universidade de Trás-os-Montes e Alto Douro, UTAD, Quinta de Prados 5000-801, Vila Real, Portugal

Mariela Curetti
Plant Nutrition, Horticultural Department, National Institute of Agricultural Technology (INTA-EEA Alto Valle), General Roca 8332, Argentina

Dario Donno
Department of Agricultural, Forest and Food Sciences, University of Turin, Grugliasco, Turin, Italy
Chestnut R&D Centre, Cuneo, Italy

Chiara Ferracini
Department of Agricultural, Forest and Food Sciences, University of Turin, Grugliasco, Turin, Italy

Berta Gonçalves
Centre for the Research and Technology for Agro-Environmental and Biological Sciences, CITAB, Universidade de Trás-os-Montes e Alto Douro, UTAD, Quinta de Prados 5000-801, Vila Real, Portugal
Department of Biology and Environment, Escola das Ciências da Vida e Ambiente, Universidade de Trás-os-Montes e Alto Douro, UTAD, Quinta de Prados 5000-801 Vila Real, Portugal

Paolo Gonthier
Department of Agricultural, Forest and Food Sciences, University of Turin, Grugliasco, Turin, Italy

Monika Gupta
Department of Fruit Science, Punjab Agricultural University, Ludhiana, India

Yi-Hsuan Hsu
Department of Horticulture, National Chung Hsing University 145, Taichung 40227, Taiwan, Republic of China

Gitesh Kumar
Department of Fruit Science, Dr YS Parmar University of Horticulture and Forestry-Nauni, Solan, Himachal Pradesh 173230, India

Debashis Mandal
Department of Horticulture, Aromatic & Medicinal Plants, Mizoram University, Aizawl 796004, Mizoram, India

Sandra Martins
Centre for the Research and Technology for Agro-Environmental and Biological Sciences, CITAB, Universidade de Trás-os-Montes e Alto Douro, UTAD, Quinta de Prados 5000-801, Vila Real, Portugal

Maria Gabriella Mellano
Department of Agricultural, Forest and Food Sciences, University of Turin, Grugliasco, Turin, Italy
Chestnut R&D Centre, Cuneo, Italy

Maria Cristina Morais
Centre for the Research and Technology for Agro-Environmental and Biological Sciences, CITAB, Universidade de Trás-os-Montes e Alto Douro, UTAD, Quinta de Prados 5000-801, Vila Real, Portugal

Ivo Oliveira
Centre for the Research and Technology for Agro-Environmental and Biological Sciences, CITAB, Universidade de Trás-os-Montes e Alto Douro, UTAD, Quinta de Prados 5000-801, Vila Real, Portugal

Dolores Raffo
Fruit Production, Horticultural Department, National Institute of Agricultural Technology (INTA-EEA Alto Valle), General Roca 8332, Argentina

Vishal S. Rana
Department of Fruit Science, Dr YS Parmar University of Horticulture and Forestry-Nauni, Solan, Himachal Pradesh 173230, India

Ana Paula Silva
Centre for the Research and Technology for Agro-Environmental and Biological Sciences, CITAB, Universidade de Trás-os-Montes e Alto Douro, UTAD, Quinta de Prados 5000-801, Vila Real, Portugal
Department of Agronomy, Universidade de Trás-os-Montes e Alto Douro, UTAD, Quinta de Prados 5001-801, Vila Real, Portugal

Contributors xvii

María Cristina Sosa
Fitopatology, Comahue Institute of Agricultural Biotechnology, Faculty of Agricultural Sciences, Comahue National University, National Council for Science and Technology (UNCo-CONICET), Cinco Saltos, Río Negro, Patagonia, Argentina

J. L. Valenzuela
Departament of Biology and Geology, Campus of International Excellence (ceiA3), CIAIMBITAL, Universidad de Almería, 04120 Almería, Spain

J. Vargas
Departamento de Ingeniería Civil y Agrícola Universidad Nacional de Colombia Bogotá, República de Colombia

Laura I. Vita
Plant Physiology, Comahue Institute of Agricultural Biotechnology, Faculty of Agricultural Sciences, Comahue National University, National Council for Science and Technology (UNCo-CONICET), Km 11, 5 Ruta 151, Cinco Saltos, Río Negro, Patagonia, Argentina

Michele Warmund
Division of Plant Sciences, University of Missouri, Columbia, SC, USA

Lobsang Wangchu
Department of Fruit Science, College of Horticulture and Forestry, CAU, Pasighat 791102, Arunachal Pradesh, India

Feng Zou
Key Laboratory of Cultivation and Protection for Non-wood Forest trees, Central South University of Forestry and Technology, Changsha, People's Republic of China

ABBREVIATIONS

β-Gal	β-galactosidase
1-MCP	1-methylcyclopropene
2iPA	N6-Δ2-isopentenyl adenine
6-BA	6-benzyladenine
ABB	abscisic acid
ACC	1-aminocyclopropane-1-carboxylic acid
ACO	1-aminocyclopropane-1-carboxylic acid oxidase
ACS	1-aminocyclopropane-1-carboxylate synthase
ASGV	apple stem growing virus
ASPV	apple stem pitting virus
AUND	Apple Union Necrosis and Decline
AVG	aminoethoxyvinylglycine
BA	benzyladenine
C/N	carbon-nitrogen ratio
CA	controlled atmosphere
CAJ	concentrated apple juice
Cas9	CRISPR-associated protein 9
CI	chilling injury
COX	cyclooxygenase enzymes
CpGV	cydia pomonella granulo virus
CRISPR	clustered regularly interspaced short palindromic repeats
CRP	C-reactive protein
DAFF	Department of Agriculture, Forestry and Fisheries
DCA	dynamic controlled atmosphere
DNJ	1-deoxynojirimycin
DNOC	dinitro-*o*-cresylate
EC	electrical conductivity
EP	ethylene production
EPP	effective pollination period
EPPO	European and Mediterranean Plant Protection Organization
FAO	Food and Agriculture Organization
FAOSTAT	Food and Agriculture Organization Corporate Statistical Database
FeEDDHA	Fe-ethylene diamine(di-*O*-hydroxyphenyl)acetate

FFV	fresh fruit and vegetables
FYM	farm yard manure
GA	gibberellins
GAIN	Global Agriculture Information Network
GDH	growth degree hours
GRIN	Germplasm Resources Information Network
IBA	indole-3-butyric acid
IEC	internal ethylene concentration
ILOS	initial low-oxygen stress
IPM	integrated pest management
IR	infrared radiation
IU	international unit
KGB	Kym Green Bush
LOL	lower oxygen limit
LTP	lipid transfer protein
MAP	modified atmosphere packaging
msl	mean sea level
NAA	naphthalene acetic acid
NAAm	naphthaleneacetamide
NAGREF-PI	National Agricultural Research Foundation, Pomology Institute
NIR	near infrared
NRCS	Natural Resources Conservation Service
NUTTAB	NUTrient TABles for use in Australia
PAR	photosynthetically active radiation
PCA	principal component analysis
PE	polyethylene
PG	polygalacturonase
PME	pectinmethylesterase
PPI	producers' price index
PPV	plum pox virus
PRD	partial root-zone drying
PS	polystyrene
PVC	polyvinyl chloride
QTLs	quantitative trait loci
RAE	retinol activity equivalents
RAW	readily available water
RDI	regulated deficit irrigation
RH	relative humidity

SAM	S-adenosyl-l-methionine
SB	Spanish Bush
SDV	stem diameter variation
SIR	sterile insect release
SL	steep leader
SSA	super slender axe
SSC	soluble solid concentration
SSR	simple sequence repeat
TA	titratable acidity
TAC	total acid content
TmRSV	tomato ringspot virus
TSA	tall spindle axe
TSS	total soluble solids
UC Davis	University of California, Davis
UFO	Upright Fruiting Offshoots
ULO	ultralow oxygen
UPOV	International Union for the Protection of New Varieties of Plants
USDA	United States Department of Agriculture
USDA-ARS	United States Department of Agriculture: Agriculture Research Service
VC	vegetative compatibility
VCL	Vogel Central Leader
VIS	visible infrared spectroscopy
VPD	vapor pressure deficit
WAPA	World Apple and Pear Association
WSU	Washington State University

PREFACE

Fruits are the most delicious and attractive horticultural crops and are rich in vitamins, antioxidants, and other nutraceuticals much needed for a healthy diet. Because of this reason, fruit is popularly found in almost every fresh food basket as well as consumed in processed form of juice, nectar or RTS, etc. Past decades have witnessed a notable increase in the production of fruit crops. However, it is worth mentioning that advancement in agro-techniques, timely harvest, and proper post-harvest handling, and care have paved the way for enhanced market share of fruit crops. Though, extreme challenges of climate vagaries, erratic rainfall, drought, rapid urbanization in particular to South Asia, major fruit producing belt land degradation, long-term use of inorganic inputs, land fertility depletion, vulnerabilities of pest and diseases had put forth, and are still presenting threats to producing quality fruits in requisite quantities. Even though production has increased, but it is still not in parity with the increasing demand of global food quantity; further or consistent improvement is needed.

Global fruit production has been benefited because of latest techniques for land–water–fertility management, biotechnology, and molecular biology, advanced irrigation, safe pest and disease management, organic and biodynamic farming, mechanized harvesting, etc. Advancement in fruit processing and post-harvest handling and storage has reduced the loss and has facilitated the production of better processed product(s) favored by a wider population. Thus, this has caused a significant impact on marketing of temperate fruits. China, India, along with other Asiatic giants, America, and Latin American countries along with some European countries, like Spain and Italy—particularly for temperate fruits—are building the strong fruit production network.

In this context, the present book volume on *Temperate Fruits: Production, Processing and Marketing*, is a compendium of pomological research for a wide range of temperate fruits, namely, Apple, Pear, Quince, Peach, Plum, Cherries, Kiwifruit, Strawberries, Mulberries, and Chestnut, incorporating the latest research output on production, post-harvest handling, processing, storage and marketing.

— **Debashis Mandal**
Ursula Wermund
Lop Phavaphutanon
Regina Cronje

CHAPTER 1

APPLE

GRACIELA MARÍA COLAVITA[1*], MARIELA CURETTI[2],
DOLORES RAFFO[3], MARÍA CRISTINA SOSA[4], and LAURA I. VITA[1]

[1]*Plant Physiology, Comahue Institute of Agricultural Biotechnology, Faculty of Agricultural Sciences, Comahue National University, National Council for Science and Technology (UNCo-CONICET), Road 151, Km 12.5, Cinco Saltos, 8303 Río Negro, Patagonia, Argentina*

[2]*Plant Nutrition, Horticultural Department, National Institute of Agricultural Technology (INTA-E.E.A Alto Valle), General Roca 8332, Argentina*

[3]*Fruit Production, Horticultural Department, National Institute of Agricultural Technology (INTA-E.E.A Alto Valle), General Roca 8332, Argentina*

[4]*Plant Pathology, Comahue Institute of Agricultural Biotechnology, Faculty of Agricultural Sciences, Comahue National University, National Council for Science and Technology (UNCo-CONICET), Cinco Saltos, Río Negro, Patagonia, Argentina.*

Corresponding author. E-mail: graciela.colavita@faca.uncoma.edu.ar; gmcolavita@gmail.com

ABSTRACT

The genus *Malus* originated in Central and Minor Asia. There is evidence that apple was cultivated as early as 1000 BC. Today, about 72 million tons of quality fruit are produced annually worldwide across approximately 5 million hectares.

The reasons for this success are that apple trees grow in different agro-ecological conditions and respond to the application of technological tools that increase yields. Apple fruits are pleasing to the eye and to the taste, they provide good nutrients with low calories, and they adapt well to conservation. Its particular suitability for transport has made the apple one of the most accomplished examples of globalization of markets. This chapter covers such topics as botany, taxonomy, varieties and cultivars, rootstocks, composition, and nutritional use of the apple. There is updated information on basic aspects of breeding and crop improvement, orchard management, harvest, postharvest, high-tech cultivation, transport, and packing. Topics such as disease, pest, and physiological disorders are also taken into account.

1.1 GENERAL INTRODUCTION

Wild apple (*Malus* sp.) archeological records have been dated to prehistory. There is general agreement that apple originated in Central and Minor Asia, and the species were distributed worldwide, mostly in temperate regions. The genus *Malus* is a member of the Rosaceae family. Apple tree cultivation is assumed to have started in Greek times. Of the 7500 types of apples that exist in the world today, only 10 cultivars account for 90% of the commercial production. Apple production is a multi-billion-dollar industry; exports earn over 7.5 billion dollars globally. World apple production is approximately 72 million tons across approximately 5 million hectares. Different types of apples

have been developed in order to produce fruits with different tastes and textures. Such varieties allow apples to be eaten fresh, cooked, dried, or to be used in juice and cider production. The nutritional benefits of apples are well known. Apples are rich in flavonoids, pectin, vitamins, potassium, and fiber, and low in sodium, calories, and fat. Apples show enormous genetic variation, so breeders are constantly researching to obtain new cultivars and rootstocks. Breeders' programs work on genes related to tree vigor control, precocity of flowering, aspects related to fruit quality and storage life, and resistance to many diseases and pests that have already been identified. Apples are cultivated in temperate regions, but they can also be produced across a diversity of climates and soils. The amount of water needed at an apple orchard depends on the climate, soil characteristics, planting system, irrigation technique adopted according to the environmental conditions, and the management and productive level of the orchard. The success of apple culture and production depends on the selection of appropriate scion and rootstock, the local agro-ecosystem, and the cultural practices. Apple trees are grown in rectangular plots and the training systems are conditioned by the scion–rootstock selected. The shape and size of the tree are obtained and maintained by pruning. Apple trees set more fruit than needed for good quality production, so fruit thinning must be done to decrease limb breakage, stimulate floral initiation for the next productive cycle, increase fruit size, and improve fruit color. Apple fruits allow long-term storage, so less mature fruits can be marketed after a cold-storage period, whereas ripe fruits are sold immediately after harvest.

1.2 AREA AND PRODUCTION

The world apple production has increased by about 155% over the past 50 years, increasing from 1.9 million to 5 million hectares in 2014 (www.fao.org/faostat/en/#data). The average production volume of apple fruits was 25.4 million tons between 1965 and 1974, which rose to 84.6 million tons in 2014, representing an increase of more than 200% (Figure 1.1).

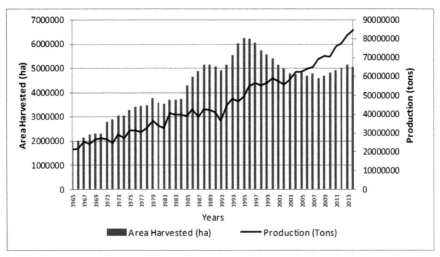

FIGURE 1.1 Annual evolution of the area harvested and the production of apple crops in the world (1965–2014).
Source: FAO (www.fao.org/faostat/en/#data).

The comparison of the area and production data makes it possible to consider the evolution over time and the yield of apple crops. Table 1.1. shows the world average values for each period of 10 years for three indicators: area, production, and yield of apple crops, as well as the percentage increase, calculated using the values for 1965–1974 as reference. The largest percentage increase was observed in production (184.5%), in relation to the area and the progressive increase in the yields. About 72 million tons of quality apple fruits are produced annually worldwide across approximately 5 million hectares.

TABLE 1.1 Average Values of Production, Worldwide Area, and Yield of Apple Crops for Periods of 10 years (Percentual Increase from 1965–1974 to 2005–2014)

Period	1965–1974	1975–1984	1985–1994	1995–2004	2005–2014	Percentage Increase
Area harvested (ha)	2,482,934	3,569,333	5,092,717	5,500,128	4,857,879	155
Production (tons)	25,392,973	34,879,330	30,844,167	56,596,046	72,249,175	184
Yield (kg/ha)	101,730.8	97,505.8	83,159.5	104,336.1	148,503.9	45

Data source: FAO (www.fao.org/faostat/en/#data)

All continents produce apples, but in the last 20 years, Asia has stood out as the largest producer in the world, expanding its apple production from 29.2 million tons to 44.4 million tons (2005–2014) (Figure 1.2).

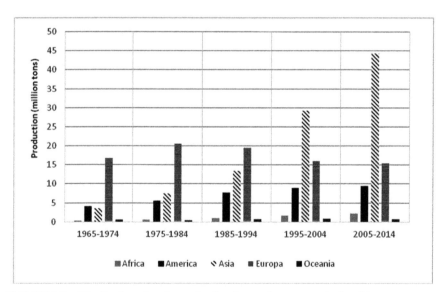

FIGURE 1.2 Distribution of apple production for continents over the last 50 years. *Source:* FAO (www.fao.org/faostat/en/#data).

The first position of Asia in the world is heavily influenced by China, currently the largest producer of apples. About 43.5 million tons come from this country, which is more than 50% of the global

Apple 5

production (Table 1.2). China's apple crops are expected to increase according to a recent United States Department of Agriculture (USDA) Global Agriculture Information Network (GAIN) report on fresh deciduous fruits. The United States is the second largest producer of apples with over 4 million tons, followed by Turkey, Poland, and Italy (Table 1.2).

TABLE 1.2 Leading Apple-producing Countries in 2016, in Tons

Country	Production	Country	Production
China	43,500,000	Chile	1,770,000
United States	4,600,000	France	1,528,000
Turkey	2,700,000	Iran	1,340,000
Poland	2,500,000	Russia	1,300,000
India	2,164,000	Argentina	983,000
Italy	1,911,000	Brazil	980,000

Sources: The World Apple and Pear Association (http://www.wapa-association.org/asp/index.asp) and United States Department of Agriculture, Fresh Deciduous Fruit: World Markets and Trade (Apples, Grapes, and Pears) (https://apps.fas.usda.gov/psdonline/circulars/fruit.pdf) (June 2017).

PolandApples (http://www.polandapples.com) (October 2017).

In the Southern Hemisphere, the largest producing countries are Brazil, Chile, Argentina, South Africa, New Zealand, and Australia (FAO, 2014; Jackson, 2003).

1.3 MARKETING AND TRADE

World apple production in marketing year 2016/17 was approximately 72.8 million tons (United States Department of Agriculture, 2017). China's entry into the apple market has had a crucial effect on the world trade in terms of this fruit since the production of this country in the early 1990s was almost the same as it currently exports. It should be expected that China's production will continue to rise because it continues to plant apples. Apple exports value over 7.5 billion dollars globally. The United States is at the top among the 10 countries that export the apple fruit with worth well over 1 billion dollars. In addition to the United States and China, other leading exporter countries of the Northern Hemisphere are Poland, Italy, France, Holland, and Belgium. In the Southern Hemisphere, Chile, South Africa, New Zealand, and, to a lesser extent, Argentina and Brazil, also stand out (Figure 1.3).

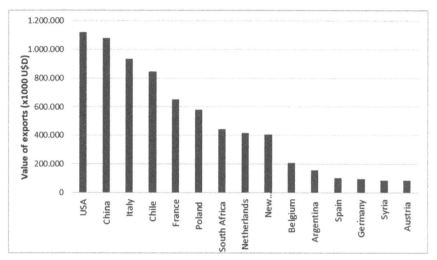

FIGURE 1.3 Value of exports from different countries (×1000 USD).
Source: FAO (www.fao.org/faostat/en/#data).

Russia is the main importer of apples. This country annually imports approximately 800 million dollars of apples, followed by countries such as Germany, the United Kingdom, the Netherlands, and China (Figure 1.4). In these countries, the domestic demand for apple is largely boosted by an increasing consciousness about the benefits of eating healthy food and by a growing purchasing power.

Apple demand is projected to continue expanding in India because of the strong economic growth of this country (Deodhar et al., 2006).

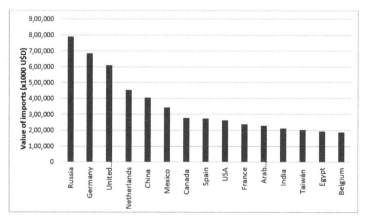

FIGURE 1.4 Value of imports from different countries (1000 USD).
Source: FAO (www.fao.org/faostat/en/#data).

Russell et al. (1992) suggested that apple price is generated by the fruit size and quality, storage method, and seasonality. Nowadays, new apple varieties are being patented and trademarked in order to be produced and marketed by a limited number of growers. The growers that produce these varieties must pay certain royalties, and these varieties are denominated "club varieties." There is a growing interest in club varieties; although club member volume is still small in comparison with the other traditional varieties, the price is much more attractive. These club varieties are driven by large supermarket chains that need to have different products on their shelves. Therefore, for the producer to compete with the traditional varieties, the fruit offered must be of superior quality.

The demand for organic apples has increased in the last decade. For example, the production of organic apples in Washington State (United States) in 2003 was 4047 ha, representing more than a tenfold increase since 1989 (Peck et al., 2005).

1.4 COMPOSITION AND USES

1.4.1 COMPOSITION OF APPLES

Information on the basic nutritional content of the apple (lipids, vitamins, sugars, and minerals) has been studied in traditional cultivars, but knowledge about metabolites has been analyzed in modern cultivars, especially those that are important in international trade (Simmonds and Howes, 2015). Consequently, data are available on the nutritional composition and energetic value of the main commercial cultivars and varieties (USDA-National Nutrient Database for Standard Reference, 2017) (Table 1.3).

TABLE 1.3 Nutrient Content Values for the Edible Portion of Main Traditional Cultivars

Nutrient Unit	1 Value/100 g/Cultivar Apple					
	Apple*	Granny Smith	Red Delicious	Golden Delicious	Gala	Fuji
Water (g)	85.56	85	85	85.5	85	84.16
Energy (kcal)	52	58	59	57	57	63
Protein (g)	0.26	0.44	0.27	0.28	0.25	0.20
Total lipid (fat) (g)	0.17	0.19	0.20	0.15	0.12	0.18
Carbohydrate, by difference (g)	13.81	13.29	14.06	13.6	13.68	15.22
Fiber, total dietary (g)	2.4	2.8	2.3	2.4	2.3	2.1
Sugars, total (g)	10.39	9.59	14.06,	10.04	10.37	11.68

8   Temperate Fruits: Production, Processing, and Marketing

TABLE 1.3   *(Continued)*

Nutrient Unit	1 Value/100 g/Cultivar Apple					
	Apple*	Granny Smith	Red Delicious	Golden Delicious	Gala	Fuji
Minerals						
Calcium, Ca (mg)	6	5	6	6	7	7
Iron, Fe (mg)	0.12	0.15	0.11	0.13	0.12	0.10
Magnesium, Mg (mg)	5	5	5	5	5	5
Phosphorus, P (mg)	11	12	12	10	11	13
Potassium, K (mg)	107	120	104	100	108	109
Sodium, Na (mg)	1	1	1	2	1	1
Zinc, Zn (mg)	0.04	0.04	0.04	0.04	0.05	0.04
Vitamins						
Vitamin C, total ascorbic acid (mg)	4.6					
Thiamin (mg)	0.017	0.019	0.015	0.018	0.017	0.013
Riboflavin (mg)	0.026	0.025	0.025	0.026	0.025	0.026
Niacin (mg)	0.091	0.126	0.075	0.094	0.075	0.070
Vitamin B-6 (mg)	0.041	0.037	0.035	0.051	0.049	0.045
Folate, DFE (µg)	3					
Vitamin A, RAE (µg)	3	5	3	3	1	2
Vitamin A, IU IU	54	100	55	51	28	38
Vitamin E (alpha-tocopherol) (mg)	0.18	0.18	0.24	0.18	0.18	0.18
Vitamin K (phylloquinone) (µg)	2.2	3.2	2.6	1.8	1.3	1.1
Lipids						
Fatty acids, total saturated (g)	0.028	0.18				
Fatty acids, total monounsaturated (g)	0.035	3.2				
Fatty acids, total polyunsaturated (g)	0.051					

*Based on analytical data for var. *Red Delicious*, *Golden Delicious* and *Granny Smith*, and cv. Gala and Fuji. Samples of 3- and 5-pound bags of apples typically contain small and extra small sizes.

Source: USDA database. National Nutrient Database for Standard Reference Release 28 and slightly revised in May 2016. Date available and last updated by the company: July 14, 2017.

The apple is considered an important food source due to its composition and the biological activities of its compounds. It is low in calories (50–60 calories per 100 g in a fresh fruit) and contains no saturated fats (Table 1.3).

1.4.2 USES OF APPLES

1.4.2.1 MARKET USES

Most apples produced worldwide are mainly marketed for domestic fresh use. In some apple producer regions without a fruit-processing industry, apples that cannot be marketed as fresh products are used for animal feed or wasted. Few countries, such as the United States, Germany, and Australia have developed an important market for processed apple products. Since 1960, the development of techniques to obtain concentrated apple juice (CAJ) at a ratio of 6:1 and the location of juice processing plants near the production regions have made it possible to supply CAJ at low cost around the world. Concentrated juice has become the primary use of apple processing, and its prices affect the prices of other processing uses (O´Rourke, 2003).

1.4.2.2 NUTRITIONAL/ NEUTRACEUTICAL USES

The apple fruit has dietary fiber that decreases the absorption of low-density lipoproteins and saves the mucous membrane of the colon from exposure to toxic substances that cause cancer. The fruit contains vitamin C (powerful natural antioxidant), β-carotene, and vitamin B (Simmonds and Howes, 2015). These vitamins increase the resistance of the body against infectious agents and prevent oxidative stress.

Apples fruits have minerals such as potassium, phosphorus, calcium, and sodium in very low concentration (Table 1.3). The nutritional value of foods depends on their chemical composition, which is related to their genetic potential and several aspects of production and processing (Kalinowska et al., 2014; Simmonds and Howes, 2015). In this sense, management of both orchard and postharvest affects the nutritional and organoleptic qualities of the fruit. Ecological agriculture improves the nutritional and organoleptic aspects and contributes to maintaining varietal diversity and producing healthy foods. Raigón et al. (2006) researched the nutritional value of the apple fruit both in organic and conventional production management and concluded that ecological apples have a higher amount of minerals, polyphenols, and sugars.

The phenolic compounds of the apple fruit can help in the prevention of chronic diseases. (Kalinowska, 2014; Markowski et al., 2015;

Simmonds and Howes, 2015; Persic et al., 2017). Numerous studies have evaluated the metabolite profile of polyphenols both in old commercial cultivars and in local or indigenous cultivars (Onivogui, 2014; Raigón et al., 2006; Wojdylo et al., 2008; Kalinowska et al., 2014; Persic et al., 2017). The concentration of individual phenolic compounds varies according to the maturity of the fruit, tissue analyzed (peel or flesh), environmental factors, time of harvest, storage conditions, infections suffered, and genetic variations. The phenolic content of fresh apple ranges from 110 to 357 mg 100 g^{-1}, and it is higher in the peel than in the flesh (Wolfe and Liu, 2003; Kalinowska et al., 2014).

1.4.2.3 PROCESSED PRODUCT USES

The principal by-product of the apple juice industry is pomace, a mixture of skin, pulp, and seeds (Vendruscolo et al., 2008). Apple pomace is used as animal feed and for the extraction of pectins and fibers (Bhushan et al., 2008; Vendruscolo et al., 2008; Miceli-García, 2014). In foods, pectins are used as gelling, thickener, texturizer, emulsifier, and stabilizer agents. In many applications, apple pectins show better gelling properties than citrus pectins.

Dietary fiber is defined as the plant components in the diet that cannot be digested by enzymes of the human digestive tract. Fiber is composed of complex polysaccharides, such as pectin, hemicelluloses, cellulose, and lignin, which are components of cell walls. Apple has a higher fiber content than the rest of the fruits (Trowell, 1976). The fibers can be extracted from by-products of the food industry—from apple pomace and depectinated apple pomace—purified, and added to processed foods (Renard and Thibault, 1991). In bakery products, Wang and Thomas (1989) showed that apple pomace is more desirable to consumers than apple fiber.

Apple pomace has been used as a substrate for producing citric acid through a fermentation system with *Aspergillus niger*. However, ethanol has been produced from fermentation processes of the pomace culture with yeast strains *Saccharomyces cerevisiae* and *Pichia fermentans*.

1.4.2.4 HEALTH AND ANTIOXIDANT RESEARCH

Fruits and vegetables are essential sources of a nondigestible components and phytochemicals that may act synergistically to offer health benefits (Magaia, 2013). As mentioned above, apple is one of the most popular and healthy fruits whose benefits include nutrition and health, and it prevents human illnesses. Regarding nutrition, apple fruits have low energetic value and contain vitamins and minerals, high content of pectin and fiber, high potassium, low sodium, and zero fat. The consumption of apples helps to avoid diarrhea and constipation, control headaches and nervous tension, and reduce colds and upper respiratory diseases. Apples are easily accepted and digested by babies and cause less colic than other fruits. Some governments promote apple consumption, not only because of its biological attributes but also because a fresh apple contributes to tooth cleaning. Recently discovered polyphenols and flavonoids are important to prevent cancer and reduce the risk of heart disease and stroke (Kalinowska et al., 2014; Simmonds and Howes, 2015). Apples are rich in antioxidant phytonutrients (vitamin C, flavonoids, and polyphenolics), with 5900 antioxidant strength (ORAC value) in every 100 g apple fruit. Additionally, apples are also rich in tartaric acid, which gives a tart flavor to them.

1.5 ORIGIN AND DISTRIBUTION

Apple originated in Central and Minor Asia, the Caucasus, the Himalayan, and western China. The Old Silk Road (an ancient network of trade routes connecting Asia and Europe) was important in the evolution of cultivated apple. Travelers on foot and pack-horse trains must have started taking this route in the Bronze Age (Luby et al., 2001). Roman authors (1000 BC) documented that selected cultivars from random hybridizations were disseminated by grafting. In fact, there is evidence that apple was cultivated in Israel as early as 800 BC. In addition, Greek horticulturalists were familiar with apple-cultivation techniques, as shown by Theophrastus (c. 371–c. 287 BC), who wrote about the art of grafting apple trees (Juniper et al., 1998).

An apple contains 15 indigenous, "primary" species. Most of them are self-incompatible, containing $2n = 2x = 34$ chromosomes, which can be crossed with one another, and their interspecific hybrids are fertile or at least partially fertile. However, some cultivars and wild forms are triploid ($2n = 3x = 51$ chromosomes), and a few are tetraploid ($2n = 4x = 68$ chromosomes) (Zohary et al., 2012). Cultivars such as Jonagold and Mutsu are triploids and are prized for their fruit size (Brown, 2012).

The domesticated apple varieties derive from *Malus domestica* Borkh or *Malus pumila* Mill (Zohary et al., 2012). There are currently around 7500 types of apples in the world. Different types of apples are developed in order to produce fruits with different tastes and textures. These differences allow apples to be used for different purposes, such as to be eaten raw, cooked, dried, or to produce juice and cider. Intensive apple production is found in highly concentrated regional clusters. More than half of China's apples are harvested in Shandong, Shaanxi, Henan, Shanxi, Hebei, and Gansu (Gale, 2011). In some countries, the production clusters are explained by climatic conditions, such as in Maule/Chile and Rio Negro/Argentina (Figure 1.5). Italian apple production serves to illustrate another important cluster, concentrated in the Alpine region of Trentino and Alto Adige. These regions are characterized by cool highland climate with high sunshine, favorable for apple quality with regard to color, texture, and the balance of sugars and acidity. Statistics show that the share of the total national apple production of this region increased from 50% in 2000 to more than 70% in 2010, while overall production in Italy remained stable. Climate is not the only reason for forming apple clusters. Other reasons are traditions and local knowledge, tourism, easy access to advisory services, skilled labor and specialized input and technology supply, and distance to markets, storage, grading, and processing facilities. While in some countries, a single region may account for more than half of the national apple production, for example, Argentina, Chile, South Africa, Italy, and the United States, in other countries, more than one major apple cluster may exist (Figure 1.5). In Poland, nearly 50% of apples are grown in Mazowiecki, while a second important cluster is found in Lubelskie, with 17% of the total harvest. In France, in addition to the Provence, Alpes, Côtes d'Azur region, an important share of apples comes from the Northern Pays de la Loire. In Turkey, apple production extends to several provinces, including Isparta, Denizli, Antalya, and Karaman.

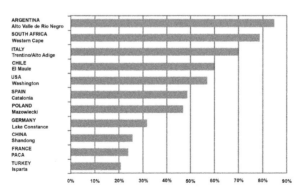

FIGURE 1.5 Share of national apple production in major clusters.
Source: Reprinted with permission from Garming (2015) (http://www.agribenchmark.org/agri-benchmark/did-you-know/einzelansicht/artikel//a-quarter-of.html).

Apple trees cover 450,000 ha in the European Union (EU) (Figure 1.6), with Poland being the largest producer of apples, followed by Italy, Romania, France, Germany, Spain, and Hungary. Together, these seven countries cover more than 80% of the total area of the EU dedicated to apple production (Eurostat, 2012).

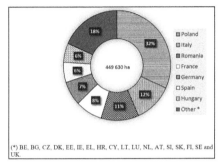

(*) BE, BG, CZ, DK, EE, IE, EL, HR, CY, LT, LU, NL, AT, SI, SK, FI, SE and UK.

FIGURE 1.6 Most important EU countries in terms of area for apple production, 2012. Source: Eurostat (online data code: orch_apples2).

1.6 TAXONOMY AND BOTANY

1.6.1 TAXONOMY

The apple that we identify today as *Malus domestica* is a descendant of *Malus sieversii*. There is agreement that *Malus sieversii* (Lodeb.) M. Roem. is the progenitor species of apple hybridizing with *M. prunifolia* Borkh, *M. baccata* Borkh, and *M. sieboldii* (eastern species) and with *M. turkmenorum* and *M. sylvestris* Mill (Western species). Western European cultivars were cut off from their parents, and they evolved in relative isolation (Ferree and Warrington, 2003).

Current apple varieties and cultivars are part of the *Malus domestica* species. The interspecific apple hybrid complex is usually designated as *Malus × domestica* Borkh or *Malus domestica* Borkh (Brown, 2012). The domesticated apple is a member of the family Rosaceae, subfamily Maloideae, and genus *Malus*. The genus *Malus* includes ~55 species, although 79 species have been recognized (Table 1.4), which are grouped into infrageneric groups (section and series), and each species can be divided into intraspecific groups (cultivar). In accordance with the requirements of the 2012 edition of the International Code of Nomenclature for Algae, Mushrooms and Plants (USDA-ARS, 2012; USDA-NRCS, 2012), there was a nomenclatural change that includes the Maloideae in a new subfamily, Amygdaloideae. Apples and pears belong to the same subfamily, both having a haploid number (x) of 17 chromosomes ($x = 17$).

Taxonomic position according to USDA-ARS (2012) and USDA-NRCS (2012):

Kingdom: Plantae (plants)

Subkingdom: Tracheobionta (vascular plants)

Superdivision: Spermatophyta (seed plants)

Division: Magnoliophyta (flowering plants)
Class: Magnoliopsida (dicotyledons)
Subclass: Rosidae
Order: Rosales
Family: Rosaceae (rose family)

Subfamily: Amygdaloideae
Tribe: Maleae
Subtribe: Malinae
Genus: *Malus* Species: *Malus domestica* Borkh.

TABLE 1.4 Species and Hybrid Species Currently Recognized in the Genus *Malus*, According to the Taxonomy Database of the US Department of Agriculture Germplasm Resources Information Network-GRIN) (USDA-ARS 2012)

S. No.	Latin Name	Common Name
1.	*Malus* × *adstringens* Zabel	
2.	*Malus angustifolia* (Aiton) Michx.	Southern crabapple
3.	*Malus* × *arnoldiana* (Rehder) Sarg. ex Rehder	
4.	*Malus* × *asiatica* Nakai	
5.	*Malus* × *astracanica* hort. exDum. Cours.	
6.	*Malus* × *atrosanguinea* (hort. ex Spath) C. K. Schneid.	
7.	*Malus baccata* (L.) Borkh	Siberian crabapple
8.	*Malus baoshanensis* G. T. Deng	
9.	*Malus brevipes* (Rehder) Rehder	
10.	*Malus chitralensis* Vassilcz.	
11.	*Malus coronaria* (L.) Mill.	Sweet crabapple
12.	*Malus crescimannoi* Raimondo	
13.	*Malus* × *dawsoniana* Rehder	
14.	*Malus domestica* Borkh	
15.	*Malus doumeri* (Bois) A. Chev.	
16.	*Malus florentina* (Zuccagni) C. K. Schneid.	Hawthorn-leaf crabapple
17.	*Malus floribunda* Siebold ex Van Houtte	Japanese flowering crabapple
18.	*Malus fusca* (Raf.) C. K. Schneid.	Oregon crabapple
19.	*Malus shalliana* Koehne	Hall crabapple
20.	*Malus* × *hartwigii* Koehne	
21.	*Malus honanensis* Rehder	
22.	*Malus hupehensis* (Pamp.) Rehder	Chinese crabapple
23.	*Malus ioensis* (Alph. Wood) Britton	Prairie crabapple
24.	*Malus kansuensis* (Batalin) C. K. Schneid.	

Apple 15

TABLE 1.4 *(Continued)*

S. No.	Latin Name	Common Name
25.	*Malus komarovii* (Sarg.) Rehder	
26.	*Malus leiocalyca* S. Z. Huang	
27.	*Malus maerkangensis* M. H. Cheng et al.	
28.	*Malus × magdeburgensis* Hartwig	
29.	*Malus mandshurica* (Maxim.) Kom. ex Skvortsov	Manchurian crabapple
30.	*Malus × micromalus* Makino	Kaido crabapple
31.	*Malus × moerlandsii* Door.	
32.	*Malus muliensis* T. C. Ku	
33.	*Malus ombrophila* Hand.-Mazz.	
34.	*Malus orientalis* Uglitzk.	
35.	*Malus orthocarpa* Lavallee ex anon.	
36.	*Malus × platycarpa* Rehder	Big fruit crabapple
37.	*Malus prattii* (Hemsl.) C. K. Schneid.	Pratt apple
38.	*Malus prunifolia* (Willd.) Borkh	Plum leaf crabapple
39.	*Malus pumila* Mill.	Paradise apple
40.	*Malus × purpurea* (A. Barbier) Rehder	
41.	*Malus × robusta* (Carriere) Rehder	Siberian crabapple
42.	*Malus sargentii* Rehder	Sargent's apple
43.	*Malus × scheideckeri* Spath ex Zabel	
44.	*Malus sieversii* (Ledeb.) M. Roem.	
45.	*Malus sikkimensis* (Wenz.) Koehne ex C. K. Schneid.	
46.	*Malus × soulardii* (L. H. Bailey) Britton	Soulard crabapple
47.	*Malus spectabilis* (Aiton) Borkh	Asiatic apple
48.	*Malus spontanea* (Makino) Makino	
49.	*Malus × sublobata* (Dippel) Rehder	
50.	*Malus sylvestris* (L.) Mill.	European crabapple
51.	*Malus toringo* (Siebold) de Vriese	Toringo crab
52.	*Malus toringoides* (Rehder) Hughes	Cutleaf crabapple
53.	*Malus transitoria* (Batalin) C. K. Schneid.	
54.	*Malus trilobata* (Poir.) C. K. Schneid.	
55.	*Malus tschonoskii* (Maxim.) C. K. Schneid.	Pillar apple
56.	*Malus yunnanensis* (Franch.) C. K. Schneid.	
57.	*Malus zhaojiaoensis* N. G. Jiang	
58.	*Malus zumi* (Matsum.) Rehder	O-zumi

1.6.2 BOTANICAL DESCRIPTION

Apple trees are deciduous, small- to medium-sized, and have a single trunk and highly branched canopy. Cultivated trees are 2.5–5.5 m high, depending on the rootstock and training system, whereas wild trees can reach up to the height of 9–10 m (Rieger, 2006). From an architectural point of view, the apple tree is characterized as follows: branch orthotropic with rhythmic growth; monopodial branching at least during the first years following germination, then sympodial when flowering becomes terminal on long shoots. The root system consists of a principal deep root with horizontal permanent scaffold roots with ramifications descending to the water table layer. Except for the roots of seedling rootstocks, all the roots of apple trees are adventitious in origin (Jackson, 2003). Under favorable conditions, the root system penetrates to a depth of approximately 1.5–2 m, and its diameter is slightly greater than that of the canopy.

Leaves are alternate, simple, ovate-elliptical, and 3–8 cm in length. They have a finely serrated margin, green and dark green color, with white pubescent. Flowers are usually terminal on spurs (although they can grow laterally from one-year-old wood), and they are grouped into 4–6 inflorescences called corymbs (Jackson, 2003; Rieger, 2006). Flowers are white to pink, have 5 petals, 5 sepals, 20 stamens in 3 whorls (10 + 5 + 5) with yellow anthers, and a pistil with 5 styles united at the base (Jackson, 2003). The pedicel and calyx are usually woolly, and the calyx is persistent in the fruit. The ovary is inferior, embedded in the floral cup, the central flower opening first (Photo 1.1). The central flower (called the "King blossom") has the capacity to produce a larger fruit than the other flowers of the corymb. Flowers are produced terminally from mixed buds (containing both leaves and flowers) and on spurs.

PHOTO 1.1 Apple inflorescence (note that central flower opens first). (Courtesy Graciela María Colavita)

Apple fruits and related fruits are classified as a special fruit type, called "pome." Pomes are shaped like ellipsoid–obovoid balloons, and their sizes vary from smaller to more than 7 cm diameter and 170–350 g in weight. The color depends on the cultivar, and it can be red, green, yellow, or bicolor, displaying stripes or red blushes on a yellow or green background. The edible portion of the fruit is derived from the floral cup (hypanthium), not from the ovary as in most other fruits. The floral cup is formed with the basal portions of the calyx, the corolla, and the stamens. In apple fruits, the seeds are relatively small, black, and developed in five cavities with two seeds each. Fruiting begins 3–5 years after grafting. Depending on the cultivars, apples reach maturity in 80–180 days after bloom (Jackson, 2003; Rieger, 2006). The bearing habit of apple is based on the location and type of the buds that produce flowers and fruits. Flower buds are borne laterally on shoots or spurs, and the terminal bud grows at the tip of a branch. Flower buds are larger and plumper than growth buds, with a downy surface. Bark is variable, generally smooth when young, and later becomes scaly. The lateral branches (scaffolds) are the main supporting branches of the tree. Twigs are moderate in thickness, brownish gray, with white pubescens. Growth buds are smaller and more pointed than flower buds, and they grow flush with the branch. Bourse shoots are vegetative short shoots that arise from beneath, where a flower bud emerges. Water

sprouts grow vigorously throughout the grow season, resulting in excessive upright growth. These very long, upright branches create shade within the tree and are unproductive. The spurs are fruiting branches, with the appearance of small and stubby compressed stems. Spurs are shoots that usually grow very slowly, often less than 1.5 cm per year, and they typically terminate in a flower bud containing numerous flowers. Depending on where the fruit bud is produced, apples and pears are divided in three groups: spur-bearers, tip-bearers, and partial tip-bearers. Spur-bearers are the largest group and they produce fruit buds on two-year-old wood, and as spurs (short, branched shoots) on the older wood. Tip-bearers produce fruit buds at the tips of long shoots generated the previous year. Partial tip-bearers are apple cultivars that produce fruit on the tips of the previous year's shoots.

1.7 VARIETIES AND CULTIVARS

Apple varieties suit nearly every taste and cultural demand. New varieties are usually obtained from mutations in known varieties, occurring mostly spontaneously, which improve the original variety. They are marketed through traditional commercial models. Another way to obtain a new variety is through directed crosses with differential and specific characters. These crosses require strong marketing programs throughout the chain. Sales impact is usually reduced at the beginning, and the commercial strategies are complex and specific with a marked tendency to restrictive systems of the club varieties type (Calvo et al., 2016).

There are hundreds of apple varieties and some of them have several strains, each with its own characteristics. Some of the most important varieties are shown in Photo 1.2.

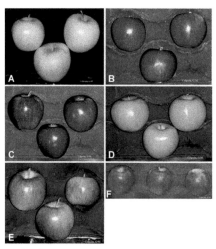

PHOTO 1.2 Apple varieties and cultivars: (A) Golden Delicious, (B) Gala, (C) Red Delicious, (D) Granny Smith, (E) Fuji, and (F) Cripps Pink. Courtesy G. M. Colavita.

1.7.1 GOLDEN DELICIOUS

Golden Delicious was discovered in the United States at the end of the 19th century. It is one of the most widespread varieties in the world (Benitez, 2001). Its skin is pale yellow, sometimes with a red blush, and a yellowish white flesh (Abbott et al., 2004; WAPA, 2017) (Photo 1.2A). This apple is firm at harvest but tends to soften during storage. The flavor is sweet, spicy, and moderately acidic (Abbott et al., 2004).

1.7.2 GALA

Gala was originated in New Zealand in 1939, by a crossing between Kidd's Orange Red and Golden Delicious (WAPA, 2017). Its background color is light green, but it becomes almost white when the optimum harvest time approaches and then yellower as the maturity progresses (Photo 1.2B). The intensity and percentage of the area covered in red depend on the clone (Benítez, 2001). In hot climates, Gala has a propensity not to develop red color and to drop prematurely (Layne et al., 2002). The flesh is fine-textured, yellowish, crispy, juicy, sweet-tasting, suitable for consumers who dislike strongly acidic apples (Benítez, 2001; WAPA, 2017).

1.7.3 RED DELICIOUS

Red Delicious originated in the 19th century is a sweet, crunchy and juicy apple that varies in color from red to solid red. Red Delicious apples have an elongated shape with five lobes (USApple, 2017). Coverage color is

Apple

more intense in sun-exposed fruits (Photo 1.2C). The pulp is yellowish white, finely grainy, juicy, perfumed, and sweet, with good flavor, and tending to lose firmness and flouriness under certain conditions. Production entry is quite late in the original standard clones. Different Delicious clone alternatives were disseminated: Starkrimson, Red Chief, Hi Early, Oregon Spur, and Red King Oregon (Benítez, 2001).

1.7.3.1 RED CHIEF

Red Chief is a Starkrimson mutation, originally from the United States. The fruit has the characteristics of Red Delicious. The skin is red, bright, and slightly striated. The pulp is fine, consistent, juicy, slightly acidulated, and it has good taste quality (Benítez, 2001).

1.7.3.2 TOP RED

Top Red is a Starking mutation, obtained in Wenatchee, the United States. It is a medium- to large-sized fruit, and its skin is red with striations on a strong pink background. Its pulp is firm, and it has good taste quality (Benítez, 2001).

1.7.4 GRANNY SMITH

Granny Smith is an Australian variety discovered in 1868 as a chance seedling by Mary Ann Smith of Ryde, New South Wales (WAPA, 2017). The skin of Granny Smith is green, and the flesh is white (Photo 1.2D). Granny Smith apple is tart, crispy to crunchy, and juicy. This variety of apple stores well, although it is susceptible to superficial scald (Abbott et al., 2004). Sometimes it may develop a pink blush on the sun-exposed side, which has been observed in high insolation regions or areas with wide thermal amplitude (Benítez, 2001).

1.7.5 JONATHAN

Jonathan apple is a medium-sized sweet apple. Its red color sometimes has green tones. It was originated in Woodstock, the United States, before 1826. The fruits texture is soft, and the flesh is juicy, sweet, and pleasant flavor (National Fruit Collection, 2015).

1.7.6 JONAGOLD

Jonagold was obtained in 1953 in New York, and it is a blend of Jonathan and Golden Delicious apples (Benítez, 2001; WAPA, 2017). Jonagold fruits have honey-tart flavor and are crispy, juicy flesh. It has a yellow–green skin color and a red–orange blush (USApple, 2017). Currently, the most widely planted clones in producing countries are Morrens® Jonagold, New Jonagold, and Jonica® Schneica (Benítez, 2001).

1.7.7 ELSTAR

Elstar was originated in Holland in 1955. It is a cross between Golden Delicious and Ingrid Marie (Benítez, 2001). The skin is characterized by a golden yellow overlaid with intense red, and its taste is sweet, generally described as syrupy (WAPA, 2017). It usually presents some russeting in the calicinal and peduncular cavities. The pulp is yellowish white, consistent, and crispy. Flavor improves considerably after a month of cooling, which is the main attribute of this variety (Benítez, 2001).

1.7.8 BRAEBURN

Braeburn was originated in New Zealand in the 1950s. Its probable parents are Lady Hamilton and Granny Smith. Its color varies from orange to red over a yellow background (WAPA, 2017). Braeburn apple is firm, flattened, and irregular, especially when produced by young trees. The epidermis is striated, red, covering 25%–75% of a dark green background. The pulp is yellowish, firm, crispy, juicy, and sweet (Benítez, 2001).

1.7.9 FUJI

Fuji was developed in Japan in the late 1930s, and it is a cross between Ralls Janet and Red Delicious. Fuji apple is bi-colored, typically yellow and red striped (USApple, 2017) (Photo 1.2E). The flesh is juicy, firm, and crisp, with a sweet and spicy flavor. It has high sugar and low acidity (Yoshida et al., 1995).

1.7.10 HIMEKAMI AND IWAKAMI

Himekami and Iwakami were first produced in 1985 at the Morioka Branch, Fruit Tree Research Station, Japan, and they are a blend of Fuji and Jonathan apples. Himekami's skin is pinkish red and it tends to develop water core. Iwakami's skin is striped red and it is slightly acidic (Soejima et al., 1998).

1.7.11 CRIPPS PINK

Cripps Pink was developed in the 1970s by John Cripps by crossing Lady Williams with Golden Delicious (Cripps et al., 1993; WAPA, 2017) (Photo 1.2F). It combines the firmness, storage potential, and low susceptibility to bitter pit of the former, with the good organoleptic quality and low scald incidence of the latter. Cripps Pink is a late-maturing and long-storing cultivar. Its maturity is reached 8–9 weeks later than that of Red Delicious, and it is superior in size and quality. This cultivar has low susceptibility to superficial scald (Cripps et al., 1993). Cripps Pink apples are often found under the retail name Pink Lady® (USApple, 2017).

1.7.12 IDARED

Idared was introduced in 1942, as a result of a cross between the Jonathan and Wagener apples (WAPA, 2017). Idared has a bright red skin and a tangy–tart flavor (Benítez, 2001).

1.7.13 CAMEO

Cameo was found out as a chance seedling in Washington State at the end of the 1990s (USApple, 2017). It has bright red stripes over an orange or cream-colored background and a sweet–tart taste (WAPA, 2017).

1.7.14 CORTLAND

Cortland originated in 1890 at the New York State Agricultural Experiment Station. It is a cross between McIntosh and Ben Davis. Cortland peel is yellow–green with red tones. It is a sweet apple with only a hint of tartness (USApple, 2017; WAPA, 2017).

1.7.15 MCINTOSH

McIntosh is an old variety of apple discovered in 1811. These fruits are characterized by intense red color skin and white, juicy, tangy–tart taste flesh (USApple, 2017; WAPA, 2017).

1.7.16 SPARTAN

Spartan is a crunchy and sweet apple. It was produced in Summerland, Canada, by a crossing McIntosh with Newtown Pippin (WAPA, 2017).

1.7.17 SHAMROCK

Shamrock is a medium-sized fruit similar to Granny Smith. Its flesh is light green, and it has a thick texture with a strong flavor similar to that of McIntosh. It was developed in Summerland, BC, Canada, by crossing Spur McIntosh with Spur Golden Delicious (Orange Pippin, 2018).

1.7.18 EMPIRE

Empire was obtained by crossing Red Delicious with McIntosh at the New York State Agricultural Experiment Station. It is a crispy and juicy apple with a sweet–tart flavor. Empire apples have an intense red color, overlying a light green background (USApple, 2017; WAPA, 2017).

1.7.19 FREEDOM

Freedom was obtained by crossing Macoun with Antonovka in the New York State Agricultural Experiment Station, Cornell University, Geneva, the United States (Orange Pippin, 2018). Freedom is a medium-large red fruit with crisp, juicy, and sweet flesh. These fruits can be eaten fresh or used for processing, cider, or juice.

1.7.20 BALDWIN

Baldwin is an old American apple variety from Massachusetts. This apple is dark red, medium-large size, and its skin has prominent lenticels. Its flesh is dense grained and acidic with an aromatic flavor. The fruit's good quality can be maintained for long periods in a cold storage (Merwin, 2008).

1.7.21 MOLLIES DELICIOUS

Mollies Delicious was developed in 1948 by G.W. Schneider at the New Jersey Agricultural Experiment Station. These fruits are large, conical in shape with a pinkish-red color (Orange Pippin, 2018).

1.7.22 BRAMLEY'S SEEDLING

Bramley's Seedling has been a traditional English cooking apple since the 1830s. The fruit is large, oblate, green with red stripes, and very tart with great content of vitamin C (Merwin, 2008).

1.7.23 KENT

Kent apple is greenish-yellow and has a crisp, white, juicy flesh with aromatic flavor. Kent was obtained in England, United Kingdom, by crossing Cox's Orange Pippin with Jonathan (Orange Pippin, 2018).

1.7.24 SUNTAN

Suntan apple is a medium-large fruit with red–orange stripes on a yellow background. The pulp is white–yellow, crispy, and juicy. Suntan was obtained in England, United Kingdom, by crossing Cox's Orange Pippin with Court Pendu Plat (Trillot et al., 1993).

1.7.25 GREENSLEEVES

Greensleeves is a pale green–yellow, skinned apple with a crisp flesh which becomes sweeter. It can be susceptible to apple scab, powdery mildews, and apple canker (Royal Horticultural Society, 2018).

1.7.26 JESTER

Jester is an attractive, crisp, and refreshing apple, with nice yet

Apple

bland flavor. Jester was obtained in England, United Kingdom, by crossing Worcester with Starkspur Golden Delicious (Orange Pippin, 2018).

1.7.27 JUPITER

Jupiter apple was developed in the 1960s by East Malling Research Station. The fruits are yellow covered by an orange–red flush, and the flesh has a cox-like flavor. The trees can produce heavy crops and be affected scab, canker, powdery mildews, and blossom wilt (Royal Horticultural Society, 2018).

1.7.28 LORD LOMBOURNE

Lord Lombourne originated in England in 1907 by crossing James Grieve with Worcester Pearmain. The skin is red, bright, and slightly striated, with some russet (Orange Pippin, 2018)

1.7.29 FIESTA

Fiesta was obtained by East Malling Research in the United Kingdom by crossing Cox's Orange Pippin with Idared. Fiesta apples are small- or medium-sized apples with red to orange skin color and white to cream flesh (Trillot et al., 1993).

1.7.30 GLOSTER 69

Gloster 69 was obtained in 1951 in Germany at the Fruit Research Station, Jork, Hamburg, by crossing Pomme Cloche with Rich-A-Red Delicious. It is a large apple with good flavor and crisp flesh. Gloster 69 is bicolor rather than fully red, morphologically resembling Red Delicious (Trillot et al., 1993).

1.7.31 PINOVA

Pinova was obtained in Germany in 1993 by crossing Clivia (Cox's Orange Pippin × Duchess of Oldenburg) with Golden Delicious. It is also known as Pinata. It has bright red, striped skin over a yellow–green background (Trillot et al., 1993).

1.7.32 MERAN

Meran originated in Italy in 1976 as a chance seedling. It has a yellow–green skin color with red blush depending on the environment. Its surface is slightly greasy with few lenticels. Meran fruit is firm, crispy, and juicy (Trillot et al., 1993).

1.7.33 DELBARESTIVALE

Delbarestivale is an early/mid-season apple from France, obtained by crossing Golden Delicious with Stark Jonagrimes. It is also known as Estival or Delcorf. Its skin is yellow–green with red streaks. The pulp is white, crispy, and juicy, with the suitable balance between sugars and acidity (Benítez, 2001).

1.7.34 RUBINETTE

Rubinette was obtained in 1966 by Walter Hauenstein in Rafz, Switzerland, by crossing Cox's Orange Pippin with Golden Delicious. Rubinette is a small-sized apple with deep red striations and an orange flush over a yellow–green skin, covered with white lenticels (Trillot et al., 1993).

1.7.35 AMRI

Amri is a medium-sized popular apple in India. It is a large, round, red, and sweet fruit, which keeps its condition for a long time (Virk and Sogi, 2004).

1.7.36 AMBRICH

Amrich apple fruits are conical with red skin. The pulp is white, slightly acid, and juicy. It was obtained in India by a crossing between Rich-A-Red and Amri (Schmidt and Kellerhals, 2012).

1.7.37 AMBRED

Ambred was obtained from crossing Red Delicious with Amri. The fruit is conical, and its skin has bright red stripes on a yellow base. The flesh is white, crisp, and it has a good storage quality (Schmidt and Kellerhals, 2012).

1.7.38 CHAUBATTIA ANUPAM AND CHAUBATTIA PRINCESS

Chaubattia Anupam and Chaubattia Princess were obtained in India by crossing Red Delicious with Early Shanbury. Chaubattia Anupam fruit skin has shining red streaks with a red blush on a pale background. Chaubattia Princess was obtained earlier than Chaubattia Anupam, and it has deep red streaks on a pale background. It is very sweet and has good keeping quality (Kumar, 2006).

1.7.39 MAJDA

Majda is a Slovenian variety. It was created in 1947 by crossing Jonathan with Golden Noble. Its taste is very acidic, and therefore it is suitable for lovers of acidic flavor. Its pulp does not get dark when it is cut into slices, and it is very suitable for drying (EkoDrevesnica Ocepek, 2018).

It is a general information of all varieties most commercial varieties and cultivars (e.g., Red Delicious, Golden Delicious, Jonagold, Gala, Elstar, and Braeburn) are susceptible to apple scab and other are scab-resistant (e.g., Fiesta, Goldrush, Goldstar, Florina,

Rubinola, and Topaz) (Liebhard et al., 2003; Petkovsek et al., 2007; Leccese et al., 2009). Cultivated apples are generally diploid, but some of the main cultivars are triploid (Trillot et al. 1993). Triploids have much larger fruits and are of value to the industry (Brown and Maloney, 2003), but they are not suitable as pollinators. Almost all apple varieties need to be cross-pollinated, and therefore, the grower should plant different apple cultivars together in the same orchard. The compatibility between pollinizer and pollinated tree depends on the blooming time. Other characteristics of some commercial cultivars, such as harvest time, main uses, fruit-bearing type, and their self-compatibility are summarized in Table 1.5.

TABLE 1.5 Uses, Fruit-Bearing Type, Self-Compatibility, According to Early, Mid- or Late-Season Cultivars

Cultivar	Uses			Fruit Bearing		Compatibility	
	Fresh Eating	Cooking	Juice Making	Spur Bearer	Partial Tip Bearer	Self-compatible	Self-incompatible
Early Season							
Elstar	X	X	X	X			X
Gala	X			X		X	
McIntosh	X	X		X			X
Mid-season							
Cortland	X	X	X	X		X	
Red Delicious	X			X			X
Bramley's Seedling		X	X		X		X
Golden Delicious	X	X	X	X		X	
Idared	X	X		X			X
Jonagold	X	X	X	X			X
Spartan	X		X	X			X
Late Season							
Braeburn	X	X	X	X		X	
Fuji	X			X			X
Granny Smith	X	X	X		X	X	
Pinova	X	X		X			X
Rubinette	X		X	X			X
Cripp's Pink	X				X		X

The club varieties which are trademarked only the club members are allowed to grow them. The success of new club varieties will depend on both the quality of the new variety and the marketing efforts of the club (Brown and Maloney, 2009). Club varieties include Pink Lady®, Envy™, Kanzi®, Jazz™, Ambrosia®, SweeTango®, Honeycrisp®, Evelyna®, Modí®, Jeromine, and Opal® (Toranzo, 2016).

1.8 BREEDING AND CROP IMPROVEMENT

Cultivated apple is an interspecific hybrid designated *Malus × domestica* Borkh. There are about 30 species of *Malus*, mainly from Asia. Four species correspond to America and three to Europe. Apple breeding is currently based on the use of species of *Malus*, but with new approaches that value genetic diversity. Apple crop is grown mostly throughout temperate climate zones, but over the last years, the crop has expanded its world distribution. This expansion toward new apple-producing areas demands cultivars with low chilling-hour requirements. On the other hand, areas with adequate chilling requirement present spring frosts during bloom and require the development of cultivars with later bloom to avoid the frosts (Brown, 2012). Environmental adaptability is researched in breeding stations, where the selection is performed in adverse climatic conditions. Varietal selection is based on precise territorial requirements without considering commercial aspects. Climate change has recently added new challenges for apple researchers, due to the emergence of new diseases, changes in climatic conditions, hotter weather, sunburn, and drought.

Historically, breeding programs have been developed at research stations in apple-producing areas worldwide and frequently in collaboration with universities. Genetic breeding programs are associated with germplasm banks, which are an important source of materials to breeders. These banks contribute knowledge about the characteristics of the available material. In particular, the germplasm banks of Asia, center of origin of apple, publish significant information on conservation and characterization of materials. For example, the knowledge of heterozygosis and the possibility of auto fecundation is valuable that could contribute to developing resistant genotypes. However, data on the harvest and conservation of materials are no less important contributions of breeding programs (Sansavini et al., 2005; Brown, 2012).

Several organizations in the world work in cooperation with genetic resources. In the United States, the US Department of Agriculture's Agricultural Research

Service is responsible for the clonal repository of apples in Geneva, NY. This service acquires materials, mainly *Malus sieversii*, from centers of origin, making these materials available to researchers and offering detailed information on apple germplasm (Brown, 2012). Similarly, the European Cooperative Program for Plant Genetic Resources has a working group on *Malus*. Thirteen countries belong to the group and many of them are responsible for its germplasm collection. The collections are used to study resistance to diseases, among other topics (Brown, 2012). Over the past few decades, Europe has increased the number of breeding programs to reduce the dependence on foreign genetic material, mainly American cultivars. Between 1991 and 2001, over 500 new apple cultivars were released; in this quest for new cultivars, Europe ranks first, followed by North America, Asia, and Oceania (Sansavini et al., 2005).

1.8.1 BREEDING OBJECTIVES

Scion breeding programs focus on various aspects of apple production:

1. Fruit quality: aspect, skin color, structure, juiciness, sensorial quality, and pulp flavor, as well as new types of fruit

2. Diseases and phytophagous resistance: apple scab, powdery mildew, European canker, fire blight, and aphids
3. Nutritional components
4. Postharvest traits that allow long storage
5. Uniformity of maturity of fruit; and
6. Shelf life.

Rootstock breeding programs are focused on resistance to abiotic and biotic stress and on the control of plant vigor:

1. Productivity and structure: vigor, production of harvest (kg/tree);
2. Fructification aspects: columnar, compact, spur, and monocarpic inflorescences; and
3. Environmental adaptability: high temperatures in summer sun scald, low temperatures in spring, and salinity of water–soil (Brown, 2012; Sansavini et al., 2012).

Some traits and sources depend on the objectives of breeding programs, as follows:

1. Quality: The typology of apple fruit depends on fruit destination: fresh consumption, medium- to long-term storage, industrial use (juice, cider, slicer, lyophilized products,

and alcohol extracts). Today, no single idiotype exists, although market tendencies are researched to know whether consumer preferences are related to the geographic area. Supermarkets and big distributors in Europe offer consumers fruits of high organoleptic quality, which are relatively aromatic, have crispy and very juicy flesh, good balance between sugars and acidity, and good maturation state.

2. Apple scab (*Venturia inaequalis*): To obtain resistance to apple scab, several *Malus* species have been used. The strategy was a modified backcross program with susceptible cultivars serving as recurrent parents. Eighteen scab-resistant cultivars containing the *Vf* gene derived from *Malus floribunda* 821 were released by a Purdue Rutgers Illinois cooperative scab-resistant apple breeding program developed by Purdue University, Rutgers, The State University of New Jersey, and the University of Illinois (Janick, 2002). In addition, Russian seedling R 1270-4A and *Malus sieversii* are new sources of resistance, especially for pyramiding resistance. Nowadays, other sources of resistance are being studied (Brown, 2012).

3. Powdery mildew (*Podosphaera leucotricha*): Several sources of resistance include Pl_i from *Malus* × *robusta* and PL_2 from *M. zumi; PLd* from D12 (open-pollinized crabapple seed); Pl_m and *U211* (quantitative resistance) (Brown, 2012).

4. Fire blight (*Erwinia amylovora*): The main source of resistance is *Malus robusta* 5. It is used for both scion and rootstock breeding. Delicious and Liberty are cultivars with good resistance.

5. Vitamin C and antioxidants: Various breeding stations use parental lines with distinctive biochemical parameters such as a high content of vitamin C or polyphenols and antioxidants (Brown, 2012).

6. Red pigmentation: New Zealand and Holland work jointly to create apple with pink flesh or strongly mottled red color, to demonstrate a high content of flavonoids (Sansavini, 2008; Brown, 2012). In the future, these functional foods can be differentiated and commercially classified by specific nutraceutical properties.

At present, there are many breeding programs in the world. For example, Brown and Maloney (2003) reported 57 apple-breeding projects in 25 countries. In the United States, there are full-time programs dedicated exclusively to apple crops and others that investigate apple and other crops. The program of Cornell University at the New York State Agricultural Experimental Station in Geneva has released 63 apple cultivars. Other important pome fruit breeding programs at the Washington State University (WSU), in the Tree Fruit Research and Extension Center in Wenatchee, has produced new improved apple varieties, especially suited to Pacific Northwest growing conditions. The WSU program (http://tfrec.cahnrs.wsu.edu/breed/) was one of the 12 core United States breeding programs of the SCRI RosBREED project, "Enabling Marker-Assisted Breeding in Rosaceae."

China has many apple breeding programs in provincial fruit research institutes, resulting in around 180 new cultivars released from 1950 to 1995. Other important breeding research centers around the world are located in France, Germany, England, Italy, Japan, the Netherlands, New Zealand, Canada, Australia, Czech Republic, and 19 other countries (Brown and Maloney, 2003).

Currently, the internationalization of markets imposes major goals on breeding programs. This implies that breeding programs must comply with the fruit quality standards required by consumers and the consumption patterns resulting from the globalization process. These standards in apple are diversified in relation to the geographic area. Worldwide, apple fruit completely covers the annual consumption arc, so that, for example, the European apple can compete with imported fresh fruits from the Southern Hemisphere (Pellegrino and Guerra, 2008).

In many regions of the world, breeding programs not only strive for fruit quality, but also for genotypes resistant and tolerant to the main biotic adversities, which are common objectives in apple crops.

Today, the goals of breeding programs include the following points:

1. Disease resistance such as scab (*Venturia inaequalis*), powdery mildew (*Podosphaera leucotricha*), tolerance of rosy apple aphid (*Dysaphis plantaginea*), and fire blight (*Erwinia amylovora*). These varieties require minimal manual labor and have good fruit quality (Lespinasse, 1992). Other breeding programs are focusing on apple resistance to canker (Turechek, 2004). New research about organic

production is conducted to produce resistant varieties and quality fruit with low-spray management in orchards (Brown, 2012).

2. Low allergenicity apples. It is known that a lipid transfer protein (LTP) is involved in fruit allergy. However, further research is required to confirm the role of LTP (Brown, 2012).

3. Processing and fresh-cut markets. There is high demand for fruits with desirable organoleptic characteristics (flavor, firmness, and nonbrowning).

4. Rootstocks from breeding programs focusing on resistance to abiotic and biotic stresses.

1.8.2 METHODOLOGY OF GENETIC BREEDING

Kumar et al. (2012) explained that three stages are followed in a standard breeding program. In the first stage, parents are identified from the pool of available candidates. Then these parents are crossed, and the best offspring is selected. This stage implies knowledge of the morphogenetic background of the traits for which the breeding program is conducted. In the second stage, the material that was selected is many times propagated onto clonal rootstock. Finally, in the third stage, the best selections are tested at a larger scale, often on commercial orchards.

The genetic breeding methods are briefly described as follows. Controlled crossing is done between especially preselected parental lines and by a subsequent selective process. The improvement in apple is still based on the conventional method, that is, crossing of the parental line that carries the chosen trait. It is an intraspecific crossing between interfertile varieties. Sometimes, interspecific crossing is done when the sources of the trait do not manifest in *Malus × domestica*, but in other species of the genus *Malus*. This is the case of repeatable resistance to apple scab. The most cited species are *M. floribunda*, apart from *M. baccata f. jackalii, M. micromalus* (Pellegrino and Guerra, 2008).

The primary aim of the Geneva Rootstock Program (USA) was to develop dwarfing rootstocks with resistance to pathogens for some producer areas. Now, it is focused on producing rootstock genotypes that retain the disease resistance of Geneva rootstocks, but improving horticultural properties (Johnson et al., 2001).

Induced mutagenesis, together with conventional breeding, accelerates the breeding process. Mutagenesis consists in inducing changes in genetic material. These changes can be translated into characters of agronomic interest, such as resistance to pathogens, tolerance to water stress, early flowering, and increase in fruit size. The induced process can be done with chemical methods (e.g., colchicine) or physical methods (e.g., ionizing radiation such as C 60, X-rays, laser rays, or other sources), followed by the selection procedure.

Genetic transformation in apple rootstocks of a popular series known as Malling is done to improve disease resistance. Particularly, in the United States, the Geneva program aims to improve resistance to the fire blight (*Erwinia amylovora*) by means of evaluations of rootstocks Malling 26 with transgenes. The best selections are distributed throughout North America among cooperating nurseries for evaluations of the trees (Johnson et al., 2001).

Clonal selection is a breeding method based on phenotypic variants of the known varieties. The variant will repeat randomly in nature, usually deriving from a positive or negative mutation. In addition, these mutations involve regression to the original genetic matrix. Clonal selection may be subtractive, that is, may aim at the identification and elimination of the negative clones, or may be regenerative and improved when the positive variants of the selected character are individualized. Then the stability of the mutated character must be demonstrated throughout many years in orchard evaluations, where the improved character retention can be verified. Then, the original genetic matrix must be contrasted with the clonal selection. When the process is favorable and really finalized, new clones are obtained.

In vitro somaclonal variation is a tool for inducing variation in fruit improvement. Somaclonal variation leads to the selection of several variants with increased resistance to pests and diseases. Although the mechanisms of somaclonal variation are still not completely understood, several explanations are proposed. These include karyotypic changes, point mutations, sister chromatid exchange, somatic gene re-arrangement, changes in organelle DNA, DNA amplification, insertion or excision of transposable elements, DNA methylation, epigenetic variation, and segregation of chimera tissue. Somaclones exhibiting resistance to *Erwinia amylovora* have been obtained from apple. Besides, resistant somaclones to *Phytophthora cactorum* were also selected in apple rootstocks M26 and MM106 (Predieri, 2001).

1.8.3 BIOTECHNOLOGICAL AND MOLECULAR APPROACHES

In the last years, biotechnology and the integration of molecular tools in breeding programs have improved their speed and efficiency. These improvements have become possible due to the development of transformation systems, multiple genetic maps, and a large number of markers, extensive expressed sequence tag libraries and, most recently, the whole genome sequencing of apple (Sansavini et al., 2005).

Nowadays, qualitative and quantitative genetic traits are being studied through molecular genotyping. Identification of profiles characterizing individual genotypes is obtained by polymorphic DNA markers, especially amplified fragment length polymorphisms and simple sequence repeats. Studies of functional markers of candidate genes range from stress resistance to quality (Sansavini et al., 2005; Brown, 2012).

The bioinformatics analysis of the data will allow researchers to identify the fruit quality of quantitative trait loci and to follow the transmission of their alleles along the pedigrees (Gianfranceschi and Soglio, 2003; Monforte et al., 2004). This original and rigorous, multidisciplinary approach includes the interactive work of apple breeders, geneticists, molecular biologists, statisticians, and experts in bioinformatics.

1.9 SOIL AND CLIMATE

1.9.1 SOIL AND CLIMATE REQUIREMENTS

Most fruit crops need medium-texture soil with at least 50 cm in depth (Sansavini et al., 2012). Apples prefer soil with pH values between 6.5 and 8.5 and less than 10%–12% of carbonate presence (Sansavini et al., 2012). High concentration of calcium carbonate leads to high pH and high carbonate levels in soil solution, and these factors generate nutritional deficiency (Sozzi, 2007).

Apple is adapted to several climates, but it is considered best adapted to the cool-temperate climate between 35° and 50° latitude (Rieger, 2006). The most relevant climate factors that restrict fruit crops are winter's minimum temperature, growth season length and temperatures, light intensity, and annual rainfall and its distribution (Westwood, 1993).

While in most deciduous trees the photoperiod decrease determines leaf fall and the beginning of endodormancy (Sansavini et al., 2012), in apple trees, this process is regulated by temperature (Heide and Prestrud, 2005). Endodormancy release is obtained with low-temperature exposition for variable times depending on species and cultivars. Such ecological need is known as cold requirement (Sansavini et al., 2012). Apple is one of the most cold-resistant fruit crops, and dormant trees tolerate temperatures of up to −40 °C (Ryugo, 1988). Cold resistance depends on tree phenology, and blooming is a critical stage. Open flowers and young fruitlets are affected by brief exposure at −2 °C (28 °F) or less. The late blooming of apple decreases the likelihood of spring frost damage, which can reduce production by up to 90% (Collett, 2011).

Most traditional apple cultivars have a high cold requirement (1000–1600 chilling hours), so they cannot be cultivated properly at low latitudes (Ryugo, 1988). If this requirement is not fulfilled, budding is unequal and poor. Some cultivars with a low cold requirement were obtained (100,300 chilling hours), which allows apple cultivation in tropical regions, such as Brazil (Sansavini et al., 2012). Because of this cold requirement, the most important areas for apple production are found between 35° and 50° latitudes in both hemispheres. Apples planted in regions of latitudes lower than 35° must have a low cold need or must be cultivated at a high altitude, which allows them to accumulate winter cold need and go out of dormancy (Westwood, 1993). Temperature decreases as altitude increases. Thermal vertical gradient presents an average value of 0.64 °C every 100 m in atmospheric stability conditions (Sansavini et al., 2012). After winter dormancy, apple requires heat accumulation to sprout. Apple requirement is about 6900 GDH (Growth Degree Hours) according to Richardson model (Sansavini et al., 2012). In most cultivars, apples reach ripening between 120 and 150 days after full bloom. In general, apples require 500–600 mm of rainfall for irrigation during growing season. Rainfall is considered an adverse factor because most apple production regions are irrigated, and rainfall increases fungal and bacterial disease incidence (Jackson, 2003). Moreover, during bloom, rainfall can set a limit to dispersal of pollen grains, wash them from stigmas, and even hinder insect pollinator flight (Sansavini et al., 2012).

1.9.2 IMPACT OF CLIMATE CHANGE

Climate change is a modification in climate status verified in mean or variance of climatic parameters (temperature, precipitation, among others) during a long time period (decades). Climate change can impact agrosystems in many ways. Atmospheric CO_2 rises, temperature and rainfall change, and the frequency of extreme events can vary plant phenology, crop areas, insect and pathogen dispersion, their attack time, etc. The atmospheric CO_2 rise has a direct effect on tree productivity because a photosynthesis increase of 30%–50% is expected for C3 species (Sansavini et al., 2012).

Over the past 50 years, an annual temperature close to 0.8 °C has been registered in Europe (Sansavini et al., 2012), while in South Africa, spring temperatures have shown an increment of up to 0.45 °C per decade (Grab and Craparo, 2011). In a changing global environment, temperate fruit crops could be at risk due to temperature change (Campoy et al., 2011). Records all over the world have shown that tree phenology has advanced at spring time, and it is associated with temperature increase (Grab and Craparo, 2011). Air temperature in some regions of the mountain state of Himachal Pradesh has shown increasing trends (Basannagari and Kala, 2013). For example, apple crops have shown decreased productivity at low elevation (<2500 m.s.l.) in regions of Himachal Pradesh and improved yield at higher elevations (Rana et al., 2011). Based on the phenology register conducted during decades, an advance of four weeks in apple blooming was determined for Western Europe, while blooming was belated for two weeks in Eastern Europe. Spring starts earlier in Western Europe because of the flow of warmer air masses from the Atlantic, while in Eastern Europe there is a different trend due to the influence of the Siberian high (Ahas et al., 2002). In Germany, apple blooming was found to occur 10 days earlier, and harvest time and leave fall were also affected (3–9 and 2–3 days, respectively). Blooming length has become shorter in the last 10 years: from 12–15 days to 810 days (Blanke and Kunz, 2009). Bloom advance exposes flowers to a higher risk of spring frost damage (Sansavini et al., 2012). Apple blooming advance was also observed in northeastern United States (Wolfe et al., 2005), South Africa (Grab and Craparo, 2011), and Australia (Darbyshire et al., 2013). Consistent with global warming, apple bloom tends to occur about −0.1 to −0.3 days per year earlier (Ahas et al., 2002; Wolfe et al., 2005; Grab and Craparo, 2011).

1.10 PROPAGATION AND ROOTSTOCK

Apples reproduce naturally by seed, and most cultivars depend on pollination to produce fertile seeds, although some cultivars can produce unfertile (apomictic) seeds. Most cultivars cannot self-pollinate due to a gametophytic self-incompatibility, and they present great heterozygosity. Thus, when apple propagates from seed, trees and fruits are highly variable (Webster and Wertheim, 2005).

To maintain apple cultivars with desirable characteristics, asexual propagation techniques are required. Commercial apple trees are composed of two genetically distinct portions. One portion is the rootstock, which consists of the root system, whereas the other is the scion, which includes all the canopy tissues above the ground. The reason for breeding certain types of cultivars is associated with the desired qualities of the fruiting scion portion or of the rootstock. In scion-breeding programs, the objectives are to increase the quality or marketability of the fruit, reduce production costs, and increase resistance to pests and adaptation to extreme abiotic conditions.

Commercial apple trees are obtained by grafting. Grafting is a horticultural technique that connect two pieces of living tissue together in such a way that they unite and grow as one. Budding is a special form of grafting in which the initial scion wood component is reduced to a single bud (Jackson, 2003). Several grafting techniques are possible. Which one to use depends on the conditions under which the grafting is to be done. When trees are dormant, the most common propagation techniques are dormant chip budding and cleft grafting. The former is primarily used for creating new trees, whereas the latter is used for topworking older established apple and pear trees. It is best adapted to branches 2.5–5.0 cm in diameter. The grafts are made within 0.5–0.9 m of the trunk or main branches and preferably not higher than 0.7–1.0 m from the ground. Another commonly used grafting technique is whip graft, which is performed by interlocking a small tongue and notch in the obliquely cut base of the scion with corresponding cuts in the stock. Whip graft is used on young apple and pear trees when the branches are not larger than 1.5 cm in diameter and when the understock is about the same diameter as the scion of the new cultivar.

1.10.1 STOCK–SCION RELATIONSHIP

Rootstocks affect the vigor of vegetative growth (Figure 1.7), fruit size and growth pattern, precocity, yield, and susceptibility to biotic and abiotic stresses (Webster and Wertheim, 2005; Fioravanço et al., 2016).

Seedling apple trees have juvenile characteristics for a long period of time and are slow to start fruit production. To produce a commercial apple tree, it is necessary to make a graft between the rootstock and the scion. To achieve a good graft, the cambium of the rhizome must coincide with the scion cambium as well as possible roodstock (Jackson, 2003). For grafting to be successful, only dormant wood can be used; thus, scions must be cut while dormant and grafted immediately or preserved under refrigerated conditions until grafting can be done.

The vigor of a compound tree reflects the vigor of its component rootstock and scion (Tworkoski and Miller, 2007). The use of a dwarfing rootstock as a management tool to control scion vigor is primarily dependent on the reduced vigor of the root system. The main effect of root system size on shoot growth is probably via supply of cytokinins but may also involve nutrient and water supply. The graft union per se may also restrict the upward movement of water and nutrients.

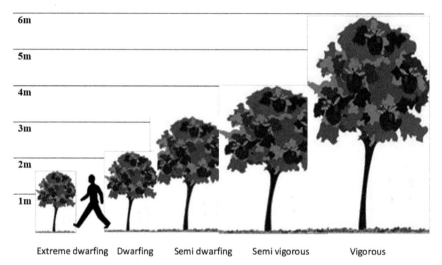

FIGURE 1.7 Height and vigor reached by the trees according to the rootstock.
Source: Apple Rootstock Characteristics and Descriptions, 2015, USDA National Institute of Food and Agriculture, New Technologies for Ag Extension project (http://articles.extension.org/pages/60736/apple-rootstock-characteristics-and-descriptions).

1.10.2 ROOTSTOCK TYPES

1.10.2.1 DWARF ROOTSTOCKS

These rootstocks are very precocious, with high yield efficiency and limited root volume, so they benefit from tree support and supplemental irrigation in dry seasons. The dwarf rootstocks are M.9 (Malling 9), M.26 (Malling 26), M.27 (Malling 27), G.65 (Geneva 65), Mark V.1 (Vineland 1), V.2 (Vineland 2), V.3 (Vineland 3), O.3 (Ottawa 3), and P.2.

1.10.2.2 SEMIDWARF ROOTSTOCKS

Semidwarf rootstocks generate a plant that does not require much ladder work. The semidwarf rootstocks are B. 118 (Budagovsky 118), G.202 (Geneva 202), G.30 (Geneva 30), M.7 (Malling 7), MM.106 (Malling-Merton 106), and J-TE-H.

1.10.2.3 SEMIVIGOROUS ROOTSTOCKS

These rootstocks produce trees three-quarters the standard size or larger. They are slow to start cropping and costly to prune and pick. The semivigorous rootstocks are MM.111 (Malling-Merton 111), Antonovka 313, and P.18.

1.10.2.4 SEEDLING ROOTSTOCKS

Seedling rootstocks are usually produced from seeds obtained from apple juice plants. Because seedling rootstocks are not clonal, there is tree-to-tree variation. Trees on seedling rootstocks are considered full-sized trees (100% standard). They generally have good survival, are free-standing, relatively nonprecocious, produce few root suckers and burr knots, and are relatively cold hardy and resistant to diseases.

1.10.3 INTERSTEM TREES

High-density orchard systems aim to achieve high yield in the early age of trees and control vegetative growth. Therefore, cultivars can be dwarfed by grafting on dwarf rootstocks, but dwarf rootstocks provide poor anchoring and are more sensitive to water and nutritional stress. On the other hand, vigorous rootstocks provide resistance against drought and better anchorage. To control the vegetative growth by means of a dwarfing rootstock and obtain the advantages of a vigorous rootstock, it is necessary to graft a dwarf interstem or interstock between the scion and the rootstock (Karlıdağ et al., 2014). Interstem trees consist of three parts: a vigorous rootstock (usually MM.111 or MM.106), the interstem (a dwarfing rootstock, usually M.9 or M.27), and the scion cultivar. Although this approach seems well conceived, results are usually disappointing. Interstem trees are generally more expensive than trees on a rootstock because an additional year is required in the nursery.

1.10.4 SCION–ROOTSTOCK INCOMPATIBILITY

In some cases, the graft union does not result in a long-lived composite plant. In this situation, the graft is incompatible. In general, the more closely related the plants, the higher the chance of compatibility, although permanent unions between one genus and another may occur. Incompatibility has the following symptoms: initial failure to form graft or bud unions, poor growth of the scion often followed by premature death in the nursery, premature yellowing of the foliage, and breaking off of trees at the graft union, especially when they have been growing for many years and the break is clean and smooth (Karlıdağ et al., 2014).

1.10.5 MICROPROPAGATION AND TISSUE CULTURE

Breeding a new apple cultivar by traditional methodologies takes approximately 15 years. Both scions and rootstocks are vegetatively propagated by tissue culture. These techniques provide efficient regeneration systems from somatic tissues, which are essential to developing the transfer of individual genes in the process of genetic engineering.

Shoot culture involves the use of explants. The tissues used for explants are meristems or shoot tips because of their genetic stability. Meristem culture is used to obtain virus-free plants by culturing terminal and axillary buds, and the survival of explants depends on contamination with microbes, oxidative conditions, the explants used for culture initiation, the physiological stage of mother plants, and the season when the explants are collected (Dobránszki and Teixeira da Silva, 2010).

Explants are surface sterilized, and then cultured in a medium containing mineral salts, vitamins, carbohydrates such as sucrose or sorbitol, and plant growth regulators (cytokinin, auxin, and possibly some gibberellins), and solidified with agar (Murashige and Skoog, 1962; Castillo et al., 2015). The medium commonly used for apple micropropagation is the Murashige and Skoog medium (1962). Shoot cultures are maintained in the culture medium in rooms with programmed photoperiod and thermoperiod.

Rooting of in vitro shoots depends on three phases: dedifferentiation, induction, and morphological differentiation. Microshoots can be rooted under both in vitro and ex vitro conditions (Gonçalves et al., 2017).

Rooting in ex vitro conditions is done directly by dipping the bases of shoots into indolbutiric acid and fungicide. The shoots are then planted in pots with sterilized sand under a transparent cover and maintained for four weeks at 22 °C. They are transferred to compost and grown in a glasshouse.

The acclimatization from in vitro to ex vitro conditions is an important step because it is the beginning of the autotrophic life of the plants (Dobránszki and Teixeira da Silva, 2010).

In vitro propagation of apple and pear may be relatively more expensive than conventional vegetative propagation. The advantages of mass propagation are rapid multiplication, disease-free plants, clonal identity, and propagation of cultivars which are difficult to root. However, under some circumstances, in vitro produced rootstocks or scions produce excessive numbers of burr knots and suckers, possibly associated with enhanced juvenility. Sequential shoot subculture may greatly increase rejuvenation (i.e., development of juvenile characteristics including ready rooting, but also spinyness and slowness to bear fruit). Regeneration systems for *Malus* have been developed from a wide range of other explants, including embryo, nucellus, cotyledon, hypocotyl, immature seed, flower part, and anther tissues (Jackson, 2003).

1.11 LAYOUT AND PLANTING

Apple trees grow well in a wide range of soils. However, in limited soil, extra effort and cost are required to ameliorate the site before planting. Tillage in orchards is usually conducted at a depth of about 20–25 cm before planting, in order to produce appropriate conditions for a better growth of the trees (Ferree and Warrington, 2003). In poor soils, the use of organic materials (hays and straw, grape pomace, manures or sewage bio-solids, and composted combinations of several of these materials) as soil amendments is recommended (Glover et al., 2000). These materials should be applied near the root zone prior to planting. Nitrogen fertilization alone or in combination with root pruning increased root length density and tips in young apple trees (Fang et al., 2017). Phosphorus is also a recommended practice at planting because it stimulates root growth.

The goals of orchard planting systems are early, high, and sustained yields with excellent fruit quality. Accordingly, there are tools to control the size of trees, such as dwarf rootstocks, pruning, and training systems. The chosen distance between trees and rows must take into account the vigor inherent to the rootstock and the final size of the tree. Apples trees are grown in rectangular plots or in hedgerows at different distances (Table 1.6). It is necessary to consider the distance between rows to allow the passage of tractors and other equipment that are used in the orchard.

TABLE 1.6 Recommended Spacing of Planting According to the Vigor of the Rootstocks Used

Rootstocks	Spacing
Seedling	6×6 m^2 or wider
Semidwarf	4.5×4.5 m^2
Dwarf trees	3×3 m^2 or slightly closer

Tree density is the most important factor affecting early yields in any system, and high planting densities achieve large yields during the first years. Improving the vegetative development of trees is the main objective in the first years so that full canopy closure can be reached. In the most common densities of the plantation (1500 and 3000 trees ha^{-1}) tree-canopy closure can be reached in the third or fourth season, meanwhile, in ultrahigh tree densities (>4000), canopy closure can be obtained in the first year (Ferree and Warrington, 2003).

.The results of several studies indicate that a north–south row orientation is preferred to an east–west row orientation. A north–south orientation provides better light interception and a more even distribution throughout the tree canopy (Cain et al., 1972; Jackson, 1980).

1.12 IRRIGATION

Irrigation is fundamental for apple production, because it increases yield, improves fruit quality, stabilizes production between years, allows production in environments not suitable for cultivation, and increases fertilization efficiency. The water requirement of an apple orchard depends on the cultivar, rootstock, planting design, irrigation system, soil management, weather conditions, and tree performance and production. Apple trees have a high-water requirement, about 6000 m^3 ha^{-1} yr^{-1}, because of their long productive cycle and high transpiration (Dzikiti et al., 2018).

If there is a cover crop between the trees, the water requirement is about 7.2 mm day^{-1} (Brouwer and Heibloem, 1986). The trees should be watered weekly during dry periods, especially during the first two years after planting. After this period, irrigation may be done every two weeks. Precision quantification of irrigation moment and water requirement is crucial to achieving good fruit yield and quality. The principal water requirement occurs when shoots and fruits are in active growth. Determination of soil humidity with tensiometers is also important in order to avoid excessive vegetative growth. Other important points are related to tree size, canopy volume, and root depth. There are three critical growth stages for water requirement: flowering; rapid shoot growth and spring root growth (6–7 weeks after budburst); and the fruit fill stage (4–8 weeks prior to harvest). After harvest, the amount of water must be sufficient to allow trees take up postharvest fertilizers. During dormancy phase irrigation is not generally needed. During the dormancy phase, irrigation is not necessary. The rooting system influences the volume and frequency of irrigation required. The root profile may extend down from 0.6 m (dwarf rootstocks) to 1.5–3 m (vigorous rootstock) depending on the local conditions (Westwood, 1993). Irrigation systems frequently used in apple cultivation are furrows, sprinklers, and drippers. Furrows are small, parallel channels that carry water to irrigate the trees. The furrows can be used in most soils, but they are not recommended in sandy soils because percolation losses can be high. In the sprinkler irrigation system, water is distributed through a system of pipes by pumping. It is then sprayed into the air through sprinklers, as small water drops which fall to the ground. Drip irrigation involves dripping water onto the soil at very low rates (2–20 l h^{-1}) from a system of small diameter plastic pipes fitted with outlets called emitters or drippers. In drip irrigation, water is usually applied every 1–3 days to provide the root system with favorable soil moisture. The selection of the irrigation system depends on the plant and fruit growth stage, root system, soil structure, weather conditions, water availability, and the need to use the irrigation system for other purposes such as fertigation or frost control (Table 1.7).

There are two possible scenarios for irrigated orchards:

1. Nonlimiting water, so the optimal amount of water to be applied will be the amount that results in the highest yield and fruit quality.
2. Limited water (e.g., drought) so maintaining production is desirable. Regulated Deficit Irrigation involves the management of apple orchards with limited water. The objective is to save water by controlling water stress without affecting fruit production or quality (Leib et al., 2006; Fallahi et al., 2010).

TABLE 1.7 Characteristics of Different Irrigation Systems

Irrigation System	% Water Efficiency (Vol. for Production/Vol. Irrigated)	Soil Characteristics	Application Rate	Suitable for Fertigation	Relative Costs
Furrow	50–70	Not suitable for sandy soils	8–12 ml ha^{-1}	No	Low
Sprinklers	60–80	Suitable for all soils	6 mm h^{-1}, 3–11 ml ha^{-1}	Yes	Medium

TABLE 1.7 *(Continued)*

Irrigation System	% Water Efficiency (Vol. for Production/Vol. Irrigated)	Soil Characteristics	Application Rate	Suitable for Fertigation	Relative Costs
Mini-sprinklers/microjets	7090	Suitable for all soils	5 mm h^{-1}, 6–10 ml ha^{-1}	Yes	Medium
Surface drip	80–95	Not suitable for sandy soils	3–8 mm h^{-1}, 4–8 ml ha^{-1}	Yes	High
Subsurface drip	80–95	Not suitable for sandy soils	2 mm h^{-1} 4–7 ml ha^{-1}	Yes	High

Another irrigation tool that saves water is partial root-zone drying (PRD), in which water is withheld from part of the root zone, while another part is well watered. However, O'Connell and Goodwin (2007) found some detrimental effects of PRD over fruit size, shoot growth, and yield in Pink Lady apples.

The irrigation strategies called Readily Available Water (RAW) involves the amount of water that is easily available to the tree. This is between a water potential of –0.002 to –0.03 MPa (field capacity) until the potential in which plants start to have difficulty sucking up water at –1.5 MPa (permanent wilting point). RAW is approximately 6% of total soil volume for sandy loam and clay soils and 8% of total soil volume for clay loam soils. Monitoring soil moisture is important to identify how much water should be applied

and when irrigation should be done (Allen et al., 1998; Naor et al., 2008).

In commercial orchards the measurements of stem water potential, canopy temperature, sap flow, and stem diameter variation have indicated to be adequate for precise irrigation scheduling. In apple trees, Ψstem is a sensitive indicator (Fernández and Cuevas, 2010). Depending on the plant variable response, the threshold value for Ψstem varied between –1.04 and –1.46 MPa (De Swaef et al., 2009). Based on controlled water stress, irrigation strategies were developed to improve irrigation efficiency and control the vegetative growth of commercial fruit trees. Regulated deficit irrigation was shown to improve fruit quality and reduce water consumption without negatively affecting fruit production (Ebel et al., 1995). Fruit color is often improved due to the lower vegetative growth of apple trees. In regions with high frost risks during spring, sprinkler irrigation is one of the most important techniques to reduce damage. This frost control consists in covering the plant with a layer of ice in continuous formation until the end of the frost.

1.13 NUTRIENT MANAGEMENT

Nitrogen (N) is a nutrient that needs to be applied universally to commercial orchards on a yearly basis to maintain productivity. It is the most widely applied nutrient and almost all soils are deficient in it. Nitrogen is usually applied via soil, but it could also be used as foliar applications (Tagliavini and Toselli, 2005). For soil fertilization, the selection of the nitrogen fertilizer must take into account soil pH. If soil pH is 5.5 or less, $(NO_3)_2Ca$ is preferred, whereas if soil pH is 5.5 or more, urea or $SO_4(NH_2)_2$ should be chosen (Westwood, 1993). For foliar application, urea at 0.6 kg 100 l^{-1} is used to improve tree growth and increase N reserves. Urea sprays at a higher concentration (2%–4%) can cause leaf burn and chlorosis in apple (Khemira et al., 2000). Late application of N is an ideal way to increase the reserves of the tree for early spring activities. N absorbed and translocated from the leaves in fall depends on the temperature. If the leaves senesce slowly, most of the N is translocated (Faust, 1989).

Phosphorus (P) requirement of apple trees is small with respect to N and Potassium (K). In mature orchards, only a few studies have reported P deficiency. Neilsen et al. (2008) found an increase of 20% in yield in several cultivars of apple with P-fertigation at bloom time. Anyway, P fertilization is standard practice at planting because it stimulates root growth.

Potassium (K) demand in apple is high, especially during fruit growth (Faust, 1989). Potassium is the most important component of fruit; however, any excess should be avoided and an adequate K:Ca balance should be achieved for good fruit quality and storage capacity (Brunetto et al., 2015).

Calcium (Ca) is an important nutrient for fruit quality. Ca prevents fruit disorders and increases tolerance to pathogens (Brunetto et al., 2015). Due to its mobility mediated by transpiration, it is difficult to increase calcium content in the fruit under conditions with high transpiration. To improve Ca absorption during early spring, it is recommended to avoid the fertilization with NH_4^+, K^+, or Mg^{+2} during the 4th–6th weeks after bloom. Foliar application with $CaCl_2$ at 0.1–0.6 g l^{-1} concentration (500–1000 l ha^{-1}) is also recommended in some cultivars that are susceptible to physiological disorders during fruit conservation (Marangoni and Baldi, 2008). Rates of CaCl$_2$ penetration are greatly affected by the stage of fruit development and it occurs in the early fruit development (in the Northern Hemisphere, between middle June and middle July) (Schlegel and Schönherr, 2002).

Boron (B) is a key micronutrient. Insufficient B has been associated with poor fruit set and fruit disorders. The B needed in flowers is transported from reserves in the wood. Boron sprays are effective in increasing the B content of flowers. Sprays before fall are more effective than in spring (Faust, 1989; Sánchez and Righetti, 2005).

Last but not least, zinc (Zn) plays an important role in fruit tree nutrition. It is the most limiting micronutrient for tree fruit production, especially when grown in alkaline soils. Zinc is included in many fertilizer programs as foliar applications since Zn fertigation fails to improve leaf Zn concentration (Neilsen et al., 2004). Neither dormant nor fall foliar applications contribute to the Zn content of young tissues form in the next season (Sánchez and Righetti, 2002).

To determine the nutrient doses for fertilization in an apple orchard, the absorption and distribution of nutrients within the trees must be taken into account (Marangoni and Baldi, 2008). An example of nitrogen and potassium is shown in Figure 1.8.

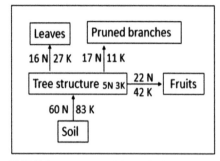

FIGURE 1.8 Quantities of nitrogen (N) and potassium (K) (kg ha^{-1}) absorbed annually and their distribution in the tree for a six-year-old orchard of apple cv. Gala/M9 with a production of 40 t ha^{-1}.

Nutrients in leaves, shoots and pruned branches return to soil after their degradation. Therefore, a simple approach is to apply nutrients that are exported by fruit harvest (Marangoni and Baldi, 2008). Along with yield data, macronutrient concentration in fruit is important to determine nutrient export (Table 1.8).

TABLE 1.8 Usual Concentration of Macronutrients in Apple Fruit at Harvest (g t^{-1} fresh weight)

N	P	K	Ca	Mg
300–500	70–140	700–1500	40–80	40–65

Fertilization could be applied to soil or via foliar. Soil fertilization could be distributed broadcast or diluted in irrigation water, which is called fertigation. The most widely used fertilizers for fertigation are urea, phosphoric acid, and potassium nitrate (Sansavini et al., 2012). The inclusion of K in fertigation was shown to increase leaf K concentration and improve fruit color development in Gala (Malaguti et al., 2006). Care must be taken in preparing the solution for fertigation. The pH of the solution should be between 5.5 and 7 and its electric conductivity must be kept below 1.5–2.0 dS m^{-1} (Marangoni and Baldi, 2008; Sansavini et al., 2012).

The efficiency of fertilization with N is usually low (<30%), even in fertigation, so it is important to be careful with the timing and technology of the application (Neilsen and Neilsen, 2002; Tagliavini and Marangoni, 2002). Neilsen and Neilsen (2001) reported that the efficiency of N use increased when irrigation was scheduled to meet the evaporative demand rather than a fixed rate. Demand for N in the young apple trees was low. The N availability should be monitored to improve N use efficiency and to avoid excessive N uptake (Brunetto et al., 2015).

Foliar cannot replace fertilization nutrition via soil but can be useful under certain circumstances (Sansavini et al., 2012). Absorption is fast and finished after 48 hours, reaching an efficiency of 75%–92% at least for nitrogen (Sansavini et al., 2012).

Some micronutrients, such as Zn and B, can be supplied exclusively via foliar applications.

Solution pH is important and low values of pH (5.5–6.5) are optimum for the absorption of urea and Zn (Sansavini et al., 2012). Weather conditions are also important because they determine the evaporation of solution and absorption of nutrients. It is not recommended to pulverize nutrients in high evapotranspiration conditions (temperature >25 °C, relative humidity <70%, and/or wind speed >8 km h^{-1}) (Sansavini et al., 2012).

Organic systems depend on nutrient cycling to assure proper nutrition, and N is the most limiting macronutrient (Granatstein and Sánchez, 2009). Permanent covers in an organic apple orchard for six years increased the organic matter and N content in soil in comparison with soil cultivation, but the apple leaf N declined and showed that additional organic fertilizer was required (Sánchez et al., 2007). Some other soil management systems, like compost application and use of mulches, also increased soil organic matter in organic apple orchards (Neilsen et al., 2014).

Soil analysis is used to determine the nutrient availability, but leaf analysis better indicates the nutritional status of the trees. Nutrient concentrations change with the age of the leaves. Nutrients with low mobility via the phloem, such as Ca, Mn, and Fe, tend to increase their concentration in leaves, whereas mobile nutrients, such as N, P, and K, can be remobilized to fruits or other centers of active growth (Faust, 1989; Sansavini et al., 2012). There is a certain plateau after shoot growth in which nutrient concentration is relatively stable and appropriate for determining the nutritional status of the tree (Sansavini et al., 2012). Nutrient standards for mid-shoot leaves is shown in Table 1.9.

TABLE 1.9 Nutrient Standard for Leaves of Apple in July–August (Northern Hemisphere)

N (%)	P (%)	K (%)	Ca (%)
2.1–2.8	0.16–0.28	1.2–1.8	1.2–2.0

Leaf analysis offers a limited diagnosis of deficiency in some nutrients such as Zn, Fe, and B. Another nutrient that could be erroneously diagnosed is N, because trees overfertilized with this nutrient are vigorous and can present a low concentration of N in leaves (Sansavini et al., 2012). Another important point when leaf analysis is interpreted is the relationships between nutrients. For example, with the N/Ca ratio of 10, apple can be stored for a long time, whereas an N/Ca ratio of 30 indicates that fruit will suffer breakdown (Faust, 1989). More recently, nutrient interactions using dual ratios in the Diagnosis and Recommendation Integrated System were developed for apple, and their efficacy in making correct diagnosis was validated (Sofi et al., 2017).

A correct determination of the nutritional status of the trees includes not only nutrient concentrations in leaf analysis but also the observation of the orchard. Important aspects of the orchard for the diagnosis are shoot growth, leaf size and color, symptoms of nutritional deficiency, and fruit load and size (Sansavini et al., 2012). Some symptoms of nutritional deficiencies are listed below:

N: lower shoot growth and leaf size; widespread pale green leaves

P: not usual; lower fruit set; early leaf drop

K: necrosis of leaf margins; smaller fruits with less color, sugar, and acidity

Ca: new leaves are small, light green, almost white; fruit disorders

Mg: chlorosis between veins; in severe deficiency, necrosis, and leaf drop

Fe: new leaves with chlorosis between veins

Mn: internal leaves with chlorosis between veins

B: shorter shoots; chlorosis and distortion of leaves; low fruit set; small and misshapen fruits

Zn: shorter shoots; small narrow leaves; rosette leaves; especially in calcareous soils.

1.14 TRAINING AND PRUNING

The fruiting patterns of the apple tree are classified into four groups according to the age of the fruitful wood.

1. Type I: there is one central axis, which is vertical and unbranched with short spurs along its entire length. The fruiting occurs predominantly on two- to three-year-old wood.
2. Type II (called spur type) fructifies in the lateral spurs of strong and upright branches in three- and four-year-old wood.
3. Type III has a weak main axis with several long branches and can fruit in one-, two-, and three-year-old wood.
4. Type IV has domed branches with fruit buds along the main branch and produces fruits on one- and two-year-old wood (Lespinasse, 1977).

The fruits produced on wood of more than five years are not of good quality. It is necessary to periodically renew the wood bearing fruit buds in order to obtain good quality fruits of and avoid biennial behavior, especially in cultivars belonging to types I and II (Dzhuvinov and Gandev, 2015). Cultivars respond differently to pruning and training due to differences in their overall level of vigor, growth, and fruiting. The main objective of training systems is to allow a greater interception and distribution of light. The choice of training system depends on the conditions that determine the growth potential of a site. Training utilizes various techniques, including pruning, to direct tree growth or form, and the development of the structural framework of the tree.

1.14.1 TRAINING TYPE

The most used training system for apple production is central leader. This system works well in areas of high solar radiation and long sunny days (Photo 1.3) (Collett, 2011). In the SolAxe system, branch bending is more intense than in central leader (Photo 1.4). This system does not adapt to mechanized pruning. There are other conduction systems that open the tree canopy by V (Tatura) or Y shapes (Photo 1.5).

PHOTO 1.3 Central leader in cv. Brookfield, San Patricio del Chañar, Argentina.

PHOTO 1.4 SolAxe in cv. Cripps Pink, General Roca, Argentina.

PHOTO 1.5 V Tatura in cv. Fuji, Cte. Guerrico, Argentina.

1.14.2 TYPE OF PRUNING

The purpose of pruning is to achieve structures that allow the best fruit production in the shortest possible time. The practice of pruning modifies the distribution of the intercepted sunlight to improve the formation and differentiation of fruit buds. Pruning is the most commonly used method to alter apical dominance, and there are general responses. Pruning changes the balance between the upper part of the tree and the part below the ground, while reducing the overall amount of dry matter accumulation.

Types of pruning according to planting age	
Planting pruning	Done immediately after planting to stimulate the growth of vigorous buds. The type of pruning depends on the training system, the density of the plantation, and the characteristics of the plant.

Training pruning	Done only once, when the tree is young, to achieve an adequate structure that quickly occupies the assigned space, while simultaneously producing fruit.
Fruit pruning	Carried out every year to regulate production. It helps to maintain stable productions and quality fruit every year, constantly rejuvenating the tree (stimulating growth and the production of good fruit structures), retaining and limiting the volume and the height of the plant, and increasing light penetration inside the canopy.

Types of pruning according to time of the year

Winter pruning	Generally carried out every year, when most of the leaves are not active. Its purpose is to train the plant and manage the fructification. Applied in areas with heavy winter frosts, to delay fruiting as much as possible.
Green pruning	Done when the plants still have leaves. According to the objective pursued, it is performed in the following times of the year: • Early spring pruning: to retain growth. • Late spring pruning: to eliminate unwanted growths (suckers) that compete with the normal development of the fruit.

• Summer pruning: to increase the light income inside the plant, which enhances the red color in apples.

• Late summer or fall pruning: carried out after harvest to increase light distribution in the canopy (especially in the inner and lower parts), to retain growth once the tree has reached the desired height and occupied the corresponding space, and to eliminate large branches or retain plants that are very high.

1.14.3 CANOPY MANAGEMENT

Apple trees growing on their own roots are, in general, too large for the economic production of good-quality fruits. Much of their canopy volume is unproductive, which adds to management costs without generating financial profits. Therefore, they are inefficient and expensive to produce and manage. For example, picking and pruning large trees require the use of ladders as well as high volumes of pesticide and powerful spraying machinery due to the size and density of the canopies. By contrast, the best quality of fruits is directly related to the light environment inside the tree canopy, which is achieved in part by dwarf rootstocks (Figures 1.9 and 1.10).

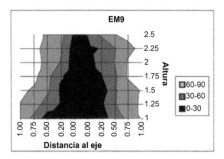

FIGURE 1.9 Photosynthetically active radiation PAR (%) inside the canopy of apple cv. Mondial Gala/EM9.
Source: Raffo et al. (2006).

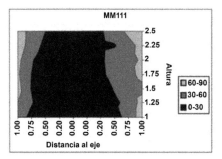

FIGURE 1.10 PAR (%) inside the canopy of apple cv. Mondial Gala/MM111.
Source: Raffo et al. (2006).

A relatively porous canopy is required to achieve an optimum balance between yield and quality; this is an important objective of the training and pruning systems, which also controls the relationship between vegetative growth and fruiting.

1.14.4 MECHANIZED PRUNING

The incorporation of mechanical pruning into flat production systems generates two very divergent but highly desirable situations:

1. The incorporation of family labor and/or people with physical conditions unsuitable for pruning and harvesting with stairs.
2. The incorporation of technology of both remote and proximal sensors, which is much more applicable to these flat driving systems.

The implementation of conduction systems such as "fruit wall" makes it possible to improve the economic and ecological aspects of fruit production. The advantages of the low and narrow fruit walls are related to the management, the characteristics of the crop, and environmental aspects. With small pedestrian trees, it is not necessary to use platforms or ladders, which is safer for workers. All the tasks of orchard management, from pruning to thinning and harvesting, can be done more quickly and easily from the ground. Any type of mechanization (e.g., flower-cutting machines, pruning, tunnel spraying, and weed control) is easier to apply in a low fruit wall than in traditional trees.

The main techniques to control the vigor and facilitate the formation and maintenance of the fruit wall involve increasing the number of axes per plant and green pruning during the summer. Other tools available to maintain the fruit walls are root pruning, application of growth regulators, and controlled water stress.

Multi-axis tree driving (ME) is a powerful tool for the formation of the fruit wall. The energy is divided into the formation of multiple structural branches, so each branch has less vigor. Consequently, in trees with six axes, for example, the secondary structures are usually absent and only the trunks remain as structural wood (Dorigoni and Micheli, 2015).

The Bibaum® trees (a registered trademark in Italy for trees with two axes preformed in the nursery) and the ME has a natural tendency to form a narrow fruit frame, with little secondary structure.

1.15 INTERCROPPING AND INTERCULTURE

Soil management techniques include intercropping with cover crops, use of herbicides under apple trees, tillage and mowing between rows, manure or compost applications, and use of different kinds of mulch (Westwood, 1993; Marangoni and Baldi, 2008; Sansavini et al., 2012).

Cover crops between rows improve soil health and apple tree performance (Sánchez et al., 2007; Marangoni and Baldi, 2008; Sansavini et al., 2012). Inclusion of cover crops in soil management of an orchard produces the following positive effects:

1. It improves soil fertility, due to an improvement in structure and increases in organic matter, soil aeration, nutrient retention, and availability in superficial layers of soil and nitrogen content, if legume species are used.
2. It reduces soil compaction and erosion.
3. It increases water infiltration.
4. It allows passage of agricultural machines in flood-irrigated orchards.
5. It decreases soil temperature.
6. It produces a more favorable soil–root interface.
7. It increases root growth of apple trees in superficial layers of soil.
8. It improves fruit quality (color and sugar content).

On the other hand, cover crops present competition to the apple tree root for water and nutrients, have their own requirements to be satisfied by irrigation and fertilization, and are not recommended in regions with low availability of water (<500 mm yr^{-1}) (Ryugo, 1988; Atucha et al., 2011; Sansavini et al., 2012). Cover crops also reduce the vegetative vigor of apple trees. Some other limitations of cover crops include a higher frost risk because of the reduction in heat accumulation in the soil. In addition, tree flower pollination could be hindered because bees could be attracted by cover crop flowers if they are not cut in apple bloom time. Apple fruit weight may also be reduced (Ryugo, 1988; Marangoni and Baldi, 2008).

The species most widely used as cover crop in orchards belong to the family of *Poaceas,* even though they are very competitive in water and nutrients. Sometimes, mixes of species are used, such as *Lolium perenne* plus *Festuca arundinacea* or *Poa pratensis* (Sozzi, 2007). In poor soils, legumes are preferred due to their ability to fix nitrogen from the atmosphere. Some characteristics of different species are listed in Table 1.10.

TABLE 1.10 Characteristics of Different Vegetal Species Used as Cover Crops in Apple Orchards

Species	Settlement
Festuca arundinacea	Excellent
Festuca rubra	Excellent
Lolium multiflorum	Excellent
Lolium perenne	Excellent
Po pratense	Good
Trifolium pratense	Good
Trifolium repens	Excellent

Alternative management of orchard involves a compromise: cover crops controlled by brush cutter or mowing machines between rows and weed control closed to the trees in the rows (Faust, 1989; Westwood, 1993). Weed control is necessary to avoid water and nutrient competition, and it is especially important in spring time and crucial in young plantations. Bands of 1.0–1.5 m in width must be kept without spontaneous vegetation.

Weed control could be achieved by manual, mechanical or chemical techniques. Manual control involves labor cost and is usually expensive. Mechanical weed control is difficult close to young trees, so in this case chemical control is preferred (Marangoni and Baldi, 2008). Even with herbicide application, caution must be taken to avoid pulverization over tree foliage or small trunk in these young trees (Ryugo, 1988; Westwood, 1993). Shelter around new plants (one- to three-years old) is used to protect them from mechanical or phytotoxicity damage (Marangoni and Baldi, 2008).

Herbicides could be divided into two groups: those of residual and preventive action (oxifluorfel, diuron, etc.) and those with direct action over weeds (glifosate, fluazifop, etc.). Herbicides of the first group inhibit the germinative capacity of seeds or are absorbed by roots, and they have a long persistence in soils. However, they are detrimental to orchard productivity and soil fertility in the long term and they are dangerous to the environment, especially if percolated to the water table. Herbicides of the second group (desiccant) are more widely used in orchards because they are safer to the environment, even though they are less effective in perennial species (Atucha et al., 2011; Sansavini et al., 2012).

Herbicides could be applied at different times during the season. The first application is usually done in spring. Dry grass after treatment generates a mulching effect that reduces the emergence of new weeds. A second application is usually necessary during summer. Fall application of systemic herbicides like glyphosate is useful to control perennial species (*Cynodon dactylon* or *Convolvulus arvensis*) in highly infested orchards. Efficient applications are achieved when weeds are in the first stages of development and weather conditions are appropriate (Marangoni and Baldi, 2008).

Tillage is used to reduce evaporation and water and nutrient competition, but it destroys superficial tree roots (>12 cm) and soil organic matter (Faust, 1989; Sansavini et al., 2012). Mowing reduces the transpiration of cover crops and restricts water loss. Dry residues coming from cover crop mowing provide nitrogen and decrease nutrient competition. Mower machines can be equipped with re-entry devices that can not only make it possible to get closer to the trees but can also cause some

damage to small trunks (Marangoni and Baldi, 2008). Another important problem related to mechanization is the soil compaction generated by the passage of agricultural machines between the rows, especially in wet or naked soil conditions. Soil compaction affects water infiltration, soil aeration, nutrient availability, root and tree growth, and fruit production and quality (Sozzi, 2007).

Other soil management techniques include manure or compost application, incorporation of annual crops to the soil, and mulch use in the rows. These practices improve soil properties, like organic matter, nutrient content, and water infiltration (Neilsen et al., 2003; TerAvest et al., 2010; Atucha et al., 2011; Sansavini et al., 2012; Forge et al., 2013; Neilsen et al., 2014). Mulch use also contributes to reducing weed competition, keeping water in the soil, decreasing soil temperature, and increasing apple tree vigor and yield (Andersen et al., 2013; Forge et al., 2013). Mulch may be organic or inorganic. Organic mulch may be made from dry cuts of cover crops or pruning residues. Cover crops produce 4–7 tons of dry matter per hectare each year (80% above ground). Pruning residues could provide 2–6 tons of dry matter per hectare if they are left in the orchard. Inorganic mulch is usually made of black polyethylene film (Sansavini et al., 2012).

1.16 FLOWERING AND FRUIT SET

Flowering in apple trees occurs once the trees reach the adult or mature phase after they overcome the juvenile period (Ryugo, 1988). The flowering process in fruit trees implies the initiation and development of flower buds in the summer and fall of one season. Then, the flowering process itself occurs in early spring the next season (Faust, 1989).

1.16.1 INITIATION AND DEVELOPMENT OF FLOWER BUDS

Induction of flower buds is a qualitative change governed by hormonal balance and/or nutrient distribution. To receive the inductive stimulus, buds must have a critical number of nodes of 16–20 (Tromp, 2000). Induction of flower buds of apple trees in the Northern Hemisphere occurs between middle June and middle July (Westwood, 1993). The initiation of flower buds is strongly stimulated by exposure to light. Practically no development of flower buds occurs in shaded areas of apple trees where there is less than 30% of sunlight (Faust, 1989), while flowering is maximal at exposures above 60%. Gibberellins (GA) were reported to inhibit the initiation of flower buds in apple. The seed of developing fruit is high in GA-like substances, so the presence of young fruits is the most important factor inhibiting flower bud initiation. This inhibiting effect takes place between three and eight weeks after bloom (Tromp, 2000). Bud development during fall is a prerequisite to achieving a good bloom in the following year. A delay in harvest greatly decreases the bloom in the following year. Premature defoliation before normal leaf fall produces similar results and greatly decreases successive yield. Buds continue to develop during winter. Fall application of nitrogen greatly accelerates the development of buds and plays an important role in poor buds (Faust, 1989). Flower initiation during spring has been considered an important condition for good flowering. The late events in bud development during the fall are also important. Early phases of flower bud development are governed by hormonal events, whereas the late development of flower buds depends on the availability of carbohydrates and nitrogen. Growers must not only reduce the fruit load to prevent flower inhibition but also protect the leaves until late fall to maintain their photosynthetic efficiency.

1.16.2 FLOWERING TIME

Flowering time depends on cultivar and weather conditions during winter. Unfavorable conditions to

Apple

complete dormancy result in a late, weak and extended bloom period. Flowering tends to last about 15 days. Extended blooms are undesirable because they cause an extended harvest that requires several pickings (Faust, 1989).

1.16.3 ANTHESIS AND FLOWERING TYPES

Apple belongs to the Rosacea family, whose flowers are hermaphrodite with five sepals, five petals, numerous stamens, and a pentacarpelar pistil (Sansavini et al., 2012). Inflorescences of apple usually have 5–6 flowers with similar pedicels (Ryugo, 1988; Westwood, 1993) in which the apical flower develops and opens first, followed by the lateral flowers in the cluster (determined inflorescence). The apical flower of the cluster is called the "king" flower of the apple (Photo 1.5).

1.16.4 POLLEN VIABILITY AND GERMINATION

Cold temperature during early spring diminishes pollen production and decreases the vitality of the pollen grains formed. On the other hand, high temperature during spring often results in sterile pollen. A temperature response index based on the mean daily temperature was developed in order to estimate the time required for pollen tube growth in apples (Faust, 1989). Boron and calcium improve pollen germination and enhance pollen tube growth (Sanzol and Herrero, 2001; Sansavini et al., 2012).

1.16.5 POLLINATION AND FERTILIZATION

Pollination is the transfer of pollen grains from the anthers to the stigma. The effective pollination period (EPP) is determined by the number of days of ovule longevity minus the time lag between pollination and fertilization (Ryugo, 1988). EPP allows us to identify the factors limiting the fruit set associated with stigmatic receptivity, pollen growth, and ovule development (Sanzol and Herrero, 2001).

High temperatures during flowering not only accelerate pollen tube growth, but also stigma and ovule degeneration. Low temperatures during flowering extend the EPP through a longer viability of ovules. In apple, ovule longevity is normally 10–15 days, and pollen requires 1–2 days to reach ovules at 15 °C. Ovule longevity and EPP can be extended through cultural practices like summer nitrogen application (Sanzol and Herrero, 2001; Sansavini et al., 2012). Fruit trees that need insects for the pollination have colorful flowers and nectar presence. In commercial orchards, beehives are located during bloom (5–8 beehives per hectare). Bees do not fly very well under rain conditions, strong winds (more than 2–40 km h⁻¹), or temperatures below 10 ºC (Ryugo, 1988). Bees fly from tree to tree within the row, rather than from one row to another one (Sansavini et al., 2012).

1.16.6 POLLINIZER

Apple and pear trees are self-incompatible. If one cultivar is expected to pollinate another, the two cultivars must bloom together. There are lists of the bloom times of cultivars that can be used as guidelines for planting plans (Faust, 1989).

To be pollinizers, varieties must have good pollen production, be compatible, and bloom just before the commercial cultivar. In the Rosacea family, self-incompatibility is regulated by one multi-allelic locus S. To avoid a poor fruit set, a combination of cultivars with different allelic components must be chosen (Sansavini et al., 2012). A list of incompatibility (S-) genotypes of apple cultivars is presented in Broothaerts et al. (2004). Triploid apple cultivars have low value as pollinizers due to their low viable pollen (Faust, 1989). A minimum percentage of pollinizer trees in an orchard is 11%, one pollinizer every nine trees. For practical management, pollinizer trees are planted in complete rows (one row of pollinizer every 2–5 rows).

1.16.7 FRUIT SET

To achieve a good fruit set, a sequence of physiological events is needed. The first requisite for a good fruit set is the development of strong flower buds. The second one is a certain temperature to assure good pollination and fertilization. The third requirement is a high level of photosynthate supply for young fruits (Faust, 1989; Schaffer and Andersen, 1994).

1.17 FRUIT GROWTH, DEVELOPMENT, AND RIPENING

The embryo development in fruit takes place in parallel with the ovary development, which ultimately produces the fruit. The apple fruit is composed of the mature ovary associated with the receptacle. The fruit provides an adequate medium for seed maturation and allows a natural mechanism for its dispersion (Sozzi, 2007). The fruits go through a significant increase in size after anthesis. The growth pattern followed by apple fruits is represented by a smooth sigmoid curve. The development of the apple fruit can be divided into three stages: phase I or period of exponential growth (cell multiplication); phase II, or period of linear growth (enlargement or cell expansion); and phase III, or the final stage of growth and maturation period (Faust, 1989) (Figure 1.11). The cell division period lasts 3–4 weeks. A significant difference in fruit size has been observed among apple varieties, depending on the length of both the cell division and expansion stages (Sozzi, 2007; Eccher et al., 2014).

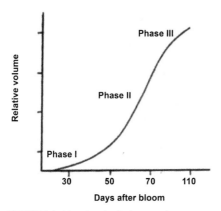

FIGURE 1.11 Apple fruit growth curve.

The endogenous factors that control fruit growth are hormonal balance, variety and rootstock, flower position in the inflorescence and in the tree, competition with other developing organs. As regards hormonal factors, seeds are a source of different phytohormones that promote growth. Seed development regulated fruit growth, final size and shape because it ensures an adequate hormone production (Eccher et al., 2014). A variety can directly influence the fruit growth rate through the branch type and orientation, and the insertion angle of the branches. By means of these characteristics, the variety can affect light interception, and therefore, the fruit growth. Rootstock choice affects the flowering abundance, the fruit set, size, and quality (Sozzi, 2007). Flower position in the inflorescence and in the tree also affects fruit size: in apple var. Red Delicious and its clones, the "king" or central flower produces a bigger fruit than the lateral flowers, particularly due to its larger size cells (Sozzi, 2007).

Competition between developing organs for photoassimilates and cell turgor pressure also directly affects the final fruit size. Sugars such as sorbitol and sucrose are the source of the required energy for fruit growth (Zhang et al., 2004). Hormones and seed number determine the fruit sink strength in attracting nutrients. Smaller fruitlets are weaker sinks because of their lower auxin content and smaller number of viable seeds. If the tree is unable to satisfy the nutritional requirement of all the developing organs, it regulates its crop load throughout the abscission of young fruitlets called "physiological drop" (Eccher et al., 2014).

The exogenous factors that affect fruit growth are climatic and edaphic factors, including temperature, light exposure, and the availability of water and nutrients. Early fruit thinning increases the size of the nonthinned fruits mainly through the increase in the number of cells (Sozzi, 2007).

Fruit maturation is the step in which the apple reaches its full size, acquires the capacity to mature, and finally ripens (Eccher et al., 2014). The apple is a climacteric fruit, generating an ethylene burst that coincides with increased respiration rates during maturation and ripening (Watkins et al., 2004; Giovannoni, 2004). Since apple fruits start their rise in the respiratory rate in the climacteric phase and therefore its ripening stage, this process may slow down but never be blocked (Fadanelli, 2008).

Ethylene is a gaseous phytohormone of very simple composition (C_2H_2), which has numerous effects on plant growth, development, and senescence. However, its incidence in climacteric fruit maturation represents, by far, its most important role (Alexander and Grierson, 2002; Sozzi, 2007).

Apple fruit respiration is conditioned throughout the period of growth by its phenological stage. Some conditioning factors affect the fruit after harvest: variety, fruit temperature, environment temperature, and gaseous composition of the storage atmosphere (Fadanelli, 2008), temperature being the most relevant factor (Table 1.11).

These biochemical changes during fruit ripening include an increase in ethylene production and respiration. Other processes also occur, including chlorophyll degradation, carotenoid and anthocyanin biosynthesis, and an increase in flavor and aroma components. Along with these processes, there is also a decrease in starch content and acidity, an increase in sugars and cell wall-degrading enzymes (Prasanna et al., 2007).

Color changes during apple maturation and ripening are largely the result of chlorophyll breakdown, increased carotenoid concentration, and increased anthocyanin concentration in red varieties. The loss of chlorophyll results in decreased green color. This process can occur via several routes, one of which is catalyzed by a family of enzymes called chlorophyllases. The activity of these enzymes is stimulated by ethylene and increases in respiration during the climacteric rise. The loss of chlorophyll and the increase in carotenoids are most noticeable

TABLE 1.11 Apple Respiratory Rate at Different Temperatures

Temperature (°C)	0	5	10	15	20
Respiratory rate (ml CO_2 $kg^{-1} h^{-1}$)	3–6	4–8	7–12	9–20	15–30

in apple varieties where yellowing of the peel occurs. The carotenoid synthesis increases during the early stages of ripening in apples, but this process can lag behind chlorophyll loss. Differences in the regulation of chlorophyll loss and carotenoid synthesis between varieties lead to variety-specific patterns of background color development. A whitish background color stage is evident in apples where chlorophyll loss occurs at a greater rate than carotenoid synthesis. Seasonal temperature variation may affect this process. Tree nutritional status may also affect color changes during ripening. High nitrogen availability and uptake delay the maturation process, as well as the synthesis of anthocyanin. Light is required for anthocyanin accumulation in most apple cultivars, and no relationship between red color development and climacteric ethylene production has been demonstrated. Regulation of anthocyanin synthesis varies between cultivars and strains. Low temperature exposure is an important factor regulating anthocyanin synthesis in many, but not all, red apple varieties (Mattheis, 1996).

During ripening, apples develop a profile of volatile compounds (basically esters) that produce a characteristic aroma. In ripe apples, butyl acetate, 2-methylbutyl acetate and hexyl acetate are volatile compounds that dominate the flavor (Zhu et al., 2008). Ethylene gas regulates ester biosynthetic pathways (Yang et al., 2016).

Apple starch is composed of two glucose polymers: amylose and amylopectin. Differences in amylose to amylopectin ratios exist between different apple varieties, and the ratio can change during ripening. As starch breakdown progresses, fruit sugars increase. Sugar released during starch breakdown is utilized for respiratory metabolism (Mattheis, 1996) for as well as organic acids. Both of them are found mostly inside vacuoles and they are the largest contribution to flavor. Fructose, sucrose, and glucose predominate among sugars and malic acid among organic acids. Apple flavor is influenced by the absolute and relative contents of sugars. In general, organic acids decline during maturation, possibly due to their use as a respiratory substrate or to their conversion into sugars (Sozzi, 2007).

Apple fruits soften moderately during ripening, causing a crunchy texture (Sozzi, 2007). The firmness decrease pattern in most apple cultivars presents a nonlinear curve. It is characterized by slowly softening in the first phase, accelerating during the second phase, and then softening again in the final phase (Johnston et al., 2002).

The factors that affect apple fruit firmness include those related to production, such as variety, maturity state at harvest, cultural practices, light exposure during development, and environmental stress before harvest. Fruit firmness is also conditioned by tissue structural characteristics, including the chemical composition of the cell wall and their macromolecule organization, enzymatic activity, turgor pressure and shape, size and distribution of the cells and intercellular spaces (Sozzi, 2007). During storage, early-season cultivars soften more rapidly than later-season cultivars. This maybe due to early-season cultivars have larger cells and intercellular spaces than later maturing cultivars (Johnston et al., 2002). The cell wall is a complex structure consisting of cellulose microfibrils embedded in a matrix of hemicellulose and pectin. Softening of apple fruit occurs due to solubilization and depolymerization of cell wall components by hydrolases and by loss of sugars, mainly galactose and arabinose (Sozzi, 2007; Bapat et al., 2010).

1.18 FRUIT RETENTION, FRUIT DROP, AND FRUIT THINNING

1.18.1 FRUIT RETENTION AND FRUIT DROP

There are three moments of fruit fall in apple trees. The first drop of nonfertilized flowers occurs shortly after petal fall (Racskó et al., 2006). The second drop ensues during the 6th–8th weeks after bloom and is known as the "June drop" because it occurs in June for the Northern Hemisphere (Faust, 1989). The third drop occurs at preharvest time (Racskó et al., 2006). The first drop is the largest one (90%–98% of initial flowers) so it is called the "cleaning" drop (Sansavini et al., 2002; Racskó et al., 2006). It occurs in all varieties but is more pronounced in some like var. Red Delicious. This drop depends on weather conditions during bloom (Sansavini et al., 2002), and it is usually heavy after self-pollination but less so after cross-pollination (Faust, 1989), because fruits with less than three seeds tend to shed (Racskó et al., 2007).

The second drop is more conspicuous because the size of the dropping fruit is much larger. The second drop is related to the loading capacity of the trees (Racskó et al., 2006). Vigorous shoot growth is also responsible for fruit drop (Racskó et al., 2007).

The third drop depends on weather conditions (e.g., wind) and varietal susceptibility (Racskó et al., 2006). A common compound for avoiding preharvest fruit drop is naphthalene acetic acid (NAA). The ethylene biosynthesis inhibitor aminoethoxyvinylglycine (AVG) and the competitive antagonist for ethylene receptors 1-methylcyclopropene

(1-MCP) both reduce preharvest fruit drop (Arseneault and Cline, 2016). Several strategies with NAA 20 mg l⁻¹, AVG 125 mg l⁻¹ and 1-MCP 396 mg l⁻¹ have been studied in var. Golden Delicious (Yuan and Carbaugh, 2007; Yuan and Li, 2008).

Standard management strategies for fruit retention include using a combination of varieties as pollinizers, introducing beehives during bloom, and applying boron and different substances (Faust, 1989; Sanzol and Herrero, 2001; Sansavini et al., 2012; Ramírez and Davenport, 2013). Applications of AVG or Daminozide soon after bloom usually decrease June drop (Greene, 1980; Faust, 1989; Luckwill and Silva, 2015). Response may depend on the time of application, the position of the flower within the cluster, and the age of the wood on which the cluster is situated. Unexpectedly, a detrimental effect on fruit production was observed two years after daminozide application (Luckwill and Silva, 2015).

1.18.2 FRUIT THINNING

Fruit trees tend to set a large number of fruits, which trees cannot support in good conditions. Removing excess fruits makes it possible to adjust fruit load to a desirable level (Ryugo, 1988). Thinning improves the characteristics of fruit quality, such as the size and color of the

fruit. Concurrently, the taste of the fruit (sugar, acid content, and firmness) will be increased (Link, 2000). Another purpose of thinning in apple is to partially remove the source of gibberellins in the seeds, which prevents flower bud formation (Faust, 1989).

To obtain high-quality apples, a good relation between leaves and fruit has been established for different times over the season: 1–4 leaves per fruit at petal fall, 1015 around June drop, and 20–40 leaves at harvest (Racskó et al., 2007). Minimum leaf to fruit ratio depends on growing habits (Sansavini et al., 2012). Standard cultivars of apple require a leaf to fruit ratio of 40:1 at harvest, while spur-type cultivars of apple achieve good fruit size with a lower leaf to fruit ratio (25:1) (Ryugo, 1988).

Thinning can be done by chemical applications, manually or mechanicaly (Sansavini et al., 2012; Greene and Costa, 2013).

Bloom time thinning is undesirable in many locations until the frost danger is over (Faust, 1989; Sansavini et al., 2012). The most widely applied blossom-thinning chemical was sodium 4,6, dinitro-o-cresylate (DNOC). It is caustic, so it burns stigmas and prevents germination of pollen deposited on the stigma. DNOC applications started to generate environmental concerns because of its heavy metal contents

(Dennis, 2000). A range of new flower thinners has emerged after DNOC production was discontinued (Wertheim, 2000). The chemicals frequently use in apple for blossom-thinning are 2-chlorophosphonic acid (ethephon), ammonium thiosulfate, oil and lime sulfur (Greene and Costa, 2013).

Postbloom thinners are the auxins NAA, naphthaleneacetamide (NAAm), the cytokinin 6- benzyladenine (6-BA) (Greene et al., 2016). NAA is applied at concentrations up to 20 ppm and NAAm up to 100 ppm. The most common dose of carbaryl is 750–1000 ppm. Benzyladenine is usually applied at 50–100 ppm, but even higher concentrations may be needed for difficult-to-thin cultivars (Wertheim, 2000). Chemical thinners are applied when fruit diameters are 9–12 mm, between 10 and 25 days after bloom (Ryugo, 1988).

The possible mode of action of auxins includes several simultaneous processes: a reduced basipetal auxin transport plus a photoassimilate deficiency, which enhance enzyme activity in the abscission layer in peduncles, while fruit growth is retarded and more stress ethylene is released (Untiedt and Blanke, 2001). Both NAA and NAAm can induce the formation of "pigmy" fruit in var. Delicious (Dennis, 2000), while 6-BA not only thins apples but also increases fruit size via cell division stimulation (Greene et al., 2016).

The insecticide Carbaryl is effective to thin apple, but its thinning mechanism is still unknown (Wertheim, 2000). The use of Carbaryl is prohibited in Europe and auxins are not allowed in North European countries (Greene and Costa, 2013).

More recently, new chemical thinners have emerged and are under evaluation. They are the abscisic acid and 1-aminocyclopropane-1-carboxylic acid and the photosynthesis inhibitor metamitron (Brevis). Metamitron is an effective thinner when applied at 8–14 mm of king fruit diameter (Greene and Costa, 2013).

Multiple applications of single chemicals or mixtures are sometimes used to achieve the desired degree of thinning, especially in difficult-to-thin cultivars like spur-type Delicious (Ryugo, 1988; Dennis, 2000)

1.19 HARVESTING AND YIELD

Fruit quality is an essential parameter in the agro-food business, and it depends on the optimal harvest window and adequate storage conditions during the postharvest period (Vielma et al., 2008). Apple is harvested at the preclimacteric stage, when respirations is low. The optimal harvest window depends on the variety. For example, the harvest period for var. Red Delicious, Granny Smith and cv. Fuji last from 10 to 15 days, var. Golden Delicious from

20 to 25 days, and cv. Gala from 6 to 8 days (Fadanelli, 2008). Not all the fruit is in optimal conditions for harvest at the same time. For this reason, it is necessary, in some cases, to pick it in two or more steps. The quantity and frequency of each step will depend on the characteristics of the variety maturing process; var. Red Delicious apple clones and cv. Gala clones may be picked in more than one step (Campana, 2007).

The yields of different varieties depend on several factors, such as variety-rootstock system, crop age, planting density, training system, cultural practices, and environmental conditions. Some varieties, such as cv. Gala, Fuji and Cripps Pink, have a high productive potential with yields of 55–60 t ha^{-1}, whereas other varieties, like var. Red Delicious, have a lesser potential with yields of 40–45 t ha^{-1} (Fadanelli, 2008).

1.19.1 MATURITY INDICES

Several methods may be used to determine optimal harvest time. Fruit maturity evaluation can be done by destructive and nondestructive methods. Destructive methods involve physical and chemical analysis such as flesh firmness, starch content, soluble solid content, and titratable acidity (Skic et al., 2016). However, no single maturity parameter is adequate across all varieties (Watkins et al., 2004).

Apple pulp softens during ripening. In 1925, Magness and Taylor developed an instrument, called penetrometer, to measure the fruit firmness in order to estimate maturity at harvest. The principle is still used, although several adaptations in size, shape, and function have been made. The penetrometer measures the resistance offered by the fruit pulp to the penetration of a rigid tip or plunger. This tip is an interchangeable piece of known diameter and convex ends. It also has a ring-shaped groove that indicates the depth to which it must be introduced into the pulp to perform the measurement. A tip of 11 mm (7/16") is used in apples. In the industry, the pulp resistance is expressed in pound-force (lb$_f$) and kg cm^{-2} (kg$_f$) by a graduated scale but in scientific papers, firmness is expressed in Newton (N) (Campana, 2007). The conversion is N = lb$_f$ × 4.448222; and N = kg$_f$ × 9.807 (Mitcham, 1996). To perform the measurement, the operator must insert the tip in a perpendicular direction to the fruit surface without skin at a constant speed. The most critical feature of testing is the speed of the force. The faster the pressure is applied, the higher the reading will be. The penetrometer can be mounted on a drill press, which allows greater control of the direction and speed of penetration, reducing measurement variability. In addition, electronic equipment that facilitates measurements is available. However, the size and energy requirement of this equipment restrict its use in the orchard, and it is therefore used exclusively in laboratories and packinghouses (Bramlage, 1983; Harker et al., 1996; Campana, 2007).

Starch accumulates in fruits during its growth and it is progressively hydrolyzed to soluble sugars during the ripening period. The level of hydrolysis of the starch correlates with the state of advancement of the maturation process and is very useful to determine harvest start (Campana, 2007) and maturity. The presence of starch can be detected from the staining produced by the action of a solution of iodine and potassium iodide when exposing the pulp of the fruit, cut transversally at the equatorial zone. The colorless area corresponds to the sectors in which the polysaccharide has been degraded, while the blue zones indicate that there is still starch presence. The colorless area expands from the center of the fruit, and the colored area retracts toward the periphery of the fruit until it disappears completely. The figure obtained in this color reaction is compared with a chart or pattern of specific degradation for each variety or with generic letters of international use such as those published by the Center Technique Interprofessionnel des Fruits et Légumes in 2002 (Campana, 2007).

The visual pattern provided by this test is unique for each apple cultivar, as is the relationship between starch loss and ethylene production (Mattheis, 1996).

The accumulation of sugar released from the glucose polymer chains results in increased fruit soluble solid concentration (SSC). Varieties harvested with low starch scores (such as var. Red Delicious) exhibit higher SSC values after storage because the starch remaining at harvest continues to be degraded to sugar. Sugar released during starch breakdown is used for respiratory metabolism (Mattheis, 1996).

The SSC in fruits is composed mainly of sugars, although it includes organic acids, amino acids, phenolic compounds, soluble pectins, and various minerals. During maturation, SSC tends to increase, and it is determined by refractometry. The refractometer is an optical, analog, or digital instrument whose operation is based on the phenomenon of light refraction when light passes through a solution. Light beam deviation will depend on the number of solutes contained in the fruit juice (Mitcham et al., 1996; Campana, 2007). The refractometer has a scale for the refractive index and another for equivalent °Brix or SSC percentage.

Apple fruit flavor results from the combination of sugars, organics acids, astringent compounds, and aromatic substances. During ripening, organic acids decrease. The gradual evolution and importance of these acids in apple fruit flavor have led to the use of an acidity index, combined with other parameters, to determine the optimal harvesting moment (Campana, 2007). Titratable acidity (TA) can be determined by titrating a known volume of apple juice with 0.1 N NaOH (sodium hydroxide) to an endpoint of pH = 8.2. The TA (express as the percentage of malic acid) is calculat as follows: ml NaOH × N (NaOH) × 0.067 (malic acid meq. factor) × 100 ml juice titrated (Mitcham et al., 1996).

A nondestructive index is the time period between full bloom (75% open flower) and the commercial harvest. This index behaves as a good maturity indicator and provides a rough estimate of harvest date (Collett, 2011; Calvo et al., 2012). Internal ethylene concentration has become a widely used maturity index to determine harvest decisions because a rise in ethylene production is associated with the initiation of ripening. The internal ethylene concentration can be affected by the cultivar, growing region, orchard within a region, growing season conditions, and nutrition (Watkins et al., 2004). Other nondestructive index indices may be considered peel color wich is defined by comparison with specific color tables for each variety or by using a colorimeter (Calvo et al., 2012). There are also nondestructive systems to determine the apple ripening stage and optimal harvest time. Optical methods such as visible and near infrared (VIS/NIR) spectroscopy, time-resolved reflectance spectroscopy, hyperspectral backscattering imaging, laser-induced backscattering, and chlorophyll fluorescence are the most promising nondestructive systems. These methods make possible to classify the fruit and control its quality, and they may be used directly in the orchard (Zhang et al., 2014; Skic et al., 2016).

Commercial applications of maturity indices use days after full bloom as a first guide and complement it with starch index and pulp firmness. Maturity standards and values at which the harvest of the different varieties begins and finishes should take into account whether the fruit will be stored or shipped immediately (Kader, 2007). Many apple-producing regions have developed maturity standards that guide the beginning of the harvest. These standards can be voluntary or legally imposed. With some differences according to apple varieties, harvest can be started with pulp firmness values of 6–9 kg cm^{-2}, 9%–14.5% of SCC, TA of 3–10 g l^{-1}, and 2–4 starch degradation index (Fadanelli, 2008).

1.19.2 HARVEST

Harvest consists in detaching fruits of the plant once they have reached an appropriate maturity degree, according to their destination. It is a complex operation in which it is necessary to consider the maturity stage of the fruit, management and coordination of the factors involved in harvesting (resources, training, and control), environmental conditions, and preharvesting cultural practices (Campana, 2007). Harvest is a key process in fruit production since it determines the product quality obtained and therefore the percentage of fruit suitable for the market (Calvo et al., 2012). Apple is harvested by hand, placed in bags, carefully transferred to the field containers, and transported to the packing plant. During this process, cushioned bags with a solid frame reduce fruit damage by impact (Mitcham and Mitchell, 2002).

Blows and wounds must also be avoided, and the fruit that does not meet the desired quality conditions, such as those with sunburn browning, inappropriate sizes, insufficient color percentage, and staining, must be discarded. This type of fruit causes an unnecessary increase in harvesting, transportation, and cooling costs, and reduces quality grade and storage capacity. Therefore, proper training of the harvesters and the supervision of harvest are very important tasks (Collett, 2011; Calvo et al., 2012).

According to the variety–rootstock and the training system, harvest can be done from the ground or will require auxiliary equipment such as ladders or picking platforms that allow workers to reach the tree top and inside the canopy. Picking platforms are devices on which workers are located to carry out harvest or other tasks such as thinning or pruning. This equipment can be pulled by a tractor or self-propelled. These platforms are coupled to an articulated hydraulic arm that allows vertical and horizontal movements to access the different parts of the cup. In multiple teams, 6–8 workers are located at different platform heights where they collect the fruits as the equipment advances through the ally way. Fruits can be either collected in picking bags and transferred or directly set on bins. Another available device consists of transport belts that carry fruit through rails to bins set in the back of the machine. Mobile picking platforms offer the operator more comfort and better stability conditions than the ladder. These mechanical aids reduce operating time and decrease the number of damaged fruits during harvesting (Campana, 2007).

1.19.3 MECHANIZED HARVESTING

Mechanized harvesting can be an interesting alternative because costs can be reduced, due to an increase in performance and an improvement in the quality of harvested fruits (Calvo et al., 2012). Even though commercial robotic harvesting systems have not been applied to apple harvest yet, some research has been conducted and several companies have developed prototypes with different degrees of success in terms of efficiency and harvest speed (Zhang, 2017). A robotic harvesting system would make it possible to attend the increasing labor demand, decrease the accidents of workers and reduce harvesting cost.

The fruit location in the tree is one of the first aspects to be solved in the implementation of robotic harvest. Over the past decades, several successful projects have applied a few technologies. Artificial vision is one of the most commonly used strategies since it is capable of detecting fruits based on color, size, texture, and other parameters. Other sensing technologies such as laser scanners are also used to determine the relative position of an apple from a robot (Zhang, 2017).

Another aspect to be considered is finding the appropriate tree structure where robotic harvesting systems may be implemented.

Separating fruit from the tree is the main and most challenging point in the implementation of this type of system. Almost all robots use a mechanical arm for fruit collection. This technology must be adjusted to allow fruit deposition in containers or rails (Zhang, 2017).

The fruit location in the tree is one of the first aspects to be solved in the implementation of robotic harvest. Over the past decades, several successful projects have applied a few technologies. Artificial vision is one of the most commonly used strategies since it is capable of detecting fruits based on color, size, texture and other parameters. Other sensing technologies such as laser scanners are also used to determine the relative position of an apple from a robot (Zhang, 2017).

1.20 PROTECTED/HIGH-TECH CULTIVATION

1.20.1 FRUIT PROTECTANTS

To establish the impact of the environment on crops, it is important to understand that a cultivated plant is not usually found in its natural habitat. Dr Robert Evans stated, "The best time to protect a crop is before it is planted, and site selection is the most effective passive risk avoidance strategy" (Lehnert, 2013). Another important topic is the selection of the most adequate scion/rootstock.

However, it is often necessary to use technologies to protect the crop from environmental stress variables.

The sun-exposed side of the fruits can suffer a physiological disorder called solar injury or sunburn (Schrader, 2003; Felicetti, 2008), making the fruit unmarketable. Sunburn may not only occur on fruits, but also on the unprotected bark of the young tree trunks. Reducing fruit surface temperature and/or UV-B light exposure are strategies to mitigate sunburn. Accordingly, some technologies have been developed: evaporative cooling, sprayable sunburn protectants, and protective netting (Racskó and Schrader, 2012). Evaporative cooling is an overtree irrigation system that decreases fruit surface temperature and reduces sunburn, improving red fruit color in some cultivars (Gindaba and Wand, 2005).

Sunburn protectants can be based on sun radiation reflective particle films and UV-blocking wax matrices. The first particle films developed at the end of the 1990s was Kaolin-based (Glenn et al., 2002). Kaolin is a chemically inert clay that reflects solar radiation and reduces the heating and oxidation of the fruit (Colavita et al., 2011). The particle film reflects infrared radiation (IR), photosynthetically active radiation (PAR), and ultraviolet radiation. The reduction in PAR at the leaf level is compensated by diffusion of PAR in the interior of the tree, and consequently, canopy photosynthesis can be increased (Glenn, 2012). Other sun radiation reflective particle films are based on calcium carbonate and calcium oxide.

Protective netting attenuates solar irradiance by shading, thereby reducing temperature and wind velocity while increasing humidity. Attenuation of solar irradiance may affect fruit color and therefore new multifilament nets are being developed to protect the trees/fruits from solar radiation and prevent negative effects on fruit skin color. Moreover, another important advantage of networks is that they generate a barrier and reduce the attack of insects. For maximum protection during severe weather periods, it is necessary to apply a combination of strategies.

1.20.2 HIGH-TECH CULTIVATION

The advancement in dwarfing rootstocks, which permits intensive planting systems, has improved tree forms, resulting in better fruit quality and yield (Robinson, 2003). Milkovich (2015) stated, "The planting density will be around 1300 trees per acre while ultra-high-density systems may reach 3000 trees."

These intensive planting systems are in progress with mechanization technologies from simple platforms to robotic tractors. The goal of mechanization technologies is robotization of pruning, thinning, and picking. These machines use combined packages that include robotic hardware and analysis software. The energy for mechanical equipment is frequently solar powered and operated remotely via GPS and remote sensing with satellite and drone reports. The advancement in drone technology provides detailed aerial maps of farms, enabling farmers to take data-driven site-specific action. The drones are equipped with multispectral sensors that make it possible to measure parameters associated with plant health, yields, water stress and nutritional deficiencies (Ghaffarzadeh, 2017).

1.20.3 PRECISION CROP MANAGEMENT AND PROTECTION

In agronomic science, it is important to analyze the factors that vary both in space and time for achieving high crop production. Site-specific management systems apply agronomic knowledge to manage technology, cultural practices, and inputs in the spatial and temporal variability of the farm. They focus on the particular characteristics of the field, the soil types, and the management system, instead of the production uniformities. Site-specific farming needs a large amount of reliable data (Atherton et al., 1999). To respond to this situation, the management concept "precision agriculture" was developed, which is a multidisciplinary analysis that relates computer science with agricultural engineering and horticulture (Zhou et al., 2012).

This concept of agriculture uses different electronic sensors such as thermography, chlorophyll fluorescence, and hyperspectral sensors (Mahlein et al., 2012). These sensors collect data that are processed by algorithms, and they make it possible to define management, productivity, and diseases of crops. Early prediction of potential yield in apple orchards is made using image processing of flowering distribution (Aggelopoulou et al., 2011) or of color features (Zhou et al., 2012).

Phytopathogenic fungi increase leaf transpiration, and this alteration may be sensed by digital infrared thermography. This technology is being studied as a useful tool for assessing scab disease on apple leaves (Oerke et al., 2011).

1.21 PACKAGING AND TRANSPORT

Packaging is an important process to maintain apple quality during storage and transportation and until it reaches the final consumer. It prevents quality deterioration and facilitates distribution and marketing (Singh et al., 2014). Packaging must be designed to protect apples from injuries and adverse environmental conditions during transport, storage, and marketing. The injuries can be due to impact, compression, abrasion, and wounds, and the storage environment is influenced basically by temperature and relative humidity. The packaging design should reduce the loss of water and facilitate the distribution of cold air within the package during refrigerated storage (Opara, 2013). In addition, the packaging must include clear information for buyers (variety, weight, number of units, quality grade, producer's name, country and area of origin, and trace-back system) (El-Ramady et al., 2015).

Apple fruits are packaged in 18–20 kg carton, depending on the variety, and stacked into pallets. Each pallet contains 49 cartons. Each fully loaded pallet is then placed back in storage (Schotzko and Granatstein, 2004). The apples are classified according to the number of apples that the box can contain, and the size ranges from 48 to 216 (the larger number corresponds to the smaller apple diameter). Inside the box, the fruits are packaged in 4–5 trays of recycled paper or soft polystyrene, which prevents damage from impacts, compression and vibration

(Watkins et al., 2004, Subedi et al., 2017). Sometimes apple retailers use polyethylene bags of 1.4, 2.3, or 4.5 kg. Bags with 2–6 apples are most often sold in retail stores (Watkins et al., 2004).

Grade standards are established to facilitate communication and transaction between buyers and sellers. These grades are based on cover color, shape, insect and disease damage and bruises (Schotzko and Granatstein, 2004).

The traditional model for the apple and pear supply flow is similar and it involves farmers, packers, commission agents, auctioneers, wholesalers, traditional retailers of all types, exporters, shipping companies, and customers (Bhaskare and Shinde, 2017). However, an insignificant proportion of apple production is sold directly from the orchard to the consumer. It is represented by roadside stands, farmer's markets, etc. (Schotzko and Granatstein, 2004). Several apple supply chains can be identified according to different countries and regions. In direct selling fruits are conditioned in plastic bins for transportation and are then sent to refrigerated storage without packaging (Cerutti et al., 2011).

When apples are destined to the export market, the fruit is processed after harvest and conditioned in the packinghouse for proper transportation (Sozzi, 2007). The basic transportation requirements are similar to those needed for storage, which include temperature and humidity control and adequate ventilation. In apple and pear packaging, it is important to avoid shock and vibration without affecting air circulation (Luna-Maldonado et al., 2012). Temperature control is needed to maintain the cold chain throughout the postharvest handling system and ensure apple fruit quality and safety. All other postharvest technologies are complementary temperature management (Kader, 2003). Temperature loggers traveling with the produce are used to monitoring the supply chain performance. Temperature information can be downloaded, and these data are used in a quality prediction tool and serve as an audit function (Bollen et al., 2003)

Apple is shipped domestically by truck or railcar (Kader, 2003). Large refrigerated vehicles for road transport are generally semitrailers. The trailer has its own cooling system that provides adequate temperature where the load is located. Apple pallets must be loaded and insured in such a way that an air space is provided around the load. This allows cooled air to flow between the load and the walls, below the product, and finally return to the cooling unit. This design allows refrigerated air to intercept the heat before it affects the temperature of the product (Thompson, 2007).

Overseas transportation is by ship (Kader, 2003). Maritime refrigerated containers and refrigerated ships are used to transport apple fruit through the ocean. Refrigerated ships have a larger capacity and less refrigeration equipment than containers. However, containers may be transferred directly to a cooled dock keeping the cold chain. Refrigerated ships are loaded in open fruit terminals and pallets are exposed to environmental conditions before reaching the ship's hold. Refrigerated ships usually transport products which are sold in large volumes by big companies. Refrigerated containers and ships often have automatic systems for monitoring internal temperatures (Thompson, 2007).

Transportation and distribution logistics must be adjusted by wholesalers and retailers when a large volume of refrigerated ships arrives at the fruit terminals. Dealers must have a good marketing plan to avoid product saturation and loss of fruit quality during transportation so that consumers obtain a quality product (Thompson, 2007).

1.22 POSTHARVEST HANDLING AND STORAGE

After apples are harvested, value is added in successive processes, such as cleaning, washing, selection, grading, disinfection, drying, packing, and storage. These processes improve product appearance and ensure that the fruits comply with established quality standards (El-Ramady et al., 2015).

After harvest, the fruit is transported in bins to the packing house, where it can be processed immediately or stored in cold rooms for further processing. Storage can last from a few days to a year. Depending on the technical and commercial strategy, apple fruits may be treated with products before their conservation, such as scald inhibitor, fungicide, and sometimes calcium chloride for bitter pit control during storage (Mitcham and Mitchell, 2002; Calvo et al., 2012).

The packing process involves a sequence of operations performed by machines and by trained personnel in which different stages can be identified (Figure 1.12). Apples are fed onto the packing line either by hydro handling or by dry feeding (Simcox et al., 2001). The tank water must be kept clean, by filtering the organic waste that accumulates during the process (El-Ramady et al., 2015).

Presorting removes small size or severely damaged fruit (Simcox et al., 2001; Mitcham and Mitchell, 2002). Apples are then soaked in water that contains detergents and phytosanitary products and are transported via rotating brushes, under a rinse-water spray, to a dryer (Simcox et al., 2001).

Waxing is done to improve apple external appearance, reduce dehydration, and replace the natural wax removed in the wash. Drying is carried out by forced air with fans to accelerate the fruit drying (Calvo et al., 2012).

Apple grading and sizing are based on external quality aspects such as color, shape, size and spots, and the presence of wounds and injuries (Calvo et al., 2012). Selection usually involves a combination of human and electronic selectors. Electronic selectors of color and size are very common in big packinghouses (Mitcham and Mitchell, 2002). According to the customer's request, adhesive labels can be placed in each fruit at the beginning of this operation (Calvo et al., 2012).

Apples are packed into different containers such as boxes with trays or into bags or bins determined by the packinghouse design, apple type, and client needs. Packing can be manual or semi-automatic. The boxes are labeled and inventoried to maintain their traceability and workers manually lift boxes from a delivery conveyor and place them on wooden pallets. Palletizing is usually carried out manually, but there are technologies that allow automation. Finally, the fruits are cooled to slowed down their metabolism.

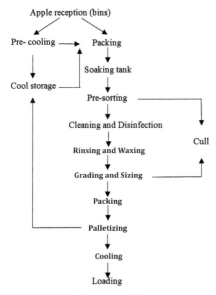

FIGURE 1.12 Apple packing process.

Apple is a typical climacteric fruit, showing a dramatic increase in respiration and ethylene production rate during ripening. Postharvest treatments have managed to delay physiological and biochemical changes and to extend apple storage and marketable life (Paul and Pandey, 2014; El-Ramady et al., 2015).

Apple fruit responds to both temperature and atmosphere modifications. Rapid temperature reduction and the accurate maintenance of low temperatures can provide a quality product (Watkins et al., 2004). Cooling delays are associated with a short postharvest life, pulp softening, increase in physiological disorders and diseases. Therefore, the fruit should be cooled as soon as possible after harvest (Mitcham and Mitchell, 2002). Precooling consists in rapidly lowering the temperature of freshly harvested apple prior to processing or storage. This operation is critical in maintaining the apple quality and allowing the producers to maximize postharvest life (El-Ramady et al., 2015).

The systems used for precooling are forced-air cooling, which pushes air through the vents in storage containers, and hydro-cooling, which uses chilled water to remove pear fruit heat from the field. There are two main types of hydro-cooler: water shower and immersion. Water chlorination is important to prevent pathogen accumulation (El-Ramady et al., 2015).

Several systems, methods, and products have been developed for apple and pear storage. Some of them are regular air storage, modified atmosphere, controlled atmosphere, among others, as described below. Regular air storage is based on storage temperatures around 0 °C (32 °F) without changing other factors to reduce metabolism. Temperature management has so far been the most effective tool to extend the postharvest life of the fruits because it reduces respiration, respiration heat, ethylene production, water loss, and susceptibility to attack by pathogens. The optimum temperature of conservation varies according to the cultivars. It is important to maintain

90%–95% relative humidity levels to reduce losses due to dehydration (Calvo et al., 2012).

Modified atmosphere storage refers to any atmosphere with a gaseous content different from air (Calvo et al., 2012). Modified atmosphere consists in modifying the levels of O_2 and CO_2 inside a package with some type of polymer film. In passive modified atmospheres, the interaction between the respiration of the fruits and the gas permeability of the film generates changes in the gas balance (O_2/CO_2) within the package. In an active modified atmosphere, the desired gas balance is introduced intentionally before the container is sealed (Mattos et al., 2012).

Controlled atmosphere (CA) storage consists in reducing the O_2 content and increasing CO_2 with respect to air composition. The proportion of both gases is controlled throughout the storage period of the fruit at pre-established levels (Calvo et al., 2012). Often CA is used when apples are stored for periods of more than three months, although CA has shown its benefits in periods shorter than 1.5 months in Gala apples. Most AC stored fruit is in bins and comes directly from the orchard; however, some apples are sorted, calibrated, and filled in the same bins before taking them to CA rooms. It is critical to maintain adequate levels of O_2 and CO_2, as well as optimum harvest maturity for each variety. In general, early harvest fruit is a better candidate for long-term storage in CA than late-harvest fruit. Apple fruit with an advanced stage of maturity can be more susceptible to the development of disorders associated with CA storage (Mitcham and Mitchell, 2002). Low storage temperatures increase fruit susceptibility to low O_2 injury (Watkins et al., 2004). General recommendations for CA storage of some varieties are described below:

Varieties like Fuji, Gala and Golden Delicous can be stored at 0–1 °C temperature with 1%–2% of O_2 and 2%–3% CO_2 (exception: 0.5% in the case of Fuji). Suitable temperature for CA storage in case of cv. Red Delicous and Jonagold is 0 °C with 2%–3% CO_2 and O_2; however, it is 0.7%–2% O_2 for Red Delicious. Cv. Braeburn and Granny Smith can suitable stored at 1 °C temperature with 0.5% CO_2 and 1.5%–2% O_2. However, for the apple cultivar Empire, the best temperature for CA storage is reported to be 20 °C along with 2%–3% CO_2 and 2% O_2 (Watkins et al., 2004).

Storage at low oxygen levels reduces apple ethylene production, respiration rate, enzymatic oxidation, and chlorophyll degradation, maintains skin color, flesh firmness, sugars and vitamins, and reduces storage disorders such as superficial scald. Ultralow oxygen (ULO) protocols require oxygen levels to be set below conventional CA, initial low-oxygen stress (ILOS) employs a temporary induction of low-oxygen stress, and fermentation and dynamic controlled atmosphere (DCA) consists in lowering the O_2 levels to the minimum limit tolerated by the fruit. The O_2 levels are set according to different forms of biological feedback of low oxygen stress such as chlorophyll fluorescence, ethanol measurements, and respiratory quotient measurements (CO_2 production/O_2 consumption). One of the main benefits of storage protocols of ULO, ILOS, and DCA is superficial scald control (Wright et al., 2015; Bessemans et al., 2016).

Ethylene is the main hormone involved in apples and pears ripening, therefore, controlling ethylene production make possible to extend the storage life of fruits (Valero et al., 2016). The use of 1-MCP inhibits ethylene action and it is a gas (at standard temperature and pressure) with a formula of C_4H_6. This molecule occupies ethylene receptors such that ethylene cannot bind and elicit action. The affinity of 1-MCP for the receptor is approximately 10 times greater than that of ethylene (Blankenship and Dole, 2003), and also blocks ethylene biosynthesis. The compound is nontoxic, odorless, stable at temperature environment; in addition, it is easy to apply and highly effective (Balaguera-López et al., 2014). Application of 1-MCP

delays or inhibits softening and yellowing (Watkins et al., 2004; Calvo et al., 2012). This compound reduces certain disorders such as internal decay and is an effective alternative to chemical products for controlling superficial scald in apples. Treatment effectiveness may be affected by variety, concentration, temperature, and treatment duration, time between harvest and treatment, apple maturity and storage time (Calvo et al., 2012).

1.23 PROCESSING AND VALUE ADDITION

Apples are consumed both fresh and processed. Processed apples are marketed in a broad spectrum of food: apple sauce, apple slices, baby food, apple butter, juice, cider, brandy, distilled drinks, pies and cakes, among others (Jackson, 2003; Brown, 2012).

Fresh apple is the most widely consumed fruit in the world. Average per capita consumption of apple is over 13 kg yr^{-1}. Countries such as Turkey and Italy consume more per head than any other country. In the United States, the consumption of apple products is greater than that of other fruits, that is, apples lead consumers' choice with 12.6 kg per head per year. This consumption includes juice (6.3 kg), fresh fruit (4.81 kg), and canned, dried, and frozen products (1.48 kg) (USDA,

2017). The availability of organic apple and fresh-cut apple (minimally processed) is currently increasing and expanding markets, mainly in developed countries. These products are aimed at the fast food industry, school lunch programs promoted by governments, and consumers seeking innocuous foods.

The fruit destined to industry lacks adequate quality for fresh consumption. Worldwide, apple juice is the second most important product. Apple juice consumption was higher than consumption per person of orange juice (USDA, 2017). In the production of juice, apples are ground, pressed, filtered, and then pasteurized (Jackson, 2003).

Apple juice may be marketed in an unconcentrated or concentrated form. The availability of CAJ reduces transportation costs and increases global commerce from areas of production to remote markets (Jackson, 2003; Collett, 2011; Brown, 2012).

Almost all the world juice production is exported as concentrated juice. China (50%) is the most important producer and exporter of apple juice in the world. Poland is the second largest exporter (20%), followed by the United States (Jackson, 2003).

In the world apple juice market, China has stronger international competitiveness than other countries. However, China continues to expand, and therefore, it needs the development of new markets and cultivars (Juan et al., 2013). The United States, Japan, and Europe are the main importers of apple juice from China (Jackson, 2003). Argentina, Brazil, Chile, and Germany are other suppliers of juice to the United States (Fonsah, 2008). The increase in the world supply of CAJ, led by China, implies higher quality requirements. In response to requirements of juice with a higher percentage of acidity, China is producing new cultivars of high acidity (Juan et al., 2013).

Phenolic compounds in fruit affect the quality of apple juice and apple slices. It is known that these compounds influence the oxidation process and thus cause browning. For this reason, apple cultivars with a balanced phenolic composition are usually the best choice for the production of both apple juice and apple slices. Because apple juice is an important economic driver, some of the new cultivars have few phenolic compounds.

In recent years, the ready-to-eat fruit market has grown rapidly. Fresh apple slices are one of these ready-to-eat products, but these commodities are more perishable than the original fruit. Tissue integrity is more easily altered during processing, and therefore, it is more sensitive to browning and smoothing. (Bhushan et al., 2008; Chiabrando and Giacalone, 2011). To preserve product quality, different methods have been developed. One of the most widely used methods is dipping in calcium ascorbic acid and storing in modified atmosphere chambers (Aguayo et al., 2010). Other industrialized products of apple include apple sauce, cider, and pop wine (Land, 2010).

The fruit processing industries are increasing by demand for food products as a result of the growing human population. This trend gives rise to huge quantities of fruit waste, up to 40% of the total fruit processed. The solid residues, called apple pomace, consist of a mixture of skin, pulp, and seeds derived mainly from the production of CAJ (Bhushan et al., 2008; Vendruscolo et al., 2008). Researchers have proposed ways to use the waste to obtain value-added products. The advancement in technology has led to alternative options for direct extraction of bioproducts. Valuable compounds can be extracted from apple pomace to increase the use of apple:

1. The production of pectin is considered the most reasonable way of utilization from both economic and ecological perspectives.
2. Pectins are used in foods as a gelling agent, thickener, texturizer, emulsifier, and stabilizer. Furthermore, applications in the pharmaceutical industry were also reported.
3. Apple aroma is a byproduct recovered in the manufacturing and sold separately. It is equivalent to 1.5% of concentrated juice production and is used by industries that manufacture drinks and perfumes.
4. Bioactive compounds, such as dietary fiber, protein, and natural antioxidants, are value-added products that can be used as functional food ingredients.
5. Enzymes, organic acids, biofuels are other valuable products.

The biotechnological applications of apple pomace are very interesting not only for the low-cost substrate and the production of value-added compounds but also as an alternative to the accumulation of apple waste (Bhushan et al., 2008, Vendruscolo et al., 2008; Miceli-García, 2014).

1.24 DISEASES, PESTS, AND PHYSIOLOGICAL DISORDERS

Pests and diseases are currently changing worldwide. An increase in both has been registered due to climate change and to control programs that are stricter on chemical use because of environmental and health concerns. Research on disease and pest control is important and implementation of Integrated Pest Management (IPM) appears to be the best solution (Damos et al., 2015). Existing and innovative tools are combined as strategies for optimizing IPM in many producer regions of the world. The IPM concept was developed to reduce the application of chemical pesticides. It involves replacing chemical pesticides with ecological ones, reducing the pressure of pests with cultural methods, using resistant or tolerant cultivars, and applying precision agriculture in space (Mayus, 2012). The IPM is currently oriented toward a system of fruit production with low residues, aimed at meeting the increasing requirements mainly of the most important European supermarket chains, which limit the residue levels and the number of active ingredients in fresh fruit.

This section briefly describes the diseases that affect apple trees and fruit, both in the orchard and during postharvest, the physiological disorders of the fruit, and the main pests.

1.24.1   ORCHARD DISEASES

The effect of diseases on apple production may be direct or indirect. There are pathogens affecting the fruit directly, making it unattractive, inedible or unmarketable. Other pathogens weaken fruit trees by wounding or invading the leaves, trunk, and branches and can lead to tree death (Jones and Aldwinckle, 2002; Turechek, 2004).

1.24.1.1   FUNGAL DISEASES

The most common fungal diseases in apple orchard are apple scab, powdery mildew, cankers, and rusts. These diseases are briefly described below.

1.24.1.1.1   Name: Apple Scab

Causal agent: Fungi *Venturia inaequalis* (*Spilocaea pomi*)

Infection: *V. inaequalis* overwinter as immature pseudothecia arising from sexual reproduction, in previous lesions of fallen leaves on the orchard floor. In areas with mild winters, the fungus can also overwinter on wood and in bud scales as conidia arising from asexual reproduction. In most areas, the ascospores are the main inoculum. In early spring or late winter, the primary infections are initiated by ascospores released at green tips of apple tree phenological stage. Release of ascospores during a period of 8 to 12 weeks peaks around bloom. Polycyclic phase is produced by conidia originating from lesions on leave and fruits (Mac Hardy, 2000; Carisse et al, 2010).

Symptoms: On leaves, lesions are developed first as light green areas and then become covered with conidia, giving an olive-dark-green

Apple 79

and velvety appearance. Lesions on fruits are small and become corky with age (Carisse et al., 2010).

Seasonality: Apple scab is one of the most damaging apple diseases worldwide. *V. inaequalis* causes annual epidemics, which vary largely in severity according to weather conditions. If it rains too much in the spring, the presence of ascospores leads to epidemics. The temperature during the leaf wetness period affects the infection efficiency (Mac Hardy, 2000). Besides, during rainy spring, the conidia originate new lesions on leave and fruits.

Management: In most apple producing areas of the world, apple scab is the most important disease and its control depends almost exclusively on the frequent use of fungicides; in some areas, up to 75% of all pesticides used are applied to control apple scab. Production of high-quality fruit for fresh market sales requires very high levels of control and cultivars bred for scab resistance constitute the most effective strategy.

1.24.1.1.2 Name: Powdery Mildew

Causal agent: Fungi *Podosphaera leucotricha (Oidium farinosum)*

Infection: The fungus over-winters as mycelium into the scale of dormant buds. In spring, the overwintered fungus is evident as "primary" mildew on leaves newly emerged from buds infected during the previous growing season. Conidia released from these primary mildew colonies disperse in air and initiate an epidemic of "secondary" mildew on growing shoots (Jones and Aldwinckle, 2002).

Symptoms: In early spring, primary symptoms occur on apple leaves, flowers, fruits, buds, shoots and twigs, flowers and fruits, which are covered by a powdery mass of mycelia and conidiophores with conidia.

Seasonality: Once the primary infection has been produced, the secondary epidemics continues to spread because the infection process does not require surface wetness.

Management: The disease requires control programs with fungicides in susceptible cultivars, such as Gala and Rome Beauty. Control failure of both scab and powdery mildew in early spring will result in secondary infections occur-ring throughout the growing season. Consequently, fungicide applica-tions until late growing season will be necessary for acceptable control of diseases (Jones and Aldwinckle, 2002; Montesinos et al., 2000; Turechek, 2004).

1.24.1.1.3 Name: Canker disease, die-back

Causal Agents: Fungus *Diplodia seriata, Botryosphaeria obtusa, D.*

stevennsis, D. mutila, B. dothidea (Family Botryosphaeriaceae)

Infection: Fungus survives from season to season as perithecia and pycnidia in cankers or mummified fruit. Cankers act as a primary inoculum source for other related diseases such as leaf spots, fruit rots, and die back. Cankers often form at sites of pruning, insect infestation, diseased tissue, or wounds from horticultural practices or adverse weather conditions.

Symptoms: On apple wood, the disease shows different degrees of severity. The fungus also causes leaf spots and fruit rots. Black rot and white rot of apple have been found in apple orchards. Delicious and Pristine cultivars are susceptible, while Golden Delicious, Braeburn, and Fuji are relatively resistant. Seasonality: Warm and wet weather favors spore dispersal, infection, and disease development.

Management: It consists in removing and burning overwintering cankers and inspecting and pruning trees during the winter. Use of fungicides for protection of prune wounds is another primary control practice. In orchards with affected fruit, the manual removal of fruit mummies is necessary. A preharvest fungicide program is used to control postharvest fruit rots (Jones and Aldwinckle, 2002; Brown-Rytlewski and McManus, 2000; Montesinos et al., 2000; Turechek, 2004; Cloete, 2011).

1.24.1.1.4 Name: European Canker

Causal agent: *Neonectria galligena* syn. *Neonectria ditissima* (*Cylindrocarpon heteronema*)

Infection: Fungus overwinters as mycelium in twigs and callus tissue of cankers. Spore production is initiated during cool and wet periods and spread by wind, rain, insects and birds (Agrios 2005).

Symptoms: Symptoms are present in late spring or early summer. Young developing cankers appear as reddish-brown lesions on small branches, generally around a leaf scar, spur or pruning wound. The lesions elongate into elliptical, sunken areas with ring-shaped cracks. Cankers on the main stem of older trees reduce their vigor and the value or productivity.

Seasonality: Production and release of spores are largely climate dependent and most common in spring and fall when there is sufficient moisture and the temperature is above 5 °C.

Management: European canker is managed primarily by pruning and protectant fungicides. It is necessary to prune and burn all diseased wood in early summer. Because infection occurs through leaf scars and leaves fall over a long period, two fungicide treatments are necessary each fall to protect new leaf scars.

Apple

1.24.1.1.5 Name: Rust disease

Causal agent: *Gymnosporangium* sp., *G. juniperi-virginianae, G. globosum, G. clavipes*

Rust is usually a minor problem, even in regions where several species of rust diseases are prevalent, because many fungicides used for apple scab also control rusts. Although the old literature generally mentioned cedar-apple rust by *G. juniperi-virginianae*, recent reviews describe different species as causes of rust. In the United States, cedar-apple rust infects both leaves and fruit, hawthorn rust (*G. globosum*) only infects leaves, and quince rust (*G. clavipes*) infects fruit but does not cause leaf lesions (Rosenberger, 2016).

1.24.1.2 OOMYCETES DISEASES

1.24.1.2.1 Name: Root, Crown and Collar Rot

Causal agent: Soil-inhabiting pathogens, *Phytophthora cactorum* (class Oomycetes) is the most important species. The disease occurs in nearly all apple-growing regions of the world (Jones and Aldwinckle, 2002; Turechek, 2004).

Infection: The pathogen overwinters in the soil as mycelium in infected wood or as oospores (sexual origin) or chlamydospores. Oospores remain viable in the soil for years. When soils are wet, oospores germinate to form mycelia and produce reproductive structures called sporangia. These sporangia are filled with zoospores that infect susceptible plant tissues or wounds of roots.

Seasonality: Disease development depends on the inherent susceptibility of the variety/rootstock, environmental conditions, degree of infection, and physiological and nutritional health of the tree Zoospores move over longer distances in runoff and irrigation water. The risk of infection increases with the period of soil saturation (Montesinos et al., 2000; Turechek, 2004).

Symptoms: reddish-brown discoloration of the inner bark and wood from the crown of a declining tree is observed after cutting away the outer bark layer. A sharp line demarcates the reddish-brown (diseased) and the white (healthy) portion of the crown. Similar symptoms can be found on roots. Tree symptoms may become noticeable in early spring, while in mid to late summer foliar symptoms of chlorosis followed by a reddening usually become evident. Developing fruits typically remain small.

Management: Disease control is generally difficult, and the strategies are based on the selection of vegetal material and implantation site, management of irrigation, drainage and soil, protection of roots, and stimulation of defense mechanisms in plants.

1.24.1.3 COMPLEX DISEASES

1.24.1.3.1 Name: Replant Disease

Causal agents: Both abiotic and biotic factors cause replant problem and possible factors vary between different regions or between orchards of the same region. Biotic factors include nematodes, oomycetes, bacteria, and fungi species, while abiotic factors involve soil structure, nutrition, and the release of allelochemicals from roots and/or residue decomposition (Montesinos et al., 2000; Yin et al., 2016). Other minor pathogens of soil include *Phymatotrichopsis omnivora*, *Fusarium* spp., *Armillaria* spp., *Rhizoctonia* spp., *Rosellinia necatrix*, *Xylaria mali*, *Nectria* spp., and *Pythium* spp. (Montesinos et al., 2000; Turechek, 2004).

Symptoms: Disease can produce up to 20% of losses in soil previously planted with apples or related species. The symptoms include root rot, chlorosis, lack of vigor, and death of apple trees.

1.24.1.4 BACTERIAL DISEASES

Bacterial diseases are devastating in rainy regions and difficult to manage. The main bacterial disease world is described below.

1.24.1.4.1 Name: Fire Blight

Causal agent: *Erwinia amylovora*.

Fire blight is a highly destructive disease that produces severe losses throughout Europe and the United States. The disease is absent in South America, South Africa, and Australia, where it is quarantine. The pathogen attacks many plants in the Rosaceae family, including apples, pears, and quince, among others (Turechek, 2004).

Infection: Bacteria survive on healthy plant surfaces, such as leaves and branches, on flower stigmas or woody surfaces as epiphytic bacteria. Infection can initiate in flower stigmas or wounds of scion/rootsocks graft. Floral infections of apple can lead to rootstock infections through xylem movement. Dissemination is by pollinating insects from flower to flower or by propagation material in nurseries.

Symptoms: Fire blight affects blossoms, fruits, shoots, branches and limbs, and the rootstock near the graft union on the lower trunk. Symptoms on blossoms appear 1–2 weeks after petal fall. The floral pieces become water soaked, grayish-green, and then black–brown. During periods of high humidity, small droplets of bacterial ooze, initially creamy-white that become amber-tinted, are formed on affected tissues. Similar but faster symptoms are observed on shoots. Leaves on shoots often show blackening before becoming fully necrotic. Shoots give trees a burnt and blighted appearance. The bark of infected rootstocks may show water-soaking, a purplish to black discoloration, cracking, and ooze.

Seasonality: Infections can expand throughout the summer. Canker expansion slows in late summer as temperatures cool and growth rates of trees and shoots decline.

Management: Sanitation through the removal of expanding and overwintering cankers is essential for control in susceptible cultivars. It is necessary to develop fire blight-resistant cultivars of fruit and ornamental trees. Chemical control against blossom infections is unsatisfactory, due to resistance to antibiotics or limited efficacy of alternative control agents. In summer, established infections are controlled principally by pruning 20–30 cm below the visible end-expanding canker. Pruning tools must be disinfected to prevent cut-to-cut transmission (Johnson, 2000).

1.24.1.5 VIRUS DISEASES

Below is a list of the most common viruses affecting apple.

1.24.1.5.1   Name: Apple Mosaic Virus

Causal agent: Apple Mosaic Virus. Hosts: apple, pear, and more than 30 mostly wood hosts (e.g., plum, peach, and Rosa).

Transmission: Vegetative propagation through budding and grafting with infected trees. Natural spreads by root grafting.

Symptoms: In the spring, leaves show pale or yellowish spots (mosaic) and banding areas along their veins. Chlorotic irregular spots may become necrotic later in the season. The disease can result in the loss of up to 40% of the crop in very susceptible cultivars (e.g. Golden Delicious, Jonathan, and Granny Smith).

Management: It includes certified virus-tested and virus-free planting material to propagate trees. Affected trees and roots must be removed and destroyed (Pscheidt and Ocamb, 2018).

1.24.1.5.2   Name: Apple Union Necrosis and Decline (AUND)

Causal agent: Tomato ringspot virus (TmRSV)

Transmission: nematodes as *Xiphinema americanum* (sensu lato). Hosts: apple, peach, nectarine, prune, ornamental trees, vegetables, and weeds.

Symptoms: AUND is caused by incompatible graft unions involving a resistant scion to TmRSV grafted onto a susceptible but tolerant rootstock (e.g., Delicious/MM.106). On old infected trees, budbreak may be delayed; leaves are pale green, smaller, and sparser, and shoot growth is reduced. Separation of graft union can occur in severe infections and lead to tree death. Under bark near union, there is a distinct necrotic line, and the graft zone is spongy and orange.

Management: It involves virus-free and nematode-resistant rootstock and growing trees in sites with no history of the disease and preferably preplant fumigation to control nematodes.

1.24.1.5.3 Latent virus diseases

Name/causal agents: Three common latent viruses in apple are Apple Chlorotic Leaf Spot (Apple chlorotic leaf spot virus [ACLSV]), Apple Stem Pitting (Apple stem pitting virus [ASPV]), and Apple Stem Grooving (Apple stem grooving virus). Transmission: graft to apple, crabapple and pear. The virus is transmitted from virus-infected scions grafted onto the susceptible rootstock.

Symptoms: Viruses live in their host plant without causing symptoms. Some new cold tolerant and fire blight resistant rootstocks are hypersensitive to one or more latent viruses. ACLSV causes pear ring pattern mosaic, chlorotic spots, and distortion, and it has been found in all pome and stone fruit species. The ASPV infected scion can cause a downward curving of leaves and pitting and weakening of the rootstock. The ASTGV on rootstocks of *M. sylvestris* Virginia Crab can develop apple decline, a brown necrotic line at the graft when infected, and complete breakage of the tree at the graft union. The scion tissue above the graft union may swell.

Management: It involves only virus-free certified planting material.

1.24.1.5.4 Fruit apple affected by virus

Name/causal agents: Flat Apple Virus (Cherry rasp leaf virus [CRLV]).

Transmission: The disease is spread from tree to tree by *Xiphinema americanum* nematodes in the soil.

Symptoms: Apple fruits become flattened from both the stem and the calyx. Leaf rolling up from the midrib may occur in infected Red Delicious trees.

Name/causal agents: Russet Ring Virus

Symptoms: Rings of russeted areas on the skin of Golden Delicious and Newtown apple fruits.

It develops more in cool growing seasons.

Name: Star crack virus

The Golden Delicious and Gravenstein varieties in the Pacific Northwest are susceptible. Affected fruits are distorted, and cracks may develop anywhere. Cork-like tissue forms on the surface of fruit on healed cracks.

Management: It includes virus-tested, virus-free planting material, and removal of infected trees as soon as symptoms appear. Fumigation of replant sites may delay CRLV development; management also includes controlling the nematode vector.

1.24.1.6 PHYTOPLASMA DISEASES

Name: Apple Proliferation

Causal agent: *Candidatus Phytoplasma mali* (Mollicutes)

Transmission: Phytoplasmas are transmitted by infected budwood or insect vectors (psyllidae) (Montesinos et al., 2000). Two species, Cacopsylla melanoneura and Cacopsylla picta (Hemiptera: Psyllidae), are vectors of apple proliferation disease by *Candidatus Phytoplasma mali* (Oettl and Schlink, 2015).

Symptoms: Phytoplasmas are systemic pathogens that cause witches' broom symptoms. Apple Proliferation is considered one of the most economically important apple diseases in Europe, causing economic losses of 10%–80%.

Management: Primary control involves producing trees under propagation programs, which ensure and certify conditions free of virus diseases and other systemic pathogens. Registered scions and rootstocks are also used and tested regularly (Montesinos et al., 2000; Menzel et al., 2003).

1.24.2 POSTHARVEST DISEASES

Postharvest diseases of apple fruit are caused worldwide by fungi, producing important economic losses of up to 30% depending on the postharvest technology. The main diseases are briefly described below.

1.24.2.1 NAME: BLUE MOLD

Causal organisms: *Penicillium expansum*, *P. crustosum* and *P. solitum*, *P. verrocosum* among other species.

Cycle and epidemiology: Blue mold represents the principal disease of apple fruit in storage worldwide. *Penicillium* spp. can saprophytically survive as conidia in orchards (plant debris, soil, tree fruit, or tree bark). Blue mold is rarely seen as fruit rot in the orchard, except on fallen fruit on the orchard floor. *P. expansum* is a ubiquitous fungus and its conidia are a common constituent of air microflora, especially in fruit stores and packinghouses, and of water flumes and dump tanks in packhouses. It survives saprophytically on plant debris in orchards, packaging stores, cold stores, and especially on fruit remains adhering to fruit bulk bins. Most species of *Penicillium* are spread by wind or contaminated water in water drenchers or dump tanks on fruit packaging lines. Decay is associated with fruit wounds.

Symptoms: Infected fruit shows an aqueous-softly rot, with definite circle edge and dark to light brown color. Rotted tissues can be easily separated from healthy tissues. Decay evolves quickly with apple fruit ripening. Blue mold typical symptoms (conidia and conidiophores) are observed in lesions.

1.24.2.2 NAME: GRAY MOLD

Causal organism: *Botrytis cinerea* (Bondoux, 1993; Spotts et al., 1999).

Cycle and epidemiology: Conidia are spread by wind and rain any time of the year. Inoculum sources of *B. cinerea* in the orchard are ubiquitous and hard to eliminate. Conidia are produced during wet weather throughout the year and they colonize dying flower parts during bloom. These infections either develop into dry calyx rot visible in the orchard or remain as latent infection and subsequently develop in the store.

The risk of botrytis eye rot in store can be assessed preharvest from previous orchard rot history and from the rainfall incidence at preharvest. When a high risk of calyx rot has been determined, earlier marketing of the fruit should be scheduled to minimize losses in store. At harvest and postharvest, the infection starts through fruit skin injuries or through the stem. *B. cinerea* moves from one fruit to the next causing a "nesting" effect, and at this stage of decay, small black sclerotia can be observed.

Symptoms: Grey mold occurs whenever apples are in cold storage and it is considered second in importance. The symptoms are variable depending on the variety and the source of infection. Botrytis rot associated with wounds is regular in shape, firmish, pale-mid brown in color often with darker areas around the calyx and lenticels. Botrytis rot associated with calyx infections varies in color from pale-dark brown and is irregular in shape.

1.24.2.3 NAME: MOLD CORE

Causal agents: *P. expansum*, *Penicillium* sp. (aquous mold core) or *A. alternata*, *Cladosporium* sp. (dry mold core).

Cycle and epidemiology: In both cases, the infection occurs in the opened calyx and affects the seeds, carpels, and pulp of var. Red Delicious apple fruit and its clones (Bondoux. 1993; Montesinos et al., 2000).

1.24.2.4 MINOR POSTHARVEST DISEASES

These minor diseases are Alternaria rot (*Alternaria alternata, Alternaria* spp.), Cladosporium rot (*Cladosporium herbarum, Cladosporium* sp.), bull's eye rot caused by *Gloeosporium perennans,* brown rot by *Monilia fructigena* and *M. fructicola,* bitter rot by *Colletotrichum fructicola, C. theobromicola, C. paranaense, C. melonis*, black rot and white rot by species of Botryosphaeriaceae (Bondoux, 1993; Montesinos et al., 2000).

Management: Preventive actions are based on cultural practices and suitable crop management at harvest, such as reduction of injuries when handling the fruit, reduction of fungal inoculum, and balanced nutrition of plant. Harvest times have a marked effect on decay resistance. Chemical control has limitations due to the appearance of resistant strains of *Botrytis* spp. and *Penicillium* spp., and to restrictions imposed by the markets. Effective management of the postharvest diseases includes preharvest and postharvest applications of new biological and more innocuous fungicides against pathogens registered for use on fruit, and resistance inducers in fruit

trees. In addition, storage conditions (temperature and gases) reduce fruit maturity, thus indirectly controlling postharvest diseases (Bondoux, 1993; Jones and Aldwinckle, 2002; Jijakli and Lepoivre, 2004; Quaglia et al., 2011).

1.24.3 PESTS

The number of pests in apple orchards varies according to the different regions where it is grown. The importance of each species is different, but codling moth is of major importance throughout the world. Some of the main pests are briefly described below.

1.24.3.1 NAME: CODLING MOTH

Scientific name: *Cydia pomonella*, Order Lepidoptera: Tortricidae

Relevance: Codling moth is a major pest of global pome fruit production. It was widely distributed due to commerce and exportation of apple plants and fruits. It is a key pest, which has become a serious quarantine pest for export apples.

Infestation: Codling moth presents from one to four generations depending on the latitude, altitude, and climate. The stings are caused by early larvae that die shortly after puncturing the apple skin. Codling moth larvae that are feeding on the core when they exit to expel excrements or to pupate make a larger hole from the entry point.

Damage: The larvae cause small wounds to the skin, construct galleries into the fruit pulp, and go to seeds. Injury to apple fruit is characterized by a tunnel emanating from the apple side or calyx and extending to the core, or small holes, with a small amount of dead tissue.

Management: importer countries require exporter and producer countries to comply with regulations that guarantee the health of the fruit. Samplings are required to ensure that fruits are codling moth-free both in orchard prior to packing and during storage. Monitoring by pheromone tramps and evaluation of damage on fruits are performed for insecticide application. Codling moth is rarely seen when regular spray programs are applied. Programs to eradicate and suppress pest populations involve the use of *Cydia pomonella* granulo virus (CpGV) as a bioinsecticide and sterile insect release. Yet, the success of the latter hinges on the performance of the released insects and the prevailing environmental conditions (Beers et al., 1993; Pasqualini and Angeli, 2008).

1.24.3.2 NAME: LIGHT BROWN APPLE MOTH

Scientific name: *Epiphyas postvittana*, Order Lepidoptera: Tortricidae.

Relevance: It is a leafroller pest with a wide host range that includes horticultural, woody and herbaceous plants. Light brown apple moth is native to Australia.

Infestation: The life cycle is longer during colder temperatures. In warmer climates, 4–5 overlapping generations can occur, whereas in colder climates only two generations occur. The first-generation larvae occur on fruit and leaves in summer and the second-generation larvae over winter, probably in leaf litter.

Damage: The larvae construct leaf rolls, or nests, which damage the leaves, surfaces of the fruits, and sometimes tunnel into the fruit flesh.

Management: It involves inspecting fruits for damage, either while developing on the tree, at harvest or during packaging. Damaged fruit indicates that populations have been high, and that treatment will be required for the next generation or the next season. Major detection and regulatory efforts have been implemented because of concerns about economic and environmental impacts. This pest was often intercepted on imports of fruit and other plant parts, and it has the potential to become a successful invader in temperate and subtropical regions worldwide. The detection of light brown apple moth in several countries (i.e., the United States) led to immediate quarantine and eradication programs. The importance of the harmful insect prompted the development of classical biological control programs together with other interventions in integrated pest management or integrated pest eradication (Suckling and Brocker-hoff, 2010).

1.24.3.3 ORIENTAL FRUIT MOTH

Scientific name: *Cydia molesta* syn. *Grapholita molesta*, Order Lepidoptera: Tortricidae

Relevance: Oriental fruit moth is currently a very important pest in commercial orchards worldwide.

Damage: It affects mainly young shoot tips where it causes galleries in stone fruit and apple fruit. In apple fruit, the moth also causes galleries in pulp to seeds.

Infestation: Fruit moth presents from four to five generations yearly. Infestation on apple is important when high populations exist on stone fruit.

Management: Detection and identification of larva are relevant for pest control, conducted by means of sexual confusion technique, continuous monitoring of damage in shoots and fruit in orchards, and insecticide sprays.

1.24.3.4 SAN JOSÉ SCALE

Scientific name: *Quadraspidiotus perniciosus*, Order Rhyncotha: Family Diaspididae

Damage: San José scale is another important pest worldwide. It is a very harmful and phytophagous insect that attacks trunk, branches, and fruits. The affected tree is dried, and the fruit damaged with red circles cannot be marketed. Young trees may be killed before fruiting.

Management: San Jose scale is the most common and damaging pest in apple orchards. Dormant season insecticide treatments are the key to controlling this pest. Natural enemies can contribute significantly to controlling it when not disrupted by insecticides.

1.24.4 PHYSIOLOGICAL DISORDERS

The development of physiological disorders during postharvest ripening and storage of apple fruit depends on a set of preharvest factors. These factors may predispose apple fruit to ripening-related disorders and determine the response of fruit to low temperatures and postharvest conditions. The pre-harvest factors include:

1. Fruit maturity
2. The position of the fruit in the tree
3. Characteristics of the fruiting site
4. Load of the tree
5. Nutrition during the development of the fruit
6. Water relations; and

7. Temperature.

Among these, fruit maturity at harvest is one of the most important preharvest factors. Maturity indices are used for optimizing harvest times, diminishing the preharvest impact of fruit maturity, and extending the conservation period. Therefore, growing conditions and fruit development have many direct and indirect effects on the incidence of postharvest disorders (Ferguson et al., 1999).

The apple physiological disorders affected by preharvest factors include cracking, water core, sunburn and sunscald, bitter pit, lenticel blotch pit, cork spot, superficial scald, lenticel breakdown, core flush, and senescent breakdown (Bondoux, 1983; Meheriuk et al., 1994; Ferguson et al., 1999).

1.24.4.1 BITTER PIT

This disorder results from the unavailability of calcium toward the end of the fruit-growing period by the mineral disproportion of calcium, potassium, magnesium, and nitrogen in the fruit pulp. Typical symptoms are cork and dark, small spots isolated and mainly distributed on the calyx zone in more susceptible varieties and cultivars: Red Delicious and its clones, Granny Smith, Braeburn and Fuji (Bondoux, 1983; Meheriuk et al., 1994).

1.24.4.2 CORK SPOT

This disorder is due to a nutritional imbalance that causes low levels of calcium. Tissues predisposed to cork include a low level of calcium or high ratio of potassium/calcium. In the orchard, when the water is insufficient and high evapotranspiration occurs, there is competition between leaves and fruits. Then, isolated cell groups die, giving a cork appearance. Typical symptoms are necrotic tissue in flesh and cork in subepidermic tissues. On var. Granny Smith fruit, the cork spot is deeper than on var. Red Delicious fruit (Bondoux, 1983; Meheriuk et al., 1994).

1.24.4.3 JONATHAN SPOT

This disorder is caused by toxins accumulating in the lenticels. It is characterized by black spots on the apple skin. The possible predisposing factors are insufficient use of nitrogen fertilizer, late harvest, delayed cooling, and prolonged storage. In addition to Jonathan, Wealthy and Rome Beauty are moderately susceptible to this disorder (Bondoux, 1983; Meheriuk et al., 1994).

1.24.4.4 RUSSETING

Russeting is a disorder of the fruit skin that results from microscopic cracks in the cuticle and the subsequent formation of a periderm. Russeting is a commercially important surface defect on apples. It has a rough, brown, corky appearance on the skin of apples. It may appear on only a small portion of each fruit or may cover its surface (Khanal et al., 2012). Severe russeting may be accompanied by fruit cracking. Russeting has been associated with cool and wet weather, frost, pesticides, viruses, fungi, and bacteria (Pscheidt and Ocamb, 2008).

1.24.4.5 SUNBURN AND SUNSCALD

Sunburn is characterized by color changes (yellow-dark brown) on fruits skin exposed to sunlight for a long time. The sun-exposed size of fruits frequently is developed under high solar radiation stress. This situation cause increase in lipid peroxides with is positively correlated to sunburn damage (Colavita, 2008).

Sunscald is a physiological disorder of apple that develop during low temperature storage. Sunscald is exclusively observed on sun-exposed section during postharvest periode. During storage the sunburn areas turns dark brown, and black (Vita et al., 2016). The apple cultivars and varieties that are most susceptible to both disorders are Granny Smith, Fuji, and Golden Delicious, while Gala and Red Delicious and their clones are less susceptible (Racsko and Schrader, 2012).

1.24.4.6 SUPERFICIAL SCALD

This is an apple postharvest disorder related to several preharvest factors—such as mineral unbalance (calcium, potassium, and nitrogen) during fruit development—growth conditions, and harvest maturity of the fruit. Irregular and brown spots on fruit skin are observed after three or four months of cold storage, and these symptoms become more severe if the fruit is kept at environment temperature. The symptoms are produced by oxidation of α-pharnesene, natural compounds of the fruit cuticle. Red Delicious and its clones and Granny Smith are more susceptible to superficial scald disorder. Control of this postharvest disorder by antioxidant treatments is limited due to requirements of food without chemical residues, but treatments with 1-MCP and storage in dynamic and controlled atmospheres are effective control strategies (Bondoux, 1983; Meheriuk et al., 1994).

1.24.4.7 OTHER DISORDERS

The postharvest disorders are induced by storage conditions such as low temperature and high CO_2 or by treatments with antiscalding before storage. However, preharvest modifications of environmental conditions and orchard practice can affect the incidence and expression of disorders resulting from storage conditions. These postharvest disorders include freezing injury, water loss, low O_2 and high CO_2 injury, and mealiness (Bondoux, 1993; Ferguson et al., 1999).

KEYWORDS

- *Malus domestica*
- **temperate fruit tree**
- **varieties and cultivars**
- **rootstock**
- **diseases and pest**
- **orchard management**
- **postharvest**
- **nutrient**

REFERENCES

Abbott, J.A.; Saftner, R.A.; Gross, K.C; Vinyard, B.T.; Janick, J. Consumer evaluation and quality measurement of fresh-cut slices of "Fuji," "Golden Delicious," "Gold Rush" and "Granny Smith" apples. *Postharvest biology and technology* 2004, 33, 127–140.

Aggelopoulou, A.D.; Bochtis, D.; Fountas, S.; Swain, K.C.; Gemtos, T.A.; Nanos, G.D. Yield prediction in apple orchards based on image processing. *Precision Agriculture* 2011, 12, 448–456.

Agri benchmark: A quarter of world apple production grows just in three provinces of China. http://www.agribenchmark.org (accessed Oct 5, 2017).

Agrios, G.N. *Plant pathology, 5th ed.*; Elsevier Academic Press: San Diego, USA, 2005.

Aguayo, E.; Requejo-Jackman, C.; Stanley, R.; Woolf, A. Effect of calcium ascorbate treatments and storage atmosphere on antioxidant activity and quality of fresh-cut apple slices. *Postharvest Biology and Technology* 2010, 57, 52–60.

Ahas, R.; Aasa, R.; Menzel, A.; Fedotova, V.G.; Scheifinger, H. Changes in European spring phenology. *International Journal of Climatology* 2002, 22, 1727–1738.

Alexander, L.; Grierson, D. Ethylene biosynthesis and action in tomato: a model for climacteric fruit ripening. *Jounal Experimental Botany* 2002, 53, 2039–2055.

Allen, R.G.; Pereira, L.S.; Raes, D.; Smith, M. Crop evapotranspiration—guidelines for computing crop water requirements. FAO Irrigation and drainage paper 56. Ed. Online 1998. Food and Agriculture Organization, Rome. http://www.fao.org/docrep/X0490E/X0490E00.htm (accessed Nov 15, 2017).

Andersen, L.; Kühna, B.F.; Bertelsena, M.; Bruusb, M.; Larsen, S.E.; Strandberg, M. Alternatives to herbicides in an apple orchard, effects on yield, earth worms and plant diversity. *Agriculture, ecosystems and environment* 2013, 172, 1–5.

Apple (*Malus* spp.) Union Necrosis and Decline, Pacific Northwest Plant Disease Management Handbook, webpage. (accessed Jan 17, 2017).

Apple Rootstock Characteristics and Descriptions, USDA National Institute of Food and Agriculture, New Technologies for Ag Extension project 2015, http://articles.extension.org/pages/60736/apple-rootstock-characteristics-and-descriptions

Apple. Apple Union Necrosis and Decline (adapted by A.R. Biggs), 2011. https://apples.extension.org/apple-union-necrosis-and-decline (accessed Jan 17, 2017).

Apple Union necrosis and decline, Michigan State University IPM webpage. (accessed Jan 17, 2017).

Arseneault, M.H; Cline, J.A. A review of apple preharvest fruit drop and practices for horticultural management. *Scientia Horticulturae* 2016, 211, 40–52.

Atherton, B.C.; Morgan, M.T.; Shearer, S.A.; Stombaugh, T.S.; Ward, A.D. Site-specific farming: A perspective on information needs, benefits and limitations. *Journal of Soil and Water Conservation* 1999, 54, 455–461.

Atucha, A.; Mervin, I.A.; Brown, M.G. Long-term effects of four groundcover management systems in an appleorchard. *HortScience* 2011, 46, 1176–1183.

Balaguera-López, H.E.; Salamanca-Gutiérrez, F.A.; García, J.C.; Herrera-Arévalo, A. Ethylene and maturation retardants in the postharvest of perishable horticultural products. *Revista Colombiana de Ciencias Hortícolas,* 2014, 8, 302–313.

Bapat, V.A.; Trivedi, P.K.; Ghosh, A.; Sane, V.A.; Ganapathi, T.R.; Nath, P. Ripening of fleshy fruit: molecular insight and the role of ethylene. *Biotechnology advances*, 2010, 28, 94–107.

Basannagari, B.; Kala, C.P. Climate Change and Apple Farming in Indian Himalayas: A Study of Local Perceptions and Responses. *PLoS ONE* 8(10): e77976. 2013. doi:10.1371/journal.pone.0077976

Beers, E.H.; Brunner, J.F.; Willet, M.; Warner, G. Orchard pest management. A resource book for the Pacific Northwest. Good Fruit Grower, Yakima, Washington, USA. 1993.

Benítez, C.E. *Cosecha y Poscosecha de Peras y Manzanas en los Valles Irrigados de la Patagonia, 1st ed.*; Instituto Nacional de Tecnología Agropecuaria, Argentina, 2001.

Bessemans, N.; Verboven, P.; Verlinden, B.E.; Nicolaï, B.M. A novel type of dynamic controlled atmosphere storage based on the respiratory quotient (RQ-DCA). *Postharvest Biology and Technology*, 2016, 115, 91–102.

Bhaskare, R.; Shinde, D.K. Development of Cold Supply Chain for a Controlled Atmosphere Cold Store for Storage of Apple. *International Journal of Engineering Science* 2017, 14207.

Bhushan, S.; Kalia, K.; Sharma, M.; Singh, B.; Ahuja, P.S. Processing and Fresh-Cut Markets Processing of Apple Pomace for Bioactive Molecules. *Critical Reviews in Biotechnology* 2008, 28, 285–296.

Blanke, M.; Kunz, A. Effect of climate change on pome fruit phenology at Klein-Altendorf-based on 50 years of meteorological and phenological records. *Erwerbs-Obstbau*, 2009, 51, 101–114.

Blankenship, S.M.; Dole, J.M. 1-Methylcyclopropene: a review. *Postharvest biology and technology*, 2003, 28, 1–25.

Bollen, F.; Praat, J.P.; Garden, G. Information management for apple quality in the supply chain. *Acta Horticulturae* 2003, 604, 753–759.

Bondoux, P. *Enfermedades de conservación de frutos de pepita, manzana y peras,* 1st edn.; Mundi-Prensa: Madrid, Spain, 1993.

Bramlage, W.J. Measuring Fruit Firmness with a Penetrometer. *Post-Harvest Pomology Newsletter* 1983, vol.1, No.3.

Broothaerts, W.; Nerum, I.V.; Keulemans, J. Update on and review of the incompatibility (S-) genotypes of apple cultivars. *HortScience* 2004; 39, 943–947.

Brouwer, C.; Heibloem, M. Irrigation Water Management: Irrigation Water Needs, FAO, Natural Resources and Environment Department |Online| 1986. http://www.fao.org/docrep/s2022e/s2022e00.htm#Contents

Brown S. *Apple.* In *Handbook of Plant Breeding, Fruit Breeding;* Badenes, M.; Byrne, D., Eds.; Springer, Boston, MA, 2012.

Brown, S.K.; Maloney, K.E. Making sense of new apple varieties, trademarks and clubs:

current status. *New York fruit quarterly*, 2009, 17, 9–12.

Brown, S.K.; Maloney, K.E. *Genetic Improvement of Apple: Breeding, Markers, Mapping and Biotechnology.* In: *Apples: Botany, Production, and Uses;* Ferree, D.C.; Warrington, I.J. Eds.; CABI Publishing, Cambridge, United States, 2003.

Brown-Rytlewski, D.E.; McManus, P.S. Virulence of *Botryosphaeriadothidea* and *Botryosphaeriaobtusa* on apple and management of stem cankers with fungicides. *Plant Diseases* 2000, 84, 1031–1037.

Brunetto, G.; Bastos de Melo, G.W.; Toselli, M.; Quartieri, M.; Taglianini, M. The role of mineral nutrition on yields and fruit quality in grapevine, pear and apple. *Revista Brasileira de Fruticultura* 2015, 37, 1089–1104.

Cain, J.C. Hedgerow orchard design for most efficient interception of solar adiation: effects of tree size, shape, spacing, and row direction. Search Agriculture, New York State Agricultural Experiment Station, Geneva, 1972, 2, 1–14.

Calvo, G.; Colodner, A.; Candan, A.P. Cosecha y poscosecha de frutos de pepita. Ed. Instituto Nacional de Tecnología Agropecuaria. Rio Negro, Argentina, 2012.

Calvo, P.; Barda, N.; de Ángelis, V.; Suarez, P. Presentación de nuevas variedades de manzanas. *Revista Fruticultura y Diversificación* 2016, 77, 24–29.

Campana, B.M.R. *Índices de madurez, cosecha y empaque de frutas.* In *Árboles frutales, ecofisiología, cultivo y aprovechamiento.* Sozzi, G.O. Ed.; Facultad de Agronomía, Universidad de Buenos Aires, Argentina, pp. 707–766. 2007.

Campoy, J.A.; Ruiz, D.; Egea J. Dormancy in temperate fruit trees in a global warming context: A review. *Scientia Horticulturae* 2011, 130, 357–372.

Canadian Food Inspection Agencia. *The Biology of Malus domestica Borkh (Apple). Appendix 1: Species and hybrid species currently recognized in the genus Malus, according to the taxonomy database of the U.S. Department of Agriculture Germplasm Resources Information Network GRIN) (USDA-ARS 2012).* Ed Online 2014. http://www.inspection.gc.ca/plants/plants-with-novel-traits/applicants/directive-94-08/biology-documents/malus-domestica/eng/1404417088821/1404417158789?chap=8 (accessed Nov 5, 2017).

Canadian Food Inspection Agency. Biology Document BIO2014-01. *The Biology of Malus domestica Borkh (Apple).* Ed Online 2014. http://www.inspection.gc.ca/plants/plants-with-novel-traits/applicants/directive-94-08/biology-documents/malus-domestica/eng/1404417088821/1404417158789 (accessed Dic 7, 2017).

Carisse, O., Tremblay, D.M., Jobin, T., Walker, A.S, (2010). Disease Decision Support Systems: Their Impact on Disease Management and Durability of Fungicide Effectiveness. Fungicides, Odile Carisse (Ed.), InTech, Available from: http://www.intechopen.com/books/fungicides/disease-decision-support-systems-their-impact-on-diseasemanagement-and-durability-of-fungicide-effe

Castillo, A.; Cabrera, D.; Rodriguez, P.; Zoppolo, R. In vitro micropropagationof CG41 applerootstock. *Acta Horticulturae* 2015, 1083, 569–576.

Cerutti, A.K.; Galizia, D.; Bruun, S.; Mellano, G.M.; Beccaro, G.L.; Bounous, G. *Assessing environmental sustainability of different apple supply chains in Northern Italy.* In *Towards life cycle sustainability management;* Finkbeiner M., Ed.; Springer, Dordrecht, 2011 pp. 341–348.

Chiabrando, V.; Giacalone, G. Effect of antibrowning agents on color and related enzymes in fresh-cut apples during cold storage. *Journal of Food Processing and Preservation* 2012, 36, 133–140.

Cloete, M.; Fourie, P.H.; Ulrike, D.A.M.M.; Crous, P.W.; Mostert, L. Fungi associated with die-back symptoms of apple and pear trees, a possible inoculum source of grapevine trunk disease pathogens. *Phytopathologia Mediterranea* 2011, 50, 176–190.

Colavita, G.M. Evaluación de la incidencia de asoleado en la producción de manzanas en la región del Alto Valle del río Negro. *Fruticultura y Diversificación* 2008, 58, 16–23.

Colavita, G.M.; Blackhall, V.; Valdez, S. effect of kaolin particle films on the temperature and solar injury of pear fruits. *Acta Horticulturae* 2011, 909, 609–615.

Collett, L. *About The Apple – Malus domestica*. Oregon State University Extension Service. http://extension. oregonstate.edu/lincoln/sites/default/files/about_the_apple.lc_.2011.pdf

Cripps, J.E.L.; Richards, L.A.; Mairata, A.M. Pink Lady'apple. *HortScience* 1993, 28, 1057.

Damos, P.; Colomar, L.A.; Ioriatti, C. Integrated Fruit Production and Pest Management in Europe: The Apple Case Study and How Far We Are From the Original Concept? *Insects 2015*, 6, 626–657.

Darbyshire, R.; Webb, L.; Goodwin, I.; Barlow, E. Evaluation of recent trends in Australian pome fruit spring phenology. *International Journal of Biometeorology* 2013, 57, 409–421.

De Swaef, T.; Steppe, K.; Lemeur, R. Determining reference values for stem water potential and maximum daily trunk shrinkage in young apple trees based on plant responses to water deficit. *Agricultural Water Management* 2009, 96, 541–550.

Dennis F.G. The history of fruit thinning. *Plant Growth Regulation* 2000, 31, 1–16.

Deodhar, S.Y.; Landes, M.; Krissoff, B. Prospects for India's Emerging Apple Market/FTS-319-01. Electronic Outlook Report from the Economic Research Service (Online) January 2006. United States Department of Agriculture. http://www.flex-news-food.com/files/USDAAppleMarketIndia_11012006.pdf

Dobránszki, J.; Teixeira da Silva, J.A. Micropropagation of apple—A review. *Biotechnology Advances* 2010, 28, 462–488.

Dorigoni, A.; Micheli, F. Come ottenere un frutteto semi-pedonabile. *L'Informatore Agrario* 2015, 35, 38–42.

Dzhuvinov, V. and Gandev, S.I. Evaluation of fruit bearing habit of apple, sweetcherry, walnut and strawberrycultivars in Bulgaria. *Acta Horticulturae* 2015, 1139, 177–182.

Dzikiti, S.; Volschenk, T.; Midgley, S.J.; Lötze, E.; Taylor, N.J.; Gush, M.B.; Ntshidi, Z.; Zirebwa, S.F.; Doko, Q.; Schmeisser, M.; Jarmain, C.; Steyn, W.J.; Pienaar, H.H. Estimating the water requirements of high yielding and young apple orchards in the winter rainfall areas of South Africa using a dual source evapotranspiration model. *Agricultural Water Management* 2018, 208, 152–162.

Ebel, R.C.; Proebsting, E.L.; Evans, R.G. Deficit irrigation to control vegetative growth in apple and monitoring fruit growth to schedule irrigation. *HortSciense* 1995, 30, 1229–1232.

Eccher, G.; Ferrero, S.; Populin, F.; Colombo, L.; Botton, A. Apple (Malus domestica L. Borkh) as an emerging model for fruit development. *Plant Biosystems-An International Journal Dealing with all Aspects of Plant Biology* 2014, 148, 157–168.

EkoDrevesnica Ocepek, 2018 [http://ekodrevesnica.si/jablane.html#majda] (accessed Apr 25, 2018).

El-Ramady, H.R.; Domokos-Szabolcsy, É.; Abdalla, N.A.; Taha, H.S.; Fári, M. *Postharvest management of fruits* and vegetables storage. In *Sustainable agriculture reviews*, Springer International Publishing, Switzerland, 2015, p 65.

Eurostat, Statistics Explained. Agricultural production – orchards. http://ec.europa.eu/eurostat/statistics-explained (accessed Oct 4, 2017).

Fadanelli, L. *Post-raccolta. Coltivazone.* In: *Il Melo*; Angelini, R. Ed.; Bayer CropScience, S.r.l., Ed.; ART S.p.A.: Bologna, Italy, 2008, p 274.

Fallahi, E.; Neilsen, D.; Neilsen, G.H.; Fallahi, B.; Shaffi, B. Efficient irrigation for optimum fruit quality and yield in apples. *HortScience* 2010, 45, 1616–1619.

Fang, C.; Zhou, K.; Zhang, Y.; Li, B.; Han, M. Effect of root pruning and nitrogen fertilization on growth of young 'fuji' apple (*Malus domestica* Borkh) trees. *Journal of Plant Nutrition* 2017, 40, 1538–1546.

FAOSTAT data base (http://www.fao.org/faostat/en/#data)

Faust, M. *Physiology of temperate zone fruit trees*. John Wiley & Sons, New York, USA, 1989.

Felicetti, D.A.; Schrader, L.E. Changes in pigment concentrations associated with the degree of sunburn browning of 'Fuji' apple. *Journal American Society Horticultural Sience* 2008, 133, 27–34.

Ferguson, I.; Volz, R.; Woolf, A. Preharvest factors affecting physiological disorders of fruit. *Postharvest Biology and Technology* 1999, 15, 255–262.

Fernández, J.E.; Cuevas, M.V. Irrigation scheduling from stem diameter variations: A review. *Agricultural and Forest Meteorology* 2010, 150, 135–151.

Ferree, D.C.; Warrington, I.J. Apples, Botany, Production and Uses. CABI Publishing, 2003.

Fioravanço, J.C.; Czermainski, A.B.C.; Oliveira, P.R D. Yiel defficiency for nine apple cultivars grafted on two rootstocks. *Ciencia Rural* 2016, 46, 1701–1706.

Fonsah E.F.; Muhammad A. The Demand for Imported Apple Juice in the United States.

Journal of Food Distribution Research 2008, 39, 57–61.

Food Availability data (data products food availability per capita data system) and Adjusted Food Availability data (data products food availability per capita. United States Department of Agriculture (httpswww.usda.gov) Economic Research Service.

Forge, T., Neilsen, G.; Neilsen, D.; Hogue, E.; Faubion, D. Composted dairy manure and alfalfa hay mulch affect soil ecology and early production of 'Braeburn' appleon M.9 rootstock. HortScience 2013, 48, 645–651.

Gale, F.; Huang, S.; Gu, Y. Investment in Processing Industry Turns Chinese Apples Into Juice Exports, 1st edn; DIANE Pennsylvania, 2011.

Garming, H. A quarter of world apple production grows just in three provinces of China. Agribenchmark.org (Online) July 2015. http://www.agribenchmark.org/agri-benchmark/did-you-know/einzelansicht/artikel//a-quarter-of.html (accessed Nov 11, 2017).

Ghaffarzadeh K., Agricultural Robots and Drones 2018-2038: Technologies, Markets and Players The future of farming; ultra precision farming; autonomous farming. https://www.idtechex.com/research/reports/agricultural-robots-and-drones-2017-2027-technologies-markets-players-000525.asp (accessed Jun 20, 2020).

Gianfranceschi, L.; Soglio, V. The European project HiDRAS: innovative multidisciplinary approaches to breeding high quality disease resistant apples. In XI Eucarpia Symposium on Fruit Breeding and Genetics 663, 2003, pp. 327–330.

Gindaba, J.; Wand S.J.E. Comparative effects of evaporative cooling, kaolin particle film, and shade net on sunburn and fruit quality in apples. *Hort Science* 2005, 40, 592–596.

Giovannoni, J. Genetic regulation of fruit development and ripening. *Plant Cell* 2004, 16, 170–180.

Glenn D.M. The mechanisms of plant stress mitigation by kaolin-based particle films and applications in horticultural and agricultural crops. Hortscience 2012, 47, 710–711.

Glenn, D.M.; Prado E.; Erez A.; McFerson J.; Puterka G.J. A reflective processed-kaolin particle film affects fruit temperature, radiation reflection and solar injury in apple. *Journal American Society Horticultural Science* 2002, 127,188–193.

Glover, J.D.; Reganold, J.P.; Andrews, P.K. Systematic method for rating soil quality of conventional, organic and integrated apple orchards in Washington State. *Agricultural Ecosystem and Environment* 2000, 80, 29–45.

Gonçalves, M.; Meneguzzi, A.; Camargo, S.; Grimaldi, F.; Candido Weber, G.; Rufato, L. Micropropagation of the new applerootstock "G. 814." *Ciência Rural* 2017, 47, 1–5. https://doi.org/10.1590/0103-8478cr20160615

Grab, S.; Craparo, A. Advance of apple and pear tree full bloom dates in response to climate change in the southwestern Cape, South Africa: 1973-2009. *Agricultural and Forest Meteorology* 2011, 151, 406–413.

Granatstein, D.; Sánchez, E.E. Research knowledge and needs for orchard floor management in organic tree fruit systems. *International Journal of Fruit Sciences* 2009, 9, 257–281.

Greene D., Crovetti A.J., Pienaar A.J. Development of 6-Benzyladenine as an Apple Thinner. *HortScience* 2016, 51, 1448–1451.

Greene, D.; Costa, G. Fruit Thinning in pome- and stone-fruit: state of the art. *Acta Horticulturae* 2013, 998, 93–102.

Greene, D.W. Effect of silver nitrate, aminoethoxyvinylglycine and gibberellins A 4+7 plus 6-benzylaminopurine on fruit set and development of 'Delicious' apples. *Journal of the American Society for Horticultural Science* 1980, 105, 717–720.

Harker, F.R.; Maindonald, J.H.; Jackson, P.J. Penetrometer measurement of apple and kiwifruit firmness: operator and instrument differences. *Journal of the American Society for Horticultural Science* 1996, 121, 927–936.

Heide, O.M.; Prestrud, A.K. Low temperature, but not photoperiod, controls growth cessation and dormancy induction and release in apple and pear. *Tree Physiology* 2005, 25, 109–114.

Jackson, J.E. *Biology of Apples and Pears.* Cambridge University Press. The Edinburgh Building, Cambridge, United Kingdom, 2003, p. 488.

Jackson, J.E. Light interception and utilization by orchard systems. *Horticultural Reviews* 1980, 2, 208–267.

Janick, J. History of the PRI apple breeding program. *Acta Horticulture* 2002, 595, 55–60.

Jijakli, M.H.; Lepoivre, P. *State of the Art and Challenges of Post-harvest Disease Management in Apples.* In Mukerji K.G. (eds) *Fruit and Vegetable Diseases. Disease Management of Fruits and Vegetables*, Mukerji, K.G. Ed.; Academic Publishers, Nederlands, 2004, p 59.

Johnson, K.B. 2000. Fire blight of apple and pear. *The Plant Health Instructor.* DOI: 10.1094/PHI-I-2000-0726-01 *Updated 2015.*

Johnson, W.C.; Aldwinckle, H.S.; Cummins, J.N.; Forsline, P.L.; Holleran, H.T.: Norelli, J.L.; Robinson, T.L. The new USDA-ARS/Cornell University apple rootstock breeding and evaluation program. *Acta Horticulturae* 2001, 557, 35–40.

Johnston, J.W.; Hewett, E.W.; Hertog, M.L. Postharvest softening of apple (Malus domestica) fruit: a review. *New Zealand Journal of Crop and Horticultural Science*, 2002, 30, 145–160.

Jones, A.L.; Aldwinckle, H.S. *Plagas y Enfermedades del manzano y del peral.* APS Mundi Prensa, Madrid, Spain, 2002.

Juan D., Runping W., Xiaoyue X.; Fuzhong L. Research on the Influencing Factors of China Apple Juice Trade. *Research Journal of Applied Sciences, Engineering and Technology* 2013, 6, 691–696.

Juniper, B.E.; Watkins, R.; Harris, S.A. The origin of the apple. *Acta Horticulturae,* 1998, 484, 27–34.

Kader, A. *Postharvest technology of horticultural crops, 3rd edn.;* University of California; Oakland; USA, 2007.

Kader, A.A. A perspective on postharvest horticulture (1978-2003). *HortScience* 2003, 38, 1004–1008.

Kalinowska, M., Bielawska A., Lewandowska-Siwkiewicz H., Priebe W. and Lewandowski, W. Apples: Content of phenolic compounds vs. variety, part of apple and cultivation model, extraction of phenolic compounds, biological properties. *Plant Physiology and Biochemistry* 2014, 84, 169–188.

Karlıdağ, H.; Aslantaş, R.; EŞİTKEN, A. Effects of interstock length on sylleptic shoot formation, growth and yield in apple. *Tarım Bilimleri Dergisi* 2014, 20, 331. https://doi.org/10.15832/tbd.63462

Khanal, B.P.; Grimm, E.; Knoche, M. Russeting in apple and pear: a plastic periderm replaces a stiff cuticle. AoB Plants. 2013; 5: pls048. doi:10.1093/aobpla/pls048.

Khemira, H.; Sánchez, E.E.; Riguetti, T.L.; Azarenko, A.N. Phytotoxicity of urea and biuret sprays to apple foliage. *Journal of Plant Nutrition* 2000, 23, 35–40.

Kumar, N. *Temperate fruits. Apple.* In *Breeding of horticultural crops: principles and practices.* New India Publishing. 2006, pp. 122–123.

Kumar, S.; Chagné, D.; Bink, M.; Volz R.; Whitworth, C.; Carlisle, C. Genomic Selection for Fruit Quality Traits in Apple (*Malus × domestica* Borkh). PLOS ONE 7(5): e36674, 2012. https://doi.org/10.1371/journal.pone.0036674

Land A.; Abadias M.; Sárraga C.; Viñas I.; Picouet P.A. Effect of high pressure processing on the quality of acidified Granny Smith apple purée product. *Innovative Food Science & Emerging Technologies* 2010, 11, 557–564.

Layne, D.R.; Jiang, Z.; Rushing, J.W. The Influence of Reflective Film and ReTain on Red Skin Coloration and Maturity of Gala Apples. *HortTechnology* 2002, 12, 640–645.

Leccese, A.; Bartolini, S.; Viti, R. Antioxidant properties of peel and flesh in 'GoldRush'and 'Fiorina'scab-resistant apple (Malus domestica) cultivars. *New Zealand Journal of Crop and Horticultural Science* 2009, 37, 71–78.

Lehnert, R. Frost protection strategies. http://www.goodfruit.com/frost-protection-strategies/ Feb 1, 2013 (accessed Jun 20, 2020).

Leib, B.G.; Caspari, H.W.; Redulla, C.A.; Andrews, P.K.; Jabro, J.J. Partial rootzone drying and deficit irrigation of 'Fuji' apples in a semi-arid climate. *Irrigation Science* 2006, 24, 85–99.

Lespinasse, J.-M. *La Conduite du Pommier: Types de Fructification, Incidence sur la Conduite de l'Arbre.* Paris, France: INVUFLEC, 1977.

Lespinasse, Y. Breeding apple tree: aims and methods. In Proceedings of the Joint Conference of the EAPR Breeding & Varietal Assessment Section and the EUCARPIA Potato Section, Landerneau, France, 12-17 January 1992, 103–110, INRA.

Liebhard, R.; Koller, B.; Patocchi, A.; Kellerhals, M.; Pfammatter, W.; Jermini, M.; Gessler, C. Mapping quantitative field resistance against apple scab in a 'Fiesta' × 'Discovery' progeny. *Phytopathology* 2003, 93, 493–501.

Link, H. Significance of flower and fruit thinning on fruit quality. *Plant Growth Regulation* 2000, 31, 17–26.

Luby, J.; Forsline, P.; Aldwinckle, H.; Bus, V.; Geibel, M. Silk-road apples – Collection, evaluation and utilization of Malussieversii from central Asia. *HortScience* 2001, 36, 225–231.

Luckwill, L.C.; Silva, J.M. The effects of daminozide and gibberellic acid on flower initiation, growth and fruiting of apple cv golden delicious. *Journal of Horticultural Science* 2015, 54, 217–223.

Luna-Maldonado, A.I.; Vigneault, C.; Nakaji, K. 2012. *Postharvest Technologies of Fresh Horticulture Produce*. In *Horticulture;* Luna Maldonado, A.I., Ed.; InTech, Rijeka, Croatia, 2012, p 161.

Mac Hardy, W.E. Current status of IPM in apple orchards. *Crop Protection* 2000, 19, 801–806.

Magaia, T.; Uamusse, A.; Sjöholm, I.; Skog, K. Dietary fiber, organic acids and minerals in selected wild edible fruits of Mozambique. *SpringerPlus*, 2013, 2:88.

Mahlein, AK.; Oerke, E.C.; Steiner, U.; Dehne H.W. Recent advances in sensing plant diseases for precision crop protection. *European Journal of Plant Pathology* 2012, 133, 197–209.

Malaguti, D.; Rombolà, A.D.; Quartieri, M.; Lucchi, A. Effects of the rate of nutrients by fertigation and broadcast application in `Gala` and `Fuji` apple. *Acta Horticulturae* 2006, 721, 165–172.

Marangoni, B.; Baldi, E. *Concimazione e irrigazione*. In: *Il Melo*; Angelini, R. Ed.; Bayer CropScience, S.r.l., Ed.; ART S.p.A.: Bologna, Italy, 2008, p. 142.

Markowski, J.; Baron, A.; Le Quéré, J.; Płocharski W. Composition of clear and cloudy juices from French and Polish apples in relation to processing technology. *LWT-Food Science and Technology*, 2015, 62, 813–820.

Mattheis, J. *Fruit maturity and ripening*. In *Tree fruit physiology: Growth and development, a Comprehensive Manual for Regulating Deciduous Tree Fruit Growth and Development;* Maib, K.M.; Andrews, P.K.; Lang, G.A.; Mullinix, K. Ed.; Good Fruit Grower, Yakima Washington, p 165, 1996.

Mattos, L.M.; Moretti, C.L.; Ferreira, M.D. *Modified atmosphere packaging for perishable plant products*. In *Polypropylene*; Fatih Dogan Ed.; InTech, Rijeka, Croatia, 2012.

Mayus, M.M., Strassemeyer, J., Heijne, B., Alaphilippe, A., Holb, I., Rossi, V. and Pattori, E. PURE: WP5-Milestone MS14: descriptions of most important innovative non-chemical methods to control pests in apple and pear orchards. 2012. https://www.researchgate.net/publication/239849906_PURE_WP5_Milestone_MS14_descriptions_of_most_important_innovative_non-chemical_methods_to_control_pests_in_apple_and_pear_orchards/references

Mc Guckian, R. Guidelines For Irrigation Management For Apple and Pear Growers. http://apal.org.au/wp-content/uploads/2013/04/Apple_pear_guidelines_Irrigation_Management.pdf (accessed Nov 15, 2017).

Meheriuk, M.; Prange, R.K.; Lidster, P.S.W.; Porritt, S.W. *Postharvest disorders of apples and pears*. Agriculture and Agri-Food Canada, Otawa, Canada, 1994.

Menzel, W.; Zahn, V.; Maiss, E. Multiplex RT-PCR-ELISA compared with bioassay for the detection of four apple viruses. *Journal of Virological Methods* 2003, 110, 153–157.

Merwin, I. Some antique apples for modern orchards. *New York Fruit Quart*, 2008, 16, 11–17.

Miceli-Garcia, L.G. Pectin from apple pomace: extraction, characterization, and utilization in encapsulating alpha-tocopherol acetate. Dissertations, Theses, & Student Research in Food. University of Nebraska-Lincol, United States of America.

Milkovich M. The orchard of the future: Higher tree densities, more automation. Fruit Growers News, 2015. https://fruitgrowersnews.com/article/the-orchard-of-the-future-higher-tree-densities-more-automation/

Mitcham, B.; Cantwell, M.; Kader, A. Methods for determining quality of fresh commodities. *Perishables Handling Newsletter* 1996, 85, 1–6.

Mitcham, E.J.; Mitchell, F.G. *Postharvest handling systems: Pome fruits*. In *Postharvest Technology of Horticultural Crops*; Kader, A., Ed.; University of California, USA, 2002, p 333.

Monforte, A.J.; Oliver, M.; Gonzalo, M.J.; Alvarez, J.M.; Dolcet-Sanjuan, R.; Arus, P. Identification of quantitative trait loci involved in fruit quality traits in melon (Cucumismelo L.). *Theoretical and Applied Genetics* 2004, 108, 750–758.

Montesinos, E.; Melgarejo, P.; Cambra, M.; Pinochet, J. *Enfermedades de los frutales de pepita y de hueso*. Sociedad Española de Fitopatología. Ed. Mundi Prensa, Madrid, Spain, 2000.

Murashige, T.; Skoog, F. A revised medium for rapid growth and bioassays with tobacco tissue cultures. *Physiologia Plantarum*, 1962, 15, 473–497.

Naor, A.; Naschitz, S.; Peres, M.; Gal, Y. Responses of apple fruit size to tree water status and crop load. *Tree Physiology* 2008, 28, 1255–1261.

National Fruit Collection. Jonathan. 2015. [http://www.nationalfruitcollection.org.uk/full2.php?id=3077&&fruit=apple] (accessed Apr 25, 2018).

Neilsen, D.; Millard, P.; Herbert, L.C.; Neilsen, G.H.; Hogue, E.J.; Parchomchuk, P.; Zebarth, B.J. Remobilization and uptake of N by newly planted apple (Malus domertica) trees in response to irrigation method and timing of N application. *Tree physiology* 2001, 21, 513–521.

Neilsen, D.; Neilsen, G.H. Efficient use of nitrogen and water in high-density apple orchards. *HortTechnology* 2002, 12, 19–25.

Neilsen, G.; Forge, T.; Angers, D.; Neilsen, D.; Hogue, E. Suitable orchard floor management strategies in organic apple orchards that augment soil organic matter and maintain tree performance. *Plant Soil* 2014, 378, 325–335.

Neilsen, G.H.; Hogue, E.J.; Forge, T.; Neilsen, D. Surface application of mulches and biosolids affect orchard soilproperties after 7 years. *Canadian Journal of plant science* 2003, 83, 131–137.

Neilsen, G.H.; Neilsen, D.; Hogue, E.J.; Herbert, L.C. Zinc and boron nutrition management in fertigated high density apple orchards. *Canadian Journal of plant science* 2004, 83, 823–828.

Neilsen, G.H.; Neilsen, D.; Peryea F. Response of soil and irrigated fruit trees to fertigation or broadcast application of nitrogen, phosphorus, and potassium. *HortTechnology* 1999, 9, 393–401.

Neilsen, G.H.; Neilsen, D.; Toivonen, P.; Herbert, L. Annual bloom-time phosphorus fertigation affects soil phospforus, apple tree phosphorus nutrition, yield and fruit quality. *HortSicence* 2008, 43, 885–890.

O'Rourke, D. *World Production, trade, consumption and economic of apples*. In: *Apples: Botany, Production, and Uses*. Ed Curtis Ferree, D.; Warrington, Y.J. CAB International. 2003.

O'Connell, M.G.; Goodwin, I. Responses of Pink Lady apple to deficit irrigation and partial rootzonedrying: physiology, growth, yield, and fruit quality. *Australian Journal of Agricultural Research* 2007, 58, 1068–1076.

Oerke, EC.; Fröhling, P.; Steiner, U. Thermographic assessment of scab disease on apple leaves. *Precision Agriculture* 2011, 12, 699–715.

Oettl, S.; Schlink, K. Molecular Identification of Two Vector

Species, Cacopsyllamelanoneura and Cacopsyllapicta (Hemiptera: Psyllidae), of Apple Proliferation Disease and Further Common Psyllids of Northern Italy. *Journal of Economic Entomology* 108 2015, 5, 2174–2183.

Onivogui, G.; Zhang, H.; Mlyuka, E; Diaby, M.; Song, Y. Chemical composition, nutritional properties and antioxidant activity of monkey apple (*Anisophyllealaurina* R. Br. ex Sabine). *Journal of Food and Nutrition Research* 2014, 2, 281–287.

Opara, U.L. A review on the role of packaging in securing food system: Adding value to food products and reducing losses and waste. *African Journal of Agricultural Research* 2013, 8, 2621–2630.

Orange Pippin. Apples-Kent. 2018. [https://www.orangepippin.com/apples] (accessed Apr 25, 2018).

Pacific Northwest Plant Disease Management Handbook, webpage. Apple Mosaic Virus, E.V. Podleckis, et. al., eXtension webpage, 2011. (accessed Jan 17, 2017). Apple Mosaic, University of California IPM webpage. (Accessed: 1/17/17).

Pasqualini, E.; Angeli, G. *Coltivazione, Parassiti animali.* In: *Il Melo*; Angelini, R. Ed.; Bayer CropScience, S.r.l., Ed.; ART S.p.A.: Bologna, Italy, 2008, p. 162.

Paul, V.; Pandey, R. Role of internal atmosphere on fruit ripening and storability-A review. *Journal of food science and technology*, 2014, 51, 1223–1250.

Peck, G.; Andrews, P.; Richter, C.; Reganold, J. Internationalization of the organic fruit market: The case of Washington State's organic apple exports to the European Union. *Renewable Agriculture and Food Systems,* 2005, 20, 101–112.

Pellegrino, S.; Guerra, W. *Ricerca, Innovazione varietale.* In: *Il Melo*; Angelini, R. Ed.; Bayer CropScience, S.r.l., Ed.; ART S.p.A.: Bologna, Italy, 2008, p. 378.

Persic, M.; Mikulic-Petkovsek, M.; Slatnar, A.; Veberic R. Chemical composition of apple fruit, juice and pomace and the correlation between phenolic content, enzymatic activity and browning. *LWT - Food Science and Technology* 2017, 82, 23–31

Petkovsek, M.M.; Stampar, F.; Veberic, R. Parameters of inner quality of the apple scab resistant and susceptible apple cultivars (*Malus domestica* Borkh). *Scientia Horticulturae*, 2007,114, 37–44.

Polandapples, http://www.polandapples.com (October, 2017).

Prasanna, V.; Prabha, T.N.; Tharanathan, R.N. Fruit ripening phenomena–an overview. *Critical reviews in food science and nutrition* 2007, 47, 1–19.

Predieri, S. Mutation induction and tissue culture in improving fruits. *Plant cell, tissue and organ culture* 2001, 64, 185–210.

Pscheidt, J.W.; Ocamb, C.M. (Senior Eds.). 2018 Apple Mosaic Virus, Pacific Northwest Plant Disease Management Handbook. © Oregon State University. (Accessed: 1/17/17). URL: https://pnwhandbooks.org/node/2121

Quaglia, M.; Ederli, L.; Pasqualini, S.; Zazzerini, A. Biological control agents and chemical inducers of resistance for postharvest control of *Penicillium expansum* Link. on apple fruit. *Postharvest Biology and Technology* 2011, 59, 307–315.

Racskó, J.; Leite, G.B.; Zhongfu, S.; Wang, Y.; Szabó, Z.; Soltész, M.; Nyéki, J. Fruit drop: The role of inner agents and environmental factors in the drop of flowers and fruits. International Journal oh *Horticultural Science* 2007, 13, 13–23.

Racskó, J.; Nagy, J.; Soltész, M.; Nyéki, J.; Szabó, Z. Fruit drop: I. Specific characteristics and varietal properties of fruit drop. *International Journal of Horticultural Science* 2006, 12, 59–67.

Racskó, J.; Schrader, L.E. Sunburn of Apple Fruit: Historical Background, Recent Advances and Future Perspectives. *Critical Reviews in Plant Sciences* 2012, 31, 455–504.

Raffo Benegas, M.D.; Rodríguez, R.; Rodríguez, A. Distribución lumínica en diferentes combinaciones portainjerto/variedad en manzana cv. Mondial Gala y su efecto sobre la calidad de la fruta y parámetros vegetativos. *RIA* 2006, 35, 53–69.

Raigón, M.D.; García Martínez, M.D.; Guerrero, C.; Esteve P. Evaluación de calidad de manzanas ecológicas y convencionales. Book of abstract VII Congress, 158, SEAE Zaragoza, 2006.

Ramírez, F.; Davenport, T.L. Apple pollination: A review. *Scientia horticulturae* 2013, 162, 188–203.

Rana, R.S.; Bhagat, R.M.; Kalia, V.; Lal, H. *The impact of climate change on a shift of the apple belt in Himachal Pradesh*. In *Handbook of Climate Change and India, Development, Politics and Governance*; Navroz Dubash Ed.; Taylor & Francis, London 2011. Published online on: 02 Nov 2011. Accessed on: 24 Apr 2018 https://www.routledgehandbooks.com/doi/10.4324/9780203153284.ch3

Renard, C.M.; Thibault, J.F. Composition and Physio-chemical properties of apple fibres from fresh fruits and industrial products. *Lebesm.-Wiss u Technol* 1991, 24, 523–527.

Rieger, M. *Introduction to fruit crops*. Food Products Press, Haworth Press, Inc., New York-London-Oxford 2006.

Robinson, T.L. *Apple-orchard planting systems*. In: *Apples: botany, production and uses*; Ferree, D.C.; Warrington, I.J., Ed.; Wallingford: CABI Publishing, 2003, p 345.

Rosenberger, D. Rust Issues. Update on Pest Management and Crop Development. Cornell University, NYS Agricultural Experiment Station (Geneva) and Ithaca. Scaffolds Fruit Journal 2016, 25:7.

Royal Horticultural Society. 2018. [https://www.rhs.org.uk/Plants/85590/Malus-domestica-Greensleeves-(PBR)-(D)/Details] (accessed Apr. 25, 2018)

Russell, T.; Huthoefer, L.S.; Monke, E. Market windows and hedonic price analyses: an application to apple industry. Journal of Agricultural and Resource Economics 1992, 17, 314–322.

Ryugo, K. *Fruit culture: its science and art*. John Wiley & Sons, New York, USA, 1988.

Sánchez E.E. and Righetti T.L. Effect of postharvest soil and foliar application of boron fertilizer on the partitioning of boron in apple trees. *HortScience* 2005, 40, 2115–2117.

Sánchez, E.E.; Giayetto, A; Cichón, L.; Fernández, D; Aruani, M.C.; Curetti, M. Cover crops influence soil properties and tree performance in an organic apple (*Malus domestica* Borkh) orchard in northern Patagonia. *Plant and Soil* 2007, 292, 193–302.

Sánchez, E.E.; Righetti, T.L. Misleading Zinc deficiency diagnoses in pome fruit and inappropriate use of foliar zinc sprays. *Acta Horticulturae* 2002, 594, 363–368.

Sansavini, S. *Miglioramento genético*. In: *Il Melo*; Angelini, R. Ed.; Bayer CropScience, S.r.l., Ed.; ART S.p.A.: Bologna, Italy, 2008, p 340.

Sansavini, S.; Costa, G.; Gucci, R.; Inglese, P.; Ramina, A.; Xiloyannis, C. (eds). General Arboriculture. Patron Editore, Bologna, Italy, 2012, p 536.

Sansavini, S.E.; Belfanti, F.; Costa, F.; Donati, F. European Apple Breeding Programs Turn to Biotechnology. *Chronica Horticulturae* 2005, 45, 16–19.

Sanzol, J.; Herrero, M. The "effective pollination period" in fruit trees. *Scientia Horticulturae* 2001, 90, 1–17.

Schaffer, B.; Andersen, P.C. *Handbook of environmental physiology of fruit crops*.

CRC Press, Inc. Taylor & Francis Group, 1994.

Schieber, A.; Hilt, P.; Streker, P.; Endreß, H.U.; Rentschler, C.; Reinhold, C. A new process for the combined recovery of pectin and phenolic compounds from apple pomace. *Innovative Food Science and Emerging Technologies* 2003, 4, 99–107.

Schlegel, T.K.; Schönherr, J. Penetration of calcium choride into apple fruits as affected by stage of fruit development. *Acta Horticulturae* 2002, 594, 527–533.

Schmidt, H.; Kellerhals, M. Progress in Temperate Fruit Breeding: Proceedings of the Eucarpia Fruit Breeding Wädenswil/ Einsiedeln, Switzerland from August 30 to September 3, 1993 (Vol. 1). Springer Science & Business Media, 2012.

Schotzko, R.T.; Granatstein, D.A. Brief Look at the Washington Apple Industry: Past and Present SES 04-05. Wenatchee, WA: Washington State University, 2004.

Schrader, L.E.; Sun, J.; Felicetti, D.; Seo, J.H.; Jedlow, L.; Zhang, J. *Stress induced disorders: effects on Apple fruit quality*. WA Tree Fruit Postharvest Conf., Wenatchee, WA. 7pp. 2003. http:// postharvest.tfrec.wsu.edu/PC2003A.pdf.

Simcox, N.; Flanagan, M.E.; Camp, J.; Spielholz, P.; Synder, K. *Musculoskeletal risks in Washington State apple packing companies*. University of Washington, USA, 2001.

Simmonds, M., Howes, M. *Profile of Compounds in Different Cultivars of Apple (Malus × domestica)*. In: *Nutritional Composition of Fruit Cultivars*; Preedy, V.R., Ed.; Royal Botanic Gardens, Kew, Richmond, Surrey, UK, 2015; p 796.

Singh, V.; Hedayetullah, M.; Zaman, P.; Meher, J. Postharvest technology of fruits and vegetables: an overview. *Journal of Postharvest Technology* 2014, 2, 124–135.

Skic, A.; Szymańska-Chargot, M.; Kruk, B.; Chylińska, M.; Pieczywek, P.M.; Kurenda, A.; Zdunek, A.; Rutkowski, K.P. Determination of the optimum harvest window for apples using the non-destructive biospeckle method. *Sensors* 2016, 16, 661.

Soejima, J.; Bessho, H.; Tsuchiya, S.; Komori, S.; Abe, K.; Kotoda, N. Breeding of Fuji apples and performance on JM rootstocks. *Compact Fruit Tree* 1998, 31, 22–24.

Sofi, J.A.; Rattan, R.K.; Kirmani, N.A.; Chesti, M.H.; Bisati, I. Diagnosis and recommendation integrated system approach for mayor and micronutrient diagnostic norms for apple (*Malus domestica* Borkh) under varying ages and management practices of apple orchards. *Journal of Plant Nutrition* 2017, 40, 1784–1796.

Sozzi, G.O. *Árboles frutales, ecofisiología, cultivo y aprovechamiento, 1st ed.;* Facultad de Agronomía, Universidad de Buenos Aires, Argentina, 2007.

Spotts, R.A.; Cervantes, L.A.; Mielke, E.A. Variability in postharvest decay among apple cultivars. *Plant Diseases* 1999, 83, 1051–1054

Subedi, G.D.; Gautam, D.M.; Baral, D.R.; Paudyal, K.P. Evaluation of packaging materials for transportation of apple fruits in CFB boxes. *International Journal of Horticulture* 2017, 7, 54–63.

Suckling, D.M.; Brockerhoff, E.G. Invasion Biology, ecology, and management of the light brown apple moth (tortricidae). *Annual Review of Entomology* 2010, 55, 285–306.

Tagliavini, M.; Marangoni, B. Major nutritional issues in deciduous fruit orchards of northern Italy. *HortTechnology* 2002, 12, 26–31.

Tagliavini, M.; Toselli, M. *Foliar application of nutrients*. In *Encyclopedia of soil in the environment;* Hillel, D. Ed.; Elsevier Ltd, Oxford, U.K. 2005, p. 53.

TerAvest, D.; Smith, J.L.; Carpenter-Boggs, L.; Hoagland, L.; Granastein, D.; Rengold J.P. Influence of orchard floor management and compost application timing on nitrogen partitioning in apple trees. *Hortscience* 2010, 45, 637–642.

Thompson, J.F.; Mitchell, F.G.; Kasmire, R.F. *Enfriamiento de Productos Hortofrutícolas.* In *Tecnología Postcosecha de Cultivos Hortofrutícolas*; Kader, A.; Pelayo-Saldivar, C., Eds.; Universidad d California, USA, 2007; p 111.

Toranzo, J. *Producción mundial de manzanas y peras.* Programa Nacional Frutales. Estación Experimental Agropecuaria Alto Valle. INTA Ediciones, Argentina, 2016.

Trillot, M.; Masseron, A.; Tronel, C. Pomme: les variétés Ctifl, Paris, 1993.

Tromp, J. Flower-bud formation in pome fruits as affected by fruit thinning. *Plant Growth Regulation* 2000, 31, 27–34.

Trowell, H.C. Definition of dietary fiber and hypothesis that it is a protective factor in certain diseases. *American Journal of Clinical Nutrition* 1976, 29, 417–427.

Turechek, A.M. Apple Diseases and their Management. In *Diseases of Fruits and Vegetables;* Naqvi, S.A.M.H. Ed.; Springer Netherlands, 2004; Vol. 1; pp. 1–108.

Tworkoski, T.; Miller, S. Rootstock effect on growth of apple scions with different growth habits. *Scientia Horticulturae* 2007, 111, 335–343.

United States Department of Agriculture (USDA), Fresh Deciduous Fruit: World Markets and Trade (Apples, Grapes, & Pears). https://apps.fas.usda.gov/ psdonline/circulars/fruit.pdf (accessed December, 2017).

United States Department of Agriculture (USDA). United States Standards for Grades of Apples. [https://www.ams. usda.gov/sites/default/files/media/ Apple_Grade_Standard%5B1%5D.pdf] 2002 (accessed Dic 17, 2017)

United States Department of Agriculture (USDA)-National Nutrient Database for Standard Reference Release 28 slightly revised May, 2016. Basic Report 09003, Apples, raw, with skin. https://ndb.nal. usda.gov/ndb/foods/show/2122 Report Date: October 24, 2017

Untiedt, R.; Blanke M. Effects of fruit thinning agents on apple tree canopy photosynthesis and dark respiration. *Plant Growth Regulation* 2001, 35, 1–9.

USApples Association. 2017. [http://usapple. org/all-about-apples/apple-varieties/] (accessed Dic 15, 2017)

USDA, 2017. Fresh Deciduous Fruit: World Markets and Trade (Apples, Grapes, & Pears).

USDA-ARS. Germplasm Resources Information Network—(GRIN) [Online Database]. [Online] Available: http:// www.ars-grin.gov/cgi-bin/npgs/html/ tax_search.pl [2012].

USDA-NRCS. The PLANTS Database. [Online] Available: http://plants.usda.gov [2012].

Valero, D.; Guillén, F.; Valverde, J.M.; Castillo, S.; Serrano, M. *Recent developments of 1-methylcyclopropene (1-MCP) treatments on fruit quality attributes.* In *Eco-friendly technology for postharvest produce quality;* Siddiqui, M.W., Ed.; Academic Press, London, UK, 2016, p 185.

Vendruscolo, F.; Albuquerque, P.M.; Streit, F.; Esposito, E.; Ninow, J.L. Apple pomace: a versatile substrate for biotechnological applications. *Critical Reviews in Biotechnology* 2008, 28, 1–12.

Vielma, M.S.; Matta, F.B.; Silval, J L. Optimal harvest time of various apple cultivars grown in Northern Mississippi. *Journal of the American Pomological Society* 2008, 62, 13–20.

Virk, B.S.; Sogi, D.S. Extraction and characterization of pectin from apple (Malus Pumila. Cv Amri) peel waste. *International Journal of Food Properties* 2004, 7, 693–703.

Vita, L.I.; Gonzalez, N.F.; Colavita, G.M. Effect of Preharvest Solar Radiation on Apple Skin Oxidative Disorders during

Cold-storage. *Acta Horticulturae* 2016, 1144, 325–332.

Wang, H.J.; Thomas, R.L., Direct Use of Apple Pomace in Bakery Products. *Journal of Food Science* 1989, 54, 3.

WAPA-The World Apple and Pear Association. 2017. [http://www.wapa-association.org/asp/index.asp] (accessed Dic 15, 2017)

Watkins, C.B.; Kupferman, E.; Rosenberger, D.A. Apple. The Commercial Storage of Fruits, Vegetables, and Florist and Nursery Stocks. Agriculture Handbook Number 66 (HB-66). Ed. Gross, K.C.; Wang, C.Y.; Saltveit, M. USDA, 2004.

Webster, A.D.; Wertheim, S.J. *Rootstocks and interstems.* In: *Fundamentals of Temperate Zone Tree Fruit Production*; Tromp J, Webster AD, Wertheim SJ, Eds; Backhuys Publishers, 2005, pp. 156–175.

Wertheim, S.J. Developments in the chemical thinning of apple and pear. *Plant Growth Regulation* 2000, 31, 85–100.

Westwood, M.N. *Temperate-Zone Pomology, Physiology and Culture, 3rd edn.*; Timber Press, Portland, Oregon, USA, 1993.

Wojdylo, A.; Oszmiánski, O.; Laskowski, A. Polyphenolic compounds and antioxidant activity of new and old apple varieties. *Journal of Agricultural and Food Chemistry* 2008, 56, 6520–6530.

Wolfe, D.W.; Schwartz, M.D.; Lakso, A.N.; Otsuki, Y.; Pool, R.M.; Shaulis, N.J. Climate change and shifts in spring phenology of three horticultural woody perennials in northeastern USA. *International Journal of Biometeorology* 2005, 49, 303–309.

Wolfe, K.L.; Liu, R.H.J. Apple Peels as a Value-Added Food Ingredient Agric. *Food Chemistry*. 2003, 51, 1676−1683

Wright, A.H.; Delong, J.M.; Arul, J.; Prange, R.K. The trend toward lower oxygen levels during apple (*Malus × domestica* Borkh) storage. *The Journal of Horticultural Science and Biotechnology* 2015, 90, 1–13.

Yang, X.; Song, J.; Du, L.; Forney, C.; Campbell-Palmer, L.; Fillmore, S.; Zhang, P.W.; Zhang, Z. Ethylene and 1-MCP regulate major volatile biosynthetic pathways in apple fruit. *Food Chemistry* 2016, 194, 325–336.

Yin, C.; Xiang, L.; Wang, G.; Wang, Y.; Shen, X.; Chen, X.; Mao, Z. How to plant apple trees to reduce replant disease in apple orchard: a study on the phenolic acid of the replanted apple orchard. *PLoS ONE* 2016, 11(12):e0167347.

Yoshida, Y.; Fan, X.; Patterson, M. 'Fuji' apple. *Fruit varieties journal* 1995, 49, 194–197.

Yuan, R.; Carbaugh, D.H. Effects of NAA, AVG, and 1-MCP on Ethylene Biosynthesis, Preharvest Fruit Drop, Fruit Maturity, and Quality of 'Golden Supreme' and 'Golden Delicious' Apples. *Hortscience* 2007, 42, 101–105.

Yuan, R.; Li, J. Effect of Sprayable 1-MCP, AVG, and NAA on Ethylene Biosynthesis, Preharvest Fruit Drop, Fruit Maturity, and Quality of 'Delicious' Apple. *Hortscience* 2008, 43, 1454–1460.

Zhang, B.; Huang, W.; Li, J., Zhao, C.; Fan, S.; Wu, J.; Liu, C. Principles, developments and applications of computer vision for external quality inspection of fruits and vegetables: A review. *Food Research International* 2014, 62, 326–343.

Zhang, L.Y.; Peng, Y.B.; Pelleschi-Travier, S.; Fan, Y.; Lu, Y.F.; Lu, Y.M.; Gao, X.P.; Shen, Y.Y.; Delrot, S.; Zhang, D.P. Evidence for apoplasmic phloem unloading in developing apple fruit. *Plant Physiology* 2004, 135, 574–586.

Zhang, Q. Tecnología de robotización en la producción frutícola. Curso Internacional Fruticultura de precision. Innovación en mecanización. 9-11 mayo 2017; General Roca, Patagonia, Argentina.

Zhou, R.; Damerow, L.; Sun, Y.; Blanke, M.M. Using colour features of cv. 'Gala' apple fruits in an orchard in image processing to predict yield. *Precision Agriculture* 2012, 13, 568–580.

Zhu, Y.; Rudell, D.R.; Mattheis, J.P. Characterization of cultivar differences in alcohol acyltransferase and 1-aminocyclopropane-1-carboxylate synthase gene expression and volatile ester emission during apple fruit maturation and ripening. *Postharvest Biology and Technology* 2008, 49, 330–339.

Zohary, D.; Hopf, M.; Weiss, E. Domestication of Plants in the Old World, 4th ed.; OUP Oxford, 2012.

CHAPTER 2

PEAR

GRACIELA MARÍA COLAVITA[1*], MARIELA CURETTI[2],
MARÍA CRISTINA SOSA[3], and LAURA I. VITA[1]

[1]*Plant Physiology, Comahue Institute of Agricultural Biotechnology, Faculty of Agricultural Sciences, Comahue National University, National Council for Science and Technology (UNCo-CONICET), Road 151, Km 12.5, Cinco Saltos, 8303 Río Negro, Patagonia, Argentina.*

[2]*Plant Nutrition, Horticultural Department, National Institute of Agricultural Technology (INTA-E.E.A. Alto Valle), General Roca 8332, Argentina.*

[3]*Plant Pathology, Comahue Institute of Agricultural Biotechnology, Faculty of Agricultural Sciences, Comahue National University, National Council for Science and Technology (UNCo-CONICET), Cinco Saltos, Río Negro, Patagonia, Argentina*

Corresponding author. E-mail: graciela.colavita@faca.uncoma.edu.ar; gmcolavita@gmail.com

ABSTRACT

There is evidence that the cultivation of pear (*Pyrus* sp.) began approximately 3000 years ago in the Anatolia peninsula and Central Asia regions. Today, the area devoted to world pear production is approximately 1.5 million ha, yielding 25 million tons pears annually. Pear trees grow in temperate regions and respond to the application of technological tools that increase production. Pears have a pleasant sweet taste, are a source of fiber and vitamin C, and have low sodium content, which make it a healthy snack if eaten fresh. Pears adapt well to conservation and transport, and they can be included in processed foods. This chapter discusses the botany, taxonomy, varieties and cultivars, rootstocks, composition, uses, and nutritional aspects of the pear. In addition, it includes updated information on basic aspects of breeding and crop improvement, orchard management, harvest, postharvest, high-tech cultivation, transport, and packing. Topics such as disease, pest, and physiological disorders are also reviewed.

2.1 GENERAL INTRODUCTION

The genus *Pyrus* is a member of the rose family (Rosaceae), and pear tree cultivation is assumed to have started in the Caucasus Mountains, Asia Minor, and Central Asia about 3000 years ago. There are 23 wild species classified as Asian and European according to their geographical situation. The European pear (*Pyrus communis*) is cultivated in North and South America, Africa, Europe, and Australia, whereas the Asian pear (*Pyrus pyrifolia*) is grown in southern and central China, Japan, and Southeast Asia. The annual world pear production is approximately 25 million tons, and China is the top producer country. In the top 10 producer countries, an area of 1.4 million ha is used for growing this fruit. Different cultivars of pear trees are bred mostly to obtain different tastes and textures. These differences allow pears to be used for different purposes: pears may be eaten raw, cooked, dried, or as juice and cider. The pear has many nutritional benefits such as it is a source of fiber,

contains vitamin C and minerals, and does not have fat, sodium, or cholesterol. Pears present genetic variations, so breeders conduct constant research to obtain new cultivars and rootstocks. In particular, they study the genes related to tree vigor control, precocity of flowering, aspects associated with fruit quality and storage life, as well as resistance to many diseases and pests that have already been identified. Pears are cultivated in temperate regions, but they can also be produced in arid climates with warm or hot temperatures. Pear trees are less resistant to cold than apple trees, so the most relevant climate factor restricting fruit crops is winter minimum temperature. In addition, pears may grow on a wide variety of soils but soil with the pH values between 6.5 and 8.5 and less than 10% of carbonate presence is preferred. Irrigation is fundamental for pear production, because it increases yield, improves fruit quality, stabilizes production between years, allows production in environments not suitable for cultivation, and increases fertilization efficiency. Pear trees should be watered weekly during dry periods, especially during the first two years after planting, and may then be irrigated every two weeks. The amount of water needed at pear orchards depends on the climate, soil characteristics, planting systems, irrigation techniques, and the management and productive levels of the orchards. The success of pear culture and production depends on the selection of appropriate scion and rootstock, the local agro-ecosystem, and the cultural practices. Pear trees are grown in rectangular plots, and the training systems depend on the scion–rootstock selected. The shape and size of the pear trees are obtained and maintained by pruning. Fruit thinning must be done to increase fruit size, improve fruit color, decrease limb breakage, and stimulate floral initiation for the next production cycle. Pear fruits allow long-term storage without a significant loss of quality.

2.2 AREA AND PRODUCTION

In recent years, the world pear production has followed a pattern of expansion similar to that registered for apple production, although on a much smaller scale (Jackson, 2003). The area dedicated to pear production has increased by about 178% over the past 50 years, expanding from 0.56 million ha to 1.57 million ha in 2014 (www.fao.org/faostat/en/#data). The average production volume of the pear fruit was 7.5 million tons between 1965 and 1974, while it rose to 25.8 million tons in 2014, representing an increase of more than 200% (Figure 2.1).

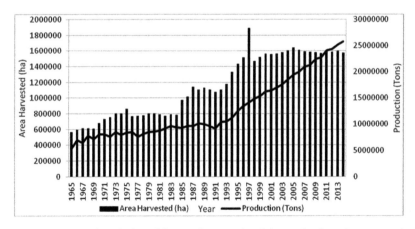

FIGURE 2.1 Annual evolution of the area harvested and the production of pear crops in the world (1965–2014).
Source: FAO (www.fao.org/faostat/en/#data).

The comparison of the area and production data makes it possible to consider the evolution over time and the yield of pear crop. Table 2.1 shows the world average values for each period of 10 years for three indicators: area, production, and yield of pear crops, as well as the percentual increase, calculated using the values for 1965–1974 as reference. The largest percentual increase was observed in production (207%), in relation to the area and the progressive increase in the yields. About 22.6 million tons of high-quality pears are produced annually worldwide across approximately 1.6 million ha.

TABLE 2.1 Average Values of Production, Worldwide Area, and Yield of Apple Crop for Periods of 10 Years. Percentual Increase from 1965–1974 to 2005–2014

Period	1965–1974	1975–1984	1985–1994	1995–2004	2005–2014	Percentual Increase
Area harvested (ha)	676,401	792,057	1,117,107	1,569,325	1,592,623	135
Production (tons)	7,357,346	8,665,256	9,961,606	15,605,663	22,594,467	207
Yield (hg/ha)	109,206	109,522	89,392	99,786	141,986	30

Source: FAO (www.fao.org/faostat/en/#data).

The areas of pear production are distributed in all continents. As observed for apple, over the past 20 years, Asia has stood out as the largest producer of pears in the world, expanding its pear production from 4.4 million tons to 16.8 million tons (2005–2014) (Figure 2.2).

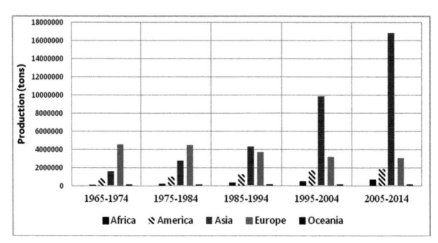

FIGURE 2.2 Distribution of pear production for continents in the last 50 years.
Source: FAO (www.fao.org/faostat/en/#data).

In Western and Central Europe, the main producing countries of pears are Italy, Spain, Germany, France, Belgium, the Netherlands, Switzerland, Portugal, and Austria, with a decline in outputs in the past few decades. This relative decline is due to the increased production in China, Argentina, Turkey, Chile, South Africa, and Iran. (Jackson, 2003) (Table 2.2).

The first position of Asia in the world is heavily influenced by China, currently the largest producer of both pear and apple. After China (the largest producer), the United States, Italy, Argentina, Turkey, and Spain rank relatively high in pear production (Table 2.2). In North America,

the main pear-producing country is the United States, where Washington, California, and Oregon states lead the nation's pear production (Geisler, 2018), whereas in South America, the main production area is Northern Patagonia, Argentina.

The main pear-producing countries in the Southern Hemisphere (Argentina and South Africa) export pears to the Northern Hemisphere in the counter-season because of their high-quality fruits (FAO, 2014; Jackson, 2003). Although China is the most important producer in the world, its exports are lower than those of the Netherlands, Argentina, and Belgium (Table 2.2).

TABLE 2.2 Leading Pear-producing and Exporter Countries in 2013–2014 (in tons)

Producer Country	2013	Exporter Country	2014
China	15,696,676	The Netherlands	420,592
United States	716,001	Argentina	409,364
Italy	668,726	Belgium	328,902
Argentina	650,092	China	295,160
Turkey	415,643	South Africa	206,958
Spain	383,130	United States	195,501
South Africa	308,883	Italy	172,813
India	306,000	Portugal	139,104
The Netherlands	294,300	Spain	129,067
Belgium	274,500	Chile	116,723

Source: WAPA (http://www.wapa-association.org/asp/page_1.asp?doc_id=446).

2.3 MARKETING AND TRADE

World-pear production in marketing year 2016–2017 was around 25 million tons, 2 million tons more than that in 2012–2013 (Figure 2.3).

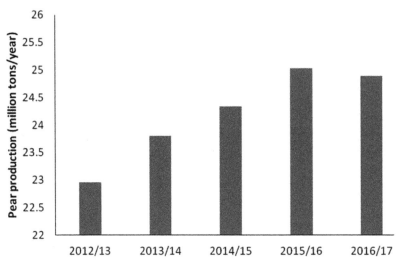

FIGURE 2.3 Evolution of pear fruit production from year 2012–2013 to 2017–2018. *Source:* USDA, Foreign Agricultural Service, 2018.

China is the largest pear-producing country; its production has risen steadily since 2013, and it is expected to continue its upward trend. China is the largest pear-producing country; its production has steadily increased since 2013. In 2018, China's production was slightly above 19 million tons (Statista, 2018) and is expected to continue its upward trend. Production in European Union countries was approximately 2.3 million tons, with Italy, Spain, the Netherlands, and Belgium being the most important contributors. Production in the United States was about 0.63 million tons. In the Southern Hemisphere, production amounted to 1.4 million tons, with Argentina (0.72 million tons), South Africa (0.38 million tons), Chile (0.18 million tons), and Australia (0.10 million tons) being the most important contributors (WAPA, 2017) (Figure 2.4).

China is at the top of the 10 countries that export pear fruit, with worth well over 0.48 million tons. In addition to China and the European Union (the latter exporting 0.30 million tons), other exporter countries in the Northern Hemisphere are

Pear 113

the United States (0.12 million tons) and Belarus (0.08 million tons). In the Southern Hemisphere, the largest exporter country is Argentina (0.28 million tons), followed by Chile and South Africa (Figure 2.5).

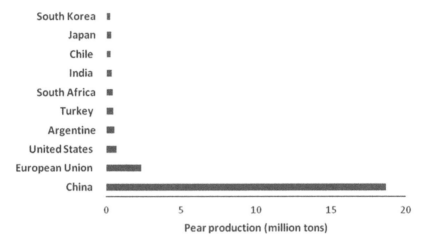

FIGURE 2.4 Production in global leading pear-producing countries in 2017 and 2018.
Source: USDA, Foreign Agricultural Service, 2017.

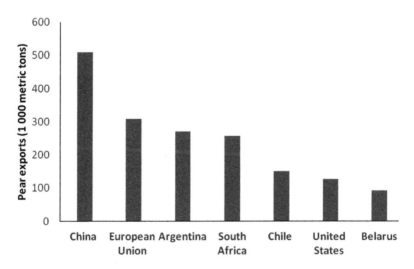

FIGURE 2.5 Main pear exporter countries worldwide.
Source: USDA, Foreign Agricultural Service, 2017.

Russia is the main importer of pear, with an annual import of approximately 0.25 million tons, followed by the European Union, which imports 0.20 million tons. Other importer countries are Brazil, Indonesia, and Belarus (Figure 2.6).

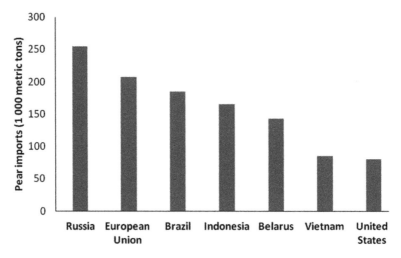

FIGURE 2.6 Main pear importer countries worldwide.
Source: USDA, Foreign Agricultural Service, 2017.

2.4 COMPOSITION AND USES

Pear is considered an important food source due to its nutritional composition and the biological activity of its compounds. Pears are a source of nutrients, fibers, antioxidants, phytochemicals, and vitamins and possess significant amounts of minerals. Besides, they are fat- and sodium-free (James-Martin et al., 2015).

The nutritional composition of common varieties from Asia, Australia, and Europe, widely distributed in the world, are summarized in Table 2.3, using both Australian nutrition composition data from NUTrient TABles for use in Australia (now called the Australian Food Composition Database) (NUTTAB, 2010), Food Standards Australia, and New Zealand's and United States Department of Agriculture (USDA) databases, National Nutrient Database for Standard Reference 2016. A comparative analysis of William Bartlett, d'Anjou, Beurré Bosc, Nashi, and Packham's Triumph pears shows a few differences among cultivars. For example, the calcium content is higher in d'Anjou pears than in the other varieties (Table 2.3).

Besides the nutritional quality, pears are low in calories; 100 g of fresh fruit provides from 50 to 66 kilocalories (kcal) depending on the variety, with an average of 60 kcal (Table 2.3). These properties favor the consumption of pear in diets.

TABLE 2.3 Nutritional Composition of Pear Cultivars (Raw, Unpeeled, Per 100 g Edible Portion)

Nutrient Unit	1 Value/100 g/Cultivar Pear						
	Williams Bartlet[1]	Red d´Anjou[1]	Green d´Anjou[1]	Beurre Bosc[2]	Nashi[2]	Packham´s Triumph[2]	Pear[4]
Water (g)	84.14	84.24	83.31	81	86.7	83.8	83.87
Energy (kcal)	63	62	66	63	50	54	59.67
Protein (g)	0.39	0.33	0.44	0.3	0.4	0.3	0.36
Total lipid (fat) (g)	0.16	0.14	0.10	0	0.1	0	0.14
Ash	0.30	0.35	NR[3]	0.3	0.2	0.3	0.29
Carbohydrate, by difference (g)	15.01	14.94	15.79	14.3	11.1	12.5	13.94
Fiber, total dietary (g)	3.1	3	3.1	3.4	2.1	2.4	2.85
Sugars, total (g)	9.69	9.54	9.73	10.4	10.6	9.1	9.85
Sucrose (g)	0.43	0.26	NR	2.4	0.2	0.3	0.74
Glucose (g)	2.50	2.74	NR	1.7	4.4	2.1	2.69
Fructose (g)	6.76	6.48	NR	6.3	6.0	6.8	6.47
Sorbitol (g)	NR	NR	NR	3.9	NR	3.3	3.6
Minerals							
Calcium, Ca (mg)	9	11	11	6	5	6	8
Iron, Fe (mg)	0.19	0.19	0.24	0.20	0	0.16	0.2
Magnesium, Mg (mg)	6	7	7	7	8	6	6.84
Phosphorus, P (mg)	11	13	13	NR	NR	12	12.25
Potassium, K (mg)	101	123	127	100	130	102	113.8
Sodium, Na (mg)	1	1	0	2	NR	1	1.25
Zinc, Zn (mg)	0.08	0.13	0.10	0.10	NR	0.10	0.11
Cooper, Cu (mg)	0.078	0.070	NR	NR	0.08	0.05	0.07

TABLE 2.3 *(Continued)*

Nutrient Unit	1 Value/100 g/Cultivar Pear						
	Williams Bartlet[1]	Red d´Anjou[1]	Green d´Anjou[1]	Beurre Bosc[2]	Nashi[2]	Packham´s Triumph[2]	Pear[4]
Manganese, Mn (mg)	0.037	0.056	NR	NR	NR	0	0.05
Selenium, Se (µg)	0.1	0.1	NR	NR	NR	0	0.1
Vitamins							
Vitamin C, total ascorbicacid (mg)	4.4	5.2	4.4	4	2	4	4

[1] Based on analytical data for Bartlet, Red d´Anjou, and Green d´Anjou cvs. USDA database. National Nutrient Database for Standard Reference Release 28, slightly revised May, 2016. Date available: 07/14/2017.

[2] Based on analytical data for Beurre Bosc, Nashi, and Packham´s Triumph. NUTTAB 2010, Database Food Standards Australia and New Zealand (accessed Jan 12, 2018).

[3] NR: not reported.

[4] Average value of six pear varieties.

The sugar content varies considerably among Japanese, Chinese, and European pears. Table 2.3 shows that Japanese pears (Nashi) contain the highest total content of sugar and glucose. Pears have sorbitol as the main translocated sugar, which is converted into glucose, fructose, and sucrose (Silva et al., 2014). In comparison with other fruits (including apples), pears are particularly rich in fructose and sorbitol (Table 2.3) (Reinald and Slavin, 2015).

Pears are an excellent source of dietary fiber; a medium-sized fruit contains approximately 3 g of fiber (71% insoluble fiber and 29% soluble fiber) (Table 2.3). Lignins, the noncarbohydrate part of dietary fiber, has antioxidant properties (USA Pears, 2018). In addition, like all fruits, pears are a source of minerals, with significant amounts of potassium (100–120 mg), calcium, phosphorus, iron, magnesium, zinc, copper, and manganese (Table 2.3).

There are differences in the content of total phenolic acids among pear cultivars. Researchers found that the pear peel has 6–20 times higher levels of total flavonoids and triterpenes than the pear flesh.

Anthocyanins, phenolic, and flavonoid compounds are associated with antioxidant and anti-inflammatory capacity in pears (James-Martin et al., 2015; Reinald and Slavin, 2015). The antioxidant activity of pears depends on the cultivar, orchard, harvest time, storage duration, and storage conditions (Li et al., 2012). Kevers et al. (2011) reported that more than 25% of the total content of phenolic and ascorbic acid is found in the pear peel. Besides, Silva et al. (2010) found that properly stored Rocha pears maintained their antioxidant levels and fruit quality during long-term storage, with little or no effect of current commercial postharvest treatments. Pears stored under good conditions maintain their antioxidant properties. The phenolic acids of pears include chlorogenic, arbutin, ferulic, and citric acids. Chlorogenic acid regulates glucose and lipid metabolism and improves metabolic biomarkers related to diabetes, cardiovascular health, and obesity. Citric acid increases iron absorption, whereas ferulic acid showed anti-inflammatory, anti-atherogenic, antidiabetic, anti-aging, neuroprotective, and hepatoprotective effects. However, further research on humans is needed because it is not clear how dietary intake generates biological effects (James-Martin et al., 2015). The main vitamin found in pear fruit is vitamin C, about 4 mg/100 g of fruit. Vitamin C is a powerful natural antioxidant (Reinald and Slavin, 2015). In fact, it prevents oxidative damage and is essential for tissue repair and cell growth, for normal metabolism and proper immune function against infectious diseases (USA Pears, 2018).

2.5 ORIGIN AND DISTRIBUTION

The cultivation of the pear tree began in the Caucasus Mountains, Asia Minor, and Central Asia about 3000 years ago (Vavilov, 1952). The first mention of pears was in ancient Mesopotamia, around 2750 B.C. Pears came to the new world with the first English and French settlers on the east coast of the United States and Canada, and spread westward with the pioneers in the 1700s (Elkins et al., 2007).

The genus *Pyrus* includes 23 wild species classified as Asian (or Eastern) and European (or Western) according to their geographical situation. The most widely used species at present are *P. betulifolia*, *P. calleryana*, *P. pyrifolia* and *P. ussuriensis* (Asian pears), and *P. communis*, *P. amygdafoliformis*, and *P. salicifolia* (European pears). The first domesticated species was *Pyrus pyrifolia*, a wild pear that produced edible fruits. Subsequently, hybridizations were carried out between *P. ussuriensis*, *P. pyrifolia*, and *P. serotina*. The European pear improvement derives from two species: *Pyrus communis* and *Pyrus nivalis*. Most pear cultivars released in Europe were developed via open pollination and the fruits were selected according to their softness and buttery aspect. Pears are cultivated in all temperate regions. The European pear (*Pyrus communis*) predominates in Europe, North and South America, Africa, and Australia, and the Asian pear (*Pyrus pyrifolia*) is produced in Southern and Central China, Japan, and Southeast Asia. Interest in Asian pear cultivars continues to increase in Western Europe, North America, New Zealand, and Australia, but the European pear has made little impact in Asia (Hancock and Lobos, 2008).

2.6 TAXONOMY AND BOTANY

2.6.1 TAXONOMY

Pear trees are categorized within the Rosaceae family, along with apple and quince. The *Pyrus* sp. belongs to the Maloideae subfamily and has a basic chromosome number of $x = 17$. Most cultivated pears are diploid ($2x = 34$), but there are a few polyploid cultivars. The genus *Pyrus* is composed of about 23 wild species that are native to Europe, temperate Asia, and the northern mountainous regions of Africa (Silva et al., 2014).

Taxonomic position according to USDA Agricultural Research Service, Germplasm Resources Information Network (GRIN):

1. Kingdom: Plantae (plants)
2. Subkingdom: Tracheobionta (vascular plants)
3. Superdivision: Spermatophyta (seed plants)
4. Division: Magnoliophyta (flowering plants)
5. Class: Magnoliopsida (dicotyledons)
6. Subclass: Rosidae
7. Order: Rosales
8. Family: Rosaceae (rose family)
9. Subfamily: Amygdaloideae
10. Tribe: Maleae
11. Subtribe: Malinae
12. Genus: *Pyrus*

The two major commercially cultivated species are *Pyrus communis* L. (European pear), which possibly derives from *P. caucasia* and *P. nivalis* (snow pear), and *Pyrus pyrifolia* (Asian pear), also called "Japanese" or "Oriental" pear (Barstow, 2017).

2.6.2 BOTANICAL DESCRIPTION

Pear trees are deciduous, upright, and conical, with very narrow branch angles. They are as high as 12 m with a crown that may reach 6 m in width. The bark of the trees is reddish brown in the first year and then turns grayish brown with shallow furrows and flat-topped scaly ridges. The twigs present spur shoots and the terminal buds are medium-sized (~1 cm), conical to dome-shaped, and may be slightly hairy. Pear trees have deep roots with a well-developed central axis, which allows them to have good anchorage and resistance to drought (Villarreal, 2010).

The leaves are 2–12 cm in length, alternate, simple, ovate-elliptical, coriaceous, and somewhat glossy in the beam; they have a crenate-serrated or almost entire margin and a defined tip; their petiole is as long as the sheet or shorter.

The flowers are about 2–3 cm in diameter, with white petals and frequently crimson anthers. The inflorescence is corymbose, containing 5–7 flowers, the external flowers opening first (Photo 2.1). The ovary is inferior, embedded in the floral cup. Flowers are produced terminally from mixed buds (containing both leaves and flowers) and on spurs, and they appear before or with the leaves. Most cultivars require cross-pollination for a commercial fruit set. Honey bees are the main pollinator.

The fruit is a pome narrowed at the base, which can be rounded or attenuated and prolonged in the peduncle. Pear fruits have a pyriform to obovoid, globe-like shape, and they are approximately 6–8 cm in diameter and 170–350 g in weight. There are three groups of fruits according to the number of carpels and fruit size: small fruits with two carpels; large fruits with five carpels; and fruits with three to four carpels, which are hybrids of the fruits mentioned above. The skin of the fruit is smooth and green, and it turns brownish or yellowish when maturing. The pulp is hard, very acidic, or astringent initially and at soft maturity, with scattered grit cells (termed brachysclereids). Asian pear fruits are heavily russeted, and their flesh has grit cells, refreshing, sweet taste, and crisp texture. European pear fruits are variable in shape, usually pyriform, sometimes round-conic, turbinate, or occasionally round-oblate; their color may be green, yellow, red, russet, or a combination of these; and the flesh has few or many grit cells at maturity.

PHOTO 2.1 Pear inflorescence (note the external flowers opening first) (Courtesy Graciela María Colavita).

The different vegetative structures of a pear tree are as follows:

1. Leader trunk is the primary woody structure.
2. Scaffolds are major limbs that arise from the leader and form the permanent structure of the tree.
3. Branches are small limbs older than 1 year that form spurs and shoots.
4. Spurs are a compressed leafy shoot with either a fruit or leaf bud that grows on two-year-old wood.
5. Laterals or pencils are short shoots with a fruit bud at the tip.
6. Terminals are shoots that develop from the tip of a branch.
7. Fruiting units are pieces of one-, two-, and three-year-old wood that grow from scaffolds and branches and must be renewed by pruning (Dejong, 2007).

In apple and pear trees, fruit buds are produced in spur-bearers, tip-bearers, and partial tip-bearers. Most pear cultivars are spur-bearing. Spur-bearers produce fruit buds on two-year-old wood and on spurs (short, branched shoots) on the older wood. This habit gives spur-bearers a tidy and compact appearance. Terminal bud formation can be considered a continuation of the axis after extension growth has ceased. Lateral buds develop from small sections of the apical meristem remaining in the axils of leaves (Jackson, 2003).

2.7 VARIETIES AND CULTIVARS

Throughout the work of genetic selection made in recent centuries, the European pear (*Pyrus communis* L.) has produced several hundred cultivars, but only a few of them are currently cultivated (Dondini and Sansavini, 2012), as described below.

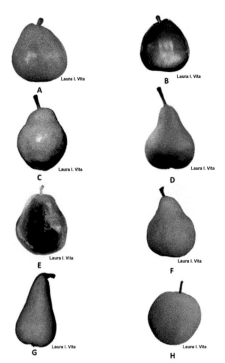

PHOTO 2.2 Pear varieties. (A) Beurré d'Anjou, (B) Red Anjou, (C) Bartlett, (D) Beurré Bosc, (E) Forelle, (F) Packham's Triumph, (G) Abate Fetel, (H) Nasi (Courtesy Laura Vita).

2.7.1 BEURRÉ D'ANJOU

Beurré d'Anjou is thought to have originated in Belgium and in the Center West of France, toward the middle of the 19th century (Benítez, 2001; WAPA, 2017). Anjou pear is short and globose piriform, medium to large, with a wide base, and a stocky neck (Photo 2.2A). It is firm and juicy, and has a rich, sweet, mellow flavor (Ingels et al., 2007; WAPA, 2017). The skin is fine, delicate, and light green not changing to yellow at maturity, with many visible lenticels. The flesh is creamy white, buttery-textured, and slightly granulated (Benítez, 2001).

2.7.2 RED ANJOU

There are two commercial clones of Red Anjou, which originated in the state of Oregon, United States. The fruit is small- and medium-sized, with red epidermis and tanned highlights (Benítez, 2001) (Photo 2.2B). Although the two strains are almost identical in appearance, the Columbia Red strain appears to ripen more consistently and has a lesser tendency to manifest physiological disorders. Both strains have maintained good overall red color, which is unusual for many red pears (Ingels et al., 2007).

2.7.3 BARTLETT

Barlett is originally from Aldermaston, Berkshire, England, where it appeared around 1799 (Benítez, 2001). This pear is known as Bartlett in North America and as Williams in other parts of the world. The fruit is pear-shaped, with a well-defined waist, and medium to large size (Benítez, 2001). The epidermis is fine and delicate, with a pale green color that turns yellow at maturity, sometimes with a pink tinge on the face exposed to the sun, and dotted with numerous, well-visible lenticels (Benítez, 2001; WAPA, 2017) (Photo 2.2C). The pulp is creamy white, fine-textured, aromatic, juicy, sweet, slightly acidulated, slightly granulated in the area close to the carpellous locules, where a few stony cells appear, and it has very good organoleptic quality (Benítez, 2001).

2.7.4 RED BARTLETT

Better known in other regions of the world as Max Red Bartlett, it was discovered as a mutation of Bartlett in the state of Washington, United States, in 1938. The fruit is pear-shaped, with a well-defined waist. The peduncle is short, woody, and somewhat curved. The skin is smooth, partially covered in purplish red that turns into carmine at maturity, more intense on the sun-exposed side, and quite even for the whole tree, although sometimes with a tendency to regression. The pulp is creamy white, fine-textured, compact and aromatic, very juicy, and sweetly and mildly acidulated. Sensation Red Bartlett, a Red Bartlett mutation, was obtained in Australia in 1940. It matures later than Max Red Bartlett, although it blooms at the same time (Benítez, 2001). It has a more consistent and brighter red color than Max Red Bartlett (Ingels et al., 2007).

2.7.5 BEURRÉ BOSC

Beurré Bosc is also called Kaiser Alexander and was discovered in the early 1800s in Belgium or France (Benítez, 2001). What is known is that Bosc pears (WAPA, 2017) were named in honor of the French horti-culturalist Louis Bosc. The fruit is large, with a long, tapering neck. The yellow skin is almost completely covered with a brown russet, which varies in intensity, depending on the climatic and cultural conditions (Photo 2.2D). Bosc flesh is yellowish white and very juicy and buttery, but it does not melt. Bosc has a sweet–spicy taste with aromatic flavor (Ingels et al., 2007; WAPA, 2017).

2.7.6 DOYENNE DU COMICE

Doyenne du Comice originated in France around 1850 (Ingels et al., 2007). The skin is sensitive to handling, fine and smooth, light green, somewhat pink on the sun-exposed side, with numerous small and earthy lenticels. Doyenne du Comice is a very sweet, creamy, juicy pear (WAPA, 2017). It develops its best flavor after a month of storage (Ingels et al., 2007).

2.7.7 FORELLE

Forelle is an old variety, which originated in Germany in the 1600s (WAPA, 2017). The Forelle pear is small, attractive, and of high quality. It has greenish skin that turns bright yellow overlaid with red and covered with grayish russet dots (Photo 2.2E). The white flesh is juicy, sweet melting, and richly flavored (Ingels et al., 2007; WAPA, 2017).

2.7.8 PACKHAM'S TRIUMPH

Packham's Triumph pear originated from a breeding program in New South Wales, Australia, at around the turn of the 20th century (Ingels et al., 2007). The fruit is pear-shaped, broad-based and short, medium, to large in size, with its surface covered with numerous protuber-ances (Benítez, 2001). Packham's Triumph has green skin that changes to yellow at maturity, with numerous dark and remarkable lenticels (Benítez, 2001; WAPA, 2017) (Photo

2.2F). The pulp is creamy white, consistent, fine-textured, aromatic, juicy, sweet, and slightly acidulated (Benítez, 2001).

2.7.9 ABATE FETEL

Originally named Abbé Fetel, this variety was found by the Prior of the Monastery of Chessny, in the Rhone Valley, South Central France, around 1866 (Benítez, 2001). It is a partially russeted Italian winter pear. The fruit is large and elongated, and it has excellent flavor (Ingels et al., 2007). Abate Fetel skin is thin and smooth, golden yellow, and may get a red blush (Benítez, 2001; WAPA, 2017) (Photo 2.2G). The skin also presents some russet in the areas near the calyx. The pulp is white, consistent, fine-textured, juicy, and neutral in flavor (Benítez, 2001).

2.7.10 CONFERENCE

Conference was obtained by Rivers, an English specialist in pomology, in 1885. It received its name in honor of the National Pear Conference of England (Calvo, 2012). The fruit has an elongated shape and smooth russet skin (WAPA, 2017). This variety has melting, juicy, and sweet pulp (Calvo, 2012).

Asian pear production consists mainly of *Pyrus pyrifolia* and *Pyrus ussuriensis* (Segrè, 2002). They are called apple-pears, salad-pears, or Nashi (pear in Japanese) (Mitcham and Mitchell, 2007). Asian pear fruit is not pear-shaped but has the appearance of an apple (Photo 2.2H). On the other hand, not only the cultivation techniques but also the criteria to determine their optimum harvest moment and their aptitude for storage are different from those of European pears (Benítez, 2001). They are crispy in texture and ready for consumption from the harvest. Asian pears do not change markedly in texture after harvest or during their storage period, as the European pears do (Mitcham and Mitchell, 2007).

2.7.11 SHINSEIKI

Shinseiki, translated as new century, was obtained by the crossing of Nijisseiki with Chojuro. The fruit has a balloon to oblate shape, and it is quite large, yellowish green, with smooth epidermis (Benítez, 2001).

2.7.12 NIJISSEIKI

Nijisseiki, translated as 20th century, originated in the City of Matsudo, in the prefecture of Chiba (Katayama, 2002). It is globose and flattened, with a yellowish green epidermis, very smooth, and sensitive to friction (Benítez, 2001).

2.7.13 CHOJURO

Chojuro means long life. It was obtained by chance in Japan at the end of the 19th century (Benítez, 2001). The fruit is firm and flat-shaped and has brown-orange skin. It must be picked when the fruit is yellowish brown; otherwise, the fruit is subject to severe bruising and skin discoloration (Beutel, 1990).

Producers are trying to position some new varieties within the "club" varieties group, such as Rode Doyenne van Doorn (Sweet Sensation), Cheeky, Carmen, Uta (Dazzling Gold), and QTee (Celina). However, so far, these varieties have not had a great market share, although they do reach good prices. Other varieties are in the process of being positioned on the market: Elliot (Selena), CRA IFF-H, Bohene, Corina, Harrow Gold HW-616, HW-606, HW-623, Network Modoc (Lowry program), among others. The number of new varieties of pears is noticeably smaller than that of apples, and breeding programs aim to produce two-colored varieties or hybrids between Asian and European pears (Toranzo, 2016).

2.8 BREEDING AND CROP IMPROVEMENT

The worldwide production of pear depends on relatively few cultivars, classified as European and Asian pears. Both pear types show well-defined characteristics. On the one hand, the European pear fruit is usually pyriform with greenish yellow skin and melting flesh. On the other hand, Asian pear fruit (Japanese and Chinese pear) usually has a rounder shape, russet brown skin, and crisp flesh (Barbosa, 2007). In many countries, the introduction of new cultivars of pear is necessary to obtain better fruit and cultivars that adapt to their environment. According to Brewer and Palmer (2010), both consumers and growers have slowly adopted new cultivars. Some of the reasons for this slow process, according to researchers, are the long length of the juvenile period, the long time span from new cultivar creation and marketability, and the lack of really novel products.

There are breeding programs in almost all continents. The first scientific breeding programs were created in the 20th century, mainly in Belgium, France, Germany, and the United States. Over the past few years, there have been advances in the development of new pear varieties, with private and public breeders working to achieve new varieties in Canada, the United States, France, Italy, Czech Republic, Romania, Germany, Japan, China, and Korea (Fischer 2009; Saito, 2016). Pear genetic improvement is aimed at adapting the plants to the changes in production technology and marketing. At first, the emphasis was placed on the improvement of agronomic (e.g., tree vigor and productivity) and pomological (e.g., fruit appearance) characteristics (Fischer, 2009), but recently, pear breeding programs have been focused on pest and disease resistance. The aim of resistance programs is to reduce chemical inputs in the orchards (Fischer, 2009). For example, resistance to the bacterial pathogen causing fire blight (*Erwinia amylovora*) is a breeding goal in pear, while resistance to fungal pathogens causing pear scab (*Venturia pirina*) is another priority (Brown, 2003).

The specific aims of breeding programs vary according to the importance of pests and diseases in each pear production region in the world. In North America and Europe, the devastating effects of fire blight and psylla on crops, along with the difficulty in controlling the pests and the high cost of pear production, have led to the development of major resistance breeding. Resistance to both species causing pear scab, *Venturia pirina*, which infects European pear, and *Venturia nashicola*, which affects Asian pear, is also important worldwide (Fischer, 2009). Besides, in many countries, another serious impediment to pear production is the mycoplasma disease, named pear decline. In New Zealand, neither pear decline nor

its vector, *Psylla pyricola*, has been recorded, but breeding programs introduced sources of resistance to *Psylla* sp. to prevent its future introduction (White, 1986).

In addition, many pear cultivars are deficient in production, fruit quality, or storability. For these reasons, pear fruit quality and appearance, together with flavor and nutritional characteristics, are important objectives in all breeding programs. There is increasing interest in adaptability to environmental factors, tree vigor control, time of harvest period, fruit longevity, and self-fertility (Fischer, 2009).

Different requirements related to quality, health, and physiological and agronomical characteristics both of crop and postharvest are considered when developing new cultivars of pear. Breeding programs develop new cultivars through interspecies hybrids, which have shown innovative combinations of traits such as texture and flavor, as well as improvement in resistance to pests and diseases (Kumar et al., 2017).

Japan conducted its first pear breeding program from 1909 to 1925, and it included crossing Japanese, European, and Chinese pears. No commercial cultivar was obtained. However, current Japanese breeding objectives mainly combine superior fruit quality and appearance, labor and cost reduction, multiple disease resistance mainly to pear scab (*V. nashicola*) and black spot (*Alternaria gaisen*) diseases, and/or self-compatibility (Saito, 2016).

According to Musacchi et al. (2005), the main goals of the pear breeding program of the University of Bologna (Italy), initiated in 1978, are to improve fruit quality and time of ripening, and to obtain red skin, resistance to *Cacopsylla pyri* and *Erwinia amylovora*, and new flesh types. Some of the 150 advanced selections evaluated show promising quality traits and harvest time.

The objectives of breeding programs in New Zealand are in general coincident with those of most other programs with respect to disease resistance, storage, and fruit characteristics. In this country, the interspecific hybridization program incorporates European and Asian species. For fruit quality traits, the species are combined to develop fruit with high flavor and without the requirement of chill induction for ripening, as well as pest and disease resistance (White, 1986).

Specific aims vary according to the agro-ecological characteristics of the region where the breeding is conducted. Some of these aims are compatible with quince rootstocks, drought, and calcium tolerance, minimizing the chilling requirement, winter frost tolerance, and dwarf tree growth (Fisher, 2009). In Brazil, São Paulo State, better fruit quality and adaptation to the subtropical–tropical climate were the main objectives of breeding programs (Barbosa, 2007).

Breeding strategies are common across pome fruits. The method of controlled crossing of selected parents began in the middle of the 19th century. Since then, combination breeding based on heredity analyses has been the main method of pear breeding (Fisher, 2009; Saito, 2015).

Pear breeding is not as advanced as apple breeding in terms of the availability of new varieties, resistance to diseases, and growth-controlling rootstocks. While pear rootstock breeding programs are developing, progress to date has been slow. In order to efficiently breed new cultivars with desirable characteristics, genetic studies and development of DNA markers for these traits have been highly promoted. To meet future breeding needs, the use of biotechnological tools, such as marker-assisted selection for each trait, quantitative trait locus analyses, and studies of genome-wide association and genomic selection are currently in progress (Saito, 2016)

2.9 SOIL AND CLIMATE

2.9.1 SOIL AND CLIMATE REQUIREMENTS

Most fruit crops need medium-texture soil with at least 50 cm in depth (Sansavini et al., 2012). Pears grafted on seedling rootstock prefer soil with pH values between 6.5 and 8.5 and less than 10%–12% of carbonate presence, whereas pears grafted on quince rootstock are more restrictive in terms of soils (pH values between 6.5 and 7.5 and less than 4%–8% of carbonate presence) (Sansavini et al., 2012). A high concentration of calcium carbonate leads to high pH and high carbonate levels in the soil solution; these factors generate nutritional deficiency (Sozzi, 2007). Pears grow well in arid climates with warm or hot temperatures. Dry summers are essential to control bacterial fire, a disease that limits pear production. The most relevant climate factors that restrict fruit crops are winter minimum temperature, growing-season length and temperatures, light intensity, annual rainfall and its distribution (Westwood,1993). Pear trees are less resistant to cold than apple trees (Westwood, 1993). Dormant pear trees tolerate temperatures of up to −26 °C (Ryugo, 1988). European pear (*Pyrus communis*) and *P. ussuriensis* are more resistant to cold than *P. pyrifolia* (Westwood, 1993). While in most deciduous trees, photoperiod decrease determines leaf fall and the beginning of endodormancy (Sansavini et al., 2012), in apple and pear trees this process is regulated by temperature drop in fall (Heide and Prestrud,

2005). Endodormancy release is obtained with low-temperature exposure for variable times depending on the species and cultivars. Such ecological need is known as cold requirement (Sansavini et al., 2012). The first reports established that cold requirement in pears was around 1200–1500 chilling hours (Ryugo, 1988; Westwood, 1993), but more recently other authors have mentioned a cold requirement of 600–800 chilling hours (Sansavini et al., 2012). If this requirement is not fulfilled, budding is unequal and poor.

2.9.2 IMPACT OF CLIMATE CHANGE

Climate change is a modification in climate status verified in the mean or variance of climatic parameters (temperature, precipitation, among others) during a long period of time (decades). Climate change can impact sagrosystems in many ways. Atmospheric CO_2 rises, temperature and rainfall change, and the frequency of extreme events can vary plant phenology, crop areas, insect and pathogen dispersion, their attack time, etc. Atmospheric CO_2 rise has a direct effect on tree productivity because a photosynthesis increase of 30%–50% is expected for C3 species (Sansavini et al., 2012). In a changing global environment, temperate fruit crops could be at risk due to temperature change (Campoy et al., 2011). Records all over the world have shown that tree phenology has advanced at spring-time, and this change is associated with temperature increase (Grab and Craparo, 2011). Bloom advance exposes flowers to a higher risk of suffering spring frost damage (Sansavini et al., 2012). Temperature changes in the Northern Hemisphere are between 0.29 and 0.34 °C/decade, whereas changes in the Southern Hemisphere are between 0.09 and 0.22 °C/decade for 1979–2005 (IPCC 2007, as cited in Darbyshire et al., 2013). Consistent with global warming, pear blooming occurred earlier in Germany (Blanke and Kunz, 2009) and South Africa (Grab and Craparo, 2011). In contrast, a delayed flowering in pear was registered in Australia (1.4 days/decade), showing that phenology change depends on each region (Darbyshire et al., 2013).

2.10 PROPAGATION AND ROOTSTOCK

Pear cultivars do not maintain their specific characteristics when grown from seed (sexual propagation). For this reason, commercial pear trees are formed from the rootstock and the scion. The former comprises the root system onto which a bud from the scion is grafted, whereas the latter is most of the above-ground tissues that produce fruit of commercial quality. Pear tree cultivars are usually asexually reproduced as clones by budding or grafting onto a genetically distinct rootstock, as opposed to being self-rooted (Chevreau and Bell, 2005). This compound tree system has the advantages that the rootstock enables specific adaptations to soil conditions and that the scion has the commercial fruit quality and yield.

2.10.1 PEAR TREE ROOTSTOCKS

Pear tree rootstocks can be classified according to the type of propagation used in the nursery: seedling rootstocks and clonal (asexual) rootstocks.

2.10.1.1 SEEDLING ROOTSTOCKS

Seedling rootstocks originate from the seed of the genus *Pyrus* sp. They are characterized by inducing precocity and promoting vigorous and heterogeneous growth, great adaptability to different climates, soil types, and irrigation systems, due to their ability to develop good radical systems, with good resistance to aphid.

In European pear cultivars, *Pyrus communis* has been extensively used as rootstock, particularly in North America and Europe. In other producing countries, such as China, India, and Japan, other rootstocks are obtained from seed belonging to other species of the genus *Pyrus* (*P. serotina, P. calleyriana, P. ussuriensis,* and *P. betulaefolia*), but they suffer from the same limitations described for *Pyrus communis* by Raffo et al. (2010).

2.10.1.2 CLONAL ROOTSTOCKS

These rootstocks are reproduced vegetatively by rooting shoots or micropropagation. Given that adventitious root formation is complicated in pears (Bell et al., 2012), clonal rootstocks are scarce worldwide. Some selections are Old Home × Farmingdale, obtained in the United States, with resistance to fire blight (*Erwinia amylovera*), BP series obtained in South Africa, and some selections of Old Home created in France.

To induce different degrees of dwarfing, quince (*Cydonia oblonga*) is used as rootstock (Musacchi et al., 2011). Clonal quince rootstocks are relevant to high-density planting but generally sensitive to fire blight and lime-induced chlorosis. Quince rootstocks are Quince C, Quince A (semi-dwarfing), EMH (more dwarfing than Quince A), Sydo, and BA (from France, more invigorating than Quince A) (Jackson, 2003; Hancock and Lobos, 2008).

2.10.2 STOCK–SCION RELATIONSHIP

Rootstocks may influence the vegetative growth, fruit size, growth pattern, precocity, and yield of the scion portion. They can also affect the tree susceptibility to biotic and abiotic stresses (Wertheim and Webster, 2005).

Grafting is the process of joining two genetically different plant tissues so that they may grow together into a single plant. The selection of grafting techniques depends on the conditions under which the graft will be performed. Whip grafting, also called bench grafting, and cleft grafting are two of the most common techniques. The vigor of a compound tree reflects the vigor of its component rootstock and scion (Tworkoski and Miller, 2007). The use of a dwarfing rootstock as a management tool to control scion vigor is primarily dependent on the reduced vigor of the root system. The main effect of root system size on shoot growth is probably via supply of cytokinins but may also involve nutrient and water supply. The graft union per se may also restrict the upward movement of water and nutrients.

2.10.3 SCION–ROOTSTOCK INCOMPATIBILITY

In some cases, the graft union does not result in a long-lived composite plant. In this situation, the graft is incompatible. In general, the more closely related the plants, the higher the chance of compatibility, although permanent unions between one genus and another may occur. Incompatibility has the following symptoms: initial failure to form graft or bud unions, poor growth of the scion often followed by premature death in the nursery, premature yellowing of the foliage, and breaking off of trees at the graft union, especially when they have been growing for many years and the break is clean and smooth (Karlıdağ et al., 2014). Quince rootstocks present symptoms of incompatibility with Williams, Beurrè d´ Anjou, Packham's Triumph, Abate Fetel, Comice, and Hardy, and all Asian cultivars.

2.10.4 INTERSTEM TREES

Interstem trees consist of three parts: a dwarf rootstock (e.g., *Cydonia oblonga*), the interstem or "filters" (a cultivar that has a great affinity with the rootstock, such as Beurrè Hardy, Passe Crassane, and Comice), and the scion cultivar. Although this approach seems well conceived, results are usually disappointing. Interstem trees are generally more expensive than those on a rootstock because an additional year is required in the nursery (Karlıdağ et al., 2014).

2.10.5 MICROPROPAGATION AND TISSUE CULTURE

Breeding a new pear cultivar by traditional methodologies takes approximately 15 years. Both scions and rootstocks are vegetatively propagated by tissue culture. These techniques provide efficient regeneration systems from somatic tissues, which are essential to developing the transfer of individual genes in the process of genetic engineering.

Shoot culture involves the use of explants. The tissues used for explants are meristems or shoot tips (dormant or actively growing buds) because of their genetic stability (Bell et al., 2012). The explants are surface sterilized, then cultured in a medium containing mineral salts, sucrose or sorbitol, cytokinin, and auxin and gibberellin (Murashige and Skoog, 1962). Shoot cultures are maintained on the culture medium in rooms with a programmed photoperiod and thermoperiod.

Plantlets are produced by excising shoots and adventitious root initiation. This is achieved by transferring the excising shoots to a medium with auxins and no cytokinin. Once the plantlets are established, they are transferred to pots of compost and grown in a glasshouse.

In vitro propagation of apple and pear may be relatively more expensive than conventional vegetative propagation. The advantages of mass propagation are rapid multiplication, disease-free plants, clonal identity, and propagation of cultivars which are difficult to root. However, under some circumstances, in-vitro-produced rootstocks or scions result in an excessive number of burr-knots and suckers, possibly associated with enhanced juvenility. Sequential shoot subculture may greatly increase rejuvenation (i.e., development of juvenile characteristics including ready rooting, but also spinyness and slowness to bear fruit) (Jackon, 2003).

2.11 LAYOUT AND PLANTING

Pear trees are planted during the winter months while the trees are still dormant, to allow proper plant rooting and induce uniform sprouting in the spring. Trees planted late in the season are exposed to temperatures above 20 °C, which promotes sprouting with an insufficient root system. Young trees are susceptible to wind damage and should be protected with a windbreak.

Although pear can tolerate a wild range of soil types, the highest yields are obtained in deep clay loam to sandy loam soils (Reil, 2007). An ideal soil pH range for pear is 6.5–7. Saline soils can exhibit a wide range of basic pH values and can have high concentrations of salts, such as sodium, chloride, and boron, which may be toxic to apple and pear

trees. Soils with high salinity values in what will be the major rooting zone can limit tree establishment. Tillage depth in orchards is usually conducted at about 20–25 cm before planting, in order to produce appropriate conditions for better growth of the trees. In poor soils, the use of organic materials (hays and straw, grape pomace, manure or sewage biosolids, and composted combinations of several of these materials) is recommended as soil amendments. These materials should be applied near the root zone prior to planting (Glover et al., 2000). Trees are planted into a hole which is large enough to accommodate the outstretched roots of the tree without bending, and the graft union should be at least 10–15 cm above the soil. Tree density affects early yields. Higher densities promote higher yields during the first years. Pear orchards are designed in single rows spaced 4 m apart. The distance between trees depends on the plantation layout. In traditional design, the distance between trees is 4 m (~600 trees/ha), in high-density designs it is 2–1.5 m (1250–1667 trees/ha) (Raffo et al., 2010), and in ultra-high-density design it is below 0.9 m (~3500 trees/ha). Some commercial pear plantations may even have 7000 trees/ha (Verna, 2014). The results of several studies indicate that a north–south row orientation is preferred to an east–west row orientation. A north–south orientation provides better light interception and a more even distribution throughout the tree canopy (Cain, 1972; Jackson, 1980).

2.12 IRRIGATION

Irrigation is fundamental for pear production, because it increases yield, improves fruit quality, stabilizes production between years, allows production in environments not suitable for cultivation, and increases fertilization efficiency. The water requirement of a pear orchard depends on the cultivar, rootstock, planting design, irrigation system, soil management, weather conditions, and tree performance and production (Brouwer and Heibloem, 1986). Pear trees should be watered weekly during dry periods, especially during the first two years after planting. After this period, irrigation may be done every two weeks. Soils must be well drained to maintain good aeration, reduce the incidence of pests and diseases, and improve tree anchorage and nutrient exploration by the roots. Precision quantification of irrigation time and water requirement is crucial to achieving good fruit yield and quality. The principal water requirement occurs when shoots and fruits are in active growth. Determination of soil humidity with tensiometers is also important in order to avoid excessive vegetative growth. Other important points are related to tree size, canopy volume, and root depth.

There are three critical growth stages for water requirement: flowering, rapid shoot growth and spring root growth (6–7 weeks after budburst), and the fruit fill stage (4–8 weeks prior to harvest). After harvest, the amount of water must be enough to allow the trees to take up postharvest fertilizers. During the dormancy phase, irrigation is not generally needed.

The rooting system will determine the volume and frequency of irrigation. Root depth is usually lower than 60 cm in pear trees, but it can be affected by soil type and moisture (Westwood, 1993).

Irrigation systems frequently used in pear plantations are furrows, sprinklers, and drippers. Furrows are small, parallel channels that carry water to irrigate the trees. The furrows can be used in most soils, but they are not recommended in sandy soils because percolation losses can be high. In the sprinkler irrigation system, water is distributed through a system of pipes by pumping. It is then sprayed into the air through sprinklers, as small water drops which fall to the ground. Drip irrigation involves dripping water onto the soil at very low rates (2–20 l/h) from a system of small diameter plastic pipes fitted with outlets called emitters or drippers. In drip irrigation, water is usually applied every

1–3 days to provide the root system with favorable soil moisture. The selection of the irrigation system depends on plant and fruit growth stage, root system, soil structure, weather conditions, water availability, and the need to use the irrigation system for other purposes such as fertigation or frost control (Table 2.4). The objective of the irrigation schedule is to determine the correct frequency and duration of irrigation for the crop (Fernández and Cuevas, 2010). With flood irrigation, there is little control over the efficiency of water application. Micro-irrigation allows greater control over the water. There are two possible scenarios for irrigated orchards:

1. Nonlimiting water: So the optimal amount of water to be applied will be the amount that results in the highest yield and fruit quality.

2. Limited water (e.g., drought): So maintaining production is desirable. Regulated deficit irrigation (RDI) is a tool for cultivating pear orchards with low water availability (Shackel, 2011). The objective is to save water by controlling water stress without affecting fruit production or quality (Marsal et al., 2002; Mc Guckian, 2013; Molina-Ochoa, 2015). However, under certain conditions, pear fruit size was found to be negatively affected by RDI (Marsal et al., 2000).

TABLE 2.4 Characteristics of Different Irrigation Systems

Irrigation System	% Water Efficiency (Vol. for Production/ Vol. Irrigated)	Soil Characteristics	Application Rate	Suitable for Fertigation	Relative Costs
Furrow	50–70	Not suitable for sandy soils	8–12 mL/ha	No	Low
Sprinklers	60–80	Suitable for all soils	6 mm/h, 3–11 mL/ha	Yes	Medium
Mini-sprinklers/ microjets	70–90	Suitable for all soils	5 mm/h, 6–10 mL/ha	Yes	Medium
Surface drip	80–95	Not suitable for sandy soils	3–8 mm/h, 4–8 mL/ha	Yes	High
Subsurface drip	80–95	Not suitable for sandy soils	2 mm/h 4–7 mL/ha	Yes	High

One irrigation strategy developed to save water is partial root-zone drying, whereby water is withheld from part of the root zone, whereas another part is well watered (Kang et al., 2002). The irrigation strategy called Readily Available Water (RAW) is the amount of water that is easily available to the tree. This is between a water potential of −0.002 to −0.03 MPa (field capacity) and the potential at which plants start to have difficulty sucking up water at −1.5 MPa (permanent wilting point). RAW is approximately 6% of the total soil volume for sandy loam and clay soils and 8% of the total soil volume for clay loam soils. Monitoring soil moisture is important to identify how much water should be applied and when irrigation should be done (Allen et al., 1998).

In commercial orchards, it has been found that measurements of stem water potential, canopy temperature, sap flow, and stem diameter variation (SDV) are suitable indicators for accurate irrigation scheduling. Stem water potential is associated with fruit production, pear size, and sugar content (Naor, 2001; Shackel, 2011). The SDV detects water stress early and can be continuously and automatically recorded. Based on controlled water stress, irrigation strategies were developed to improve irrigation efficiency and control the vegetative growth of commercial fruit trees.

Regulated deficit irrigation was shown to improve fruit quality and reduce water consumption without negatively affecting fruit production (Marsal et al., 2002). In regions with high frost risks during spring, sprinkler irrigation is one of the most important techniques to reduce damage. This frost control consists in covering the plant with a layer of ice in continuous formation until the end of the frost.

2.13 NUTRIENT MANAGEMENT

Nitrogen (N) is a nutrient that needs to be applied universally to commercial orchards on a yearly basis to maintain productivity. It is the most widely applied nutrient and almost all soils are deficient in it (Tagliavini and Toselli, 2005). Nitrogen is usually applied via soil, but it could also be used as foliar applications (Westwood, 1993). For soil fertilization, the selection of the nitrogen fertilizer must take into account soil pH. If the soil pH is 5.5 or less, then $(NO_3)_2Ca$ is preferred, whereas if the soil pH is 5.5 or more, urea or $SO_4(NH_2)_2$ should be chosen (Westwood, 1993). For foliar application, urea at 0.6 kg/100 l is used to improve tree growth and increase N reserves.

Application of N after harvest is an ideal way to increase the reserves of the tree for early spring activities (Faust, 1989; Sánchez et al., 1992).

The N absorbed and translocated from the leaves in fall depends on the temperature. If the leaves senesce slowly, most of the N is translocated (Faust, 1989).

Phosphorus (P) requirement of pear trees is smaller than N and K. Due to little requirement, a deficiency is unlikely to occur in mature orchards (Brunetto et al., 2015). Anyway, P fertilization is standard practice at planting because it stimulates root growth.

Potassium (K) demand in pear is high, especially during fruit growth. Potassium is a major nutrient that needs to be supplied in large quantities, in particular to fruit trees (Faust, 1989). It is the most abundant nutrient in fruits and it positively affects the size, firmness, skin color, sugar content, acidity, juiciness, and aroma of pear fruits (Brunetto et al., 2015).

Calcium (Ca) is an important nutrient that improves fruit quality, prevents fruit disorders, and increases tolerance to pathogens (Brunetto et al., 2015). Due to its mobility mediated by transpiration, it is difficult to increase calcium content in the fruit under conditions with high transpiration. To improve Ca uptake during early spring, any stress or interference with this process, like fertilization with NH_4^+, K^+, or Mg $^{+2}$, should be avoided during the 4–6 weeks after bloom (Sánchez, 2015).

Boron (B) is a key micronutrient. Insufficient B has been associated with poor fruit set and fruit disorders. The B needed in flowers is transported from reserves in the wood in species of the Rosaceae family. Foliar applications of B before full bloom or after harvest were reported to increase fruit set and fruit yield (Wojcik and Wojcik, 2003). Rootstocks influence the nutritional status of pears. Pears on quince absorbed less N and B and more Mg than pears on seedling rootstock (Westwood, 1993).

Iron (Fe) is also an importany micronutrient. Pear trees grafted on quince should be fertilized with Fe because this rhizome does not have the capacity to solubilize the Fe (Brunetto et al., 2015). An alternative to the replacement of iron fertilizers is the use of organic matter as manure or compost (Sánchez, 2015).

Last but not least, Zinc (Zn) plays an important role in fruit tree nutrition. It is the most limiting micronutrient for fruit tree production, especially when the trees are grown in alkaline soils (Sánchez, 2015). Zinc is included in many fertilizer programs as foliar applications. Spring applications are the only effective tool to incorporate moderate amounts of Zn into the targeted organs (Sánchez and Righetti, 2002).

To determine the nutrient doses and fertilization timing in a pear orchard, the uptake and remobilization of nutrients within the trees must be taken into account (Tagliavini et al., 1997; Quartieri et al., 2002; Neto et al., 2008). Nutrients in leaves, shoots, and pruned branches return to soil after their degradation. Therefore, a simple approach is to apply nutrients that are exported by fruit harvest (Musacchi, 2007) (Table 2.5).

TABLE 2.5 Annual Export of Macronutrients for Pear Orchards

N	P_2O_5	K_2O	CaO	MgO
70–90	5–10	65–85	135–140	12–15

Fertilization could be applied to soil or leaves. Soil fertilization could be distributed broadcast or diluted in irrigation water, which is called fertigation. The most widely used fertilizers for fertigation are urea, phosphoric acid, and potassium nitrate (Sansavini et al., 2012). The pH of the solution should be between 5.5 and 7, and its electric conductivity must be kept below 1.5–2 dS/m (Il Pero, 2007; Sansavini et al., 2012).

The N fertilization efficiency is often low (<30%), even in fertigation (Neto et al., 2008). Timing of N fertilization is crucial to determine its destiny inside the tree. Remobilization from storage was shown to provide the greatest amount of the N required for the initial growth of spur leave and flower development

(Tagliavini et al., 1997), while 50% of pear fruit N at harvest is absorbed from soil during the first two months after bloom (Quartieri et al., 2002).

Foliar applications cannot replace fertilization nutrition via soil but can be useful under certain circumstances (Sansavini et al., 2012). Occasional high nutrient requirements may not be met by uptake through the roots and need to be supplied through foliar application (Faust, 1989). Absorption is fast and finished after 48 hours, reaching an efficiency of 75%–92% at least for nitrogen (Sansavini et al., 2012). Foliar application of urea at 5% during bloom increases fruit size in Bartlett pear (Curetti et al., 2013). Some micronutrients, such as Zn and B, can be supplied exclusively via foliar applications.

Foliar fertilization usually uses between 150 and 1500 l/ha. Solution pH is important, and low values of pH (5.5–6.5) are optimum for the absorption of urea and Zn (Sansavini et al., 2012). Weather conditions are also important because they determine the evaporation of solution and absorption of nutrients. It is not recommended to pulverize nutrients in high evapotranspiration conditions (temperature >25 °C, relative humidity <70%, and/or wind speed >8 km/h) (Sansavini et al., 2012).

Permanent covers with white clover in a pear orchard were reported to increase the organic matter and N content in the soil in comparison with soil mowing (Xu et al., 2013). Soil analysis is used to determine the nutrient availability, but leaf analysis better indicates the nutritional status of the trees (Faust, 1989). Nutrient concentrations change with the age of the leaves. Nutrients with low mobility via the phloem such as Ca, Mn, Fe tend to increase their concentration in leaves, whereas mobile nutrients such as N, P, and K can be remobilized to fruits or other centers of active growth (Faust, 1989; Sansavini et al., 2012). There is a certain plateau after shoot growth in which nutrient concentrations are relatively stable and appropriate for determining the nutritional status of the tree (Faust, 1989; Sansavini et al., 2012). Nutrient standards for mid-shoot leaves are shown in Table 2.6.

TABLE 2.6 Nutrient Standards for Leaves of Pear in July–August (Northern Hemisphere)

N (%)	P (%)	K (%)	Ca (%)	Mg (%)	Fe (ppm)	Mn (ppm)	B (ppm)	Zn (ppm)	Cu (ppm)
2–2.8	0.16–0.28	1.2–1.8	1.2–2.2	0.20–0.40	40–120	15–100	15–30	10–100	>5

Leaf analysis offers a limited diagnosis of deficiency for some nutrients such as Zn, Fe, or B. Another nutrient that could be erroneously diagnosed is N, because trees overfertilized with this nutrient are vigorous and can present a low concentration of N in leaves (Sansavini et al., 2012). A correct determination of the nutritional status of the trees includes not only nutrient concentrations in leaf analysis, but also the observation of the orchard. Important aspects of the orchard for the diagnosis are shoot growth, leaf size and color, symptoms of nutritional deficiency, and fruit load and size (Sansavini et al., 2012). Some symptoms of nutritional deficiencies are listed below:

N: lower shoot growth and leaf size; widespread pale green leaves.

P: not usual; lower fruit set; early leaf drop.

K: necrosis of leaf margins; smaller fruits with subdued color, less sugar and acidity.

Ca: new leaves are small, light green, almost white; fruit disorders.

Mg: chlorosis between veins; severe deficiency, necrosis and leaf drop.

Fe: new leaves with chlorosis between veins.

Mn: internal leaves with chlorosis between veins.

B: shorter shoots; chlorosis and distortion of leaves; low fruit set; small and misshapen fruits.

Zn: shorter shoots; small narrow leaves; rosette leaves, especially in calcareous soils.

2.14 TRAINING AND PRUNING

Pears and apples are trained and pruned in a similar way (Verna, 2014). The main objectives of training systems in pear are to increase the orchard efficiency and to secure early and high returns from investments. In pear trees, scaffolds tend to grow nearly vertically; therefore, an effective training system is particularly important. Scaffolds are strong vegetative structures that grow and cause poor, delayed fruiting.

2.14.1 TRAINING TYPE

Improving the vegetative development of trees is the goal in the first years of the tree, so that full canopy closure can be reached. The closure of the tree canopy is related to the density of plantation. In common densities, closure can be achieved by the end of the third or fourth season, but in ultra-high densities, canopy closure can be reached at the end of the first year (Ferree and Schupp, 2003).

Pear trees can adapt to all kinds of shapes, but the most appropriate ones derive from central leader systems (conical or pyramidal shaped trees), double leader systems (single trees with two leaders simulating higher densities), palmette systems (central leader trees with scaffolds in the plane of the row only), and V- or Y-shaped systems (inclined canopies that improve light interception) (Sansavini and Musacchi, 1994; Musacchi et al., 2011).

Central leader is the most common training system for developing a conical shaped tree. This system is suitable for densities of up to 2000 trees/ha and for places with high solar radiation and long sunny days (Collett, 2011). Several systems originated from the central leader concept (free spindle, vertical axis, slender spindle). Central leader consists of tiers of scaffold branches along a straight central axis (Adaro et al., 2010).

In double leader systems, trees are planted at a density of around 3000 trees/ha, but the development of double leaders implies a leader density of 6000 trees/ha. The objective is to achieve high densities of the leader with a reduced number of trees.

The palmette system is suitable for densities of 700–1500 trees/ha. This system consists of a central leader with scaffolds in the plane of the row only. The bending of branches on trellises controls growth and provides a balance between fruiting and vegetative growth (Adaro et al., 2010).

The V- or Y-shaped systems consist of trees which have a vertical trunk and two opposing arms of the tree trained to either side of the

Pear

trellis, and in V-shaped trees, the whole tree is leaned to one side of the trellis, whereas the next tree in the row is leaned to the other side. These training systems facilitate maximum sunlight interception and are suitable for high and ultra-high density (2000–6000 trees/ha). Trees can be trained in different ways, such as single leader, double leaders, and in cordon.

Pruning time in pears is winter season when the trees are dormant. The pear fruits are produced from terminal buds on short spurs. This formation of spurs takes two years and occurs on older wood. The spurs have a long life and remain productive for 4–5 years or even longer, so it is important to keep them healthy and productive (Verma, 2014).

2.14.2 TYPE OF PRUNING

The purposes of pruning are to establish the basic tree architecture, limit the tree size, allow sunlight interception and air circulation, and control vegetative growth. Pruning contributes to achieving structures that allow the best fruit production in the shortest possible time. It modifies the distribution of the intercepted sunlight, thus improving the formation and differentiation of fruit buds. Pruning is the most commonly used method to alter apical dominance and change the balance between the upper part of the tree and the part

below the ground while reducing the overall amount of dry matter accumulation.

The types of pruning according to planting age are as follows:

1. Planting pruning is done immediately after planting to stimulate the growth of vigorous buds. The type of pruning depends on the training system, the density of the plantation, and the characteristics of the plant.
2. Training pruning is done only once, when the tree is young, to achieve an adequate structure that quickly occupies the assigned space while simultaneously producing fruit.
3. Fruit pruning is carried out every year to regulate production. It helps to maintain stable productions and quality fruit every year, constantly rejuvenating the tree (stimulating growth and the production of good fruit structures), retaining and limiting the volume and the height of the plant, and increasing light penetration inside the canopy.

Types of pruning according to the time of the year are as follows:

1. Winter pruning is generally carried out every year, when most of the leaves are not active. Its purpose is to train

the plant and manage the fructification. It is applied in areas with heavy winter frosts, to delay fruiting as much as possible.

2. Green pruning is performed when the plants still have leaves. Depending on the objective pursued, pruning may be performed at different times of the year as follows: early spring pruning retains growth; late spring pruning eliminates unwanted growths (suckers) that compete with the normal development of the fruit; summer pruning increases the light income inside the plant; late summer or fall pruning is carried out after harvest to increase light distribution in the canopy (especially in the inner and lower parts), to retain growth once the tree has reached the desired height and occupied the corresponding space, and to eliminate large branches or retain plants that are very high.

Types of pruning cuts:

1. Heading is done to remove the terminal portion of shoots or limbs. Heading in wood older than one year helps to increase the size of the fruits.

2. Thinning involves removing an entire shoot or limb at its point of origin from a main branch or limb. With thinning cuts, some terminal shoots are left intact, apical dominance remains, and the pruning stimulation is more evenly distributed among remaining shoots. Its main objective is to improve the light distribution within the canopy.

3. Root pruning is generally done in high-density orchards to restrict root growth and canopy size.

2.14.3 CANOPY MANAGEMENT

Pear trees growing on their own roots are, in general, too large for the economic production of high-quality fruits. Much of their canopy volume is unproductive, which adds to management costs without generating financial profits. Therefore, they are inefficient and expensive to produce and manage. A relatively porous canopy is required to achieve an optimum balance between yield and quality; this is an important objective of the training and pruning systems, which also control the relationship between vegetative growth and fruiting.

2.14.4 MECHANIZED PRUNING

The incorporation of mechanical pruning into flat production systems generates two very divergent but highly desirable situations:

1. The use of mechanical platforms instead of stairs, which incorporates family labor and/or people with physical conditions unsuitable for pruning and harvesting.
2. The use of technology of both remote and proximal sensors, which is much more applicable to these flat driving systems.

The prototype of mechanical pruning machines basically scans the tree rows with front- and rear-mounted cameras and analyzes the images to distinguish between the different vegetative structures (scaffolds, shoots, and spur) before the cutting. Then a computer program selects the branch to be cut. Hydraulic shears mounted on a robotic arm cut the branch at the precise points calculated by the system. However, the performance of mechanical pruning is not yet equal to that of manual pruning (Lehnert, 2012). The implementation of conduction systems such as "fruit wall" makes it possible to improve the economic and ecological aspects of fruit production. The advantages of the low and narrow fruit walls are related to the type of management, the crop characteristics, and environmental aspects. With small pedestrian trees, it is not necessary to use platforms or ladders, which is safer for workers. All the tasks of orchard management, from pruning to thinning and harvesting, can be done more quickly and easily from the ground. Any type of mechanization (e.g., flower-cutting machines, pruning, tunnel spraying, and weed control) is easier to apply in a low fruit wall than in traditional trees. The main techniques to control the vigor and facilitate the formation and maintenance of the fruit wall involve increasing the number of axes per plant and green pruning during the summer. Other tools available to maintain the fruit walls are root pruning, application of growth regulators, and controlled water stress. Multi-axis tree driving is a powerful tool to form the fruit wall. The energy is divided into the formation of multiple structural branches, so each branch has less vigor. Consequently, in trees with six axes, for example, the secondary structures are usually absent and only the trunks remain as structural wood (Dorigoni and Micheli, 2015). The Bibaum® (a registered trademark in Italy for trees with two axes preformed in nursery) and multi-axis trees have a natural tendency to form a narrow fruit frame, with little secondary structure.

2.15 INTERCROPPING AND INTERCULTURE

Soil management techniques include intercropping with cover crops, use of herbicides under pear trees, tillage and mowing between rows, manure or compost applications, and the use of different kinds of mulch (Ryugo, 1988; Faust, 1989; Westwood, 1993; Musacchi, 2007; Intrieri et al., 2012).

Cover crops between rows improve soil health and pear tree performance (Ryugo, 1988; Faust, 1989; Westwood, 1993; Intrieri et al., 2012). Including cover crops in the soil management of an orchard has the following positive effects:

1. Improves soil fertility, due to an improvement in structure, and increases the organic matter, soil aeration, nutrient retention, and availability in superficial layers of soil and nitrogen content if legume species are used.
2. Reduces soil compaction and erosion.
3. Increases water infiltration.
4. Allows the passage of agricultural machines in flood-irrigated orchards.
5. Decreases soil temperature.
6. Produces a more favorable soilroot interface.
7. Increases root growth of apple trees in superficial layers of soil.

8. Improves fruit quality (color and sugar content).

On the other hand, cover crops present competition to the roots of pear trees for water and nutrients, have their own requirements to be satisfied by irrigation and fertilization, and are not recommended in regions with low availability of water (less than 500 mm/year) (Ryugo, 1988; Faust, 1989; Intrieri et al., 2012). Cover crops also reduce the vegetative vigor of pear trees. Some other limitations of cover crops include a higher frost risk caused by the reduction in heat accumulation in soil and an increase in disease incidence because of the higher humidity generated. In addition, tree flower pollination could be hindered because bees could be attracted by cover crop flowers if they are not cut in pear bloom time. Pear fruit weight may also be reduced (Ryugo, 1988; Musacchi, 2007).

The species most widely used as a cover crop in orchards belong to the family of *Poaceae,* even though they are very competitive in water and nutrients. Sometimes, mixed species are used, such as *Lolium perenne* plus *Festuca arundinacea* or *Poa pratensis* (Sozzi, 2007). In poor soils, legumes are preferred due to their capacity to fix nitrogen from the atmosphere. Some characteristics of different species are listed in Table 2.7.

Pear

TABLE 2.7 Characteristics of Different Vegetal Species Used as Cover Crops in Pear Orchards

Species	Settlement	Vigor	Duration	Water Use	N–P–K Requirement	C/N Ratio
Festuca arundinacea	Excellent	High	Long	High	50–60–40	50
Festuca rubra	Excellent	Low	Long	Low	60–80–40	40
Lolium multiflorum	Excellent	High	Short	Medium	50–70–40	40
Lolium perenne	Excellent	Medium	Short	Medium	60–80–40	30
Po apratense	Good	Medium	Medium	Medium	70–80–40	35
Trifolium pratense	Good	High	Medium	High	10–90–60	18
Trifolium repens	Excellent	Medium	Long	High	10–80–60	16

Alternative management of orchard is a compromise: cover crops controlled by brush cutter or mowing machines between rows and weed control closed to the trees in the rows (Faust, 1989; Westwood, 1993). Weed control is necessary to avoid water and nutrient competition, and it is especially important in springtime and crucial in young plantations. Bands of 1.0–1.5 m in width must be kept without spontaneous vegetation.

Weed control could be achieved by manual (hoe), mechanical, or chemical techniques. Manual control involves labor cost and is usually expensive (Musacchi, 2007). Mechanical weed control is difficult close to young trees, so in this case chemical control is preferred (Musacchi, 2007). Even with herbicide application, caution must be taken to avoid pulverization over tree foliage or small trunk in these young trees (Ryugo, 1988; Westwood, 1993). Shelter around new plants (1–3 years old) is used to protect them from mechanical or phytotoxicity damage (Musacchi, 2007).

Herbicides could be divided into two groups: those of residual and preventive action (oxifluorfen, diuron, etc.) and those with direct action over weeds (glyphosate, fluazifop, etc.). Herbicides of the first group inhibit the capacity of seed germination or are absorbed by roots, and they have a long persistence in soils. However, they are detrimental to orchard productivity and soil fertility in the long term and dangerous to the environment, especially if percolated to the water table. Herbicides of the second group (desiccant) are more widely used in orchards because they are safer to the environment, even though they are less effective in perennial species (Intrieri et al., 2012).

Herbicides could be applied at different times during the season. The first application is usually done in spring. Dry grass after treatment generates a mulching effect that reduces the emergence of new weeds. A second application is usually necessary during summer. Fall application of systemic herbicides like glyphosate is useful to control perennial species (*Cynodon dactylon* or *Convolvulus arvensis*) in highly infested orchards. Efficient applications are achieved when weeds are in the first stages of development and weather conditions are appropriate (Musacchi, 2007).

Tillage is used to reduce evaporation and water and nutrient competition, but it destroys superficial tree roots (>12 cm) and soil organic matter (Faust, 1989; Intrieri et al., 2012). Mowing reduces transpiration of cover crops and restricts water loss. Dry residues coming from cover crop mowing provide nitrogen and decrease nutrient competition. Mower machines can be equipped with re-entry devices that make it possible to get closer to the trees but can also cause some damage to small trunks (Musacchi, 2007). Another important problem related to mechanization is the soil compaction generated by the transit of agricultural machines between the rows, especially in wet or naked soil conditions. Soil compaction affects water infiltration, soil aeration, nutrient availability, root and tree growth, and fruit production and quality (Sozzi, 2007).

Other soil management techniques include manure or compost application, incorporation of annual crops to the soil, and mulch use in the rows. These practices improve soil properties, especially organic matter, nutrient content, and water infiltration (Tagliavini et al., 2000; Abu-Rayyan et al., 2011; Ingels et al., 2011; Intrieri et al., 2012; Sorrenti et al., 2012; Van Schoor et al., 2012; Ingels et al., 2013; Han et al., 2015). Mulch use also contributes to reducing weed competition, keeping water in the soil, decreasing soil temperature, and increasing fruit weight and yield in pear orchards (Bertelsen, 2005; Abu-Rayyan et al., 2011; Ingels et al., 2011, 2013; Einhorn et al., 2012). Mulch may be organic or inorganic. Organic mulch could be made from dry cuts of cover crops, wood chips, or pruning residues. Cover crops produce 4–7 tons of dry matter per hectare each year (80% above ground). Pruning residues could provide 2–6 tons of dry matter per hectare if they are left in the orchard. Inorganic mulch is usually made of black polyethylene film (Intrieri et al., 2012).

2.16 FLOWERING AND FRUIT SET

Flowering in trees is achieved once the adult or mature phase is reached after the juvenile period has been overcome (Ryugo, 1988). Flowering in fruit trees can be divided into two processes: the initiation and development of flower buds occurring during the summer and fall of one season, and the flowering process itself occurring in early spring the next season (Faust, 1989).

2.16.1 INITIATION AND DEVELOPMENT OF FLOWER BUDS

Induction of flower buds is a qualitative change governed by hormonal balance and/or nutrient distribution (Faust, 1989). To receive the inductive stimulus, buds must have a critical number of nodes, between 16 and 20 (Tromp, 2000). The induction of flower buds of pear trees in the Northern Hemisphere occurs in July (Westwood, 1993), and it is strongly stimulated by exposure to light. Gibberellins (GA) were reported to inhibit the initiation of flower buds in pear. The seed of the developing fruit is high in GA-like substances, so the presence of young fruits is the most important factor inhibiting flower bud initiation. The time of the inhibitory effect of fruit in pears was estimated to

be 4–6 weeks after bloom (Faust, 1989; Tromp 2000). The development of buds during the fall season is very important to achieve a good bloom the following year. A delay in harvest greatly decreases the bloom in the following year. Premature defoliation before normal leaf fall produces similar results and greatly decreases successive yield. Buds continue to develop during winter. Fall application of nitrogen greatly accelerates the development of buds and plays an important role in poor buds (Faust, 1989). Flower initiation during spring has been promoted as an important key to obtaining good flowering. The late events in bud development during the fall are also important. The early phases of flower bud development are governed by hormonal events, whereas the late development of flower buds depends on the availability of carbohydrates and nitrogen. Growers not only have to reduce the fruit load to prevent flower inhibition but also need to protect the leaves until late fall to keep their photosynthetic efficiency (Faust, 1989).

2.16.2 FLOWERING TIME

Flowering time depends on cultivar and weather conditions during winter. Unfavorable conditions to complete dormancy generate a late, weak, and extended bloom of trees. Flowering tends to last about 15 days. Extended blooms are undesirable because they cause an extended harvest that requires several pickings (Faust, 1989).

2.16.3 ANTHESIS AND FLOWERING TYPES

The pear belongs to the Rosaceae family, whose flowers are hermaphrodite with five sepals, five petals, numerous stamens, and a pentacarpelar pistil (Costa et al., 2012). Inflorescences of pear usually have 7–8 flowers with similar pedicels (Westwood, 1993; Ryugo, 1988) in which the apical flower is less developed and opens last in the cluster (undetermined inflorescence). This type of inflorescence is named corymb (Ryugo, 1988).

2.16.4 POLLEN VIABILITY AND GERMINATION

Both the germination and the growth rate of pollen are dictated by temperature. Cold temperature during early spring diminishes pollen grains and decreases their vitality. By contrast, high temperature during spring often results in sterile pollen (Faust, 1989). Boron applications were reported to enhance pollen tube growth (Sanzol and Herrero, 2001). Calcium is also an important factor in these processes (Costa et al., 2012).

2.16.5 POLLINATION AND FERTILIZATION

Pollination is the transfer of pollen grains from anthers to the stigma (Ryugo, 1988). The effective pollination period (EPP) is determined by the number of days of ovule longevity minus the time lag between pollination and fertilization (Ryugo, 1988; Faust, 1989; Sanzol and Herrero, 2001). The EPP is a useful parameter to identify the factors limiting fruit set (Table 2.8), and it is conditioned by three processes: stigmatic receptivity, pollen growth, and ovule development (Sanzol and Herrero, 2001).

TABLE 2.8 Effective Pollination Period (EPP) Duration for Pear Cultivars[a]

Cultivar	Location or Conditions	EPP (Days)	Limiting Factor		
			Stigma Receptivity	Pollen Tube Growth	Ovule Longevity
Doyennè de Comice	UK, USA, CC[b] at 17 °C	1–2			
	USA, CC[b] at 13 °C	5–6			
	France	8	–	6	14*
Gorham	UK	6			

Pear   147

TABLE 2.8   *(Continued)*

Cultivar	Location or Conditions	EPP (Days)	Limiting Factor		
			Stigma Receptivity	Pollen Tube Growth	Ovule Longevity
Packham's Triumph	UK	7			
Bristol Cross	UK	9			
Conference	UK	7–10			
Williams	UK	9			
	France	6	–	12	6*
Agua de Aranjuez	Spain	3	4*	9–13	12–17
Tsakoniki	Greece	9–11			

Whenever the limiting factor and/or its duration was recorded, it is indicated. (b) Controlled Conditions; (*) limiting factor

High temperatures during flowering accelerate pollen tube growth (Sanzol and Herrero, 2001). For example, pollen growth is completed after 12 days at 5–6 °C in Bartlet pear, while it only requires 2 days at 15 °C. Pear ovule longevity is about 12 days. Slow pollen growth causes a low fruit set in cold springs (Ryugo, 1988; Westwood, 1993). Fruit trees that need insects for the pollination have colorful flowers and nectar presence. In commercial orchards, beehives are used during bloom (5–8 beehives per ha) (Ramirez and Davenport, 2013). Bees do not fly very well under rain conditions, strong winds (more than 25–40 km/h), or temperatures below 10 °C (Ryugo 1988; Westwood, 1993). Bees prefer cherry flowers with more than 50%–55% sugar content to pear flowers with 15% sugar content (Costa et al., 2012). Bees usually fly from tree to tree within a row, rather than from one row to another (Costa et al., 2012).

2.16.6   POLLINIZER

Apple and pear trees are self-incompatible (Faust, 1989). If one cultivar is expected to pollinate another, the two cultivars must bloom together. There are lists with bloom time of cultivars that can be used as guides for planting plans (Faust, 1989).

To be a pollinizer, varieties must have good pollen production, be compatible, and bloom right before the commercial cultivar (Faust, 1989; Westwood, 1993). In the Rosaceae family, self-incompatibility is regulated by one multi-allelic locus S. To avoid a poor fruit set, a combination of pear cultivars with different allelic components must be chosen (Costa et al., 2012). A minimum percentage of pollinizer trees in an orchard is 11%, one pollinizer every nine trees. For practical management, pollinizer trees are usually planted in complete rows (one row of pollinizer every 2–5 rows).

2.16.7 FRUIT SET

To achieve a good fruit set, a sequence of physiological events is needed. The first requisite for a good fruit set is the development of strong flower buds. The second one is a certain temperature, to assure good pollination and fertilization. The third requirement is a high level of photosynthate supply for young fruits (Faust, 1989).

Parthenocarpy is the development of fruits without fertilization (Costa et al., 2012), and it is widespread in pears. Although the fruit set is genetically determined, the rate of parthenocarpy is influenced by environmental conditions. For example, parthenocarpy in Bartlett pears depends on high temperature around bloom (Faust, 1989).

2.17 FRUIT GROWTH, DEVELOPMENT, AND RIPENING

Fruit growth comprises the increase in length, diameter, and mass. Fruit growth begins at the onset of fruit development and involves the mobilization of photosynthesis products from other parts of the plant. Pollination stimulates the enlargement of the ovarian tissue, which results in the fruit set, while the enlargement of the receptacle takes place during the later periods of fruit development (Opara, 2000). The fruit provides an adequate medium for seed maturation and allows a natural mechanism for its dispersion (Sozzi, 2007b). Pear fruits consist of more than an ovary wall; a part of the receptacle encloses the flower and becomes a part of the fruit. Both, the receptacle and the ovary wall, are edible, except for the endocarp and the seed cavity (Dennis, 1996). The growth pattern followed by pear fruits is represented by a smooth sigmoid curve (Figure 2.7). Pear fruit development may be divided into three partially overlapping stages:

Phase I: The period of exponential growth (cell division), which lasts 7–9 weeks.

Phase II: The period of linear growth (enlargement or cell expansion); and

Phase III: The final stage of growth and maturation period (Westwood, 1993).

Pears are harvested during the exponential portion of the fruit growth. If pear fruits were left on the tree, they would complete the sigmoidal curve. Later harvest achieves a greater fruit size but decreases the storage quality (Faust, 1989; Sozzi, 2007b).

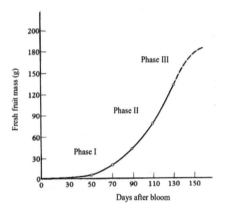

FIGURE 2.7 Pear fruit growth curve.

Pear fruit growth is under the control of a complex network of both endogenous and exogenous factors. The endogenous factors are as follows:

1. Hormonal factors: Fruit growth is associated with a pattern of hormone activity, which is similar in both parthenocarpic and seeded fruits.
2. Variety and rootstock: Variety can directly influence the fruit growth rate through the branch type and orientation and the insertion angle of the branches. By means of these characteristics, the variety can affect light interception, and therefore, fruit growth. Rootstock choice affects flowering abundance, fruit set, size, and quality (Sozzi, 2007b).
3. Flower position in the inflorescence and in the tree also affects fruit size. Pear inflorescences are racemes. The terminal flower is the last to be formed and, therefore, the smallest; thus, basal flowers tend to produce larger fruits than terminal ones (Dennis, 2000).

The exogenous factors that affect fruit growth are climatic and edaphic characteristics, including temperature, light exposure, and the availability of water and nutrients (Sozzi, 2007b). In addition, some irrigation and fertilization modify the incidence of climatic and edaphic factors in pear trees.

The final stage of pear fruit growth is the maturation period. Pear belongs to the climacteric fruits and shows an increase in respiration and ethylene production rate during ripening (Paul and Pandey, 2014). However, some Asian pears do not show climacteric behavior and produce little ethylene (Jackson, 2003).

Ethylene is a gaseous phytohormone of very simple composition (C_2H_2), which has numerous effects on plant growth, development, and senescence. However, its incidence in climacteric fruit maturation represents, by far, its most important role (Sozzi, 2007a). Ethylene biosynthesis is catalyzed by two enzymes: 1-aminocyclopropane-1-carboxylate synthase (ACS) and 1-aminocyclopropane-1-carboxylic acid oxidase (ACO). ACS restricts the rate and the production pathway to 1-aminocyclopropane-1-carboxylic acid (ACC) from S-adenosyl-L-methionine (SAM), which is then converted to ethylene by ACO (Alexander and Grierson, 2002).

Ethylene production in some pear varieties is not detectable at commercial maturity. For example, the capacity to produce ethylene in Beurré d'Anjou develops during storage, and fruits removed from storage too early do not produce ethylene in significant quantities. Quality changes associated with ripening occur at slow rates in pears not producing ethylene. Softening, color changes, and aroma development during Beurré d'Anjou pear ripening are all mediated by ethylene (Mattheis, 1996).

Generally, winter pear cultivars require exposure to low temperatures (chilling requirement) or to ethylene gas after harvest in order to develop the capacity to ripen. Low-temperature exposure during post-harvest develops the capacity of pear fruit to produce sufficient ethylene to activate and complete the ripening process (Villalobos-Acuña and Mitcham, 2008). The length of time to complete this cold temperature requirement is characteristic of the cultivar (e.g., Bosc 15 days, Comice 30 days, and Beurré d'Anjou 60 days) and depends on fruit maturity at harvest, storage atmosphere, and nutritional status of the fruit (Sugar, 2002, 2011).

These biochemical changes during fruit ripening include an increase in ethylene production and respiration. Other processes also occur, including chlorophyll degradation, carotenoid and anthocyanin biosynthesis, and an increase in flavor and aroma components. Along with these processes, there is also a decrease in starch content and acidity, an increase in sugars and the activity of cell wall-degrading enzymes (Prasanna et al., 2007).

The color of the skin depends on plastid pigments, chlorophylls, and carotenoids and on red coloration due to anthocyanins in the vacuoles (Jackson, 2003). Some pear cultivars do not change skin color during ripening (Villalobos-Acuna and Mitcham, 2008). Immature fruit color is dark green, and it disappears during ripening due to breakdown of chlorophylls *a* and *b*. Carotene declines during ripening but xanthophylls increase. The green color may fade until it has completely disappeared, and the skin fruit becomes cream to pale yellow. The green color may fade less completely, giving a greenish yellow to yellowish green color (Jackson, 2003).

The organic compounds on which pear aroma and flavor depend are synthesized during the climacteric phase (Jackson, 2003). In pears, the volatile compounds are ethyl and dimethyl esters (Paul et al., 2012). Unsaturated esters are synthesized by β-oxidation of polyunsaturated linoleic and linolenic acids (Eccher Zerbini, 2002).

During fruit growth, carbohydrate from photosynthesis is transported to small developing fruits as sorbitol. In the fruit, it is converted mainly to fructose and starch with some sucrose and glucose. The main sugars are fructose, sorbitol, glucose, and sucrose. During ripening, starch hydrolysis is accompanied by the appearance of sucrose, which is then slowly hydrolyzed to form more glucose and fructose (Eccher Zerbini, 2002; Jackson, 2003).

The sugar released during starch breakdown, along with organic acids, is utilized for respiratory metabolism (Mattheis, 1996). In general, organic acids decline during maturation, possibly due to their use as a respiratory substrate or to their conversion into sugars (Sozzi, 2007a). Malic, ascorbic, and citric acids are the predominant organic acids in Asian and European pear cultivars (Eccher Zerbini, 2002; Arzani et al., 2008).

Pear fruit softens significantly during ripening. These textural changes result from the solubilization and depolymerization that occur due to specific degradation enzymes of the cell wall (Hiwasa et al., 2003, 2004). Softening is related to the increased activity of polygalacturonase, α-galactosidase, β-galactosidase, and α-mannosidase. During pear ripening, synthesis of xylans is associated with the continued development of sclereids, which has a significant effect on texture, reducing early oversoftening (Eccher Zerbini, 2002).

2.18 FRUIT RETENTION AND FRUIT DROP

There are three waves of fruit drop in pear trees. The first drop of nonfertilized flowers occurs shortly after petal fall (Racskó et al., 2006). The second drop ensues during the 6–8 weeks after bloom and is known as the "June drop" because it occurs at the end of May or in early June for the Northern Hemisphere (Faust, 1989). The third drop occurs at preharvest time (Racskó et al., 2006).

The first drop is the largest one (90%–98% of initial flowers), so it is called the "cleaning" drop (Sansavini et al., 2002; Racskó et al., 2006). This drop depends on weather conditions during bloom (Sansavini et al., 2002), and it is usually heavy after self-pollination but less so after cross-pollination (Faust, 1989) because fruits with less than three seeds tend to shed (Racskó et al., 2007).

The second drop is more conspicuous because the size of the dropping fruit is much larger. The second drop depends on the loading capacity of the trees (Racskó et al., 2006). Carbon balance may be critical during the period before the final fruit drop. High temperatures, especially at night, have been found to cause fruit abscission (Schaffer and Andersen, 1994). Vigorous shoot growth is also responsible for fruit drop (Racskó et al., 2007).

The third drop depends on weather conditions (e.g., wind) and varietal susceptibility (Racskó et al., 2006). Some pear varieties, such as Hardy, Beurre Bosc, Conference, and Williams are susceptible to preharvest fruit drop, whereas others such as Blanquilla and Packham`s Triumph are less affected (Racskó et al., 2006). A common compound for avoiding preharvest fruit drop is naphthalene acetic acid (NAA).

2.18.1 MANAGEMENT STRATEGIES FOR FRUIT RETENTION

Standard management strategies for fruit retention include using a combination of varieties as pollinizers, introducing beehives during bloom, and applying boron and other substances (Faust, 1989; Sanzol and Herrero, 2001; Sansavini, 2012).

The ethylene biosynthesis inhibitor aminoethoxyvinylglycine and the competitive antagonist for ethylene receptors 1-methylcyclopropene (1-MCP) reduce fruit drop. The response may depend on the time of application, the position of the flower within the cluster, and the age of the wood on which the cluster is situated (Faust, 1989).

2.18.2 FRUIT THINNING

The practice of thinning is more important in apple than in pears (Greene and Costa, 2013). Removing excess fruits makes it possible to adjust fruit load to a desirable level (Ryugo, 1988). Thinning improves characteristics of fruit quality, such as size and color of the fruit. Concurrently, the taste of the fruit (sugar, acid content, and firmness) will be increased (Link, 2000). Thinning can be done by chemical applications or manually (Sansavini et al., 2012).

Chemicals can be applied at different times. Bloom time thinning is undesirable in many locations until the frost danger is over (Faust, 1989; Sansavini et al., 2012). Postbloom thinners are the auxins NAA and naphthaleneacetamide (NAAm) and the cytokinin 6-benzyladenine (6-BA) (Greene and Costa, 2013). Pears are successfully thinned with NAA at 15–20 ppm, NAAm at 10–15 ppm, or 6-BA 100–150 ppm (Ryugo, 1988; Faust, 1989; Bound and Mitchell, 2002; Maas et al., 2010; Dussi and Sugar, 2011). The most effective time for thinning pear trees is 15–20 days after bloom (Ryugo, 1988; Faust, 1989).

New chemical thinners have recently emerged and are under evaluation. These include two naturally occurring compounds, abscisic acid (ABA) and ACC, and the photosynthesis inhibitor metamitron (Brevis). Metamitron is a thinner applied 8–14 mm of king fruit diameter, and the resulting phytotoxicity does not adversely affect the fruit quality, productivity, or return bloom (Greene and Costa, 2013).

2.19 HARVESTING AND YIELD

European pears that are allowed to mature on the tree usually develop a poor texture and lack the juiciness and characteristic flavor of the variety. For this reason, they are harvested when they are physiologically mature but very firm, and they reach consumption mature before or during processing. Low temperatures during storage promote ethylene precursor synthesis inside the tissue of the pear, and when the fruit is placed at maturation temperature, ethylene production increases and the fruit matures uniformly. However, most European pears may mature without cold storage through ethylene application in the freshly harvested fruit (Mitcham and

Mitchell, 2007). Not all the fruit is in optimal conditions for harvest at the same time. For this reason, it is necessary, in some cases, to pick it in two or more steps. The quantity and frequency of each step will depend on the characteristics of the variety maturing process. Bartlett, Packham´s Triumph, and Beurré d´Anjou may be picked in more than one step (Campana, 2007).

Several destructive and nondestructive methods may be used to determine the optimal harvest time. The most common indices among the destructive methodologies are fruit firmness, starch content, soluble solids concentration, and titratable acidity. In pears, firmness is the best parameter to determine the appropriate maturity stage. In 1925, Magness and Taylor developed an instrument, called penetrometer, to measure the fruit firmness in order to estimate maturity at harvest. The principle is still used, although several adaptations in size, shape, and function have been made. The penetrometer measures the resistance offered by the fruit pulp to the penetration of a rigid tip or plunger. This tip is an interchangeable piece of known diameter and convex ends. It also has a ring-shaped groove that indicates the depth to which it must be introduced into the pulp to perform the measurement. A tip of 7.9 mm (5/16") is used in pears. In the industry, the pulp resistance is indicated in pound force (lb_f) and kg/cm^2 (kg_f) by a graduated scale but in scientitic papers, firmness is indicated in Newton (N) (Campana, 2007). The conversion is $N = lb_f \times 4.448222$; and $N = kg_f \times 9.807$ (Mitcham, 1996). To perform the measurement, the operator must insert the tip in a perpendicular direction to the fruit surface without skin at a constant speed. The most critical feature of testing is the speed of the force. The faster the pressure is applied, the higher the reading will be. The penetrometer can be mounted on a drill press, which allows a greater control of the direction and speed of penetration, reducing measurement variability. In addition, electronic equipment that facilitates measurements is available. However, the size and energy requirement of this equipment restrict its use in the orchard, and it is therefore used exclusively in laboratories and packinghouses (Bramlage, 1983; Harker et al., 1996; Campana, 2007). Different varieties have different degrees of firmness at maturity: Beurré d′Anjou, 57.8–66.7 N; Bartlett, 66.7–84.3 N; Bosc, 62.7–71.6 N; Comice 49.0–57.8 N, Packham′s Triumph, 57.8–66.7 N (Chen, 2004).

Another useful way of determining harvest maturity is the evaluation of starch content degradation. Starch accumulates in fruits during their growth and it is progressively hydrolyzed to soluble sugars during the ripening period. The level of hydrolysis of the starch correlates with the state of advancement of the maturation process and it is very useful to determine harvest start (Campana, 2007) and maturity. The presence of starch can be detected from the staining produced by the action of a solution of iodine and potassium iodide when exposing the pulp of the fruit, cut transversally at the equatorial zone. The colorless area corresponds to the sectors in which the polysaccharide has been degraded, while the blue zones indicate that there is still a starch presence. The colorless area expands from the center of the fruit, and the colored area retracts toward the periphery of the fruit until it disappears completely. The figure obtained in this color reaction is compared with a chart or pattern of specific degradation for each variety or with generic letters of international use, such as those published by the Center Technique Interprofessionnel des Fruits et Légumes in 2002 (Campana, 2007).

The soluble solid concentration (SSC) in fruits is composed mainly of sugars, although it includes organic acids, amino acids, phenolic compounds, soluble pectins, and various minerals. During maturation, SSC tends to increase, and it is determined by refractometry. The refractometer is an optical, analog, or digital instrument whose operation is based on the phenomenon of light refraction when light passes through a solution. Light beam deviation will depend on the number of solutes contained in the fruit juice. The refractometer has a scale for refractive index and another for equivalent °Brix or SSC percentage, which can be read directly (Mitcham et al., 1996; Campana, 2007).

During ripening, organic acids decrease. The gradual evolution and importance of these acids in fruit flavor have led to the use of an acidity index, combined with other parameters, to determine the optimal harvesting moment (Campana, 2007). Titratable acidity (TA) is determined by titrating a known volume of fruit juice with 0.1 N NaOH (sodium hydroxide) to an end point of pH = 8.2, as indicated by a phenolphthalein indicator, or by using a pH meter. The TA, expressed as the percentage of malic acid, can be calculated as follows: mL NaOH × N (NaOH) × 0.067 (malic acid meq. factor) × 100 mL juice titrated (Mitcham et al., 1996).

Other indices may be considered for determining the optimum harvest time and they are variety specific. Peel color is defined by comparison with specific color tables for each variety or by using a colorimeter. The period between full bloom (75% open flower) and the commercial harvest is another index which serves as a good maturity indicator and provides a rough estimate of harvest date (Calvo et al., 2012).

Commercial applications of maturity indices use days after full bloom as a first guide and complement it with starch index and pulp firmness. Maturity standards and values at which the harvest of the different varieties begins and finishes should take into account whether the fruit will be stored or shipped immediately. The goal is to determine a reliable index that can predict the best harvest date depending on the fruit (Mitcham and Mitchell, 2007).

Fruit maturity evaluation in Asian pears is usually based on background color and SSC. Background color is the one that best relates to the storage potential of the fruit. The combination of starch index, firmness, and SSC could provide a good harvest index (Mitcham and Mitchell, 2007).

There are also nondestructive systems to determine the fruit ripening stage and optimal harvest time. Optical methods such as visible and near-infrared (VIS/NIR) spectroscopy, time-resolved reflectance spectroscopy, hyperspectral backscattering imaging, laser-induced backscattering, and chlorophyll fluorescence are the most promising nondestructive systems. These methods also make it possible to classify the fruit and control its quality, and they may be used directly in the orchard (Zhang et al., 2014). The use of the VIS/NIR spectroscopy method provides estimates of soluble solids (Machado et al., 2012; Li et al., 2013), pH, and firmness for different varieties of pears.

The harvest consists in detaching fruits from the plant once they have reached the appropriate maturity degree, according to their destination. It is a complex operation in which it is necessary to consider the maturity stage of the fruit, management, and coordination of the factors involved in harvesting (resources, training, and control), environmental conditions, and preharvesting cultural practices (Campana, 2007). Harvest is a key process in fruit production since it determines the product quality obtained and, therefore, the percentage of fruit suitable for the market (Calvo et al., 2012).

Pears are harvested by hand, placed in large containers, and transported to the packing plant. During this process, cushioned bags with a solid frame reduce fruit damage by impact (Mitcham and Mitchell, 2007). Fruits must be taken gently with the palm of the hand, the index finger on the peduncle, without exerting pressure on the fruit. By a slight torsion, the fruit is detached from the branch. The fruit must be picked with a complete peduncle and without leaves or darts. Blows and wounds must also be avoided, and the fruit that does not meet the desired quality conditions, such as those with sunburn browning, inappropriate sizes, insufficient color percentage, and staining, must be discarded. This type of fruit causes an unnecessary increase in harvesting, transportation, and cooling costs and reduces the quality grade and storage capacity (Calvo et al., 2012).

Asian pears have a high susceptibility to skin superficial abrasion; therefore, the harvest must be done carefully to avoid fruit damage. The harvesters may use rubber or cotton gloves and gently place the fruit inside cushioned bags. They should also avoid peduncle injuries (Mitcham and Mitchell, 2007).

Depending on the variety–rootstock system and the training system, harvest can be done from the ground or will have to rely on auxiliary equipment such as ladders or picking platforms that allow workers to reach the tree top and inside the canopy. Picking platforms are devices on which workers are located to carry out harvest or other tasks such as thinning or pruning. This equipment can be pulled by a tractor or self-propelled. These platforms are coupled to an articulated hydraulic arm that allows vertical and horizontal movements to access the different parts of the cup. In multiple teams, 6–8 workers are located at different platform heights where they collect the fruits as the equipment advances through the ally way. Fruits can be either collected in picking bags and transferred or directly set in bins. Another available device consists of transport belts that carry fruit through rails to bins set in the back of the machine. Mobile platforms offer the operator more comfort and better stability conditions than the ladder. These mechanical aids reduce operating time and decrease the number of damaged fruits during harvesting (Campana, 2007).

2.20 PACKAGING AND TRANSPORT

Packaging is an important process to maintain fruit quality during storage and transportation until the fruit reaches the final consumer. It prevents quality deterioration and facilitates distribution and marketing (Han, 2005; Singh et al., 2014). Packaging must be designed to facilitate handling, storage, transport, and marketing. The packaging design should reduce the loss of water and facilitate the distribution of cold air within the package during refrigerated storage (Opara, 2013). It must include information for the buyers (variety, weight, number of fruits, quality grade, area of origin, and tracking system) (El-Ramady et al., 2015). The fruits are individually wrapped and placed on trays that are packed into cartons and placed in cold storage (USA Pear, 2012). The size of the pear is established according to the fruits that fit in a

standardized box designed to contain approximately 20 kg. The sizes vary from 70 (pears per box) for the largest pears to 150 for the smallest ones. Grade standards are established to facilitate communication and transaction between buyers and sellers. These grades are based on cover color, shape, insect, and disease damage and bruises (Schotzko and Granatstein, 2004). The traditional model for pear supply flow is similar to that of apples. It involves farmers, packers, commission agents, auctioneers, wholesalers, traditional retailers of all types, exporters, shipping companies, and customers (Bhaskare and Shinde, 2017).

When pears are destined to the export market, the fruit is processed and conditioned in the packinghouse after harvest for proper transportation (Altube et al., 2007). For the overseas market, the fruit must be sent to the port in refrigerated conditions before shipping. For the domestic market, periodic deliveries are made to the regional and local markets. Pear fruit that does not meet the specifications has a lower price and is allocated to the local market as soon as possible (Blanco et al., 2005).

Basic requirements during transportation are similar to those needed for storage and include temperature and humidity control and adequate ventilation. In pear packaging, it is important to avoid shock and vibration without affecting air circulation

(Luna-Maldonado et al., 2012). Temperature control is needed to maintain the cold chain throughout the postharvest handling system and ensure fruit quality and safety. All other postharvest technologies are complementary temperature management (Kader, 2003). Pear is shipped domestically by truck or railcar (Kader, 2003). Large refrigerated vehicles for road transport are generally semitrailers. The trailer has its own cooling system, which provides adequate temperature where the load is located. Pear pallets must be loaded and ensured in such a way that air space is provided around the load. This allows cooled air to flow between the load and the walls, below the product, and finally return to the cooling unit. This design allows refrigerated air to intercept the heat before it affects the temperature of the product (Thompson, 2007). Overseas transportation is by ship (Kader, 2003). Maritime refrigerated containers and refrigerated ships are used to transport pear fruit through the ocean. Refrigerated ships have a larger capacity and less refrigeration equipment than containers. However, containers may be transferred directly to a cooled dock, keeping the cold chain. Refrigerated ships are loaded in open fruit terminals, and pallets are exposed to environmental conditions before reaching the ship's hold. Refrigerated ships usually transport products which are sold in large volumes by big companies. Refrigerated containers and ships often have automatic systems for monitoring internal temperatures (Thompson, 2007). Transportation and distribution logistics must be adjusted by wholesalers and retailers when a large volume of refrigerated ships arrive at the fruit terminals. Dealers must have a good marketing plan to avoid product saturation and loss of quality fruit during transportation so that consumers obtain a quality product (Thompson, 2007).

2.21 POSTHARVEST HANDLING AND STORAGE

Postharvest handling includes cleaning, washing, selection, grading, disinfection, drying, packing, and storage. After pears are harvested, value is added in successive stages to maximize the added value (El-Ramady et al., 2015). After harvest, the fruit is transported in bins to the packinghouse, where it can be processed immediately or stored in cold rooms for further processing. Storage can last from a few days to a year. Depending on the technical and commercial strategies, pear fruits may be treated with products before storage (Calvo et al., 2012). The pear packing process has been standardized across the industry with some differences between companies (Figure 2.8). The packing process involves a sequence of operations performed by machines and trained personnel, in which different stages can be identified. Pears are emptied from the containers onto the packing line using dry systems or by immersion in water (Mitcham and Mitchell, 2007). Flotation casting is used especially for apples and pears. The bins are pushed into the water to allow the fruit to float, and the empty bin rises out of the tank (Thompson et al., 2007). Salts such as sodium lignin sulfonate, sodium silicate, or sodium sulfate must be added to the water to assure fruit flotation (Kitinoja and Kader, 2015). The tank water must be kept clean by treating with chlorine and filtering the organic waste that accumulates during the process (El-Ramady et al., 2015). Fruits are sorted to remove damaged, decomposed, or defective pears before cooling. This classification will save energy and limit the spread of the infection (Kitinoja and Kader, 2015). Pear washing is carried out with potable water and detergent to facilitate the removal of dirt. The application of phytosanitary products prevents fungal and physiological diseases (Calvo et al., 2012). Pear grading and sizing are based on external quality aspects such as color, shape, size and spots, and the presence of wounds and injuries (Calvo et al., 2012). Selection usually involves a combination of human and electronic devices. Electronic selectors of color and size are very common in big packinghouses (Mitcham and Mitchell, 2007). According to the customer's request, adhesive labels can be placed on each fruit at the beginning of this operation (Calvo et al., 2012). The fruit selected according to its quality and size is packed in different types of commercial packages. The packaging process can be manual or mechanical (Calvo et al., 2012). Most pear varieties are wrapped in paper and placed on trays inside corrugated carton boxes, although some wooden boxes are still in use. They can also be packed in polyethylene bags. Pear bagging is common, particularly in marketing in self-service stores. The polyethylene bags used inside the boxes for keeping the humidity may interfere with temperature management, but they are still used in several production systems (Mitcham and Mitchell, 2007).

Due to the high susceptibility of Asian pears to skin damage, fruits are often visually calibrated by size and shape and packaged by hand. Many Asian pears are wrapped using soft paper or plastic and packaged in plastic trays or foam nets. The fruit is often packed in one or two layers in depth and a cushion is added to the box to avoid movement during transportation (Mitcham and Mitchell, 2007). Finally, the boxes are labeled and inventoried to maintain their traceability, manually lifted from a delivery conveyor, and placed on wooden pallets. Palletizing is usually carried out manually, but some technologies allow automation (Calvo et al., 2012).

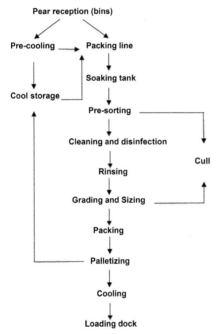

FIGURE 2.8 Pear packing process.

Postharvest temperature management is the most important factor in the maintenance of pear fruit quality. Cooling delays are associated with a short postharvest life, pulp softening, and increase in physiological disorders and diseases. Therefore, the fruit should be cooled as soon as possible after harvest (Mitcham and Mitchell, 2007). Precooling consists in rapidly reducing the temperature of freshly harvested pears prior to processing or storage. This operation is critical in maintaining the pear quality and allowing the producers to maximize postharvest life. The systems used for precooling are forced air cooling, which pushes air through the vents in storage containers, or hydro-cooling, which uses chilled water to remove pear fruit heat from the field. Two main types of hydro-cooling systems can be used: water shower and produce immersion (Vigneault et al., 2009).

Several systems, methods, and products have been developed for pear storage. Some of them are regular air storage, modified atmosphere, controlled atmosphere (CA), among others, as described below.

2.21.1 REGULAR AIR STORAGE

Regular air storage is based on storage temperatures around 0 °C (32 °F) without changing other factors to reduce metabolism. Temperature management has so far been the most effective tool to extend the postharvest life of the fruits because it reduces respiration, respiration heat, ethylene production, water loss, and susceptibility to attack by pathogens. The optimum temperature of conservation varies according to the cultivars. It is important to maintain 90%–95% relative humidity levels to reduce losses due to dehydration (Calvo et al., 2012).

2.21.2 MODIFIED ATMOSPHERE STORAGE

Modified atmosphere storage refers to any atmosphere with a gaseous content different from air (Calvo et al., 2012). Modified atmosphere consists in modifying the levels of O_2 and CO_2 inside a package. The interaction between product respiration and packaging generates an atmosphere with low levels of O_2 and/or a high concentration of CO_2 inside the package. In passive modified atmospheres, only pear respiration and film gas permeability influence the change in the gaseous composition of the package environment. In an active modified atmosphere, the environment gas composition is changed intentionally into the container before sealing (Mattos et al., 2012).

2.21.3 CONTROLLED ATMOSPHERE

CA storage consists altering and maintaining a gas composition different from that of air in the storage atmosphere. In pear, CA cold storage increases storage life and reduces the development of yellow color and physiological disorders (Villalobos-Acuña and Mitcham, 2008). It is critical to maintaining adequate levels of O_2 and CO_2, as well as optimum harvest maturity for each variety. Overripe fruit is more susceptible to damage by low O_2 or high CO_2. In general, early harvest fruit is a better candidate for long-term storage in CA than late-harvest fruit (Mitcham and Mitchell, 2007). The concentration of O_2 and CO_2 used in pear storage depends on the cultivar, but it generally ranges between 1% and 3% O_2 and 0% and 5% CO_2 (Villalobos-Acuña and Mitcham, 2008).

2.21.4 LOW O_2 STORAGE

In some pear-producing regions, research is being conducted on the use of low oxygen for pear storage and the results indicate benefits when the fruit is stored in dynamic controlled atmosphere (DCA). This method allows the storage manager to customize the O_2 concentration at the beginning of storage and change it during storage, as the lower oxygen limit (LOL) changes. The oxygen should be precisely at LOL to avoid anaerobic effects. Chlorophyll fluorescence, ethanol measurements, and respiratory quotient measurements (CO_2 production/O_2 consumption) are capable of detecting the LOL (Mitcham and Mitchell 2007; Prange et al., 2011). Atmospheric and temperature requirements for standard CA storage of selected pears cultivars have been described below.

Most of the commercially important pear European cultivars

viz. Bartlett, Beurré d´Anjou, Bosc, Comice, Forelle and Packham's Triumph, and Asian Pears like Chojuro and Nijisseiki are being stored at −1 to 0 °C temperature with varying degree of CO_2 and O_2 enrichment. Bartlett, Beurré d´Anjou, and Bosc should have 1–2% of O_2 but with 0–3%, 0–0.5%, and 0.5–1.5 % CO_2 enrichment, respectively, for CA storage. Comice and Forelle can succesfully be stored at −1 to 0 °C temperature with 1.5% of O_2 and 0.5–4 % CO_2. Packham's Triumph, Chojuro, and Nijisseiki need 1.5–2.5 %, 1–2%, and 0–1% CO_2 and 1.5–1.8%, 2%, 1–3% O_2, respectively, for controlled atmosephere storage at −1 to 0 °C temperature (Mitcham and Mitchell 2007).

2.21.5 USE OF 1-METHYLCYCLOPROPENE

The use of 1-MCP inhibits ethylene action, thus making it possible to extend the storage life of pears and delay ripening. At standard temperature and pressure, 1-MCP is a gas with a formula of C_4H_6. This molecule occupies ethylene receptors such that ethylene cannot bind and elicit action. The affinity of 1-MCP for the receptor is approximately 10 times greater than that of ethylene (Blankenship et al., 2003), and 1-MCP also blocks ethylene biosynthesis (Chiriboga et al., 2014). The compound is nontoxic, odorless, and stable at temperature environment; in addition, it is easy to apply and highly effective (Balaguera-López et al., 2014).

The treatment with 1-MCP in pears was found to be more complex than in other fruits, due in part to the high sensitivity of pears to treatment, and sometimes pears remain excessively firm and green and lose their ability to mature properly. The effectiveness of the treatment in fruit softening and physiological disorder control depends on the optimum combination of several pre- and postharvest factors (Villalobos-Acuña and Mitcham, 2008; Calvo et al., 2012). Preharvest factors include variety, environmental conditions during fruit development, orchard cultural practices, physiological state of the fruit (Chiriboga et al., 2014). The postharvest factors that may affect the application of 1-MCP are time between harvest and treatment, temperature, treatment duration, bin material, 1-MCP concentration, and storage time after 1-MCP treatment (Villalobos-Acuña and Mitcham, 2008; Calvo and Sozzi, 2009; Calvo et al., 2012).

2.22 PROCESSING AND VALUE ADDITION

Pears are commonly consumed as fresh fruit throughout the world, but they are also industrialized

(James-Martin, 2015). Dietary guidance universally promotes the consumption of fresh fruit. In the United States, a pick-your-own operation, also called U-pick, is a farm where clients may go to pick, cut, or choose their own product. This type of enterprise is an excellent channel of direct marketing (Leffew and Ernst, 2014). Despite the known benefits of eating fresh fruit, the world average per-capita consumption of fresh pear remains low (3.75 kg/year). In 2015, the consumption in the United States was still low, with 1.45 kg/year/capita (USDA, Department of Agriculture National Agricultural Statistics Service). Moreover, in the last years, the annual per capita consumption of pear has diminished from 6 kg (2013) to 2.6 kg (2015) in Argentina (Ministerio de Hacienda y Finanzas, 2016). In the European Union, the annual per capita consumption was 5.14 kg/person during the period 2009–2011 (Trentini, 2012) and 4 kg in 2013 (Van Doorn, 2015). Major producing countries (Western/ Southern EU), including Italy (11.17 kg) and Portugal (10.20 kg), have higher consumption (Van Doorn, 2015). By contrast, consumption of pears in Central and Eastern European countries is smaller. Promotion campaigns are conducted to increase pear consumption worldwide.

Little has been published on the health effects associated with the overall consumption of fruit, and particularly of pear. In this respect, the Australian apple and pear industry and government have recently funded a research review to understand the current knowledge status about the nutritional properties of pear and its components, and to provide evidence of its benefits to health (James-Martin et al., 2015).

Pear has a very sweet flavor, making it an ideal fruit in processed foods such as canned pears, baby food, glazes, and fruit bars (Reiland and Slavin, 2015). Pear fruits are mainly industrialized as canned or dried pear, juice, or fermented as pear cider. Another mode of pear processing is fresh-cut pear slices. Gorny et al. (2000) demonstrated that Bartlett is the best cultivar for fresh-cut processing.

The production process of concentrated pear juice usually includes fruit washing and crushing, pulp-pomace separation and maceration, extraction, evaporation, centrifugation, filtration, concentration, packing, and cold storage. Argentina exports about 95% of the juice production. Juice production is influenced by price fluctuations due to the worldwide supply–demand trade (Catalá et al., 2014).

Finding a niche in the markets is another way of adding value to a pear. For example, in recent years alcoholic pear beverages have become more popular. Industrialization of pear products results in a significant amount of waste that includes skins, seeds, and core. Pear seeds are another waste product from pear with potential nutritional benefits.

2.23 DISEASES, PESTS, AND PHYSIOLOGICAL DISORDERS

In the last decade, the presence and relative importance of pests and diseases have changed in some production areas of the world, mainly due to climate change. To reduce the application of chemical pesticides, the concept of integrated pest management (IPM) was developed. It involves replacing chemical pesticides with ecological ones, reducing the pressure of pests with cultural methods, using resistant or tolerant cultivars, and applying precision agriculture in space (Mayus, 2012). The IPM is currently oriented toward a system of fruit production with low residues, aimed at meeting the increasing requirements mainly of the most important European supermarket chains, which limit the residue levels and the number of active ingredients in fresh fruit.

2.23.1 ORCHARD DISEASES

Pear diseases can affect fruit directly, making it unattractive, inedible, or unmarketable; they can weaken fruit trees or lead to tree death in severe cases (Montesinos et al., 2000; Jones and Aldwinckle, 2002). Some diseases are a limiting factor for pear growth, because of their epidemic characteristics in conducive environments. One of the most destructive diseases in many parts of the world is fire blight, caused by the bacteria *Erwinia amylovora*. Fire blight produces severe losses throughout Europe and the United States. The disease is devastating and difficult to manage in susceptible commercial varieties grown in warm and humid weather. However, cultivars bred for resistance constitute the most effective strategy for fire blight control. The disease is absent in some countries of South America (Argentina, Chile, and Brazil), South Africa, and Australia (Montesinos et al., 2000; Deckers and Schoofs, 2002; Agrios, 2003). For example, Argentina, a pear-exporter country, is characterized by environmental conditions conducive to the production of high-quality fruit and low incidence of diseases (Dobra et al., 2008).

Another highly harmful disease is pear scab, caused by *Venturia pirina* fungus worldwide. *V. pirina* causes annual epidemics, which vary largely in severity and depend on weather conditions (Mac Hardy, 2000). Pear scab occurs in most pear-producing regions of the world. All commercially important European pear cultivars are susceptible to pear scab (Spotts and Castagnoli, 2010).

Most Asian pear species have been reported to be moderately resistant to *V. pirina* (Ishii and Yanase, 2000). Another species more recently reported to cause pear scab is *V. nashicola*, which was only pathogenic on Japanese and Chinese pears (Ishii and Yanase, 2000). Worldwide production of high-quality pear fruit for fresh market sales requires very high levels of control. The presence of ascospores leads to epidemics during the rainy spring season. Besides, overwintering twig pustules liberate conidia that can cause new infections. Infection prediction is an important control strategy; however, cultivars bred for resistance constitute the most effective strategy (Williams et al., 1978; Jones and Aldwinckle, 2002).

In addition, the brown spot caused by *Stemphylium vesicarium* has been an important fungal disease in pear producing areas of Europe in the past 10 years. The disease incidence is at present equal to or higher than pear scab incidence. Severe epidemics may cause global losses of 1–10% of the total production, depending on the year and pear-growing area (Llorente and Montesinos, 2006).

Other disease that affects pear wood is canker disease. It can reduce tree productivity and growth (Montesinos et al., 2000; Jones and Aldwinckle, 2002; Cloete et al., 2011). In this group, the main diseases are canker and dieback caused by *Valsa ceratosperma* in China, Japan, and Korea and recently reported in Italy (Montuschi et al., 2006; Suzaki, 2008); canker and dieback are caused by fungus of Botryosphaeriaceae such as *Botryosphaeria theobromae* in India (Shah et al., 2010); *B. obtusa, Neofusicoccum australe* and *Diplodia seriata* in South Africa (Slippers et al., 2007; Cloete et al., 2011); *Neofusicoccum luteum* in Australia; *Botryosphaeria dothidea, B. rhodina, B. obtusa,* and *B. parva* in China (Zhai et al., 2014); *D. mutila* in Uruguay (Sessa et al., 2016) and *D. seriata* in Argentina; and European canker by *Neonectria galligena* in European production areas (Jones and Aldwinckle, 2002; Montesinos et al., 2000). More recently, in the United States, *Potebniamyces pyri*, a causal agent of Phacidiopycnis rot on pear fruit, was reported as causing twig dieback and weak canker in commercial d´Anjou orchards (Xiao and Boal, 2005). Similarly, the species associated with Bull's eye rot of apple and pear fruit, *Neofabraeaperennans* and *N. alba* were reported as the causes of cankers (Henriquez et al., 2006). Furthermore, the bacteria *Pseudomonas syringae* pv. *syringae* causes blossom blast and papyraceous cankers on trunks and branches in late fall and during winter. This disease results in economic losses in pear production areas of the United States, South America, and

South Africa (Ogawa and English, 1991; Jones and Aldwincle, 2002; Moragrega et al., 2003).

Other minor pear diseases include powdery mildew, caused by *Podosphaera leuchotrica*, which occurs worldwide. The disease is prevalent on susceptible cultivars such as d′Anjou and red pear cultivars wherever pears are grown (Spotts, 1984; Dobra et al., 2008). Economic losses are due to russeted fruit and an increased need for fungicide applications to control the disease (Williams et al., 1978; Spotts, 1984). Using the core *Pyrus* germplasm collection at NCGR-Corvallis reported that European cultivars were more susceptible to powdery mildew than Asian cultivars (Serdani et al., 2005).

Oomycetes also is an important soil pathogen. In this sense, *Phytophthora cactorum* is the main causal agent of neck and root rot disease in nearly all pear-growing regions (Jones and Aldwinckle, 2002). Bartlett cv is highly susceptible (Montesinos et al., 2000). Tree symptoms may become noticeable in early spring, while in mid to late summer foliar symptoms of chlorosis followed by reddening usually become evident. *P. cactorum* sporadically induces pear fruit decay both in orchards and in postharvest storage. In Argentinean commercial orchards, preharvest rot of Golden Russet Bosc fruit caused by other species, *P. drechsleri* and *P. lacustris*,

has been recently reported (Sosa et al., 2015). Disease control is generally difficult, and the strategies are based on management practices of irrigation, drainage, soil protection of roots and wounds, and stimulation of defense mechanisms in plants in the orchard (Montesinos et al., 2000).

Pear decline is caused by a phytoplasma belonging to the apple proliferation group. Pear decline is one of the most dangerous diseases of pear trees. Pear psylla *C. pyri* and *C. pyricola* are reported as particularly the active vectors of the pear decline phytoplasma. Pear decline can overwinter in roots of affected pear trees and in *C. pyri*. To prevent the spread of pear decline, it is of fundamental importance to control the overwintering generations of the vector (which are infected and infective) (Carraro et al., 2001).

2.23.2 POSTHARVEST DISEASES

Postharvest fungal pathogens are considered the main causes of economic losses of fruits and vegetables. Most losses occur due to pathogen infections, in orchard or postharvest, which lead to postharvest decay during ripeness and senescence of fruit (Romanazzi et al., 2016). Much fruit decay takes place at the end of long- and medium-term cold storage of fruit.

The main storage diseases are blue mold and gray mold. *Penicillium expansum* is the primary cause of blue mold of pear. Other species, *P. solitum, P. commune, P. verrucosum, P. chrysogenum,* and *P. regulosum,* have also been reported to cause these diseases (Pitt et al., 1991; Sanderson and Spots, 1995). *P. expansum* is essentially a wound pathogen. Stem-end blue mold is common on d'Anjou and Bosc pears. *Botrytis cinerea,* the cause of grey mold, infects pear fruit through wounds or natural openings. Frequently, latent infections occur in the orchard and develop in storage, causing calyx and stem-end rot. This is common in d'Anjou and Bosc fruits (Kock and Holtz, 1992; Bondoux, 1993; Sardella et al., 2016).

Other pear postharvest diseases are Alternaria rot (*Alternaria alternata, Alternaria* spp.), Cladosporium rot (*Cladosporium herbarum* and *Cladosporium* sp.), and brown rot (*Monilia fructigena* and *M. laxa*) (Bondoux, 1993; Montesinos et al., 2000). In the United States, Sphaeropsis rot (*Sphaeropsis pyriputrescens*) has been described as a postharvest disease of d'Anjou pear fruit. Like in grey mold decay, the infection and development of the pathogen occur from stem-end and calyx-end of fruit, respectively (Xiao and Rogers, 2004). Phacidiopycnis rot (*Potebniamyces pyri* and *Phacidiopycnis piri*) was previously reported in Europe and India and more recently in Washington State (United States). Phacidiopycnis rot causes three types of symptoms on pears: stem-end rot, calyx-end rot, and wound-associated rot originating from infection at the stem, calyx, and wound on the fruit skin, respectively (Xiao and Boal, 2005).

Phytophthora rot by *P. cactorum* appears sporadically in pear fruit (Bondoux. 1993). In Argentina, decay in pear fruit was reported in 2010 to be caused by *P. salixsoil* in Williams, Packham's Triumph, and Red Bartlet pears after long cold storage of fruit (Dobra et al., 2011). Besides, losses were reported in connection with long storage of Packham's Triumh fruit, caused by the complex of *B. cinerea* associated with *Phytophthora* spp., as a result of latent infections in the orchard (Sosa et al., 2016).

Effective management of the postharvest diseases includes pre- and postharvest applications of new biological and more innocuous synthetic fungicides against pathogens registered for use on fruit. On the one hand, the new fungicides pyraclostrobin + boscalid applied during preharvest controlled the *Alternaria-Cladosporium* complex (44.6%) and *B. cinerea* decay (95%) in Bosc pears after four months of cold storage (Lutz et al., 2017). On the other hand, eco-friendly and safe alternatives to synthetic fungicides include yeasts as biological control agents. In the last decade, many studies have been done using yeasts with several antagonist action mechanisms against *B. cinerea* and *P. expansum* (Lutz et al, 2012). More attention is being drawn today to resistance induction in plants. Biological agents, elicitor substances, and natural compounds can stimulate and increase plant defenses against pathogens (Romanazzi et al., 2016). In addition, storage conditions (temperature and gases) reduce fruit maturity, thus indirectly controlling postharvest diseases (Bondoux, 1993; Jones and Aldwinckle, 2002; Jijakli and Lepoivre, 2004; Quaglia et al., 2011).

2.23.3 PHYSIOLOGICAL DISORDERS

The physiological disorders during the postharvest and storage of pears are related to the pre- and postharvest factors that affect the normal metabolism of the fruits. Optimal environmental conditions (temperature and gases) for postharvest fruit storage decrease the metabolic activity of fruit, which maintains their quality standards. Thus, in cold storage, the low temperatures decrease the metabolism of pear fruit, although they should not cause chilling or freezing damage. By contrast, in CA, low oxygen partial pressure decreases aerobic respiration and avoids the fermentation of fruit, and carbon dioxide partial pressure favors fruit color but increases the risk of physiological disorders.

The pear physiological disorders caused by preharvest factors include browning, sunburn and sunscald, cork spot, superficial scald, and black speck (Bondoux, 1983; Kays, 1991; Meheriuk et al., 1994; Ferguson et al., 1999). Pear browning can in general lead to considerable economic losses. Frank et al. (2007) explained that the combined action of particular preharvest factors, such as climate and orchard conditions, determine whether a pear fruit is susceptible to browning disorder or not. Browning is caused by the increment in oxidative processes due to the gas gradient inside the fruit. As a result, the membrane integrity is damaged and brown phenolic compounds are accumulated (Frank et al., 2007). This disorder is particularly important in Conference and Bartlett cultivars. Browning disorders include symptoms like cavities and flesh browning. Browning refers to both brown earth (injuries by CO_2) and core breakdown (senescence injuries), according to Giraud et al. (2001), while Larrigaudiere et al. (2004) described core browning as senescence injury.

The quality and conservation of pear fruit exposed to high solar radiation are affected by a physiological disorder caused by oxidative stress, called solar injury or sunburn. Sunburn occurs on fruit in orchard or after harvest, and it is characterized by color changes on skin exposed to sunlight (high temperature and radiation) for a long time. Kaolin particle film on fruit reflects solar radiation and diminishes superficial temperature and, therefore, the incidence of sunburn. Kaolin is an effective treatment not only to reduce sunburn in pears but also to increase fruit quality in Packham's Triumph and d´Anjou cv (Colavita et al., 2011) and productivity and tree performance in Comice cv (Sugar et al., 2005). During storage, sunscald is observed on areas of pear fruit affected by sunburn, and the affected area turns dark brown and black. Pear cultivars are less susceptible than apple cultivars to sunscald.

Superficial scald is a pear postharvest disorder related to several preharvest factors—such as mineral imbalance (calcium, potassium, and nitrogen) during fruit development—growth conditions, and harvest maturity of the fruit. Irregular and brown spots on fruit skin are observed after three or four months of cold storage, and these symptoms become more severe if the fruit is kept at environment temperature. The symptoms are produced by oxidation of α-farnesene, natural compounds of fruit cuticle. Although European and Asian pears are susceptible, d´Anjou cv is more susceptible to superficial scald disorder. Control of this postharvest disorder by antioxidant treatments is limited due to requirements of food without chemical residues, but treatments with 1-MCP and storage in DCA and CA are effective control strategies (Bondoux, 1983; Lurie and Watkins, 2012).

Identifying preharvest factors increases the likelihood of producing fruit with a lower predisposition to postharvest disorders. An understanding of these factors makes it possible to modify fruit development, optimize storage quality, and develop methods for predicting disorder risk (Ferguson et al., 1999).

2.23.4 PESTS

The species of insects and mites that are considered pests in pears vary with the climatic regions where the fruit is grown. These species differ in importance, with codling moth being the most serious species worldwide. In addition, several species of leafroller cause damage to fruit while other species requiring control measures are pear psylla, San Jose scale, and woolly aphid.

Cydia pomonella (Lepidoptera: Tortricidae), the codling moth, is a major pest of global pome fruit production that causes huge economic losses. It has become a serious quarantine pest for export apples and pears. Some countries require orchard or field bin sampling to ensure that fruits are codling moth-free prior to packing and during storage. Insecticides are applied according to results of monitoring by pheromone tramps and evaluation of damage on fruits. Programs to eradicate and suppress these pest populations involve the use of *Cydia pomonella granulovirus* (CpGV) as a bioinsecticide and sterile insect release (SIR). The success of these systems depends on the efficiency of the insects released and the environmental conditions (Beers et al., 1993).

Cacopsylla pyricola, C. pyri, and *C. bidens* (Homoptera: Psyllidae) in the genus *Cacopsylla syn. Psylla* are specialist pests of commercial pear, reproducing only on certain species of *Pyrus*. It is believed that the critical stage of infestation by pear psylla takes place during late winter as psyllas leave overwintering sites and distribute in the orchards (Soroker et al., 2004). The factors that affect psylla distribution and movement in orchards include surface colors and volatile chemical production both of plant and female. Pear *psylla* causes dark russet blotches or streaks on the skin of the fruit and brown spots on the leaves. Intense infestations can cause partial defoliation of trees, reducing vitality and preventing the formation of fruit buds. Besides, there is strong evidence that the *C. pyri* is a vector of phytoplasma pear decline associated with the development and spread of pear decline disease (Carraro et al., 2001).

Cydia molesta = *Grapholita molesta* (Lepidoptera: Tortricidae), an oriental fruit moth, is currently a very important pest in commercial orchards worldwide. Detection and identification of larvae are relevant to pest control, conducted by means of sexual confusion technique, continuous monitoring of damage in shoots and fruit in orchards, and insecticide sprays.

Quadraspidiotus perniciosus (Rhy-ncotha, Diaspididae), San José scale, is another important pest worldwide. It is a very harmful and phytophagous insect that attacks trunk, branches, and fruits. The affected tree is dried, and the fruit damaged with red circles cannot be marketed (Beers et al., 1993; Jones and Aldwinckle, 2002).

KEYWORDS

- *Pyrus*
- temperate fruit tree
- varieties and cultivars
- rootstock
- diseases and pest
- orchard management
- postharvest
- nutrient

REFERENCES

Abu-Rayyan, A.M.; Shatat, F.A.; Abu-Irmaileh, B.E. Response of fruit trees to composting of animal manures in the tree line. *Journal of Food Agriculture and Environment* 2011, 9, 492–495.

Adaro, A.; Raffo, M.D; Rodriguez, R.; Ginobili, Juan; Toranzo, J.; Zaffino, R. *Conducción y poda.* In *Pera Williams, Manual para el productor y el empacador*; Sanchez E., Ed.; INTA Argentina, 2010, p. 18.

Agrios, G.N. *Plant Pathology*, 5th ed.; Elsevier Academic Press, London, 2003.

Alexander, L.; Grierson, D. Ethylene biosynthesis and action in tomato: A model for climacteric fruit ripening. *Journal Expeimental Botany* 2002, 53, 2039–2055.

Allen, R.G.; Pereira, L.S.; Raes, D.; Smith, M. Crop evapotranspiration—guidelines for computing crop water requirements. *FAO Irrigation and Drainage Paper* 56 [Online]. Food and Agriculture Organization, Rome, 1998. http://www.fao.org/docrep/X0490E/X0490E00.htm (accessed Nov 15, 2017).

Altube, H.A.; Santinoni, L.A.; Alem, H.J. *Introducción a la fruticultura.* In *Árboles frutales, ecofisiología, cultivo y aprovechamiento*; Sozzi, G.O., Ed.; Facultad de Agronomía, Universidad de Buenos Aires, Buenos Aires, 2007, p. 3.

Arzani, K.; Khoshghalb, H.; Malakouti, M.J.; Barzegar, M. Postharvest fruit physicochemical changes and properties of Asian (*Pyrus serotina* Rehd.) and European (*Pyrus communis* L.) pear cultivars. *Horticulture Environment and Biotechnology* 2008, 49, 244–252.

Balaguera-López, H.E.; Salamanca-Gutiérrez, F.A.; García, J.C.; Herrera-Arévalo, A. Ethylene and maturation retardants in the postharvest of perishable horticultural products. A review. *Revista Colombiana de Ciencias Hortícolas* 2014, 8, 302–313.

Barbosa, W.; Pommer, C.V.; Tombolato, A.F.; Meletti, L.M.; Vega, R.F.; Moura, M.F.; Pio, R. Asian pear tree breeding for subtropical areas of Brazil. *Fruits* 2007, 62, 21–26.

Barstow, M. *Pyrus communis.* The IUCN Red List of Threatened Species 2017: e.T173010A61580281. [Online] 2018. http://dx.doi.org/10.2305/IUCN.UK.2017-3.RLTS.T173010A61580281.en (accessed 18 Jan, 2018).

Beers, E.; Brunner, J.F.; Willett, M.; Warner, G. Orchard Pest Management Online, Washington State Fruit Commission. Tree Fruit Research & Extension Center, Washington State University, http://jenny.tfrec.wsu.edu/opm/toc.php?h=2, 1993 (accesed June 5, 2017).

Bell, R.L.; Scorza, R.; Lomberk, D. Adventitious shoot regeneration of pear (*Pyrus* spp.) genotypes. *Plant Cell, Tissue and Organ Culture* 2012, 108, 229–236.

Benítez, C.E. *Cosecha y Poscosecha de Peras y Manzanas en los Valles Irrigados de la Patagonia*, 1st ed.; Instituto Nacional de Tecnología Agropecuaria, Argentina, 2001.

Bertelsen, M. Reflective mulch improves fruit size and flower bud formation of pear cv "Clara Frijs." *Acta Horticulturae* 2005, 671, 87–95.

Beutel, J.A. Asian pears. In *Advances in New Crops*; Janick, J.; Simon, J.E., Eds.; Timber Press, Portland, 1990, p. 304.

Bhaskare, R.; Shinde, D.K. Development of Cold Supply Chain for a Controlled Atmosphere Cold Store for Storage of Apple. *International Journal of Engineering Science* 2017, 7, 14207–14210.

Blanco, A.M.; Masini, G.; Petracci, N.; Bandoni, J.A. Operations management of a packaging plant in the fruit industry. *Journal of Food Engineering* 2005, 70, 299–307.

Blanke, M.; Kunz, A. Effect of climate change on pome fruit phenology at Klein-Altendorf - based on 50 years of meteorological and phenological records. *Erwerbs-Obstbau*, 2009, 51, 101–114.

Blankenship, S.M.; Dole, J.M. 1-Methylcyclopropene: A

review. *Postharvest Biology and Technology* 2003, 28, 1–25.

Bondoux, P. *Enfermedades de conservación de frutos de pepita, manzana y peras*, 1st ed; Mundi-Prensa, Madrid, 1993.

Bound, S.A.; Mitchell, L. A new post-bloom thinning agent for "Packham's Trumph" pear. *Acta Horticulturae* 2002, 596, 793–796.

Bramlage, W.J. Measuring fruit firmness with a penetrometer. *Post-Harvest Pomology Newsletter* August 1, 1983.

Brewer, L.R.; Palmer, J.W. Global pear breeding programmes: Goals, trends and progress for new cultivars and new rootstocks. *Acta Horticulturae* 2010, 909, 105–119.

Brouwer, C.; Heibloem, M. Irrigation Water Management: Irrigation Water Needs, FAO, Natural Resources and Environment Department |Online| 1986. http://www.fao.org/docrep/s2022e/s2022e00.htm#Contents

Brown S. Pome fruit breeding: Progress and prospects. *Acta Horticulturae* 2003, 622, 19–34.

Brunetto, G.; Bastos de Melo, G.W.; Toselli, M.; Quartieri, M.; Taglianini M. The role of mineral nutrition on yields and fruit quality in grapevine, pear and apple. *Revista Brasileira Fruticultura* 2015, 37, 1089–1104.

Cain, J.C. *Hedgerow Orchard Design for Most Efficient Interception of Solar Radiation: Effects of Tree Size, Shape, Spacing, and Row Direction*. Search Agriculture, New York State Agricultural Experiment Station, Geneva, 1972, 2, pp. 1–14.

Calvo, G.; Colodner, A.; Candan, A.P. *Cosecha y poscosecha de frutos de pepita*. Instituto Nacional de Tecnología Agropecuaria, Rio Negro, 2012.

Calvo, G.; Sozzi, G.O. Effectiveness of 1-MCP treatments on "Bartlett" pears as influenced by the cooling method and the bin material. *Postharvest Biology and Technology* 2009, 51, 49–55.

Campana, B.M.R. *Índices de madurez, cosecha y empaque de frutas.* In *Árboles frutales, ecofisiología, cultivo y aprovechamiento*; Sozzi, G.O., Ed.; Facultad de Agronomía, Universidad de Buenos Aires, Buenos Aires, 2007, p. 707.

Campoy, J.A.; Ruiz, D.; Egea J. Dormancy in temperate fruit trees in a global warming context: A review. *Scientia Horticulturae* 2011, 130, 357–372.

Carraro, L.; Loi, N.; Ermacora, P. The "lifecycle" of pear decline phytoplasma in the vector *Cacopsylla pyri*. *Journal of Plant Pathology* 2001, 83, 87–90.

Catalá, L.P.; Moreno, M.S.; Durand, G.A.; Blanco, A.M.; Bandoni, A. Optimal production planning of concentrated apple and pear juice plants. *Iberoamerican Journal of Industrial Engineering* 2014, 5, 10, 172–187.

Chen, P.M. *Pear.* In *The Commercial Storage of Fruits, Vegetables, and Florist and Nursery Stocks, Agriculture Handbook Number 66*; Gross, K.C.; Wang, C.Y.; Saltveit, M. Ed.; USDA, 2004.

Chevreau E.; Bell, R. *Pyrus spp. Pear and Cydonia spp. Quince.* In *Biotechnology of Fruit and Nut, Crops Biotechnology in Agriculture, 29;* Litz, R.E., Ed.; CABI Publishing, Cambridge, MA, USA, 2005, p. 543.

Chiriboga, M.A.; Schotsmans, W.C.; Larrigaudière, C.; Recasens, I. Últimos avances en la aplicación del 1-metilciclopropeno (1-MCP) en peras. Información técnica económica agraria: *Revista de la Asociación Interprofesional para el Desarrollo Agrario* 2014, 1, 34–48.

Cloete, M.; Fourie, P.H.; Ulrike, D.A.; Crous, P.W.; Mostert, L. Fungi associated with die-back symptoms of apple and pear trees, a possible inoculum source of grapevine trunk disease pathogens. *Phytopathologia Mediterranea.* 2011, 50,

176–190. http://fupress.net/index.php/pm/article/view/9004/9863 (accessed Feb 03, 2016).

Colavita, G.M.; Blackhall, V.; Valdez, S. Effect of kaolin particle films on the temperature and solar injury of pear fruits. *Acta Horticulturae* 2011, 909, 609–615.

Collett, L. *About The Apple—Malus domestica*. Oregon State University Extension Service. http://extension.oregonstate.edu/lincoln/sites/default/files/about_the_apple.lc_.2011.pdf

Costa G., Ramina A.; Bassi, D.; Bonghi, C.; Botta, R.; Fabbri A.; Massai, R.; Morini, S.; Sansavini, S.; Tonutti, P.; Vizotto G. *Ciclo ontogenetico dell`albero*. In: *Arboricoltura Generale*; Sansavini S. Pàtron editore. Bologna. 2012. p. 532.

Curetti, M.; Sánchez, E.E.; Tagliavini, M.; Gioacchini, P. Foliar-applied urea at bloom improves early fruit growth and nitrogen status of spur leaves in pear trees, cv. Williams Bon Chretien. *Scientia Horticulturae* 2013, 150, 16–21.

Darbyshire, R.; Webb, L.; Goodwin, I.; Barlow, E. Evaluation of recent trends in Australian pome fruit spring phenology. *International Journal of Biometeorology* 2013, 57, 409421.

Deckers, T.; Schoofs, H. The world pear industry and research: Present situation and future development of European pears (*Pyrus communis*). *Acta Horticulturae* 2002, 587, 37–54.

Dejong, T.M. Canopy and light management. In *Pear: Production and Handling Manual*; Mitcham, E.J.; Elkins, R.B. Eds.; University of California, Division of Agriculture and Natural Resources, Oakland, CA, USA, 2007, p. 59.

Dennis, F.G. Fruit development. In *Tree Fruit Physiology: Growth and Development*; Maib, K.M.; Andrews, P.K.; Lang, G.A.; Mullinix, K., Eds.; Yakima Washington, Good Fruit Grower, 1996, p. 107.

Dennis, F.G. The history of fruit thinning. *Plant Growth Regulation* 2000, 31, 1–16.

Dobra, A.C.; Sosa, M.C.; Dussi, M.C. Low incidence of fungal and bacterial diseases in the pear production of north Patagonia, Argentina. *Acta Horticulturae* 2008, 800, 907–912.

Dobra, C.; Sosa, M.C.; Lutz, M.C.; Rodriguez, G.; Greslebin, A.G.; Vélez, M.L. Fruit rot caused by *Phytophthora* sp. in cold-stored pears in the valley of Rio Negro and Neuquén, Argentina. *Acta Horticulturae* 2011, 909, 505–510.

Dondini, L.; Sansavini, S. European pear. In *Fruit Breeding*; Badenes, M.L.; Byrne, D.H., Eds.; Springer Science & Business Media, Boston, MA, USA, 2012; Vol. 8; p. 369.

Dorigoni, A.; Micheli F. Come ottenere un frutteto semi-pedonabile. *L'Informatore Agrario* 2015, 35, 38–42.

Dussi, M.C.; Sugar, D. 2011. Fruit thinners and fruit size enhancement with 6-benzyladenine application to "Williams" pear. *Acta Horticulturae* 2011, 909, 403–407.

Eccher Zerbini, P. The quality of pear fruit. *Acta Horticulturea* 2002, 596, 805–810.

Einhorn, T.C.; Turner, J.; Laraway, D. Effect of reflective fabric on yield of mature "d´Anjou" pear trees. *HortScience* 2012, 47, 1580–1585.

Elkins, R.B.; Moratorio, M.S.; Mc Clain, R.; Siebert, J.B. History and overview of the California pear industry. In *Pear Production and Handling Manual*; Mitcham, E.J.; Elkins, R.B., Eds.; University of California, Berkely, CA, USA, 2007, p. 3.

El-Ramady, H.R.; Domokos-Szabolcsy, É.; Abdalla, N.A.; Taha, H.S.; Fári, M. Postharvest management of fruits and vegetables storage. In *Sustainable Agriculture Reviews*; Lichtfouse, E., Ed.; Springer International Publishing, Switzerland, 2015, p. 65.

Faust, M. *Physiology of Temperate Zone Fruit Trees*. John Wiley & Sons, New York, NY, USA, 1989.

Ferguson, I.; Volz, R.; Woolf, A. Preharvest factors affecting physiological disorders of fruit. *Postharvest Biology and Technology* 1999, 15, 255–262.

Fernández, J.E.; Cuevas, M.V. Irrigation scheduling from stem diameter variations: A review. *Agricultural and Forest Meteorology* 2010, 150, 135–151.

Ferree, D.J.; Schupp, J.R. Pruning and training physiology. In *Apples: Botany, Production, and Uses*; Ferree, D.C.; Warrington, I.J., Eds.; CABI Publishing, Cambridge, MA, USA, 2003, p. 319.

Fischer, M. Pear breeding. In *Breeding Plantation Tree Crops: Temperate Species*; Jain, S.M.; Priyadarshaneds P.M., Eds.; Springer Science Business Media, LLC, New York, NY, USA, 2009, pp. 135–160.

Franck, C.; Lammertyn, J.; Ho, Q.T.; Verboven, P.; Verlinden, B.; Nicolaï, B.M. Browning disorders in pear fruit. *Postharvest Biology and Technology* 2007, 43, 113.

FruitYearbookSupplyandUtilization_GTables USDA CONSUME 2017, Table G-26—Canned pears: Supply and utilization, 1980/81 to date. usda. mannlib.cornell.edu/usda/.../2017/FruitYearbookSupplyandUtilization_GTables.xls

Geisler, M. Ag Marketing Resource Center, Iowa State University. https://www.agmrc.org/commodities-products/fruits/pears/ (accesed Jan 6, 2018).

Giraud, M.; Westercamp, P.; Coureau, C.; Chapon, J.F.; Berrie, A. *Recognizing postharvest diseases of apple and pear.* Centre Technique Interprofessionnel des Fruits et Légumes, Paris, France, 2001.

Glover, J.D.; Reganold, J.P.; Andrews, P.K. Systematic method for rating soil quality of conventional, organic and integrated apple orchards in Washington State. *Agricultural Ecosystem and Environment* 2000, 80, 29–45.

Gorny, J.R.; Cifuentes, R.A.; Hess-Pierce, B.; Kader, A.A. Quality Changes in Fresh-cut Pear Slices as Affected by Cultivar, Ripeness Stage, Fruit Size, and Storage Regime. *Journal of Food Science JFS: Sensory and Nutritive Qualities of Food.* 2000, 65, 3, 541–544.

Grab, S.; Craparo, A. Advance of apple and pear tree full bloom dates in response to climate change in the southwestern Cape, South Africa: 1973–2009. *Agricultural and Forest Meteorology* 2011, 151, 406–413.

Greene, D.; Costa, G. Fruit thinning in pome- and stone-fruit: State of the art. *Acta Horticulturae* 2013, 998, 93–102.

Han, J.H. New technologies in food packaging: Overview. In *Innovations in Food Packaging*; Han, J.H., Ed.; Elsevier Academic Press, the Netherlands, 2005, p. 3.

Han, X.; Li, Y.L.; Wang, L.F.; Liu, S.L.; Zhanga, J.G. Effect of tree-row mulching on soil characteristics as well as growth and development of pear trees. *Acta Horticulturae* 2015, 1094, 299–306.

Hancock, J.F.; Lobos, G.A. Pears. In *Temperate Fruit Crop Breeding: Germplasm to Genomics*; Hancock, J.F., Ed.; Springer, the Netherlands, 2008, p. 299.

Harker, F.R.; Maindonald, J.H.; Jackson, P.J. Penetrometer measurement of apple and kiwifruit firmness: operator and instrument differences. *Journal of the American Society for Horticultural Science* 1996, 121, 927–936.

Heide, O.M.; Prestrud, A.K. Low temperature, but not photoperiod, controls growth cessation and dormancy induction and release in apple and pear. *Tree Physiology* 2005, 25, 109–114.

Henriquez, J.L.; Sugar, D.; Spotts, R.A. Induction of cankers on pear tree branches by Neofabraea alba and N. perennans, and fungicide effects on conidial production on cankers. *Plant Disease* 2006. 90, 481–486.

Hiwasa, K.; Kinugasa, Y.; Amano, S.; Hashimoto, A.; Nakano, R.; Inaba, A.;

Kubo, Y. Ethylene is required for both the initiation and progression of softening in pear (*Pyrus communis* L.) fruit. *Journal of Experimental Botany* 2003, 54, 771−779.

Hiwasa, K.; Nakano, R.; Hashimoto, A.; Matsuzaki, M.; Murayama, H.; Inaba, A.; Kubo, Y. European, Chinese and Japanese pear fruits exhibit differential softening characteristics during ripening. *Journal of Experimental Botany* 2004, 55, 2281−2290.

Ingels, C.A.; Burkhart, D.J.; Elkins, R.B. Varieties. In *Pear Production and Handling Manual*; Mitcham, E.J.; Elkins, R.B., Eds.; University of California, Berkely, CA, USA, 2007, p. 26.

Ingels, C.A.; Lanini, T.; Klonsky, K.M.; Demoura, R. Effects of weed and nutrient management practices in organic pear orchards. *Acta Horticulturae* 2013, 1001, 175−184.

Ingels, C.A.; Lanini, W.T.; Shackel, K.A.; Klonsky, K.M.; Demoura, R.; Elkins, R.B. Evaluation of weed and nutrient management practices in organic pear orchards. *Acta Horticulturae* 2011, 909, 571−578.

Intrieri, C.; Ciavatta, C.; Xiloyannis, C.; Di Lorenzo, R. *Gestione del suolo*. In: *Arboricoltura Generale*; Sansavini S. Ed.; Pàtron editore, Bologna, Italy, 2012, p. 532.

Ishii, H.; Yanase, H. Venturia nashicola, the scab fungus of Japanese and Chinese pears: A species distinct from V. pirina. *Mycological Research* 2000, 104, 755−759.

Jackson, J.E. Light interception and utilization by orchard systems. *Horticultural Reviews* 1980, 2, 208−267.

Jackson, J.E. *Biology of Apples and Pears*, 1st ed. Cambridge University Press, The Edinburgh Building, Cambridge, United Kingdom, 2003.

James-Martin, G.; Williams, G.; Stonehouse, W.; O'Callaghan, N.; Noakes, M. Health and nutritional properties of pears (*Pyrus*): A literature review. For Hort Innovation and Apple and Pear Australia Ltd (APAL) Final report 15 October 2015, Food and Nutrition Glagship, pp. 1−25. http://rediscoverthepear.com.au/wp-content/uploads/2017/05/Pears-Health-Study-AP15010-Final-Report-Complete.pdf

Jijakli, M.H.; Lepoivre, P. State of the art and challenges of post-harvest disease management in apples. In *Fruit and Vegetable Diseases*, Springer, Dordrecht, 2004, p. 59.

Jones A.L.; Aldwinckle, H.S. *Plagas y Enfermedades del manzano y del peral*. APS Mundi Prensa 2002, p. 99.

Kader, A.A. A perspective on postharvest horticulture (1978−2003). *HortScience* 2003, 38, 1004−1008.

Kang, S.; Hu, X.; Goodwin, I.; Jerie, P. Soil water distribution, water use, and yield response to partial root zone drying under a shallow groundwater table condition in a pear orchard. *Scientia Horticulturae* 2002, 92, 277−291.

Karlıdağ, H., Aslantaş, R., & EŞİTKEN, A. (2014). Effects of interstock length on sylleptic shoot formation, growth and yield in apple. *TarımBilimleriDergisi* 20(3), 331. https://doi.org/10.15832/tbd.63462

Katayama Y. Nijisseiki Nashi in Tottori. *Acta Horticulturea* 2002, 587, 55−57.

Kays, S.J. Preharvest factors affecting appearance. *Postharvest Biology and Technology* 1991, 15, 233−247.

Kevers, C.; Pincemail, J.; Tabart, J.; Defraigne, J.O.; Dommens, J. Influence of cultivar, harvest time, storage conditions and peeling on the antioxidant capacity and phenolic and ascorbic acid contents of apples and pears. *Journal of Agricultural and Food Chemistry* 2011, 59, 6165Y6171.

Kitinoja, L.; Kader, A.A. *Small-Scale Postharvest Handling Practices: A Manual for Horticultural Crops*, 5th Ed. Postharvest Technology Research and Information Center University of California, Davis, 2015.

Kock, S.L.; Holz, G. Blossom-end rot of pears: Systemic infection of flowers and immature fruit by *Botrytis cinerea*. *Journal of Phytopathology* 1992, 135, 317−327.

Kumar, S.; Kirk, C.; Deng, C.; Wiedow, C.; Knaebel, M.; Brewer, L. Genotyping-by-sequencing of pear (*Pyrus* spp.) accessions unravels novel patterns of genetic diversity and selection footprints. *Horticulture Research* 2017, 4, 17015.

Larrigaudiere, C., Lentheric, I., Puy, J.; Pinto, E. Biochemical characterisation of core browning and brown heart disorders in pear by multivariate analysis. *Postharvest Biology and Technology* 2004, 31, 29−39.

Leffew, M.B.; Ernst, M.D. *A Farmer's Guide to a Pick-Your-Own-Operation*. University of Tennessee, Institute of Agriculture, 2014, 24 pp. https://extension.tennessee.edu/publications/Documents/PB1802.pdf

Lehnert, R. *Robotic Pruning Good Frut Growers*, 2012. https://www.goodfruit.com/robotic-pruning/

Li, J.; Huang, W.; Zhao, C.; Zhang, B.A. Comparative study for the quantitative determination of soluble solids content, pH and firmness of pears by VIS/NIR spectroscopy. *Journal of Food Engineering* 2013, 116, 324−332.

Li, X.; Zhang, J.Y.; Gao, W.Y.; Wang, Y.; Wang, H.Y., Cao, J.G.; Huang, L.Q. Chemical composition and anti-inflammatory and antioxidant activities of eight pear cultivars. *Journal of Agricultural and Food Chemistry* 2012, 60, 8738−8744.

Link, H. Significance of flower and fruit thinning on fruit quality. *Plant Growth Regulation* 2000, 31, 17−26.

Llorente, I.; Montesinos, E. Brown spot of pear: An emerging disease of economic importance in Europe. *Plant Disease* 2006, 90, 1368−1375.

Luna-Maldonado, A.I.; Vigneault, C.; Nakaji, K. Postharvest technologies of fresh horticulture produce. In *Horticulture*; Luna-Maldonado, A.I., Ed.; Intechopen Publications, London, UK, 2012, p. 161.

Lurie, S.; Watkins, C.B. Superficial scald, its etiology and control. *Postharvest biology and Technology* 2012, 65, 44−60.

Lutz, C.; Lopes, C.; Sosa, M.C.; Sangorrín, M. A new improved strategy for the selection of cold-adapted antagonist yeasts to control postharvest pear diseases. *Biocontrol Science and Technology* 2012, 22 (12): 1465−1483. ISSN 0958-3157.

Lutz, M.C.; M.C. Sosa; A. Colodner. 2017. Effect of pre and postharvest application of fungicides on postharvest decay of Bosc pear caused by *Alternaria/Cladosporium* complex in North Patagonia, Argentina. *Scientia Horticulturae* 225, 810−817

Maas, F.M.; Kanne, H.J.; Van der Steeg, P.A.H. Chemical thinning of "Conference" pears. *Acta Horticulturae* 2010, 884, 293−304.

Mac Hardy, W.E. Current status of IPM in apple orchards. *Crop Protection* 2000, 19, 801−806.

Machado, N.P.; Fachinello, J.C.; Galarça, S.P.; Betemps, D.L.; Pasa, M.S.; Schmitz, J.D. Pear quality characteristics by VIS/NIR spectroscopy. *Anais da Academia Brasileira de Ciências* 2012, 84, 853−863.

Marsal, J.; Mata, M.; Arbonés, A.; Rufat, J.; Girona, J. Regulated deficit irrigation and rectification of irrigation scheduling in young pear trees: An evaluation based on vegetative and productive response. *European Journal of Agronomy* 2002, 17, 111−122.

Marsal, J.; Rapoport, H.F.; Manrique, T.; Girona, J. Pear fruit growth under regulated deficit irrigation in container-grown trees. *Scientia Horticulturae* 2000, 85, 243−259.

Mattheis, J. Fruit maturity and ripening. In *Tree Fruit Physiology: Growth and Development*; Maib, K.M.; Andrews, P.K.; Lang, G.A.; Mullinix, K., Eds.; Good Fruit Grower, Yakima Washington, 1996, p. 117.

Mattos, L.M.; Moretti, C.L.; Ferreira, M.D. Modified atmosphere packaging for perishable plant products. In *Polypropylene*; Dogan, F., Ed.; Intechopen Publications, London, UK, 2012, p. 95.

Mayus, M.M.; Strassemeyer, J.; Heijne, B.; Alaphilippe, A.; Holb, I.; Rossi, V.; Pattori, E. PURE: WP5-Milestone MS14: Descriptions of most important innovative non-chemical methods to control pests in apple and pear orchards, 2012.

Mc Guckian, R. *Guidelines for Irrigation Management for Apple and Pear Growers*. http://apal.org.au/wp-content/uploads/2013/04/Apple_pear_guidelines_Irrigation_Management.pdf (accessed Nov 15, 2017)

Meheriuk, M., Prange R.K.; Lidster P.; Porritt, S.W. *Postharvest Disorders of Apples and Pears*. Canadian Agriculture, Canada, Otawa, 1994.

Ministerio de Hacienda y Finanzas Públicas Presidencia de la Nación. Informes de cadenas de valor. Frutícola Manzanas y Peras., ArgentinaAÑO 1 - N° 23 – Diciembre 2016, p. 58. https://www.economia.gob.ar/peconomica/docs/Complejo_fruta_pepita.pdf

Mitcham, B.; Cantwell, M.; Kader, A. Methods for determining quality of fresh commodities. *Perishables Handling Newsletter* 1996, 85, 1–5.

Mitcham, E.J.; Mitchell, F.G. *Sistemas de Manejo Postcosecha: Frutos Pomo*. In *Tecnología Postcosecha de Cultivos Hortofrutícolas*, 3rd ed.; Kader A.A., Ed.; Universidad de California, CA, USA, 2007, p. 373.

Molina-Ochoa, M.J.; Vélez-Sánchez, J.E.; Galindo-Egea A. Preliminary results of the effect of deficit irrigation duringthe rapid growth of the pear fruit on production and quality of Triunfo de Viena variety. *Revista Colombiana de ciencias hortícolas* 2015, 9, 38–45.

Montesinos, E.; Melgarejo, P.; Cambra, M.; Pinochet, J. *Enfermedades de los frutales de pepita y de hueso. Sociedad Española de Fitopatología*. Ed. Mundi Prensa, 2000, p. 147.

Montuschi, C.; Collina, M.; Antoniacci, L.; Cicognani, E.; Rimondi, S.; Trapella, R.; Brunelli, A. Preliminary studies on biology and epidemiology of Valsaceratosperma (Cytosporavitis), the causal agent of barkcankeronpear in Italy. *Pome Fruit Diseases IOBC/WPRS Bulletin* 2006, 29, 187–193.

Moragrega, C.; Llorente, I.; Manceau, C.; Montesinos, E. Susceptibility of European pear cultivars to *Pseudomonas syringae* pv. syringae using immature fruit and detached leaf assays. *European Journal of Plant Pathology* 2003, 109, 319–326.

Murashige, T.; Skoog, F. A revised medium for rapid growth and bioassays with tobacco tissue cultures. *Physiologia Plantarum*, 1962, 15, 473–497.

Musacchi, S. *Tecnica colturale*. In: *Il Pero*; Angelini R. Ed.; ART Servizi Editoriali S.p.A. Bologna, Italy, 2007, p. 339.

Musacchi, S.; Ancarani, V.; Gamberini, A.; Giatti, B.; Sansavini, S. Progress in pear breeding at the University of Bologna. *Acta Horticulturae* 2005, 671, 191–194.

Musacchi, S.; Serra, S.; Ancarani, V. Comparison among pear training systems and rootstocks for high density planting (hdp) of the cultivar "Abbé Fétel." *Acta Horticulturae* 2011, 909, 251–258.

Naor, A. Irrigation and crop load influence fruit size and water relations in field-grown "Spadona" pear. *Journal of the American Society for Horticultural Science* 2001, 126, 252–255.

Neto, C.; Carranca, C.; Clemente, J.; Varennes, A. Nitrogen distribution, remobilization and re-cycling in young orchard of non-bearing "Rocha" pear trees. *Scientia Horticulturae* 2008, 118, 299–307.

NUTTAB Online Searchable Database. *Food Standards Australia and New Zealand*, 2010. http://www.foodstandards.

gov.au/science/monitoringnutrients/Pages/default.aspx (accesed Jan 14, 2018).

Ogawa, J.M.; English, H. *Diseases of Temperate Zone Tree Fruit and Nut Crops*, University of California Division of Agriculture and National Resources, Publications, 1991; Vol. 3345.

Opara, L.U. Fruit growth measurement and analysis. In *Horticultural Reviews*; Janick, J., Ed.; John Wiley & Sons, Oxford, 2000; Vol. 24; p. 373.

Opara, U.L. A review on the role of packaging in securing food system: Adding value to food products and reducing losses and waste. *African Journal of Agricultural Research* 2013, 8, 2621–2630.

Paul, V.; Pandey, R. Role of internal atmosphere on fruit ripening and storability: A review. *Journal of Food Science and Technology* 2014, 51, 1223–1250.

Paul, V.; Pandey, R.; Srivastava, G.C. The fading distinctions between classical patterns of ripening in climacteric and non-climacteric fruit and the ubiquity of ethylene: An overview. *Journal of Food Science and Technology* 2012, 49, 1–21.

Pitt, J.I.; Spotts, R.A.; Holmes, R.J.; Cruickshank, R.H. *Penicillium solitum* revived, and its role as a pathogen of pomaceousfruit. *Phytopathology* 1991, 81, 1108–1112.

Prange, R.K.; DeLong J.M.; Wright A.H. Storage of pears using dynamic controlled-atmosphere (DCA), a Non-Chemical Method. *Acta Horticulturea* 2011, 909, 707–717.

Prasanna, V.; Prabha, T.N.; Tharanathan, R.N. Fruit ripening phenomena—an overview. *Critical Reviews in Food Science and Nutrition* 2007, 47, 1–19.

Quaglia, M.; Ederli, L.; Pasqualini, S.; Zazzerini, A. Biological control agents and chemical inducers of resistance for postharvest control of Penicillium expansum link on apple fruit. *Postharvest Biology and Technology* 2011, 59, 307–315.

Quartieri, M.; Millard, P.; Tagliavini, M. Storage and remobilisation of nitrogen by pear (Pyruscommunis L.) trees as affected by timing of N supply. *European Journal of Agronomy* 2002, 17, 105–110.

Racskó, J.; Leite, G.B.; Zhongfu, S.; Wang, Y.; Szabó, Z.; Soltész, M.; Nyéki, J. Fruit drop: The role of inner agents and environmental factors in the drop of flowers and fruits. *International Journal oh Horticultural Science* 2007, 13, 13–23.

Racskó, J.; Nagy, J.; Soltész, M.; Nyéki J.; Szabó Z. Fruit drop: I. Specific characteristics and varietal properties of fruit drop. *International Journal of Horticultural Science* 2006, 12, 59–67.

Raffo, M.D; Rodriguez, R.; Ginobili, J. *Plantación*. In *Pera Williams, Manual para el productor y el empacador*; Sanchez E., Ed.; INTA Argentina, 2010, p. 14.

Ramírez, F.; Davenport, T.L. Apple pollination: A review. *Scientia Horticulturae* 2013, 162, 88–203.

Reil, W.O. Orchard establishment. In *Pear Production and Handling Manual*, Mitcham, E.J.; Elkins, R.B, Eds.; Postharvest Center, University of California, CA, USA, 2007, p. 45.

Reiland, H.; Slavin, J. Systematic review of pears and health. *Nutrition Today* 2015, 50, 301–305.

Romanazzi, G.; Sanzani, S.M.; Bi, Y.; Tian, S.; Martínez, P.G.; Alkan, N. Induced resistance to control postharvest decay of fruit and vegetables. *Postharvest Biology and Technology* 2016, 122, 82–94.

Ryugo, K. *Fruit culture: its science and art*. John Wiley & Sons, New York, USA, 1988.

Saito, T. Advances in Japanese pear breeding in Japan. *Review Breeding Science* 2016, 66, 46–59.

Sánchez E.E. Nutrition and water management in intensive pear growing. *Acta Horticulturae* 2015, 1094, 307–316.

Sánchez, E.E.; Righetti, T.L. Misleading Zinc deficiency diagnoses in pome fruit and inappropriate use of foliar zinc sprays. *Acta Horticulturae* 2002, 594, 363–368.

Sánchez, E.E.; Righetti, T.L.; Sugar, D.; Lombart, P.B. Effects of timing of nitrogen application on nitrogen partitioning between vegetative, reproductive and structural components of matures Comice pears. *The Journal of Horticultural Science* 1992, 67, 51–58.

Sánchez, E.E.; Righetti, T.L.; Sugar, D.; Lombart, P.B. Response of Comice pear trees to a postharvest urea spray. *The Journal of Horticultural Science* 1990, 65, 541–546,

Sanderson, P.G.; Spotts, R.A. Postharvest decay of winter pear and apple fruit caused by species of *Penicillium*. *Phytopathology* 1995, 85, 103–110.

Sansavini, S.; Costa, G.; Gucci, R.; Inglese, P.; Ramina, A.; Xiloyannis, C. (Eds.). *General Arboriculture*. Patron Editore, Bologna, Italy, 2012. p. 536.

Sansavini, S.; Musacchi, S. Canopy architecture, training and pruning in the modern european pear orchards: An overview. *Acta Horticulturae* 1994, 367, 152–172.

Sanzol, J.; Herrero, M. The "effective pollination period" in fruit trees. *Scientia Horticulturae* 2001, 90, 1–17.

Sardella, D.; Muscat, A.; Brincat, J.P.; Gatt, R.; Decelis, S.; Valdramidis, V.A. Comprehensive review of the pear fungal diseases. *International Journal of Fruit Science* 2016, 16, 351–377.

Schaffer, B.; Andersen, P.C. *Handbook of Environmental Physiology of Fruit Crops*, 1st ed.; CRC Press, Boca Raton, FL, USA, 1994, p. 358.

Schotzko, R.T.; Granatstein, D.A. *Brief Look at the Washington Apple Industry: Past and Present SES 04-05*. Wenatchee, WA: Washington State University, 2004.

Segrè, A. The world pear industry: Current trends and prospects. *Acta Horticulturae* 2002, 596, 55–59.

Serdani, M.; Spotts, R.A.; Calabro, J.M.; Postman, J.D. Powdery mildew resistance in pyrus germplasm. *Acta Horticulturae* 2005, 671, 609–613.

Sessa, L.; Abreo, E.; Bettucci, L.; Lupo, S. Botryosphaeriaceae species associated with wood diseases of stone and pome fruits trees: Symptoms and virulence across different hosts in Uruguay. *European Journal of Plant Pathology* 2016, 146, 519–530.

Shackel, K. A plant-based approach to deficit irrigation in trees and vine. *HortScience* 2011, 46, 173–177.

Shah, M.D.; Verma, K.S.; Singh, K.; Kaur, R. Morphological, pathological and molecular variability in Botryodiplodia theobromae (Botryosphaeriaceae) isolates associated with die-back and bark canker of pear trees in Punjab, India. *Genetics and Molecular Research* 2010, 9, 1217–1228.

Silva, F.; Gomes, M.H.; Fidalgo, F.; Rodrigues, J.A.; Almeida, D. Antioxidant properties and fruit quality during long-term storage of "Rocha" pear: Effects of maturity and storage conditions. *Journal of Food Quality* 2010, 33, 1–20.

Silva, G.; Medeiros Souza, T.; Barbieri, R.; Costa de Oliveira, A. Origin, domestication, and dispersing of pear (*Pyrus* spp.). *Advances in Agriculture* 2014, ID 541097. http://dx.doi.org/10.1155/2014/541097

Singh, V.; Hedayetullah, M.; Zaman, P.; Meher, J. Postharvest technology of fruits and vegetables: an overview. *Journal of Postharvest Technology* 2014, 2, 124–135.

Slippers, B.; Smit, W.A.; Crous, P.W.; Coutinho, T.A.; Wingfield, B.D.; Wingfield, M.J. Taxonomy, phylogeny and identification of Botryosphaeriaceae associated with pome and stone fruit trees in South Africa and other regions of the world. *Plant Pathology* 2007, 56, 128–139.

Soroker, V.; Talebaev, S.; Harari, A.R.; Wesley, S.D. The role of chemical cues in host and mate location in the pear *Psylla Cacopsylla* bidens (Homoptera: Psyllidae). *Journal of Insect Behavior* 2004, 17, 613−626.

Sorrenti, G.; Toselli, M.; Marangoni, B. Use of compost to manage Fe nutrition of pear trees grown in calcareous soil. *Scientia Horticulturae* 2012, 136, 87−94.

Sosa, M.C.; Lutz, M.C.; Lefort, N.C.; Sanchez, A. Postharvest losses by complex of *Phytophthora* sp. and *Botrytis cinerea* in long storage pear fruit in the North Patagonia, Argentina. *Acta Horticulturae* 2016, 1144, 237−244.

Sosa, M.C.; Lutz, M.C.; Vélez, M.L.; Greslebin, A. Pre-harvest rot of pear fruit Golden Russet Bosc caused by *Phytophthora lacustris* and *Phytophthora drechsleri* in Argentina. *Australasian Plant Disease Notes* 2015, 10, 18.

Sozzi, G.O. *Fisiología de la maduración de los frutos de especies leñosas*. In *Árboles frutales, ecofisiología, cultivo y aprovechamiento*; Sozzi, G.O., Ed.; Facultad de Agronomía, Universidad de Buenos Aires, Buenos Aires, 2007a, p. 669.

Sozzi, G.O. *Fisiología del crecimiento de los frutos*. In *Árboles frutales, ecofisiología, cultivo y aprovechamiento*; Sozzi, G.O., Ed.; Facultad de Agronomía, Universidad de Buenos Aires, Buenos Aires, 2007b, p. 309.

Sozzi, G.O. ed. *Árboles frutales: ecofisiología, cultivo y aprovechamiento*, 1st ed. Buenos Aires: Universidad de Buenos Aires. Facultad de Agronomía, 2007.

Spotts, R.A. Infection of Anjou pear fruit by *Podosphaera leucotricha*. *Plant Disease* 1984, 68, 857−859.

Spotts, R.A.; Castagnoli, S. *Pear Scab in Oregon Symptoms, disease cycle and management* EM 9003 May 2010. Oregon State University, Extension Service.

Statista, The Statistics Portal. Global leading pear producing countries in 2016/2017. https://www.statista.com/statistics/739168/global-top-pear-producing-countries/ (accessed Mar, 2018).

Sugar, D. Advances in postharvest management of pears. *Acta Horticulturea* 2011, 909, 673−678.

Sugar, D. Postharvest physiology and pathology of pears. *Acta Horticulturea* 2002, 596, 833−838.

Sugar, D.; Hilton, R.J.; Van Buskirk, P.D. Effects of kaolin particle film and rootstock on tree performance and fruit quality in "Doyenne du Comice" pear. *HortScience* 2005, 40, 1726−1728.

Suzaki, K. Population structure of *Valsaceratosperma*, causal fungus of Valsa canker, in apple and pear orchards. *Journal of General Plant Pathology* 2008, 74. https://doi.org/10.1007/s10327-008-0078-4

Tagliavini, M.; Abadía, J.; Rombolà A.D.; Abadía, A; Tsipouridis, C.; Marangoni, B. Agronomic means for the control of iron deficiency chlorosis in deciduous fruit trees. *Journal of Plant Nutrition* 2000, 23, 2007−2022.

Tagliavini, M.; Toselli, M. *Foliar application of nutrients*. In *Encyclopedia of soil in the environment*; Hillel, D. Ed.; Elsevier Ltd, Oxford, U.K. 2005.

Tagliavini, M; Quartieri M.; Millard P. Remobilised nitrogen and root uptake of nitrate for spring leaf growth, flowers and developing fruits of pear (*Pyrus communis* L.) trees. *Plant and Soil* 1997, 195, 137−142.

Thompson, J.F. *Transporte. Tecnología Postcosecha de Cultivos Hortofrutícolas*, 3rd ed.; Kader A.A., Ed.; Universidad de California: California, 2007; p. 111.

Toranzo, J. *Producción mundial de manzanas y peras*. Programa Nacional Frutales. Estación Experimental Agropecuaria Alto Valle. INTA Ediciones, 2016.

Trentini, L. Pears: trade, consumers' habits and consumption in Europe. Conference, Interpera 2012 World Conference on Pears, 15 June, Lleida, Spain.

Tromp, J. Flower-bud formation in pome fruits as affected by fruit thinning. *Plant Growth Regulation* 2000, 31, 27–34.

Tworkoski, T.; Miller, S. Rootstock effect on growth of apple scions with different growth habits. *Scientia Horticulturae* 2007, 111, 335–343.

United States Department of Agriculture. Global Agricultural Information Network (GAIN) https://gain.fas.usda.gov (accessed October 4, 2017).

USA Pears. *Vitamin C, Phytonutrients, and Antioxidants, Oh My!* Milwaukie, OR: Pear Bureau Northwest [Online], 2018. http://usapears.org/vitamin-c-phytonutrients-and-antioxidants-oh-my/ (accessed Feb 16, 2018).

USA Pears. *Pear Handling Manual.* Milwaukie, OR: Pear Bureau Northwest [Online], 2012. http://trade.usapears.com/wp-content/uploads/2015/07/USA-Pears-Pear-Handling-Manual.pdf (accessed Jan 6, 2018).

USDA; ARS; GRIN. United States Department of Agriculture, Agricultural Research Service Germplasm Resources Information Network [Online], 2018. http://www.ars-grin.gov/ (accessed 3 Feb, 2018).

USDA, Foreign Agricultural Service. *Fresh Deciduous Fruit (Apples, Grapes, & Pears): World Markets and Trade,* December, 2017. https://downloads.usda.library.cornell.edu/usda-esmis/files/1z40ks800/7w62f868g/df65v832v/decidwm-12-08-2017.pdf (accessed June 28, 2020).

USDA, Foreign Agricultural Service. *Fresh Deciduous Fruit (Apples, Grapaes, & Peers): World Markets and Trade,* June 2017. https://downloads.usda.library.cornell.edu/usda-esmis/files/1z40ks800/wd375w73r/pk02cb160/decidwm-06-12-2017.pdf (accesed Oct 24, 2017).

USDA. *National Nutrient Database for Standard Reference Release 28* slightly revised May, 2016. Basic Report 09413, Pears, raw, with skin. Report Date: October 24, 2017. https://ndb.nal.usda.gov/ndb/foods (accesed Feb 16, 2018).

Van Doorn, D. Consumption of pears. Conference, Interpera November 19, 2015, Ferrara, Italy November 19, 2015. http://interpera.weebly.com/uploads/1/7/0/4/17040934/freshfel-wapa_presentation_interpera_-_d._van_doorn.pdf (accesed Feb 19, 2018).

Van Schoor, L.; Stassen, P.J.C.; Botha, A. Effect of organic material and biological amendments on pear tree performance and soil microbial and chemical properties. *Acta Horticulturae* 2012, 933, 205–214.

Vavilov, N.I. The origin, variation, immunity and breeding of cultivated plants. *Science* 1952, 115, 433–434.

Verma M.K. *Pear Production Technology. In Training manual on teaching of postgraduate courses in horticulture (Fruit Science),* Singh S.K.; Munshi, A.D.; Prasad, K.V.; Sureja, A.K., Eds.; Post Graduate School, Indian Agricultural Research Institute, New Delhi-110012, 2014, pp. 249–255.

Vigneault, C.; Thompson, J.; Wu, S. Designing container for handling fresh horticultural produce. *Postharvest Technologies for Horticultural Crops* 2009, 2, 25–47.

Villalobos-Acuña, M.; Mitcham, E.J. Ripening of European pears: The chilling dilemma. *Postharvest Biology and Technology* 2008, 49, 187200.

Villarreal, P. *Taxonomia y morfología de Pyrus communis.* In *Pera Williams, Manual para el productor y el empacador;* Sanchez E., Ed.; INTA Argentina, 2010, p. 11.

World Apple and Pear Association. *European Apple and Pear Crop Forecast,* August,

2017. http://www.wapa-association.org/asp/page_1.asp?doc_id=447 (accessed Mar, 2018).

Webster, A.D. Factors influencing the flowering, fruit set and fruit growth of European pears. *Acta Horticulturea* 2002, 596, 699–709.

Webster, A.D.; Wertheim, S.J. *Rootstocks and interstems*. In: *Fundamentals of Temperate Zone Tree Fruit Production*; Tromp J, Webster AD, Wertheim SJ, Eds; Backhuys Publishers, 2005, pp. 156–175.

Wertheim, S.J. Developments in the chemical thinning of apple and pear. *Plant Growth Regulation* 2000, 31, 85–100.

Wertheim, S.J.; Webster, A.D. Rootstocks and interstems. In: *Fundamentals of Temperate Zone Tree Fruit Production*; Tromp, J.; Webster, A.D.; Wertheim, S.J., Eds.; Backhuys Publishers, the Netherlands, 2005, p. 156.

Westwood, M.N. *Temperate-Zone Pomology, Physiology and Culture*, 3rd ed.; Timber Press, Portland, Oregon, USA, 1993.

White, A.G. European pear breeding strategy in New Zealand. DSIR *Plant Breeding Symposium* 1986, 25, 5, 150–154.

Williams, M.W.; Couey, H.M.; Moffitt, H.; Coyier, D.L. *Pear Production*. In *Agriculture Handbook* 526, U.S. DEPT. OF AGRICULTURE Science and Education Administration's Federal Research staff, 1978.

Wojcik, P.; Wojcik, M. Effects of boron fertilization on "Conference" pear tree vigor, nutrition, and fruit yield and storability. *Plant and Soil* 2003, 256, 41–-421.

Xiao, C.L.; Rogers, J.D. A postharvest fruit rot in d'Anjou pears caused by *Sphaeropsispyriputrescens* sp. nov. *Plant Disease* 2004, 88, 114–118.

Xiao, C.L.; Boal, R.J. Prevalence and incidence of *Phacidiopycnis* rot in d'Anjou pears in Washington State. *Plant Disease* 2004, 88, 413–418.

Xiao, C.L.; Boal, R.J. Distribution of *Potebniamyces pyri* in the U.S. Pacific Northwest and its association with a canker and twig dieback disease of pear trees. *Plant Diseases* 2005, 89, 920–925.

Xu, L.; Zhou, P.; Han, Q.; Li, Z.; Yang, B.; Nie, J. Spatial distribution of soil organic matter and nutrients in the pear orchard under clean and sod cultivation models. *Journal of Integrative Agriculture* 2013, 12, 344–351.

Zhai, L.; Zhang, M.; Lv, G.; Chen, X.; Jia, N.; Hong, N.; Wang, G. Biological and molecular characterization of four *Botryosphaeria* species isolated from pear plants showing stem wart and stem canker in China. *Plant Disease* 2014, 98, 716–726.

Zhang, B.; Huang, W.; Li, J., Zhao, C.; Fan, S.; Wu, J.; Liu, C. Principles, developments and applications of computer vision for external quality inspection of fruits and vegetables: A review. *Food Research International* 2014, 62, 326–343.

CHAPTER 3

QUINCE

HAMID ABDOLLAHI

Temperate Fruits Research Centre, Horticultural Sciences Research Institute, Agricultural Research, Education and Extension Organization (AREEO), Karaj, Iran

Corresponding author. E-mail: h.abdollahi@areeo.ac.ir; habdollahi@yahoo.it

ABSTRACT

History of growth and cultivation of quince tree dates back to at least 2500 years ago in the ancient cultures of Persia, Greece, and Romans. This tree is the monospecific member of genus *Cydonia* and is classified as a temperate fruit and also is a medicinal low-cost natural source of the phenolics. The quince tree is native to North Iran, Turkmenistan, and the Transcaucasian region, and it seems that its local cultivars originated mostly from selections of superior genotypes made by growers over past centuries. The particular features of quince trees and fruits tied their cultivation and use to the mythologies, and for this reason, the cultivation of quinces principally is subjected to the cultural believes of nations rather than commercial

interests of producers. Due to the original adaptation of quince to the foothills of the Caucasus Mountains and Persia, this tree can be cultivated in the cool temperate zones with moderate altitudes. Well-known commercial quince cultivars are "Isfahan," "Ekmek," "Şeker," "Limon," "Bardacık," "Eşme," "Smyrna," "Bereczcki," "Vranja," "Portugal," "Gamboa," "Orange," "Meech's Prolific," "Van Deman," "Pineapple," and "Champion," but attempts for breeding of quince cultivars for fire blight (*Erwinia amylovora*) resistance, tolerance to lime-induced iron chlorosis, growth habit, and bearing have resulted in the release of new cultivars such as "Viduja." Various aspects of quince cultivation and orchard management, including propagation and rootstock selection, tree nutrition and irrigation, and pest and disease control are principally similar to the other pome fruits, but some aspects such as flower bud initiation and differentiation, and the level of self-fertility are also the matters of controversy in the literature. Recent worldwide evaluations of the quince germplasm collections revealed that a wide potential for the modernization and commercialization of quince orchards exists that needs more consideration.

3.1 INTRODUCTION

Quince (/ˈkwɪns/; *Cydonia oblonga* [*C. oblonga*]) known as "cotoneum" in Latin, "cotogno" in Italian, "cognassier" in French, and "quitte" in German language and is a member of the Rosaceae family. The quince grows as a flowering tree or shrub and is believed that has originated from the regions around the Caspian Sea (Bell and Leitão, 2011). In these regions, the quince is called a variety of names, including "Beh" in Iran, "Behi" in Uzbekistan, "Heyva" in Azerbaijan, "Ayva" in Turkey, and "հպրստանմայ" in Armenia. The quince tree belongs to the genus *Cydonia*, the monospecific genus of subfamily Spiraeoideae (previously recognized as subfamily Maloideae) and consists of a single species *C. oblonga* (Bell and Leitão, 2011). The quince tree is deciduous and bears pome fruits. The fruits of quince are

similar to the apple and pear, and they turn bright golden-yellow at maturity. Many historic documents show that cooked quinces have been used as food or marmalade. But the quince trees are also grown for their attractive colors of blooms. The particular features of quince trees and fruits tied their cultivation and use to the mythologies and beliefs of various ethics.

According to the historical documents, the cultivation and growth of the quince trees have a very long history that date back to more than several thousand years ago. In ancient Persia, the first pieces of evidence for the cultivation of the quince trees were presented by Herodotus and in a cuneiform text of the Persepolis of the Achaemenid Empire (550–330 BC) (Curtis and Tallis, 2005). Besides these pieces of evidence, also the nomination of quince fruits in the tradition of Zoroastrian of ancient Persia (Daryaee, 2006–07) shows that quince cultivation has at least 2500 years of history in the southern areas of the Caspian Sea. Additional historical signs of the importance of this tree may return to the traditions and beliefs of the Jews and Muslims. The rabbinical traditions of the Jews indicate quince as the most ancient fruit and that its cultivation dates back to the Garden of Eden (Meech, 1888). Avicenna (syn. Ibn-Sina) the Persian polymath (980–1037 AD) listed quince as a medicinal plant and mentioned the numerous benefits of this tree for human health in his book *The Canon of Medicine* (Avicenna, 1973). Also, in the historic miniatures of Iran, mainly from the Safavid dynasty (1501–1736 AD), numerous evidences show the importance of the quince fruits in the imperial courts and ceremonies (Figure 3.1A–C).

In the European cultures, the appearance of quince fruits in artworks seems to be more recent than that in the Eastern cultures. This could be due to the presence of the center of origin of this fruit around the Caspian Sea. Aulus Cornelius Celsus, the Roman encyclopaedist who is well known for his *De Medicina* (47 CE) in the 1st century, referred to the various medicinal effects of quince fruits. Also, quince fruits are visible in the wall paintings of the Roman era about 36 BC, as well as in more recent oil paintings of Vincent Willem van Gogh, the Dutch painter of the 19th century (Figure 3.1D–F).

3.2 AREA AND PRODUCTION

Quince trees are grown as a scion cultivar and moreover as a rootstock for commercial pear orchards in all around the world. Quince has so far been considered as a minor fruit, and therefore its area of cultivation in the world in 2014 was limited to near 82,000 ha (Food and Agriculture Organization of the United Nations, 2017). According to these data, the global production of quince fruits has been estimated about 650,000 metric tons with an average yield of 7.8 metric tons/ha in 2014 that shows yield efficiency of quince trees in general is less than apples and pears. Also, the total production of quince according to the Food and Agriculture Organization (FAO) statistical data (Food and Agriculture Organization of the United Nations, 2017) demonstrate an increment from 315,000 ha in 1995 to near 650,000 in 2014. This increment of quince production is principally related to the extension procedure of the quince cultivation area by People's Republic of China from less than 1000 ha in 1990 to 32,776 ha in 2014, which made this country as the major quince producer in the world in 2014 in terms of the production area. In 2014, Uzbekistan (110,000 metric tons), People's Republic of China (109,090), Turkey (107,243), and Iran (82,522) were the major producers of this fruit. In addition, North African countries including Morocco, Algeria, and Tunisia; East Europe countries such as Serbia, Ukraine, Russia, and Romania; some South American countries, especially Argentina, Peru, and Uruguay; two countries of West Europe, Spain and Portugal, besides Azerbaijan and Armenia that are two countries near to the center of origin of quince trees, are other producers of this fruit. Quinces also have been selected and utilized as pear rootstocks in many of West European countries, including France, United Kingdom, Italy, the Netherlands, and Belgium (Tukey, 1964).

FIGURE 3.1 (A) Quince fruits in a miniature painted by Reza Abbasi (1565–1635 AD), the Persian painter of the Isfahan School during the later Safavid period (1501–1736 AD). (B) Quinces in a miniature painting by Mohammad Kassim Tabrizi (1575–1659 AD). The miniature date back to the Safavid period of Persia. (C) Offering quinces fruits in a dish in a wall miniature painting on ceramic that dates back to the first quarter of the 17th century, from a pavilion in Isfahan, the capital of Safavid dynasty of Persia. (D) Quinces in a painted garden, from the triclinium in the Villa of Livia Drusilla, Roma, Italy, belong to 30–20 BC. (E–F) Painting of quinces and other fruits by Vincent Willem van Gogh, a Dutch painter, in 1887.

3.3 MARKETING AND TRADE

It was mentioned that the cultivation and use of quince fruits follow principally the cultural interests, rather than commercial interests. For this reason, major part of quince production in the main global producers of this fruit is utilized in local markets and not for exportation. According to this, the last statistical data of quince global trade in 2013 demonstrated that among four main producers of quince fruits, including Uzbekistan, Republic of China, Turkey, and Iran, just Turkey allocated about 10% of its total production, equivalent to 13,100 metric tons, to exportation, whereas the other three major producer countries had not considerable quantities of exportation of these fruits (Food and Agriculture Organization of the United Nations, 2017). On the contrary, in 2013, apart from Turkey, the Netherlands (7512 metric tons), Spain (3043 metric tons), France (2690 metric tons), and Austria (2330 metric tons) had some exportation quantities of quinces. In the same year, the Russian Federation, Austria, Germany, Portugal, Romania, and France were the main importers of quince fruits with quantities more than 1000 metric tons.

3.4 COMPOSITION AND USES

The edible part of 100 g raw (unprocessed) quince fruit contains 83 g water and 16 g dry matter. This amount of dry matter of raw quince fruits contains about 57 kcal (238 KJ) energy, 0.40 g protein, 0.1 g lipids, 15.3 g carbohydrates, and 1.9 g edible fiber (USDA, 2016). Mean quantities of the other organic and inorganic materials in raw quince fruits, including minerals, vitamins, and composition of fatty acids, have been demonstrated in Table 3.1.

TABLE 3.1 Mean Food Values for Minerals and Vitamins Per 100 g Mature Quince (Excluding Core, Seeds, and Peels) According to the National Nutrient Database of USDA (2016)

Minerals		Vitamins		Lipids (Fatty Acids)	
Nutrient	Value (mg)	Nutrient	Value (mg)	Nutrient	Value (g)
Ca	11	Vitamin C	15.0	16:0 (palmitic acid)	0.007
Fe	0.7	Thiamin	0.02	18:0 (stearic acid)	0.002
Mg	8.0	Riboflavin	0.03	Total monounsaturated	0.036
P	17	Niacin	0.20	18:1 undifferentiated (oleic acid)	0.036
K	197	Pantothenic acid	0.08	Total polyunsaturated	0.050
Na	4.0	Vitamin B-6	0.04	18:2 undifferentiated (linoleic acid)	0.049
Zn	0.04	Folate, total	3×10^3		
Cu	0.130	Folic acid	0.00		
Se	0.6×10^{-3}	Vitamin A, RAE	2×10^{-3}		
		Vitamin A, IU	40 (IU)		

RAE, retinol activity equivalents; IU, international unit.

More than 4000 phenolic compounds, such as simple phenolics, cinnamic, coumarins, flavonoids, and anthocyanins, have been categorized in plants. These compounds prevent catastrophic damages of the free radicals to proteins, carbohydrates, lipids, and DNA molecules in plants and also have important roles in human health (Vermerris and Nicholson, 2006; Vinson et al. 2005). Quince leaves and fruits are reach and low-cost natural sources of the phenolics. Whereas the amounts of total phenolic compounds in various species of fresh fruits have been reported between 11.88 (*Pyrus communis* L.) and 585.52 (*Ziziphus jujuba* Mill.) mg GAE/100 g (Francini and Sebastiani, 2013), the mean amounts of phenolics in the pulp and peel of fresh quince fruits are 13 and 97.5 mg GAE/100 g, respectively (Silva et al. 2002a).

In the traditional herbal medicine of Iran, according to Avicenna historic texts (Avicenna, 1973), numerous therapeutic effects have been mentioned for quince leaves, besides therapeutic effects of fruits. Ghozati et al. (2016) and Amirahmadi et al. (2017) surveyed total phenolics and flavonoids of the selected quince germplasm of the southern regions of the Caspian Sea and reported that the total phenolics in the leaves vary from 30.8 to 1491 mg and in fruits from 7.5 to 213.9 mg GAE/100 g in fresh weight. Various harvests of leaves during growth season also demonstrated that the total phenolics increase up to the end of season, while the highest contents of flavonoids were mainly observed in harvested quince leaves of late June. Also, the mean contents of phenolics have been reported significantly different in the local landraces and genotypes of various regions of Iran, Spain, and Portugal (Ghozati et al. 2016; Silva et al. 2002a; Oliveira et al. 2007). Silva et al. (2002a) determined that caffeoylquinic acids including, 3, 5-dicaffeoylquinic acid, 3-, 4-, and 5-*O*-caffeoylquinic acids and one quercetin glycoside, and low amounts of rutin are the main phenolics of quince pulp. They reported a higher level of phenolics in quince peel and identified the same caffeoylquinic acids and several flavanol glycosides: kaempferol 3-glucoside, quercetin 3-galactoside, kaempferol 3-rutinoside, and several unidentified compounds as the main phenolic compounds of fruit peels. Oliveira et al. (2007) also demonstrated that various quince genotypes have a common phenolic profile composed of nine compounds including the same phenolics of fruits identified by Silva et al. (2002a) and quercetin-3-*O*-rutinoside, quercetin-3-*O*-galactoside, kaempferol-3-*O*-glucoside, kaempferol-3-*O*-rutinoside, and kaempferol-3-*O*-glycoside. Also, it was confirmed that quince leaves have higher relative contents of kaempferol derivatives than fruit pulps, peels, and seeds.

Historically, decoction or infusion of quince leaves and fruits is well known in traditional medicinal practices as a herbal medicine, and several nutraceutical effects have been related to its various parts, including, leaves, fruits, and seeds (Lim, 2012). Khoubnasabjafari and Jouyban (2011) reviewed phytochemicals and medicinal usages of different parts of quince tree. According to this, antibacterial, antialcoholic, carminative, expectorant, anticancer effects of fruits and seeds, anticough, antibronchitis, anticonstipation and healing effects on skin lesions of quince seeds, and benefit effects of seeds on migraine, nausea, common cold, and influenza, as well as antidiabetic, antioxidant, antihyperglycemic, and cardiovascular, hemorrhoids, bronchial asthma, and cough benefits of quince leaves have been listed from numerous researches on health effects of quinces.

3.5 ORIGIN AND DISTRIBUTION

The primary origin of quinces is thought to be Western Asia—probably in northern Iran, Turkmenistan, and the Caucasus—together with the countries of Armenia, Azerbaijan, and the Russian Federation (Bell and Leitão, 2011; Lim, 2012). Wild landraces of quinces that have been distributed toward West, East, and South areas are connected with their populations in the core diversity areas in Iran, Anatolia, Syria, Turkmenistan, and Afghanistan (Bell and Leitão, 2011). It seems that quinces have been migrated from the center of diversity to three or four different directions toward Central Asia, Iran, Anatolia (southern regions of the Black Sea), and East Europe (northern regions of the Black Sea) (Figure 3.2). The quince germplasm of Europe could be migrated from both northern and southern regions of the Black Sea, as in history, the name of quince genus *Cydonia* is derived from the name of a Greek city on the island of Crete, Cydonea (Sykes, 1972). This means that quinces reached Greece and later Italy and other parts of Europe via Crete Island (Anonymous, 2014). Quinces were first mentioned in Britain when Edward 1st planted four trees at the Tower of London in 1275. This fruit has also been considered as the emblem of love in many mythologies of ancient Greece and Mediterranean cultures (Phillips, 1820). The "golden apples" of ancient myths such as those featured in the 11th labor of Hercules were probably quinces (Hodgson, 2011).

In Iran, various landraces of quince with numerous seedlings have been well adapted to the semitropic and humid climate in moderate altitudes of southern regions of the Caspian Sea, and also abundant seedlings or native local cultivars are indigenous to the semi-arid regions of North-West, North-East, and central regions (Abdollahi et al. 2011; Khatamsaz 1992). Khoramdel Azad et al. (2013) by using simple sequence repeat (SSR) markers demonstrated that quince landraces of various regions of Iran are distinct according to their geographical distribution and origin. Also, several native cultivars and genotypes of quince are indigenous to such countries, including Turkey (Kuden et al. 2009; Bayazit et al. 2011), Azerbaijan (Aliyev, 1996), Greece (Ganopoulos et al. 2011), Ukraine (Yezhov et al. 2005), Czech Republic (Rop et al. 2011), Hungary (Szabó et al. 1999), Spain (Rodríguez-Guisado et al. 2009), France (Tukey, 1964), Portugal (Oliveira et al. 2007), Italy (Bellini and Giordani, 2000), Tunisia (Benzarti et al. 2015), Syria (Thompson, 1986), Turkmenistan, Afghanistan (Frantskevich, 1978), and Poland (Wojdylo et al. 2013).

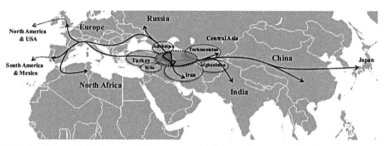

FIGURE 3.2 Hypothetical map of the center of origin and westward and eastward spread of quince genotypes and landraces from North Iran and Transcaucasian regions.

Source: Paper from Abdollahi, H. A review on history, domestication and germplasm collections of quince (Cydonia oblonga Mill.) in the world. Genet Resour Crop Evol (First Online Version) https://doi.org/10.1007/s10722-019-00769-7

Quince is tolerant to the winter frosts and 100–450 hours below 7 °C is required to break the dormancy of the dormant buds of this tree for growth and flowering in the next spring (Lim, 2012). This chilling requirement is one of the factors limiting the extension of quince cultivation in the warm condition in latitudes less than 25–30 °N and also in the tropical climate condition. Due to this limitation, majority of quince cultivations are in temperate regions of North Africa, northern half of Iran, Central Asia, Southeast China, major parts of Turkey (Figure 3.3A). In addition, long period of freezing temperatures about −26 °C and less could damage even woody parts of quince branches and trunk during winter (McMorland Hunter and Dunster, 2015). This level of tolerance limits the cultivation area of quince tree to about 50 °N latitudes, such as Turkmenistan, Uzbekistan, North Turkey, East Europe, South

Federation of Russia, South France, South Spain, and South England that have winter seasons with more tolerable freezing temperature for quince trees (Figure 3.3B–D). Comparison of built of cultivation for apple and quince trees demonstrates that due to low chilling requirement of quince, cultivation of this tree can be extended commercially in warmer regions of lower latitudes, while lower freezing tolerance of quince tree also limits its cultivation in highlands with 2000 m and higher altitudes.

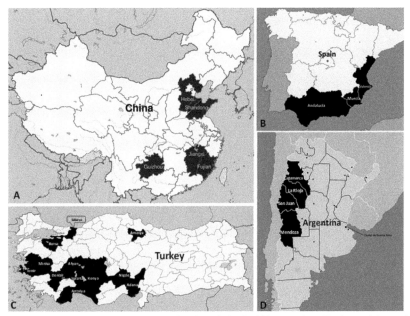

FIGURE 3.3 Major production areas of quince fruits in China, Turkey, Spain, and Argentina. (A) In China, production areas are limited to the southern regions including Guizhou, Jiangxi, and Fujian provinces where quince trees can tolerate freezing temperatures during winter seasons. Hebei and Shandong provinces in China are main producers of the pear fruit in this country, located in higher latitudes due to different climatic requirements of pear tree. (B) In Argentina, quince cultivation areas are located in foothills where the trees can get chilling requirements for blooming and bearing. (C) In Turkey, quince trees adapted to many regions, especially in the Southwest Mediterranean area. (D) In Spain, quince fruits are produced in warmer areas of southern regions, including Andalucía, Murcia, and Valencia.

3.6 BOTANY AND TAXONOMY

The genus quince, *Cydonia*, consists of a single species (Hummer and Janick, 2009). The scientific name of quince species, *C. oblonga* Mill., derived from its Greek name, Kudonian Melon, or *Cydonia* Apple, that refers to the city of Cydon in the Crete island (Meech, 1888). The previous names of this species were *Cydonia vulgaris* Pers. and *Pyrus cydonia* L. The quince tree is a member of the subtribe Malinae, tribe Maleae, subfamily Spiraeoideae (formerly *Maloideae*), family Rosaceae, order Rosales, and division Magnoliophyta (Bell and Leitão, 2011). Several subspecies or natural varieties of this species have also been identified of which two subspecies apple-shaped fruits *C. oblonga* subsp. maliformis (Mill.) Thell. and pear-shaped fruits *C. oblonga* subsp. pyriformis Medik. ex Thell.—are the most well-known subspecies (Meech, 1888). Bell and Leitão (2011) also nominated 4 subspecies, 18 varieties, and 4 forma of *C. oblonga* species, all of them are listed in *Encyclopedia of Life* (2018). *Pseudocydonia sinensis* (Thouin) (Chinese quince) developed by C. K. Schneider and *Chaenomeles* spp. (flowering quinces) are most related species to *C. oblonga* Mill., but Torkashvand (2014), by using SSR markers developed for apple and pear species, demonstrated that *P. sinensis* and *Chaenomeles* spp. are distinct from *C. oblonga*.

Wild genotypes of quince trees in Transcaucasian and Talysh forests near the Caspian Sea grow in the form of a small tree with 5–8 m in height that tends to produce numerous crown and root suckers instead of generating long shoots and branches (Figure 3.4A). The trees can have upright growing or spreading habit,

with pubescence growing in the upper third of the shoots. The color of shoots varies from dark brown to red or rarely green brown in some cultivars and genotypes (Khandan et al. 2011). The leaves and buds are tomentose when they are young but turn glabrous at maturity (Bell and Leitão, 2011). The position of vegetative buds in relation to shoots is adpressed to strongly held out (Figure 3.4B).

Blades of the leaves are complete with forms that vary from ovate to oblong, and their lengths and width are between 5 and 8 cm, and 4 and 7 cm, respectively (Alipour et al. 2014). Stipules of the leaves are caduceus, with the ovate form, and the sizes of the petioles are 0.8–1.5 cm, with the tomentose surface. Leaf veins are conspicuous on the abaxial side. The adaxial side of the leaf is glabrous when young (Figure 3.4C), and the abaxial side is pubescent. The base part of leaves is cuneate, rounded, truncate, or cordate with entire margins, and the angle of leaf apex, excluding the pointed tip, varies from acute to obtuse. Also, the undulation of leaf margins could be weak or strong (UPOV, 2003).

The single flowering shoot is 10–15 cm long with 4–5 cm diameter; the number of sepals, petals, and styles are 5. The sepals are green, and the color of petals varies from white to pink with free to overlapping margins in their arrangements.

Quince

The special and attractive pink color of quince flowers led to the nomination of color Gol-Behi in the Persian language, which especially means the color of quince bloom. The styles are free and pubescent in base. The number of stamens is 20, with yellow anthers and white to purple filaments. The abaxial part of hypanthium is tomentose, and the ovary of quince consists of five carpels (Figure 3.4D–F). Fruits of quince are botanically pome in that the outer wall develops the fruit pulp from the floral cup or hypanthium of the flower (Robinson, 2001). The color of densely villous immature fruits is either light brown or brown (Figure 3.4G). The peel of the fruits becomes pale yellow which turns to the dense golden color at maturity (Figure 3.4H and I), with pale yellow to yellow pulps. Also, the mature fruits are with persistent reflexed calyces and can weigh over 500 g or more. Postman (2009) indicated that the quince of Persia (Iran) reaches a weight of 1.5 kg, and our observations confirm this in some native cultivars of Iran, such as "Isfahan." The pedicels are short (<5 cm), thick, dark brown, and globose. The texture of mature fruits is aromatic and firm, but in some indigenous genotypes of North Iran, the fruits become extremely soft at the ripening time.

FIGURE 3.4 (A) Wild quince tree in the Talysh forest in Iran. (B) Brown color of quince shoot. (C) Young tomentose leaves of the quince tree after several days of budburst. (D) Single flower of quince with tomentose sepals and pink petals at the petal swollen stage. (E) Open pink flower of quince with yellow anthers and purple filaments. (F) Cross-section of a quince flower at the petal-swollen stage, demonstrating the tomentose hypanthium. (G) Densely villous immature fruit of quince with light brown color. (H) Quince fruit at the premature stage with pale yellow peel color. (I) Bearing habit of a domesticated quince tree with fruits along the limbs and branches.

3.7 VARIETIES AND CULTIVARS

Unlike the apple and pear production industry, in most quince producer countries, quince growers continue to produce many cultivars that have been traditional in their respective regions and cultures. This could be due to the less importance of quince cultivation than other species of temperate fruit trees. According to this, many cultivars of quinces apparently derived from the selection of the superior genotypes from native landraces and seedling populations during the past centuries. This process presumably in part has occurred during migration or transfer of quince germplasm from the center of diversity in the Transcaucasian region to its area of cultivation. Due to this, no precise history of development and release has been presented for most of quince cultivars.

In the countries on the center of diversity of quinces, a common commercial cultivar is not used in orchards, and in each country including Iran, Turkey, Azerbaijan, and Armenia, the producers prefer to use indigenous historic cultivars of the same country. In Iran, the first collection and evaluation of native cultivars between 1940 and 1950 determined that several native local cultivars including "Isfahan" (synonyms are: "Esfahan," "Gorton Esfahan," or "Gourton Isfahan,"

maybe the famous "Quince of Persia"), "Shams," "Neyshabour," and "Torsh" (synonym "Beh Torsh") are the most important cultivars of the main regions of quince production. Following the fire blight outbreak in Iran about the 1990s (Mazarei et al. 1994), especially in north and northwestern Iran, famous cultivar "Isfahan" became the most commercial and promising quince (Figure 3.5A–D).

Native quince cultivars of Turkey are "Ekmek," "Altin Ayva," "Cengelköy," '"Kalecik," "Limon," "Istanbul," "Yerli" "Bardacık," "Şeker," "Tekkes," and "Eşme." Kuden et al. (2009) reported that four cultivars "Bardakcik," "Limon," "Istanbul," and "Seker" bear the biggest fruits among quinces of Turkey. Also, the most popular quince cultivars in various regions of Turkey include "Ekmek," "Limon," and "Eşme" (Bayazit et al., 2011). Also, the quince cultivar "Smyrna" is another well-known cultivar from Turkey (Bassil et al., 2015). Dumanoglu et al. (2009), by using SSR markers, demonstrated distinct variation among clones of the "Kalecik" quince cultivar in Turkey.

There is not much available information about the quince cultivars in China, but pieces of evidence show that the quince production of China is based on one of the oriental flowering quince species that could be *Chaenomeles speciosa* (*C. speciosa*). This species has been cultivated widely in Southeast China for several centuries and also is native to the Eastern Asia. *C. speciosa* is also a traditional herb in the traditional Chinese medicine (Tang et al., 2010; Simoons, 1990). In addition, Uzbekistan is currently one of the major worldwide producers of quince fruits, and according to the National Clonal Germplasm Repository of USDA at Corvallis, Oregon, the cultivar "Tashkent" is the commercial quince of this country (NCGR-Corvallis, 2017).

Commercial quince cultivars of the countries that geographically are on the center of origin of *C. oblonga* could be interested both in breeding and germplasm conservation programs. In the Republic of Azerbaijan, indigenous folk quince cultivars are "Sary-aye," "Kara-ayva," and "Chilyachi" that grow all over this territory. The yield of the Rajably-1 and Rajably-2 cultivars is superior to that of the indigenous ones (Aliyev, 1996) (Figure 3.5E and F). The scion quince cultivars listed in the USDA germplasm repository at Corvallis, Oregon, and originated from Armenia are "Alema," "Arak-seni," "Megri" and "Seghani." Similarly, "Babaneuri" and "Dusheti" are quince cultivars from Georgia; cultivars "Kichikaradede," "Yuz-Begi," "Miradzhi," and "Zeakli" are from Turkmenistan; and the cultivar "Kashenko" is from Ukraine listed in the USDA germplasm repository at Corvallis, Oregon. The quince cultivar "Kaunching" is also a popular cultivar in orchards throughout Central Asia (NCGR-Corvallis, 2017). Other well-known quince cultivars are "Bereczcki," a Serbian/Hungarian cultivar; "Vranja," a Serbian cultivar (Figure 3.5G) (McMorland Hunter and Dunster, 2015); "Portugal" (syn. "Lusitanica") is an early ripening Portuguese quince and the same cultivar named "Gamboa" in Puerto Rico (NCGR-Corvallis, 2017; Bassil et al. 2015); and "Orange" (syn. "Apple quince") is a group rather than a distinct cultivar, and its origin is uncertain. Many orange- or apple-shaped quinces came to be called "Orange" or "Apple," and they were thought to come true from seed (NCGR-Corvallis, 2017). The origin of "Meech's Prolific" cultivar according to Bassil et al. (2015) is the USA, but it has been donated from the United Kingdom (NCGR-Corvallis, 2017). Several historic quince cultivars of France are "Le Bourgeaut," "Fontenay," and "Angers." Some of them have been selected as dwarfing clonal rootstocks for intensive pear cultivation in the East Malling Research Station (Tukey, 1964). According to some literatures, "Vranja" and "Bereczcki" are synonyms of quince cultivars, but the International Union for the Protection of New Varieties of Plants indexed them as distinct cultivars by morphological characteristics (UPOV, 2003). Also, according to the donor information lists, the origin of cultivar "Vranja" is Iran (NCGR-Corvallis, 2017).

During Colonial era in USA, a quince tree was rarely frequent in the gardens of middle classes (Postman, 2009). The development of the first quince cultivars in the USA (in California) is credited to Luther Burbank who released two cultivars "Pineapple" and "Van Deman" in the 1890s, which are important commercial cultivars (Postman, 2009) (Figure 3.5H). Other quince cultivars that originated in the USA are "Champion" (released in 1870), "Cooke's Jumbo" (syn. "Jumbo") (released in 1972), and "Rea's Mammoth" (released in 1800).

Due to a high level of similarity, it is difficult to classify quince species. Therefore, primary distinctness of the commercial cultivars is possible based on the fruit shape, as it was categorized as apple shaped (*C. oblonga* subsp. maliformis) and pear shaped (*C. oblonga* subsp. pyriformis). Also, in this species, the morphology of fruit is controlled by numerous factors such as the percentage of the fruit set and the number of seeds per fruit, and also by the environmental condition (Özbek 1978). For distinctness of quince cultivars, the International Union for the Protection of New Varieties of Plants released the guidelines to index them by both morphological and molecular markers (UPOV, 2003; Kimura et al. 2005). The national descriptor guideline for distinctness of quinces of Iran was also released by Khandan et al. (2011) with more details on the fruit and tree characteristics.

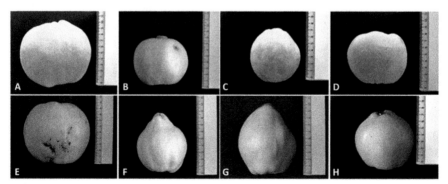

FIGURE 3.5 Forms of the quince fruits in some cultivars: (A) cultivar "Isfahan" (syn. "Esfahan"); (B) cultivar "Shams" (syn. "Shams Esfahan"); (C) cultivar "Torsh" (syn. ''Sour"); (D) new released cultivar "Viduja" (all four cultivars from Iran); (E) cultivar "Rajabli-1" (from Azerbaijan); (F) cultivar "Pyriformis" (from Caucasus); (G) cultivar "Vranja" (from Serbia); and (H) cultivar "Pineapple" (from the USA).

3.8 BREEDING AND CROP IMPROVEMENT

Unlike apples and pear trees, few attempts have been made to breed quince cultivars by using selection and in particular hybridization methods. Therefore, it seems that many quince cultivars used in different countries have been superior genotypes that were selected from indigenous landraces or local seedlings by growers and gradually cultivated and developed as commercial cultivators. Many old cultivars originated from countries located on the center of origin of quince such as the cultivars "Isfahan" and "Shams" (originated in Iran); the cultivars "Ekmek," "Şeker," "Limon," "Bardacık," "Tekkes," and "Eşme" (of Turkish origin); the cultivars "Sary-aye," "Rajably-1," and "Rajably-2" (of Azeri origin) are examples of these old native cultivars.

In order to implement modern breeding programs for quince trees, the first step could be the collection and evaluation of genetic potential and traits in the selected germplasm. Olivier de Serres (1539–1619 AC) was a French author and soil scientist who called attention to use the quinces as a useful rootstock for pears. For several centuries, French nurseries of the Angers and Orleans regions were the main producers of quince rootstocks, such as "Angers," "Portugal," and "Fontenay" for the United Kingdom and USA. Due to apparent confusion among quince rootstocks of French nurseries, R. G. Hatton at the East Malling Research Station in 1914 collected various quince rootstocks from 14 nurseries and nominated them with letters, listed as Quince A to Quince G. Among them, just Quince A, B, and C were significantly commercialized (Tukey, 1964).

Other quince collections in other parts of the world include the quince collection of Norton Priory in the United Kingdom established in the 1920s, with cultivars from Balkans, Turkey, Iran, Serbia, and Russia (Hodgson, 2011); quince collection of NCGR-USDA in Corvallis, Oregon, established in the 1980s (Postman, 2008; McGinnis, 2007). Now, this collection preserves 174,23 seed lots and clonal accessions of various genus including *Cydonia*, *Chaenomeles*, and *Pseudocydonia* (Bassil et al. 2015); the quince collection of Izmir, Turkey, established in 1964, includes many regionally developed fruit cultivars and landraces (Sykes, 1972); the quince collection of Kara Kala in Turkmenistan, which was indicated by Postman (2009) as an important collection of fruit trees in the region of Kara Kala, was part of the Vavilov Institutes during Soviet times. Some of quince accessions of this collection have been rescued in the late 1990s and brought to the NCGR-USDA collection in Corvallis Oregon; the quince collection of Iran, established in 2004 (Abdollahi et al., 2011), now contains more than 52 cultivars and accessions from various regions of Iran and Caucasus; the quince collection of Nikita Botanical Gardens, established in 1940 by Klavidya D. Dorogobuzhina, now contains more than 200 accessions, and six commercial cultivars have been developed by this collection (Yezhov et al., 2005). Smaller quince collections are growing in Italy, at the University of Pisa, with 50 accessions, mainly from the Tuscany region (Turchi, 1999); also 21 accessions in the University of Basilicata, in Italy; 14 accessions in Istituto Sperimentale per la Frutticoltura (Bellini and Giordani, 1999); 50 accessions in the National Agricultural Research Foundation, Pomology Institute (NAGREF-PI) of Greece (Thomidis et al., 2004); 19 cultivars in the National Fruit collection of the University of Reading (University of Reading, 2017); quince collection of Research Station of Újfehértó in Hungary with 55 accessions and hybrids (Maggioni et al., 2002); quince collection of Mendel University in Brno, Czech Republic with 22 accessions (Rop et al., 2011); and finally, old reported quince germplasm collection of the University of Florence in Italy with about 10 well-known quince cultivars (Scaramuzzi, 1957).

Undoubtedly, collections of quince germplasm are considered as the first step in the modern breeding of quince cultivars. The literature review demonstrates that several of the above-listed quince collections have been considered as a genetic resource base for the evaluation of genetic potential, selection of promising genotypes, and breeding new cultivars by means of hybridization schedules. In the quince collection of NCGR-USDA, about 41 out of 119 accessions represent edible cultivars for fruit production; and the others are rootstock selections, seedlings, and wild genotypes (Postman, 2008). According to a review of Postman (2008) on the potential of fire blight resistance of NCGR quince collection, resistant cultivar(s) or rootstock(s) was not noted in any *Cydonia* clones, as the quince cultivars are as susceptible to the invasion of *Erwinia amylovora* (fire blight causal agent) as the most susceptible cultivars of apple and pear trees. Also, the evaluation of susceptibility in quince germplasm of Iran under both greenhouse and orchard conditions (Abdollahi and Akbari Mehr, 2008; Mehrabipour et al., 2010; Ghahremani et al., 2014) has confirmed similar results to Postman (2008). Ghahremani et al. (2014) reported that the USDA fire blight scoring system used for evaluating the degree of fire blight in the apple and pear trees is not similarly applicable in quince germplasm. Therefore, a combination of fire blight indexes including the USDA fire blight scoring system, besides the index of varietal susceptibility and the index of frequency, could be more appropriate for the evaluation of fire blight susceptibility in quince. Bobev et al. (2009) indicated to 18 hybridization progenies of quinces in Bulgaria that exhibited a significant level of resistance to fire blight during the epidemic years 2003 and 2005. The experiences on fire blight damages in outbreak years demonstrate that the behavior of young trees in new established orchards could be different from the trees in the mature phase; therefore, complementary evaluation needs to confirm a high level of fire blight resistance in the selected genotypes.

McMorland Hunter and Dunster (2015) suggested that quince cultivar named "Iranian quince" has fruits sweeter and less gritty than many of the other cultivars. This indication suggests that breeding for fresh use or higher fruit quality could be a breeding objective with the germplasm of the southern region of the Caspian Sea. Abdollahi et al. (2018) demonstrated that various quince landraces of Northwest, Northeast, North, and Central Iran demonstrate various fruit qualities for fresh use and the highest scores

of organoleptic tests obtained from quince genotypes of Northeest and Central Iran. Stancevic and Nikolic (1992) mentioned to quince breeding program in Ex-Yugoslavia (currently in Serbia) by hybridization and release of new cultivar "Morava" derived from cross between "Leskovacka" and "Rea's Mammoth" in 1968 with medium to large fruits, and juicy, subacidic, and yellowish flesh and pleasant fruit aroma. Yezhov et al. (2005) reported that among more than 200 quince accessions in Nikita Botanical Gardens, 6 commercially important cultivars have been developed and released. Quince cultivars "Meech's Prolific," "Missouri Mammoth," "Rea's Mammoth," and "Fuller" are other cultivars raised in the USA from selected seedlings for edible purpose before years 1880 (Meech, 1888).

Quince trees are also very susceptible iron chlorosis in alkaline soils and show characteristics of yellowing in the upper leave of the shoots that generally becomes more severe about mid-summer (Alcántara et al. 2012). The leaf chlorosis reduces the bearing efficiency of quince cultivars, and for this reason, is one of the considerable breeding objectives in this species and in the case of the application of quinces as dwarfing rootstocks for pears. Ahmadi and Abdollahi (2018) screened germplasm of quince collection of Iran for evaluating the tolerance to lime-induced iron chlorosis and selected some more tolerant genotypes. Also, Abdollahi et al. (2010) showed various effects of hawthorn (*Crataegus atrosanguinea* Pojark.) and quince seedling rootstocks on the tolerance of quince cultivars and genotypes to iron chlorosis. Alcántara et al. (2012) also selected two quince clones with a higher tolerance level to iron chlorosis among 13 clones of Spain. Baninasab et al. (2015) reported a higher tolerance level of hawthorn (*C. atrosanguinea* Pojark.) to alkaline soils, followed by Quince A and seedling quince rootstocks. In another work, Ghasemi et al. (2010) demonstrated that Quince A and PQBA29 had the highest tolerance to ammonia bicarbonate in irrigation water among major commercial clonal quince rootstocks.

According to the Food and Agriculture Organization of the United Nations (2017), the global average yield of quince production is about 7.8 metric tons/ha, which is significantly less than other pome fruits including apples and pears. Alipour et al. (2014) demonstrated that growth habits of the quince genotypes originated from various regions of Iran are different and vary from extremely upright with scarce spur spread along the tree limbs to spreading form with abundant spurs along the limb. Normally, the genotypes with spreading growth habit demonstrated higher yield and bearing potentials. This means that breeding of productive cultivars could be another objective in a quince tree that has been ignored yet. Resistance to Fabraea leaf and fruit spot (*Fabraea maculata* Atk.) and a range of ripening seasons have been other minor breeding objectives of quinces that were considered recently in the NCGR-USDA collection (Hummer et al., 2012).

Over the past two decades, the quince collection of Iran has been playing the role of as one of the active quince collections, and numerous evaluations on genetic relation (Khoramdel Azad et al., 2013); fire blight susceptibility (Abdollahi and Akbari Mehr, 2008; Mehrabipour et al. 2010; Ghahremani et al. 2014), growth habit, and bearing (Alipour et al. 2014) resulted in a deeper understanding of breeding potentials of quince landraces from various regions of Iran and selection of promising and release of first quince cultivar "Viduja" with a high yield potential (SPII, 2015).

Numerous attempts have been made for using new techniques, especially using somaclonal variants obtained from *in vitro* regeneration and embryogenesis in breeding of new quince rootstocks. The principal objective of these attempts has been the *in vitro* establishment of the quinces (Khosravinezhad et al., 2016), evaluation of regeneration ability (Dolcet-Sanjuan et al., 1991; Antonelli, 1995; Aygun and Dumanoglu, 2007) and factors affecting this ability (D'Onofrio and Morini, 2005; Baker and Bhatia, 1993; Morini et al., 2004), induction of somatic embryogenesis (D'Onofrio et al. 1998; Fisichella and Morini, 2003; D'Onofrio and Morini, 2006) and caulogenesis (Chartier-Hollis, 1993; Morini et al., 2000), and *in vitro* screening of somaclones (Marino and Molendini, 2005); however, none of these somaclones have been released and used commercially practically for quince production.

3.9 SOIL AND CLIMATE

Quinces come from the foothills of the North Iran and Transcaucasian regions, the regions with long, hot summers with maximum temperature between 30 °C and 40 °C and harsh winters (McMorland Hunter and Dunster, 2015). This means that wild genotypes that domesticated cultivars derived from them are trees tolerant to high summer temperatures and also to subzero temperatures, as low as −26 °C. The comparison of the tolerance of subzero temperatures in quince trees with pears and apples shows less tolerance of quince trees to low temperatures. Agrolib (2010) explained that quince buds, woods, and roots tolerant limits to subzero temperatures are about −23 °C, −27 °C, and −12 °C, respectively. Sola (2011) indicated to the adaptation of quince cultivars to 7a, 7b, 8a, and 8b temperature zones and that the minimum winter temperature in these zones are −15 °C to −17.7 °C, −12.3 °C to −14.9 °C, −9.5 °C to −12.2 °C, and −6.7 °C to −9.4 °C, respectively. All this suggests that the quince trees are not appropriate for mountainous altitudes above 2000 m.

The chilling requirement (temperatures below 4 °C–7 °C) for the removal of endodormancy and bud break in quinces is relatively low, ranging from 100 to 500 h (Bell and Leitão, 2011; Lim, 2012). If the trees do not get this chilling requirement, the blooming will be poor and the bud break will delay for several weeks, and both of them lead to noneconomic bearing of the orchard trees. In addition, quince trees bloom later in the spring than apples and pears, and for this reason, frost is not usually a problem in this species (McMorland Hunter and Dunster, 2015). Fruits of most quince cultivars ripe in autumn, and early frosts could be harmful for fruits on the trees. For this reason, it is important to avoid the cultivation of this tree in areas with an inadequate length of growing season and with early chilling risks in autumn.

Quince is a shallow-rooted species and needs rich, moist soil with appropriate drainage, but it is also adequately tolerant to grow well on a range of soils. In most of quince producer countries, quince clonal or seedling rootstocks are used for its cultivation. In addition to quince rootstocks, hawthorn seedlings from the *C. atrosanguinea* Pojark. species are used in Iran that is considered a very tolerant rootstock to an unfavorable condition in soils of semi-arid areas, such as drought, salinity, and alkalinity. Quince trees and rootstocks are moderately to highly tolerant to low pH in soils, while its cultivation in high pH of alkaline or calcareous soils causes leaf chlorosis due to the poor uptake of iron (Bell and Leitão, 2011). This chlorosis is most commonly found on young leaves of quince trees and emerges from mid-summer. These two conditions indicate that the ideal soil pH for quince cultivation and growth is between 6.5 and 7.5.

3.10 PROPAGATION AND ROOTSTOCK

Propagation of the quince trees and rootstocks could be done both by sexual and asexual methods. In sexual methods, the seeds are the means of propagation, but the segregation of the traits leads to the diversity of the grown seedlings. This method is merely used in breeding programs and/or in the production of seedling rootstocks. Two quince

(*C. oblonga* Mill.) and hawthorn (*C. atrosanguinea* Pojark.) species are used for the production of seedling rootstocks in nurseries (Figure 3.6A–C). The later seedling rootstock (hawthorn) is only common in some parts of Iran, and the rootstocks obtained from these seedlings are more tolerant than quince seedling rootstocks to undesirable soil and water conditions. Seed germination in *Crataegus* spp. species occurs very late and often after two growth seasons from the harvest of fruits from the trees. Therefore, in nurseries, usually hawthorn seeds are cultivated in a waiting plot, and after seedling growth, they are transplanted to the propagation plots for budding (Figure 3.6D and F).

Other than quince and hawthorn seedlings, clonal rootstocks such as East Malling series, including Quince A and Quince C and other clonal rootstocks such as Quince BA29 and Quince Sydo® can be used for the propagation of commercial quince trees. Quince is an easy-rooting species and therefore the propagation of clonal rootstocks can be performed simply by hardwood and softwood cuttings or by layering (Duron et al., 1989). Quince A is the most common rootstock in many European countries the helps the tree to grow about 4.5 m in height and is categorized as the semivigorous stock. Quince C is a dwarfing rootstock, the helps trees to reach about 3 m in height (McMorland Hunter and Dunster, 2015), but this rootstock is not commercially as important as Quince A. Quince BA29 and Quince Sydo® both are slightly less dwarfing and more tolerant to contamination by different viruses than Quince A (Duron et al. 1989). Additionally, some commercial quince cultivars with upright growth habit on Quince A rootstock tend to produce fewer suckers on the limbs and over the tree canopies. This effect leads to easier training of quince cultivars, production of more spurs, and increase the efficiency of the orchard for fruit production (Figure 3.7A and B).

FIGURE 3.6 (A) Red fruits of hawthorn species *Crataegus atrosanguinea* Pojark. (B) Seeds of *C. atrosanguinea* for production of seedling rootstocks for quince. (C) Seedlings of *C. atrosanguinea* in a waiting plot. (D) Contrast of rootstocks/scions in propagated quince cultivar "Isfahan" on hawthorn rootstock. (E) Form of graft union in grafted quince cultivar "Isfahan" on hawthorn rootstock. (F) Semidwarfing effects of hawthorn rootstock on cultivar "Isfahan".

FIGURE 3.7 (A) Improvement of the growth habit of the commercial quince cultivar "Isfahan" by growth on Quince A clonal rootstock in Fars Province of Iran. (B) Modified leader training form of the commercial quince cultivar "Isfahan" on Quince A clonal rootstock in the same orchard. (Photos courtesy D. Atashkar, Horticultural Research Institute of Iran).

Moreover, conventional propagation methods, in vitro methods, have also been considered for the propagation of quince rootstocks and cultivars. First attempts for in vitro propagation of quinces maybe refer to Németh (1979) that used shoot tip culture for the multiplication and micropropagation of rootstock Quince BA29. In this report, effective multiplication and rooting of Quince BA29 were achieved on media enriched by N^6-benzyladenine (BA) and indole-3-butyric acid (IBA). Al Maarri et al. (1986) described the method for the mass production of Provence Quince through multiplication and shootlets elongation on QL (Quoirin and Lepoivre, 1977) media enriched by 2 mg/l BA and rooting on MSJ (Murashige and Skoog (1962), but with the NaEDTA and FeNaSO$_4$ replaced by FeNaEDTA at 20 mg/l) media containing 0.1–1 mg/l naphthaleneacetic acid (NAA). Giorgota et al. (2009) developed the method for in vitro propagation of "BN 70" and "A Type" quince rootstocks and "Aurii" cultivar and reported abundant multiplication of the quinces on QL and successful rooting on half-strength MS with Lee-Fossard vitamins and 1 mg/l IBA. Vinterhalter and Neškovlć (1992) demonstrated that double phase liquid–solid media are led to a higher average of the multiplication rate of quinces. But by using this method, the percentage of rooting in *in vitro* micro-cuttings was not higher than 35%. Khosravinezhad et al. (2016) confirmed more appropriateness of the mQL medium (Leblay et al. 1991) enriched by BA and N6-Δ2-isopentenyl adenine (2iP) for the micropropagation of Quince C rootstock and several promising quince genotypes of Iran.

As a practical use of in vitro culture techniques for the micropropagation of quince rootstocks and cultivars, this technique now has limited convenience for mass propagation in the commercial nurseries. But contrarily these are the useful methods for the production of virus-free prebasic and basic materials, as well as for the mass propagation of certificated pathogen-tested quince materials for further use in the nurseries (EPPO, 1999).

3.11 LAYOUT AND PLANTING

Orchard planting systems are the combination of layout and management. These two elements of orchard planting systems are variables that improve orchard production efficiency. Layout choices for the establishment of a commercial quince orchard are not as comprehensive as other species of pome fruits, such as apple and pear. High-density orchards for the commercial production of quince have not been developed, and most of suitable rootstocks for this species in main producer countries are semidwarfing. Also, most of quince cultivars are semivigorous, and the propagation of these cultivars on commercial semidwarfing quince or hawthorn (belong to *C. atrosanguinea* Pojark.) rootstocks produces trees whose heights do not exceed 4–4.5 m in the orchards. For these reasons, the most adaptable density for quince orchards is about 500 trees/ha maximum to a few less than 1000 trees/ha (Figure 3.8A–C). In these types of orchards, no supporting and wires are necessary due to appropriate anchorage of root systems in the soil.

FIGURE 3.8 (A) A semidwarfing quince orchard in the Ispartan region of Turkey, established by cultivar "Bardacik" on Quince A (photo courtesy of M. Mortaheb). (B) A semidwarfing quince orchard in the Isfahan region of Iran, established by cultivar "Isfahan" on hawthorn *Crataegus atrosanguinea* Pojark. (C) Research project on the comparison of growth and bearing of the quince cultivar "Isfahan" on various rootstocks, including Quince BA29, Quince A, Quince B, Quince C, and hawthorn *C. atrosanguinea* in the Isfahan region of Iran and horizontal palmette training of the trees by a central trunk with three horizontal scaffold branches trained along the wires (photo courtesy of A. Ghasemi).

The formation of a balance between vegetative growth and cropping is an important and considerable objective in quince orchards as some of the commercial cultivars tend to generate long spur-free or low spur density scaffold branches that lead to low bearing. Several orchard-management operations including summer pruning, use of an appropriate modified leader training system for orchard, and the adjustment of tree scaffolds angles to about 40–45° are effective in the formation of the vegetative/bearing balance of the quince trees. In addition, the evaluation of quince germplasm of the south region of the Caspian Sea demonstrated that a new breeding potential for quince trees in the form of high spur density along the scaffolds and tree limbs is existing in the local germplasm of some parts of Iran (Alipour et al. 2014). According to this, new released quince cultivar "Viduja" seems to be more appropriate for establishment of semidwarfing orchards with high bearing potential as high as 2030 tons/ha and low alternative bearing (SPII, 2015).

In addition to the selection of quince cultivar/rootstock combination for establishment, a new quince orchard, it is also important to assess the soil condition and its deficiencies. Our experience demonstrates that it is almost impossible to correct major soil limitations and deficiencies in any existing orchards. Preparatory soil-management and soil-restoration measures are fundamental for an efficient soil-management strategy. Therefore, one year prior to the establishment of a new quince orchard, the condition of the soil, especially in the alkaline terrains or waterlogged areas, should be assessed carefully so that corrective measures can be taken on time. At this stage, soil improvements can be carried out in the most efficient way. Approaches that can be taken include breaking up a hardpan layer, addition of manure and organic materials to the soil, establishment of a drainage system, and reduction of the soil pH in alkaline soils. All of these are necessary soil-management operations before cultivation and establishment of quince trees in an orchard.

Milić et al. (2010) reviewed economic aspects of quince plantation and production in the Republic of Serbia and estimated that in a quince orchard with 833 trees/ha, full fruitfulness of orchard after five years, and average of 35 tons/ha, cost estimate of soil preparation for setting up the plantation is about 24.5% and the cost of planting trees is about 41.1%.

3.12 IRRIGATION

Quince is a temperate fruit tree that is cultivated under a range of environmental conditions—from humid regions with high rainfall to semi-arid regions with mean annual precipitation (<300 mm). In all regions, quince trees prefer damp soils without a waterlogged condition. Therefore, a regular irrigation even in the regions with high rainfall is essential to optimize tree growth and development at the first year, and also to support early and abundant fruit bearing with the optimum size of the fruits. Irrigation of quince trees should also be available from the start of the first growing season, or as supplementary irrigation in dry periods (Baxter, 1997; McMorland Hunter and Dunster, 2015).

Various systems or methods of irrigation including free flooding, furrow irrigation, and basin and trickle/drip irrigation have been used for quince orchards, but now the trickle irrigation is considered as the most appropriate and economic method in many producer countries. In surface irrigation systems that root collar or graft union part of the quince trees remain in a direct contact with irrigation water for a long period, and the crown and root rot disease lead to the catastrophic loss of trees. Therefore, in conventional planting systems for quince trees, furrow irrigation, avoiding collar wetting and waterlogged of root systems, is preferred to direct flooding. Currently, trickle irrigation is the most preferred system for the quince orchard that in arid regions prepare major part of water necessity of the trees through several drippers around the trees (Figure 3.7A and B) and in humid regions with high rainfall; this system can be used for supplementary irrigation. In many quince production orchards of Iran, annual water requirement from April to late October is between 3000 and 6000 m^3/ha, depending on the mean seasonal temperature, latitude, and the evapotranspiration level. In humid regions with abundant and evenly distributed summer precipitation, the annual water requirement of quince orchards could be even less than 3000 m^3/ha. Obviously, most of annual water requirement belong to July, August, and finally early September, when cell elongation in the growing fruits needs more water. Therefore, not only the application of sufficient water throughout the growth season is necessary to maintain steady fruit growth, but also more consideration to the availability of water prior to the enlargement of the fruits has a significant importance.

According to the classification of relative tolerances of fruit plants to saline irrigation waters by Hart (1974), quince trees with other pome fruits such as apples and pears are among the fruit trees that can tolerate the irrigation waters with EC between 800 and 2300 μs/cm at 25 °C (Class 3) that is equal to 500–1000 mg/l of the total soluble solids (TSS) index. Both above-mentioned indexes of quantity and quality of irrigation water for quince orchards are true when the trees grown on quince seedling or clonal rootstocks. As per the observations, quince cultivars on hawthorn *C. atrosanguinea* Pojark. seedlings demonstrated a superior level of tolerance to the drought and salinity stresses. On the use of the hawthorn rootstock in semi-arid regions of Iran, it has been observed that the quince tree irrigation cycle can be prolonged to every 10 days during summer on this rootstock, whereas an irrigation cycle between three and seven days for quince rootstocks usually seems to be adequate at the same climatic condition.

High bicarbonate (HCO_3^-) concentration is naturally found in high pH waters, and the continued use of high bicarbonate water leads to a high soil pH and induces iron chlorosis on the leaves of quince trees. Ghasemi et al. (2010) demonstrated that Quince A and Quince BA29 rootstocks had the highest tolerance to ammonium bicarbonate in irrigation water among major commercial clonal quince rootstocks. Also, hawthorn *C. atrosanguinea* Pojark. seedlings are more tolerant than Quince A and Quince BA29 to high bicarbonate concentration in the irrigation water (Ghasemi, personal communications).

3.13 NUTRIENT MANAGEMENT

Quince trees, similar to other plants, absorb carbon, hydrogen, and oxygen from air (atmosphere) or irrigation water, but they absorb other 14 nutrient elements, including two major groups of macro- and micronutrients, from soil or nutrient solutions. The nutritional requirements of quince trees are not significantly different from other pome fruits. Therefore, these requirements are often categorized as the nutritional requirements of pome fruit trees (Ebert and Kirkby, 2009) (Table 2). The data of Table 2 show that nutritional requirements of quince trees for K, Mg, and B are moderately more than other two main species of pome fruits, apples and pears.

TABLE 3.2 Critical Concentrations of Some Macro- and Micronutrients in the Leaves of Pome Fruits for Diagnosing Fertilizer Needs

	Macronutrients (%)					Micronutrients (ppm)				References
	N	P	K	Ca	Mg	B	Zn	Mn	Cu	
Apples	2.30	0.23	1.50	1.40	0.41	42	30	98	23	Hagin and Tucker (1982)
Pears	2.30	0.18	1.50	1.80	0.33	45	50	50	20	Ebert and Kirkby (2009) Mitcham and Elkins (2007)
Quinces	2.67	0.11	2.69	1.56	0.74	57	45	43	25	Mirabdulbaghi and Abdollahi (2014) Serra (2009)

Quince trees are susceptible to iron chlorosis and show typical and progressive symptoms of yellowing in young leave (Figure 3.9) (Alcántara et al., 2012). The appearance of the leaf chlorosis in quinces is associated with the growth of trees in calcareous alkaline soils. In this soil, both high HCO_3^- concentrations and nitrate rather than ammonium are involved in the induction of leaf chlorosis (Mengel, 1994). Mirabdulbaghi and Abdollahi (2014) reported that total contents of iron in the leaves of quince cultivars and genotypes vary from 18.15 to 57.45 ppm (mg/kg) of leaf dry matter, and the mean Fe content is 33.50 ppm (mg/kg) of the leaf dry matter. Ahmadi and Abdollahi (2018) reported a higher amount of total Fe in quince genotypes and demonstrated that similar to other plants and fruit trees, the analysis of leaf total Fe does not provide any clue of Fe deficiency. This phenomenon is essentially due to the inactivation and homeostasis of iron at the subcellular level in the plant cells (Römheld and Nikolic, 2007). It has been demonstrated that in many fruit trees, such as peaches, the evaluation of the active Fe contents of the leaves provides a more reliable clue for the Fe nutritional condition (Çelik and Katkat, 2007). Due to the susceptibility of quince cultivars and rootstock to iron chlorosis, this tree has been the subject of several breeding objectives for the selection and/or improvement of the level of the iron chlorosis level both by the conventional (Ahmadi and Abdollahi, 2018; Mirabdulbaghi and Abdollahi, 2014) methods and modern techniques, such as somoclonal variation (Bunnag et al., 1996; Cinelli et al., 2004; D'Onofrio and Morini, 2002a,b; Fisichella et al., 2000; Muleo et al., 2002). Anyway, soil (by Fe-ethylene diamine(di-*O*-hydroxyphenyl)acetate) or foliar (by Fe-ethylenediaminetetraacetate) application of Fe chelates before the appearance of the irreversible symptoms of chlorosis are one of the main nutritional management issues of the quince orchards (Figure 3.9).

FIGURE 3.9 Progressive symptoms of leaf chlorosis caused by iron deficiency due to the growth of quince tree in alkaline soil (high pH level). From the stage 4 to the next, the chlorosis symptoms of iron deficiency are irreversible on the leaves and use of iron fertilizers does not lead to the disappearance of the chlorosis/necrosis symptoms.

3.14 TRAINING AND PRUNING

In quince trees, the flowering occurs singularly on the apices of a current season shootlet. The flowering shootlets of the quince tree are dispersed on various parts of scaffolds with various ages. This means that the quince trees bloom on spur-like structures on old parts of scaffolds; as well on one-year-old shoots (Figure 3.10A and B). According to this flowering and bearing habit, it could be considered that the bearing habit of quince trees is completely different from that of apple and pear trees. Additionally, it is believed that the beginning of the flower induction or initiation in this species is in the early spring, and the flowers born on the terminal part of the current growth shootlet. Therefore, the season of full bloom and the season of flower initiation both are spring (Vossen and Silva, 2015). Esumi et al. (2007a, b) demonstrated that in quince (*C. oblonga* Mill.) floral differentiation is initiated from late October to November and after that eight leaf primordia is initiated. In this stage, apical meristem is transformed to a dome-like structure and initiates sepal primordia. Auxiliary meristems of bracts or leaf primordia in this species remain undifferentiated during winter, and then in the next spring differentiates to flower meristem.

FIGURE 3.10 Singular flowering of quince tree occurs on the apices of a current season shootlet in one-year-old shoots, (A) the flowering current season shootlets and swollen buds are clear in this part of the tree and (B) spur-like structures in older parts of quince tree such as scaffolds.

This special pattern of flower bud initiation and flowering on both spur-like structures in old parts of scaffolds and on one-year-old shoots result in the appropriateness of a training form with the minimum number of cuttings for the early bearing of quince cultivars. This means that at the first years of tree growth, the flowers developed on one-year-old shoots are the main fruit bearing flowers, while after several years, both these flowers and flowers that differentiate on the spur-like structures are important for the bearing of fruit trees. According to this argument and semidwarfing growth habit of most quince cultivars and rootstocks, it is believed that a modified leader training formation with the use of ties, stick spreaders, weights, or string to a nail in the trunk is an effective method to adjust the angles of limbs/scaffolds in young quince trees (Figure 3.11A and B) and stimulation of early and high bearing (Figure 3.11C and D). Experience demonstrated that the use of stick spreaders is more effective and applied than other methods for adjusting the angles of the main branches of quince tree in commercial orchards. Other forms of training of the pome fruits including open center, central leader, and various forms of espalier training systems do not look like to be appropriate for quince trees. In relation to the central leader training systems, the lack of a suitable high dwarfing rootstock for quince cultivars is the main reason for not using this system commercially. In the case of espalier systems, the growth of numerous shoots and water sprouts in the bending site of scaffolds on the wires seems to be the main reason that makes these training systems ineffective for quince orchards.

FIGURE 3.11 (A) Schematic view of the use of stick spreaders and/or string of primary scaffolds to a nail in the trunk as an effective method to adjust the angles of scaffolds in young quince trees. (B) Real view of the use of stick spreaders to adjust the angles of primary scaffolds in young quince trees cultivar "Viduja." (C) Growth of main scaffolds with 40°–45° angles on the trunk in a young quince tree cultivar "Viduja" following the use of the stick spreaders and modified leader training system. (D) Effective bearing of a five-year-old tree (promising Iranian quince cultivar) following the modified leader training system.

Observation on the flowering and bearing habit of various quince genotypes and cultivars shows that some of them considerably tend to bloom and bear on spur-like structures, whereas some genotypes bloom principally on the one-year-old shoots. Also, there are some minor genotypes and cultivars that tend to bloom and set the fruits on both the above-mentioned structures (Figure 3.12A–C). The first group of genotypes demonstrates a pendulate/spreading form, whereas the second group remains upright when the fruits grow on the shoots.

The third group of genotypes with bearing habit on both spur-like structures and one-year-old shoots is usually very productive with a high yield potential that fruits set on both inner parts of canopy where spur-like structures form and outer parts of canopy where one-year-old shoots bloom and bear.

FIGURE 3.12 Comparison of bearing habit in various quince genotypes including quince genotypes with bearing habit on a one-year-old shoot of outer parts of the tree. (A) In this bearing habit, following the growth of fruits, the tree becomes pendulate. (B) Quince genotypes with bearing habit on spur-like structures of inner parts of the tree. (C) Productive quince genotype with bearing habit both on the spur-like structures of inner parts and a one-year-old shoot of outer parts of the tree.

The rewards of the proper training of quince trees are the growth of commercial orchards that are easy to prune and harvest, also have sustained high yields, and quality fruits. Following the modified leader training of quince trees, a few unwanted suckers during late spring and summer grow in the various parts of the canopy. The most persecutor suckers of quince trees could be those that grow on the top of canopy; they limit light penetration in the canopy and reduce flower bud formation and fruit set. Therefore, a specialist producer of quince does not substitute summer pruning with a delayed winter cutting of these undesired shoots. According to this, it seems that the main pruning operation of commercial quince orchards is a reasonable and moderate removal of growing suckers in the inner parts of the canopy structure and cutting of some of the undesired shoots on top of the tree canopy. The severity and amounts of these pruning operations are highly dependent on various factors of orchards, such as the growth habit of cultivars, rootstock vigor, soil fertility, and orchard managements.

3.15 INTERCROPPING AND INTERCULTURE

The quince tree is native to both humid and semiarid regions of Caucasus, North Iran, Turkmenistan, and Asia Minor. Therefore, it is adapted to the regions with abundant annual rainfall with regular summer precipitations where the floor of quince orchards could be managed by the use of perennial turf grasses between tree rows. In addition, quince trees are also cultivated in semiarid regions of Iran, Turkey, and Central Asia, where survival and bearing of trees are highly dependent on the irrigations, and the lack of precipitations during late spring and summer often plays as an obstacle for intercropping and use of perennial turf grasses between rows of quince trees. Floor-orchard management in quince orchards established in arid and semiarid regions of Asia is mainly based on the control of excess growth of weeds by herbicides and the alternative use of between-row cultivators to prepare a weed-free area between rows (Figure 3.8A and B). This system of floor-orchard management in quince orchards of arid and semiarid regions is essential due to the lack of available water resources and limited regular precipitations during growth season. Weed regrowth in this system occurs after several weeks or months, depending upon mixtures of the groundcover weed species and water availability.

3.16 FLOWERING AND FRUIT SET

Review of articles about flowering of quince trees demonstrates that flowering of this species occurs singularly on the apices of a current season shootlet (Figure 3.10A and B). In pome fruits like apple, flowering consists of flower induction, flower initiation, flower differentiation, and finally anthesis or blooming (Hanke et al., 2007). According to this review, flower induction is the transition of the meristem development from the vegetative to the reproductive phase, and in this stage, the meristems receive the required environmental stimuli for this process. Hanke et al. (2007) also stated that flower initiation is the adjacent period of flower induction, in which a series of histological changes that are apparently invisible occur and lead to the beginning of flower differentiation that is characterized by morphological changes of flower primordia of floral organs. Various references, maybe according to the flowering habit of quince trees, mentioned that the beginning of the flower induction or initiation in this species is in early spring and flowers born on of terminal part of a current growth shootlet (Vossen and Silva, 2015; Westwood, 1993, Bell and Leitão, 2011). Dadpour et al. (2008) investigated the differentiation of flower buds in a quince tree regardless of time of differentiation, but as previously mentioned, Esumi et al. (2007a, b) by using electron microscopy imaging demonstrated that in quince, floral differentiation is initiated from late October to November, and after that, eight leaf primordia is initiated. In this stage, apical meristem is transformed to a dome-like structure and initiated sepal primordia (Figure 3.13A–L). This means that, contrary to the previous reports, also in quince tree, similar to the apple and pear, flower induction and initiation begin in the previous growth season of bloom anthesis. On the other hand, unlike apple and pear trees, in quince, the auxiliary meristems of bracts or leaf primordia in this species remains undifferentiated during winter, and then in the next spring differentiates to flower meristem. The more interesting is that Zeller (1960) reported that the first buds of three evaluated quince varieties began flower initiation in the autumn, and the initiation and development of primordia continued throughout the winter. These contradictions indicate that the exact time of flower bud induction, initiation, and differentiation in quince tree have not been studied well, and further research is needed in this regard.

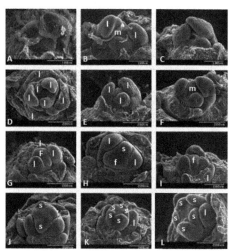

FIGURE 3.13 Electron microscopy imaging of the initiation and differentiation of the flower bud in a quince tree in Japan (Esumi et al. 2007a). (A, B) Apical meristem in July and August (C, D) Narrow apical meristem surrounded by several leaf primordia in late August and September. (E, F) Expansion of the apical meristem in October. (G, H) Initiation of the flower meristem in October. (I–K) Initiation of sepals on the fully swollen flower meristem in November. (L) Flower with five sepals in January: m = apical meristem, l = leaf primordia, f = flower meristem, s = sepal. Reproduced from Esumi et al. (2007a) with permission of the Japanese Society for Horticultural Science.

Anyway, following the completion of flower differentiation, the anthesis of quince bloom occurs in spring, considerably later than the apple and pear trees. The late anthesis of quince flowers is due to the simultaneous growth of the short flowering shootlet of the current season and development and growth of flower buds on the terminal part of shootlets. Despite the late anthesis of quince flowers in spring, the growing flower buds are unexpectedly very susceptible to spring frosts. In this case, damages of spring frost on the flowers appear as necrosis of the central part of the flower, whereas sepals and petals are more frost hardy than other parts (Figure 3.14A). Mohammadi et al. (2017) reported that quince cultivars and genotypes possess a range of resistance to the spring frost from damages exclusively on the styles, to all parts of ovary and entire stamens. But styles and anthers are more sensitive than other parts of the flower to the spring frost (Figure 3.14B and C). Mohammadi et al. (2017) also showed that the duration of flowering of quinces varies from several days to two weeks and even more, but this phenomenon is highly dependent on the climatic condition and genotypes.

FIGURE 3.14 (A) Damage of spring frost in the central part of the quince flower bud, without apparent damages on the sepals. (B) Moderate spring frost damages on style, ovary, and anther of quince flower. (C) Severe spring frost damages on the entire ovary and stamens of quince flower. In both cases, the fruit set and bearing are impossible due to the inability of the pollen tube growth in style.

Martínez-Valero et al. (2011) classified different phenological stages of quince according to the Biologische Bundesanstalt, Bundessortenamt and CHemica General Scale and Fleckinger's code. In this system, the length of phenological stages is measured in days and as cumulative degree days. According to this, 15 phenological stages from the dormant winter bud to leaf fall were identified in quince (Figure 3.15A–L). Among this phenological stages, D1 to G belong to flowering with five days for the appearance of flower bud (D1) with 10.5 °C days of duration, five days for calyx opening (D2) with 23.1 °C days of duration, four days for petals visible (E1) with 18.2 °C days of duration, three days for swollen calyx (E2) with 11.7 °C days of duration, nine days for open flower (F) with 28.2 °C days of duration, and finally four days for petal fall (G) with 12.0 °C days of duration. According to this, phenological stages of quinces from the appearance of flower bud to petal fall prolong for 31 days and needs to 103.7 °C days.

FIGURE 3.15 Phenological stages from dormant buds (Fleckinger, code A) to leaf fall (Fleckinger, code L) in quince tree according to the classification of quince phenological stages (Martínez-Valero et al., 2011).

The flowering period of quinces considerably depends on the cultivars. According to this, some commercial quince cultivars in Iran such as new released cultivar "Viduja" is early flowering and most of cultivars such as cultivar "Isfahan" is considered as medium flowering (Mohammadi et al. 2017). The early flowering habit of quince cultivars could be considered as negative characteristics that increase the risk of spring frost and loss of fruits in some areas.

Following the anthesis of quince flowers, the third whorled-androecium consists of 15–20 stamens with yellow anthers and white to purple filaments (Halamágyi and Keresztesi, 1975 from Nagy-Déri, 2011). This part of quince flowers with the styles and ovaries is highly susceptible to the late spring frosts (Mohammadi et al. 2017). According to Nagy-Déri (2011), the length of anthers in quince cultivars vary from 3259.86 μm in cultivar "Apple-Shaped" to 3779.19 in cultivar "Aromaté," with two oblate and suboblate forms. Also, it was demonstrated that some morphological aspects of pollen grains of quince cultivars are correlated to the form of anthers. Dalkiliç and Mestav (2011) showed that the number of pollen grain per flower in quince is between 12×10^3 in cultivar "Lemon" and 20×10^3 in cultivar "Ege25" in Turkey. Pollen grains of quinces are cylindrical, tricolporate, and ectoapertures that are slits. Also, the comparison of pollens in quince cultivars showed that ridges of pollens in cultivars "Lemon" resemble fingerprints, whereas in cultivar "Ekmek" they are irregular (Figure 3.16A–D). In both cultivars, perforations of pollen grains are scattered irregularly throughout the surface of the pollen, and the pollen of the cultivar "Limon" was identified as perprolate and that of "Ekmek" as prolate (Evrenosoğlu and Misirli, 2009).

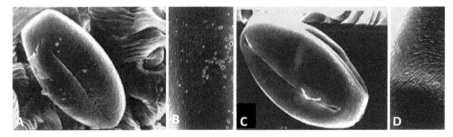

FIGURE 3.16 Comparison of cylindrical forms and ridges of pollen grains in two quince cultivars "Ekmek" (A) and "Lemon" (C). Pollen ridges of pollen grains in cultivar "Ekmek" are irregular (B) and in cultivar "Lemon" are fingerprint (D). Reproduced with permission from Evrenosoğlu and Misirli (2009).

Pollen viability and germination in quince are affected by cultivar, temperature, and media of culture. Sharafi (2011) investigated pollen viability and pollen tube longevity in the quince cultivar of Iran and reported that pollen germination varies from 37.2% (in genotype CO_2) to 65.3% (in cultivar "Isfahan"). Also, the best temperature for germination ant pollen growth tubes of quince was determined as 22 °C ± 1 °C by Yolaçtı (2006). Dalkiliç and Mestav (2011) observed the high germination percentage of the pollen grains of quince cultivars in Turkey on 1% agar and reported various effects of sucrose and boric acid on germination. In this investigation, best pollen germination observed in 10%–20% sucrose and predominantly in 50–100 mg/l boric acid.

The quince tree is an entomophilous fruit tree (Benedek et al. 2001). Simidchiev (1967) demonstrated that honeybees visit the quince flowers, and nectar production is abundant in flowers to attract honeybees for pollination. Benedek et al. (2001) concluded that the nectar production of quince flowers is low (1.07 ± 0.06 mg per bloom), and the sugar concentration in quince flowers ranges between 9% and 47.5%. Most honeybees tend to collect pollens, and usually fewer bees only collect nectar in flowers. Déri et al. (2006) reported that glucose, fructose, and sucrose are the main sugars of the nectar of quince flowers, and in most cultivars, sucrose is the dominant sugar in nectarthods. This attractiveness of nectar and pollen in quince flowers leads to about 7seven bee visits per flower per day in a good weather condition (Benedek et al. 2001). According to this, the key subject in quince pollination is the sensitivity of quince cultivars to partial or limitation of pollination during the effective pollination period and receptivity of stigma. Benedek et al. (2001) demonstrated that the complete limitation of insect pollination leads to no yield, and the partial limitation of the insect pollination period reduces the final set and yield by at least 60%–70% or more. Visiting of 4–5 or 8–10 honeybee per day per flower is necessary to achieve the required optimal fruit set, which is 20%–25% for quince. Fikret Balta and Muradoglu (2006) reported that the final fruit set of the quince cultivar "Ekmek" in the Van region of Turkey was 12.3% and 8.6% in two consecutive years, 2004 and 2005, respectively (Figure 3.17A). In other pome fruits, such as apple trees, large size of fruits permits a full crop with about 5%–8% of the flowers setting fruits (Westwood, 1993). Chaplin and Westwood (1980) reported that fruit set tendency in pear trees are between 3% and 11%. The comparison of the percentage of the optimal fruit set in quince trees with apples and pears demonstrates a higher percentage of the optimal fruit set in quince that is due to the flowering habit of quince that occurs singularly on the apices of a current season shootlet, whereas in apples, inflorescences contain five (or sometimes 6) and in pears contain seven or eight flowers (Westwood, 1993).

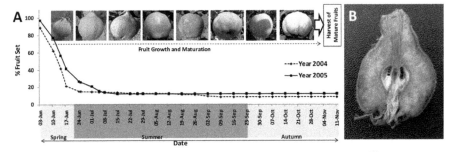

FIGURE 3.17 (A) Stages of fruit growth and maturation in quince from late spring to early autumn; and the percentage of the initial fruit set and procedure of fruit drop from late spring to harvest time of quince fruits in cultivar "Ekmek" under open-pollination conditions in the Van region of Turkey, according to Fikret Balta and Muradoglu (2006). (B) Aborted seeds in the abscised fruit of June-drop due to the lack of fertilization or incomplete fertilization of ovules.

Self-fertility of quince is a matter of controversy in the literature. Westwood (1993) indicated that quince flowers seem to be self-fertile, and Yamane and Tao (2009) categorized the quince tree as a self-compatible rosaceous species. Tatari et al. (2018) reported a different level of self-fertility in quinces. Hegedûs et al. (2008) by testing proteins and DNA-based methods indicated that S-RNases are putatively present in pistil tissues of quince cultivars. Similarly, Akbari Bishe et al. (2016) by using polymerase chain reaction-based methods amplified fragments from S-locus of this species. The first sequence of self-incompatibility allele of quince (S1) entirely was isolated from the Iranian cultivar "Torsh" (syn. "Sour") and deposited in databank by Talaei et al. (2017) that shows S-locus is entirely present in this species. Çetin and Soylu (2006) compared pollen tube growth in self-fertile and crossed flowers of several major quince cultivars of Turkey such as cultivars "Ekmek," "Tekkes," "Ege-22," "Ege-25," "Bardak," "Bencikli," "Eşme," "Limon," "Beyaz Ayva," and "Sekergevrek" and observed that the pollen tubes reached nearby the ovules at 24 h after pollination. Fruit set rates of crossed and self-fertile flowers were not significantly different, and it was concluded that the studied quince cultivars of Turkey were self-fruitful, and a quince orchard can be established with using a single cultivar. Unlike to the results of Çetin and Soylu (2006), Akbari Bishe et al. (2016) reported that in the most important commercial quince of Iran, cultivar "Isfahan" cross-pollination with cultivar "Torsh" leads to more adequate fruit set and bearing. Also, field observation in various regions of Iran confirms a higher percentage of the fruit set and more appropriate bearing of the quince cultivar "Isfahan" in orchards with a distinct cultivated cultivar as pollinizer (Figure 3.18B and C). All of these show that despite the existence of a handful of research about self-compatibility or incompatibility of quinces, the lack of research in this field is clearly comprehensible. Based on this lack of information about quince self-compatibility, comprehensive researches in this field seem to be the key point for commercial cultivars of the major quince producer countries.

Another problem of some quince cultivars is the lack of fruit set, low fertility, and bearing when planted in mountainous areas with high diurnal temperature variation. This problem occurs even when blooms are not exposed to the spring frosts, and some cultivars such as "Isfahan" are more susceptible to this phenomenon. The main reason of a low fruit set and fertility of this cultivar in high altitudes is not clear and therefore illustrative researches in this case seem to be necessary.

FIGURE 3.18 (A) High fruit set and bearing of quince trees in a well-managed orchard established on Quince A semidwarfing rootstock in the Fars Province of Iran and (B) the use of the local cultivar "pear shaped" pollinizer in the orchard (photos courtesy: E. Narimani).

3.17 FRUIT GROWTH, DEVELOPMENT, AND RIPENING

Fruits originate from hypanthium in pome fruits; therefore, similar to other pome fruits, quince fruits are false fruits. Following the fertilization of ovules and development of primary seeds, the growth in quince fruits initiates. According to the classification of phenological stages of fruit growth by Martínez-Valero et al. (2011), in quince cultivars, fruit setting duration is between 11 and 13 days, and they need 64.6 °C–69.6 °C day, the growth duration of immature fruits is between 24 and 26 days and need 175.5 °C–186.1 °C day, and finally, fruit growth duration is between 91 and 93 days and need 1152 °C–1180 °C day. It means that according to this classification, the duration of the fruit set to the initiation of fruit ripening in quince fruits is about 175 days, and also they mentioned 58–62 days for absolute fruit ripening.

In pome fruits, the first stage (S1) of fruit growth is resulted from cell division, followed by the second stage (S2) or cell expansion. In the quince tree, the pattern of fruit growth is single sigmoid, in which S1 and S2 stages are followed by a moderate final growth and ripening (S3) (Costa and Ramina, 2014). Immature quince fruits are densely villous with light brown color during the S1 growth stage (Figure 3.17A). In mid-summer, quince fruits initiate the S2 stage in which fruits enlarge through the expansion of cells. In this stage, the density of superficial pubescence of the fruits declines and green or yellow green ground color of fruit skin becomes slightly visible. Almost all quince fruits are late mature, and fruit ripening initiates in early to mid-autumn. During the fruit maturation stage (S3), in quinces, the fruit diameter slightly increases, and the peel becomes pale yellow to golden, with pale yellow to yellow pulps. Also, the mature fruits are with persistent reflexed calyces (Bell and Leitão, 2011). Maturation of quince trees is concomitant with a significant increase in the biosynthesis of a volatile compound and therefore mature quince fruits are very aromatic and fragrant. The beauty of the ripening process of quince is that it is naturally staggered and could be harvested nearly throughout the autumn (McMorland Hunter and Dunster, 2015).

Quince shows a climacteric respiratory pattern (Tuna-Gunes, 2008). Tuna-Gunes and Koksal (2005) demonstrated a climacteric increase in the respiration rate in fruits of two quince cultivars "Eşme" and "Cukurgobek" approximately up to 1.9–2.1 ml CO_2 kg^{-1} h^{-1} at three months after cold storage that was almost concomitant with the increase in the ethylene biosynthesis and internal ethylene concentration (IEC) of the fruits. In pome fruits such as quinces, the rise of climacteric respiration is concurrent with the main changes occurring during ripening (Costa and Ramina, 2014). All these show that the quince fruits demonstrate climacteric behavior and are harvested commercially at the physiological maturation stage, and the fruits continue to their physical, respirational, and biochemical changes after harvest and in the cold storage period. In pome fruits, such as quinces, main biochemical changes occur during ripening are depolymerization of cell-wall components, increase of mono and disaccharides, biosynthesis of volatile compounds, degradation of organic acids, and color development. Fruit firmness of pome fruits decreases, and fruits become edible during the maturation stage that is resulted from cellulose and pectin depolymerization in cell walls and finally cell separation (Knee, 1993). Ghozati et al. (2016) and Amirahmadi et al. (2017) reported that total phenolics in mature fruits of quince genotypes and cultivars of the southern regions of the Caspian Sea vary from 7.5 to 213.9 mg GAE/100 g in fresh weight. Moradi et al. (2017) demonstrated that the percentage of pectin in most quince cultivars of Iran varies between 5% and 10% (g/100 g FW) by the pectin extraction and weighting method, whereas Rop et al. (2011) reported the pectin range of 1.75–3.51 g/100 g FW by the photometric method. Soluble solids content (SSC) also in ripe fruits of quince cultivars of Iran reported between 14.53 and 20.08 °Brix and in the cultivar "Isfahan" was 15.2 °Brix (Moradi et al. 2017; Abdollahi et al. 2018). This range in ripe fruits of quince cultivars of Turkey reported between 10.9 and 17.7 °Brix by Rop et al. (2011) and Pinar et al. (2016).

In pome fruits such as quince, carbohydrates are transported to developing fruits as sorbitol and converted mainly to fructose and starch with some glucose and sucrose (Knee, 1993). During the maturity stage of pome fruits, increases in glucose and fructose contents are resulted from the starch hydrolysis (Chong et al. 1972). Tuna-Gunes and Koksal (2005) indicated to 2.9–3.1 g 100 ml^{-1} of fruit juice of sucrose, 3.5–5.4 g 100 ml^{-1} of fruit juice of glucose, and 8.9–9.8 ml^{-1} of the fruit juice of fructose of ripe quince fruits of "Eşme" and "Çukurgöbek." These results confirm the importance of fructose as the major soluble carbohydrate of mature quince fruits, similar to apple and pear fruits. Besides soluble and storage carbohydrates and their balance during maturity of quince fruits, organic acids have a key role during the process of maturity and also for sensory characteristics of quinces. It has shown that malic acid is the main organic acid of quince fruits (Rodríguez-Guisado et al. 2009). Rop et al. (2011) reported that the average contents of organic acids in quince fruits of various cultivars of Turkey range between 0.36 and 1.53 g/100 g FW, and Moradi et al. (2017) reported between 0.26 and 0.75 g/100 g FW in quince fruits of various cultivars of Iran and between 0.44 and 0.85 g/100 g FW in quince fruits of various cultivars of Serbia. Tuna-Gunes and Koksal (2005) indicated that malic acid, the predominant acid of quince fruits, range between 709.5 and 831.1 mg 100 ml^{-1} of fruit juice in cultivar "Eşme" and between 520.7 and 625.1 mg 100 ml^{-1} of fruit juice in the cultivar "Çukurgöbek."

In pome fruits, flavors depend upon many organic compounds that are synthesized during the climacteric phase (Knee, 1993). For this reason, the horticultural maturity of the quince fruits during the storage period leads to the release of volatile and aromatic compounds from fruits. Umano et al. (1986), Tateo and Bononi (2010), and Tsuneya et al. (1983) analyzed the volatile compound of ripe quince fruits, and Khoubnasabjafari and Jouyban (2011) reviewed these volatile compounds. Tateo and Bononi (2010) demonstrated that some of these volatiles, such as sesquiterpene, in ripe quince fruits increase from October to November.

3.18 FRUIT RETENTION AND FRUIT DROP

In quince trees, premature and preharvest fruit drops are two main periods that growing fruits face during the formation of the abscission layer in pedicels and set fruitlets or fruits lost. Croitoru and Baciu (2013)

showed that the average blooms per mature quince tree is about 450–900. This means that according to the optimal percentage of the set that has been defined between 20% and 25% by Benedek et al. (2001), the final production of 100–200 fruits per tree leads to abundant bearing and yield in commercial quince orchards. The final level of the fruit set and fruit production is dependent on the precedent fruit drops during various stages of fruit growth. Westwood (1993) divided fruit drops in pome fruits in premature and preharvest drops. The premature drop consists of two stages including the first drop of unfertilized flowers about four weeks after bloom and the June drop of fruitlets with aborted seeds about eight weeks after bloom (Figure 3.17B). These two stages of premature could overlap each other and are pursued by the premature fruit drop that occurs during the maturation of fruits on the pome trees. Fikret Balta and Muradoglu (2006) demonstrated that fruitlets consistently drop up to the end of spring in quince tree, and growing fruits remain on the tree with the minimum percentage of the premature fruit drop in this species (Figure 3.17A). According to this pattern, the first fruit drop of unfertilized flowers and June drop are not distinct in quince, but aborted seeds in fruitless with about 20 mm diameter are visible after about eight weeks after the full bloom of the tree

(3. 17B). Also, damages of quince moth (*Euzophera bigella* Zell.) could lead to fruit drop and loss of yield from mid-summer up to the maturation of the fruits on the trees.

In quince trees, flowers appear singularly on the apices of a current season shootlet. Therefore, naturally the fruits distribute appropriately in various parts and along the entire length of tree limbs and branches in quince trees. This means that unlike the apples and pears, for quince trees, fruit thinning is not necessary in many years and many conditions. The lack of need of the artificial thinning in some quince cultivars such as "Isfahan" that normally produce few spurs along the tree limbs could be more considerable than other quince cultivars. Finally, for this reason, alternative bearing usually does not appear in quince trees unlike apple or pear trees.

3.19 HARVESTING AND YIELD

Physiological maturity and ripening of quince fruits are concurrent with physiochemical changes in the fruit pulp and peel that some of them could be considered to determine the appropriate time of harvesting for cold storage or fresh use of fruits. Several indexes such as the starch pattern, TSS as a refractometer measurement of the juice (%), titratable acid per 100 g of juice, and firmness, measured with an 11.1 mm diameter penetrometer, and finally disappearance of green color of chlorophyll pigments from fruit background color and appearance of yellow colors are main maturity indexes for pome fruits such as quinces. Unlike the apple and pear trees, almost in all quince cultivars and genotypes, the fruits mature during late September to October, and early ripening summer cultivars does not exist in this species (Thomidis et al., 2004; Türk and Memicoglu, 1994).

Major physicochemical changes of quince fruits during maturity including climacteric respiration, ethylene and organic acid biosynthesis, depolymerization of cell-wall components and softening of fruits, biosynthesis of volatile compounds, and color development and conversion of cellulose and sorbitol to fructose and glucose that leads to the increase of TSS of quince fruits have been reviewed previously in this chapter. But exclusively some of these physicochemical factors are stable, distinctive, and reasonable for using as maturity indexes of quince fruits. Due to the commercial importance of apples and pears, the maturity indexes of these pome fruits have been determined previously, but in quince the data are either very limited or not available. In some quince cultivars, green color of fruits disappears before physiological maturation, and for this reason, the color index seems to be not appropriate for the determination of harvest time. Similarly, the starch pattern index, which is a reliable maturity index in some apple cultivars, in quinces could not be abundantly distinctive. It seems that unlike the above-mentioned indexes, threshold firmness and TSS are most applicable indexes for the determination of the maturity stage and harvesting time of quince fruits—both for cold storage and fresh use. The fruit firmness of mature fruits varies in different quince cultivars, as in "Eşme" 81.4 N (Tuna-Gunes and Koksal, 2005), in "Limon" 79.7 N, in "Baradacik" 71.6 N (Pinar et al. 2016), "Isfahan" 50 N, and in "Viduja" 55 N (Moradi et al. 2017) were reported at ripening time. Tuna-Gunes (2003) demonstrated that the fruit firmness of quince flesh of the cultivar "Eşme" diminishes from 124.2 N at 104 days after full bloom to 96.5 N at 118 days after full bloom, which means that in late summer, the firmness of quince fruits shows a considerable diurnal decline. A similar pattern of decline in fruit firmness was observed in other quince cultivars of Turkey in late summer and early autumn (Tuna-Gunes and Dumanoglu, 2005). Abdollahi et al. (2018) reported that the firmness of fruit in various cultivars and genotypes is correlated with their origins, as genotypes from humid regions of the Southern Caspian Sea were significantly softer than those from the central and north east regions of Iran. Field observations show that fruits of quince genotypes from north and northwest Iran, Republic of Azerbaijan, Turkey, and South Caspian Sea have relatively softer fleshes than genotypes from the central and northeast Iran. Torkashvand (2014) results demonstrated the genetic distance of further two quince landraces of Iran from other parts by using SSR markers (Figure 3.2).

TSS could be considered as the second maturity index for quince fruits. Kuzucu and Sakaldas (2008) demonstrated that TSS in fruits of the quince cultivar "Eşme" increase from 12 °Brix in late September to more than 14 °Brix in late October. Rop et al. (2011) determined SSC of ripe fruits in October in various quince cultivars of Czech Republic and showed this index is for cultivar "Champion" 16.5%, for cultivar "Leskovačka" 15.9% and for new cultivar "Morava" 15.8%. In Turkey, SSC for cultivars "Lemon," "Bardacik," and "Tekkes" were determined to be 13.1%, 14.4%, and 13%, respectively, by Pinar et al. (2016). Tuna-Gunes (2003) evaluated variations in monocarbohydrates including sucrose, glucose, and fructose separately during quince fruit ripening of the cultivar "Eşme," but these indexes cannot be used in the orchard condition for producers. Mosharaf and Ghasemi (2004) indicated that in the quince cultivar "Isfahan," TSS in harvested fruit at 149–181 days after full bloom increase from 14.8 to 15.6 °Brix and reported that the TSS index equal to 15.6 °Brix is the best time for the harvest of this cultivar for cold storage. Radović et al. (2015) determined the TSS index of cultivars "Leskovacka," "Vranja," and "Portugal," 15.6, 17.5, and 17.2 at harvest time, respectively. Türk and Memicoglu (1994) showed that TSS as a maturity index in quince depends on the altitude of production locality, and therefore this index must be determined and optimized for each cultivar in each region.

3.20 PROTECTED/HIGH-TECH CULTIVATION

Quince trees are mostly semi-vigorous. Therefore, commercial cultivars even on seedling rootstocks produce scion–rootstock combination that grows as a semidwarf tree of 3–4 m height. For this reason, historically, Quince A and Quince BA29 or hawthorn seedlings belong to *C. atrosanguinea* Pojark. have been used for the establishment of semiintensive quince orchards. Also, the development of high-intensive orchards with dwarfing rootstocks such as Quince C has not been considered in the main producer countries at the commercial level.

3.21 PACKAGING AND TRANSPORT

Following the determination of best maturity indexes for the harvest of quince, the mature fruits must be carefully picked up from the trees and packaged for wholesale or retail markets. Despite a relative high firmness level of the quince pulp, the fruit peel is very sensible to physical damages, abrasion, and bruising (Figure 3.19A). Therefore, it is important that the quince fruits are picked up in small quantities and then immediately sent for packing. Due to the sensibility of quince peels to the superficial damages, it is usually sold for retail and wholesale markets in carton box packages with smaller quantities containing 5–10 kg of quince fruits (Figure 3.19B and C).

UNECE (2014) determined that minimum requirements for sorting of all classes of quinces includes: fruits must to be intact, sound, clean, fresh in appearance, free from pests and their damages, free of abnormal external moisture, and any foreign smell, and also must to withstand transportation and handling to arrive in the satisfactory condition at the place of destination. According to this standard for marketing and commercial quality control of quinces, three categories, including

"Extra class" that fruits must be of superior quality and with the characteristic of the variety; "Class I" that fruits must be of good quality with the characteristic of the variety; and finally "Class II" that fruits do not qualify for inclusion in the higher classes but pass from the minimum requirements specified above. Also, some major producers of quince in Iran developed a local standard in which quince fruits are categorized into eight classes, including Extra-I, which is characterized as 460–560 g weight of fruits and without any internal or external damages and with excellent quality; Extra-II, which is characterized as 340–460 g weight of fruits and without any internal or external damages and with excellent quality; Superior, characterized as 240–340 g weight of fruits and without any internal or external damages and with good quality; Class-I, which is characterized as 220–240 g weight of fruits and without any internal or external damages and with acceptable quality; Orange Size, which is characterized as 180–220 g weight of fruits and without any internal or external damages, and with acceptable quality and size of an orange; Asymmetric, without standard for weight but consists all fruits that grown abnormal and without equivalent from; Dapple d-I, without standard for weight but fruits have small physical damage; and Dappled-II, without standard for weight but fruits have considerable physical damage. Other fruits with decay or damage of quince moth (*E. bigella* Zell.), partially russets, mechanical damages, sunburns, scalds, and sooty blotches could not be acceptable for sorting. This standard of sorting could be used for a quince cultivar "Isfahan" that has large fruit sizes and is not applicable to other cultivars with medium or small size of fruits.

FIGURE 3.19 (A) Physical damage on a sensible peel of harvested quince caused by bruising and pressure in a nonstandard plastic box. (B) Sorted fruits of the quince cultivar "Isfahan" in an orchard and packed in carton boxes. (C) High-quality quince fruits cultivar "Isfahan" packed in single-layer fruit carton boxes containing 15 fruits with 5–7 kg medium weight.

Quince fruits after harvest could be stored for several months in a cold storage or could be sell fresh in markets. Many factors such as harvest time affect physiochemical characteristics, quality, and shelf life of quinces. Tuna-Gunes and Koksal (2005) and Tuna-Gunes (2008) demonstrated that after 1–5 months cold storage of quince fruits at 2°C ± 1 °C with 85%–90% relative humidity of cold storage, respiratory climacterium realizes at the end of the first week during the shelf-life period. Also, sucrose and organic acid contents of quinces decrease during shelf life, and glucose and fructose tend to increase with some fluctuations. They also reported that sensory values of quince reduce permanently during three-week shelf life of quince fruits, after taking out them from the cold storage.

3.22 POSTHARVEST HANDLING AND STORAGE

Optimum and standardized conditions for quince fruits in a cold storage have been determined as −0.6 °C to 0.0 °C and 90%–95% relative humidity of the cold storage. Also, the highest freezing point of quince fruits has been reported to be −2.0 °C. During the storage period of quinces that could be prolonged up to 2–3 months, the fruits must be handled carefully as they bruise easily. The ethylene production (EP)

ranges of quince fruits in the storage period at 0 °C, 10 °C, and 20 °C are 2.3–6.1, 6.9–7.4, and 11–31.9 µl C_2H_2/Kg h, respectively. At the same temperatures, the CO_2 production range also is 2.3–5.2, 10.2–14.1, and 21.2–39 ml CO_2/kg h. 100 ppm treatment by ethylene for two days at 18 °C –21 °C and the same relative humidity of the storage period (90%–95%) can be used after the removal of quinces from the cold storage to stimulate more uniform and faster ripening of quinces before processing (Kader, 1992). For the cargo transportation of quince fruits, the optimum condition is similar to that of a cold storage, and the acceptable temperature at loading into containers is 2 °C and ventilation setting for containers must be 25 m³/h. Tuna-Gunes (2008) explained that 0 °C –2 °C for up to six months in air and for up to seven months in 2% O_2 + 3% CO_2 at 2 °C are the optimum condition for the normal and controlled atmosphere storage of quince fruits.

Rop et al. (2011) reported that in quince genotypes, biochemical and nutritional properties vary after cold storage according to the genotype. Tuna-Gunes et al. (2012) and Nanos et al. (2015) showed positive effects of 1-MCP treatment at 1000 ppb on the prevention of flesh browning and weight loss, loss of titratable acid, and green color loss in cold storage period of quince fruits. Tuna-Gunes and Koksal (2005) reviewed several biochemical changes during cold storage of quinces and demonstrated fluctuations in the activities of 1-aminocyclopropane-1-carboxylic acid oxidase, the terminator key enzyme in ethylene biosynthesis of fruits, is parallel with changes in IEC, and EP indexes in quince fruits during cold storage. Similar to the shelf-life period, the sucrose content of fruits tends to decrease, while concurrently glucose and fructose concentrations increase slightly. Tuna-Gunes and Koksal (2007) indicated that this increase in the glucose content of quince fruits during cold storage periods has positively affected the sensory quality of fruits.

Other factors including harvest time (Kuzucu and Sakaldas, 2008; Mosharaf and Ghasemi, 2004; Tatari and Mousavi, 2017; Türk and Memicoglu, 1994), packaging materials (Nikkhah and Ganji Moghadam, 2009; Kuzucu and Sakaldas, 2008) reported as effective factors for prolongation of cold storage and preservation of quality of quince fruits. Also, treatment of quinces with hot water for 36 h at 38 °C is useful for the maintenance of fruit quality in the cold storage (Rahemi and Akbari, 2004).

3.23 PROCESSING AND VALUE ADDITION

Quince processes as jam, jelly, dried fruit, nectar, and puree (Silva et al., 2002b; Babić et al., 2007, 2008; Sharma et al., 2011; Ramos and Ibarz, 1998; Voicu et al., 2009). Quince jam is the most popular processed form of this fruit. Quinces are especially rich in phenolics, including proanthocyanidins of 2–20 anthocyanin-like subunits. These aggregates are cross-linked and coagulate proteins, so they feel astringent. During the cooking process, the combination of heat and acidity causes the subunits to break off, and then oxygen from the air reacts with the subunits to form true anthocyanins. Therefore astringent, pale quince fruits become more gentle-tasting and deep pink (McGee, 2004). Sharma et al. (2011) described that quince jam should contain at least 45 parts of fruit pulp for every 55 parts of sugar and 68 °Brix or the percentage of TSSs; however, jelly should contain 65% of TSSs. They also compared nutritional composition and processed products of quinces. TSSs of jam and jelly were recorded 70 and 66 °Brix, respectively. Quince jelly contained more titratable acid than 0.72% in quince jam.

Due to the firmness of quince, this fruit is not used in a normal dried form (Stojanović et al., 2010; Guiné and Barroca, 2012), but it was proved that osmotic drying within the combined drying technology has favorable effects on reducing dried quince firmness (Babić et al. 2007, 2008). Osmotic drying of quince is also more favorable as the volume loss of osmotic dried fruits is not as much of fruits dried by using conventional drying (Koc et al. 2008). Stojanović et al. (2010) evaluated the effects of osmotic solution temperature and concentration, and the duration of the process and showed that the most intensive changes of physical properties of quince fruits occur in the first 100 min of osmotic drying.

Quince uses nectar mix with other fruits such as apple. Voicu et al. (2009) analyzed sensorial and biochemical characteristics of mix "apple, quince, and sea buckthorn nectar," and selected the optimum variant mix in which the ratio of apple and quince puree was 1:1 and the percentage of sea buckthorn juice was 3%.

3.24 DISEASE, PEST, AND PHYSIOLOGICAL DISORDERS

Various fungal, bacterial, viral, and virus-like diseases affect quince and other pome fruits world-wide. Major and minor fungal, bacterial, viral, and viral-like diseases of various parts quince trees have been listed in Table 3.3.

TABLE 3.3 Reported Diseases that Affect Various Parts of Quince Trees[a]

Disease	Pathogen	Common Name	Reference
Bacterial	*Erwinia amylovora*	Fire blight	Vanneste (1995)
	Agrobacterium rhizogenes	Hairy root	Ogawa and English (1991)
	Agrobacterium tumefaciens	Crown gall	Horst (2013)
Fungal	*Fabraea maculate*	Black Spot (Fleck)	Horst (2013)
	Phytophthora cactorum	Collar rot	Érsek et al. (2008)
	Phytophthora parsiana	Collar rot	Hajebrahimi and Banihashemi (2011)
	Monilinia fructigena	Fruit rot	Hrustić et al. (2012)
	Alternaria mali	Fruit rot	Horst (2013)
	Botrytis cinerea	Fruit rot	Horst (2013)
	Penicillium expansum	Fruit rot	Horst (2013)
	Venturia pirina	Scab	Horst (2013)
	Gymnosporangium clavipes	Rust	Horst (2013)
	Podosphaera sp.	Powdery mildew	Horst (2013)

Quince   231

TABLE 3.3   *(Continued)*

Disease	Pathogen	Common Name	Reference
Virus and virus-like	Quince sooty ringspot virus		Paunoviç and Rankoviç (1998)
	Apple chlorotic leaf spot virus		Akbaş and İlhan (2008)
	Apple stem pitting virus		Birişik and Baloğlu (2010)
	Apple stem growing virus		Birişik and Baloğlu (2010)
Nematode	*Meloidogyne* sp.	Root knot	Horst (2013)

[a] The important and moderate diseases are listed and rare diseases with local or regional importance have been ignored.

Among various diseases listed in Table 3.3, fire blight is the most important and catastrophic disease of pome fruits, especially in quince cultivars and genotypes (van der Zwet and Beer, 1999; van der Zwet and Keil, 1979). The causal agent of fire blight is *Erwinia amylovora* (Burrill) (Winslow et al., n.d.). The disease spread from the Hudson Valley of New York State, USA, and during more than two centuries was recorded from five continents and in all around the world (Gordon Bonn and van der Zwet, 2000). Quince trees are one of the susceptible hosts of *E. amylovora* and unlike apples and pears; none of the quince commercial cultivars demonstrate the considerable resistance level to the disease. Infections of fire blight initiate by the colonization of the flowers and growing shoots in spring (Figure 3.20A), growth, and progress of bacteria along branches and limbs and the formation of cankers in trunk in disease-susceptible quince cultivars (Figure 3.20B). On susceptible quince cultivars, if climatic conditions, including temperature and relative humidity, are favorable, the disease spread throughout an orchard and destroys all trees over several years (Figure 3.20C). The bacterial agent overwinter in cankers and in spring spreads from oozes on the previous year's cankers to the flowers through rain dissemination, honeybees, and other flower-visiting insects. Copper compounds such as Bordeaux mixture, copper oxychloride, two antibiotics including streptomycin (Plantomycin, Agrept, and Agristrep) or oxytetracycline (Mycoshield) (Psallidas and Tsiantos, 2000), and antagonist bacteria (Gerami et al. 2013) have been recommended for fire blight control on host plants, but the use of resistant cultivars is considered as the most effective control strategy in all host plants, as well as quince (Azarabadi et al., 2016; Abdollahi and Majidi Heravan, 2005; Abdollahi et al., 2011; Abdollahi and Akbari Mehr, 2008; Ghahremani et al., 2014; Mehrabipour et al., 2010, 2012). The disease seems to be more devastating on quince trees in semi-humid to semiarid regions with low to medium altitude where diurnal/nocturnal temperatures vary, besides minimum relative humidity that is necessary for bacterial growth, providing an appropriate condition for outbreak of the fire blight.

Phytophthora cactorum, the causal agent of the collar rot, is serious in apple and pear rootstocks, but rare reports demonstrate considerable damages of this fungus in quinces. Érsek et al. (2008) reported that meddlers on quince stocks infected by *P. cactorum* via rootstocks. Also, after humid springs, we could isolate *P. cactorum* with very sever damages in graft union of the trees in quince germplasm collection of Iran (Figure 3.20D) (unpublished data). The reaction of quince cultivars and genotypes to infection of *P. cactorum* was extremely different. Also, other common fungal agents including *Monilinia fructigena, Alternaria mali, Botrytis cinerea,* and *Penicillium expansum* listed by

Horst (2013) as the causal agents of rot on the quince fruits.

The main aerial host of *Gymnosporangium clavipes* is the quince tree. Infection from basidiospores gives rise to pycnia borne on the surface of quince fruits; they are visible from late spring to early summer, and the most favorable climatic conditions for the infection from this fungus are a wetting period with a mean temperature above 10 °C during flower bud anthesis (Jones and Aldwinckle, 1990). *G. clavipes* is widespread in North America. Fabraea leaf and fruit spot or black spot also is a fungal disease caused by *Fabraea maculata*. The disease could be particularly severe on quince in Europe. The disease lesions become brown in the center surrounded by a reddish halo. Severe defoliation of susceptible varieties reduces tree vigor and yield (Jones and Aldwinckle, 1990; Postman, 2012). *Podosphaera leucotricha* is the causal agent of powdery mildew on quince, and Postman (2012) reported a different resistance level of quince to this disease, as well as to rust (*G. clavipes*), and the Fabraea leaf and fruit spot (*F. maculate*), in quince germplasm collection of NCGR-Corvallis.

The main reported viral diseases of quince trees are quince sooty ringspot virus, apple chlorotic leaf spot virus, apple stem pitting virus (ASPV), and apple stem growing virus (ASGV) (Paunoviç and Rankoviç, 1998; Akbaş and İlhan, 2008; Birişik and Baloğlu, 2010; Birişik and Baloğlu, 2010). Birişik and Baloğlu (2010) showed that ASPV and ASGV co-infections of quince trees appear as severe leaf deformation, mosaic, and overgrowth of the trunk (Figure 3.20E). Single ASPV infection cause gummy tissues on the surface of the fruits, gritty, and gummy tissues around the seeds, fruit malformation and trunk flexion. According to the European and Mediterranean Plant Protection Organization, using certified propagation materials, including rootstocks and scions, is the main control strategy for the virus disease of quinces (EPPO, 1999).

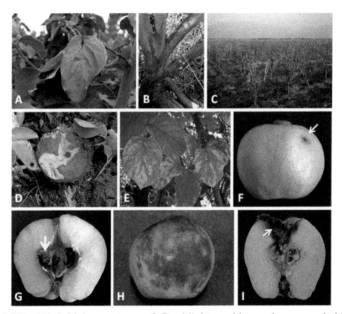

FIGURE 3.20 (A) Initial symptoms of fire blight on bloom, leaves, and thin shoots, appeared as a small round canker with distinct margins of canker. (B) Severe symptoms of fire blight on the trunk and limbs of quince cultivar "Isfahan," including paper bark and distinct margins of cankers. (C) Progressive infection of fire blight on the whole quince orchard in northwest Iran during the first years of disease outbreak between 2005 and 2006. (D) Trunk cross-section of the quince tree infected by the collar rot (*Phytophthora cactorum*). (E) Virus infection symptoms on the leaf of quince in shadow. (F) Superficial hole caused by the damage of the quince moth (*Euzophera bigella* Zell.). (G) Internal rot and growth of molds in the quince fruit following the progress and nutrition of the quince moth from seeds and ovary. (H) Considerable symptoms of the flesh browning on quince fruits as the most important physiological disorders during cold storage. (I) Decay around the stalk end of the quince fruit following the cracking of the peel and pulp around the premature fruit and growth of mold on the decayed soft area.

According to the list of Connecticut Agricultural Experiment Station, main quince pests are apple aphid (*Aphis pomi*), codling moth (*Cydia pomonella*), hawthorn lacebug (*Corythucha cydoniae*), oriental fruit moth (*Grapholita molesta*), quince curculio (*Conotrachelus crataegi*), resplendent shield bearer (*Coptodisca splendoriferella*), and San Jose scale (*Quadraspidiotus perniciosus*) (The Connecticut Agricultural Experiment Station, 2017). Also, the quince moth (*Euzophera bigella* Zell.) is the principle pest of quince orchards in Iran (Davatchi and Esmaïli, 1970). Caterpillars of the quince moth and codling moth damage the quince fruits. Pinkish or white caterpillars overwinter under bark or debris and form pupae in spring. Females emerge around bloom to lay eggs on fruits. The caterpillars of the first of the two generations enter the set fruit after anthesis, while the caterpillars of the second generation injure fruit in August and September (Figure 3.20F) and cause the penetration of fungal rots in ovary (Figure 3.20G). The amount of damage of both pests varies greatly from year to year. The pheromone traps may be used to monitor adult flights (The Connecticut Agricultural Experiment Station, 2017). Quince moth in Iran has four generation and the last one is also the most harmful generation (Rajabi and Dastgheib-Beheshti, 1979). Control of codling and quince moth in quince orchards needs an integrated pest-management strategy, including biological control for suppressing pest population to levels that would result in the acceptable crop loss. Monitoring of pest population by pheromone traps to determine the appropriate timing schedule for using botanical organic insecticides or conventional chemical spray that most of them target young larva hatching from the eggs.

Main physiological disorders of quince fruits are peel browning (Figure 3.20H), bitter pit, flesh browning, breakdown, and scald (Thomidis et al. 2004; Tuna-Gunes, 2008; Holevas and Biris, 1980). Flesh browning initially appears in the middle part of the fruit cortex and moves toward peel (Tuna-Gunes, 2008). Türk and Memicoglu (1994) reported that delay in the harvest of quince fruits and production in high altitudes increase the risk of browning during cold storage. However, the severity of flesh browning in quinces during cold storage is higher for early mature cultivars (Tuna-Gunes and Koksal, 2005). Also, quince genotypes show different susceptibility to the physiological fruit disorders, including bitter pit, breakdown, scald, and browning (Thomidis et al., 2004; Moradi et al., 2016). In a personal experience, the quince fruits with a high level of peel browning had a severe imbalance in macro-elements of fruits, including a high level of potassium and considerable deficiency of calcium and magnesium (unpublished data). Beyaz et al. (2011) determined color change during enzymatic browning of quince fruits and enzymatic browning, whereas Tuna-Gunes (2008) demonstrated that treatment with hot water at 50 °C for 5 min reduced the risk of browning on the quince fruit. Also, Holevas and Biris (1980) reported that the spray of boron and calcium reduced the percentage of pitting and internal browning of quince fruits, but major nutritional differences were not observed between healthy and disordered quince fruits. Some quince cultivars also are susceptible to stalk-end cracking that this disorder increase by wet springs following the period of fruit sets. This cracking sometimes ends to fruit rot and growth of moulds on the fruits before or during cold storage period (Figure 3.20I).

KEYWORDS

- *Cydonia oblonga* Mill.
- quince
- phenolics
- germplasm collections
- fire blight resistance
- hawthorn rootstock
- iron chlorosis
- flower initiations
- fruit browning

REFERENCES

Abdollahi, H.; Akbari Mehr, H. Evaluation of fire blight resistance in some quince (*Cydonia* oblonga Mill.) genotypes, I. isolation, evaluation and selection of causal bacteria (*Erwinia amylovora*) isolates. *Seed Plant Imp. J.* **2008**,24, 515528. DOI:10.22092/spij.2017.110835

Abdollahi, H.; Alipour, M.; Khorramdel Azad, M.; Ghasemi, A.; Adli, M.; Atashkar, D.; Akbari, M.; Nasiri, J. Establishment and primary evaluation of quince germplasm collection from various regions of IRAN. *Acta Hortic.* **2013**,976, 199–206. DOI:10.17660/ActaHortic.2013.976.25

Abdollahi, H.; Alipour, M.; Mohamadi Garamroudi, M. Assessment of physicochemical traits and their relation with the organoleptic characteristics of fruits in the quince (*Cydonia oblonga* Mill.) genotypes of various regions of Iran. *Seed Plant Imp. J.* **2018**,33, (accepted; under press).

Abdollahi, H.; Ghasemi, A.; Mehrabipour, S. Evaluation of fire blight resistance in some quince (*Cydonia oblonga* Mill.) genotypes. II. resistance of genotypes to the disease. *Seed Plant Imp. J.* **2008**, 24, 529–541. DOI:10.22092/spij.2017.110836

Abdollahi, H.; Ghasemi, A.; Mehrabipour, S. Interaction effects of rootstock and genotype on tolerance to iron deficiency chlorosis in some quince (*Cydonia oblonga* Mill.) genotypes from central regions of Iran. *Seed Plant Imp. J.* **2010**, 26, 1–14. DOI:10.22092/spij.2017.110966

Abdollahi, H.; Majidi Heravan, E. Relation between fire blight resistance and different vegetative and reproductive traits in apple (*Malus domestica* Borkh.) cultivars. *Seed Plant Imp. J.* **2005**, 21, 501–513. DOI:10.22092/spij.2017.110656

Abdollahi, H.; Tahzibi, F.; Ghahremani, Z. Correlation between fire blight resistance and morphological characteristics of pear (*Pyrus communis* L.). *Acta Hortic.*

2011, 896, 339–345. DOI:10.17660/ActaHortic.2011.896.46

Agrolib. Quince Production. https://agrolib.rs/en/quince-production (accessed Nov 25, **2017**).

Ahmadi, S.; Abdollahi H. Susceptibility to the chlorosis and its relationship to iron contents in some genotypes of Quince (*Cydonia oblonga* Mill.). *Seed Plant Imp. J.* **2018,**33 (accepted: under press).

Akbari Bisheh, H.; Abdollahi H.; Torkashvand M.; Ghasemi A. Pollen self-incompatibility and determination of appropriate pollinizer for quince (*Cydonia oblonga* Mill.) cultivar Isfahan. *Seed Plant Imp. J.* **2016,** 32, 13–26. DOI:10.22092/spij.2017.111286

Akbaş, B.; İlhan, D. First report of apple chlorotic leaf spot virus in quince (*Cydonia Oblonga* Mill.) in Turkey. *Acta Hortic.* **2008,** 781, 161–166. DOI:10.17660/ActaHortic.2008.781.24

Al Maarri, K.; Arnaud, K.; Miginiac, E. *In vitro* micropropagation of quince (*Cydonia oblonga*). *Sci. Hortic.* **1986,** 28, 315–321. DOI:10.1016/0304-4238(86)90105-6

Alcántara, E.; Montilla, I.; Ramírez, P.; García-Molina, P.; Romera, F. J. Evaluation of quince clones for tolerance to iron chlorosis on calcareous soil under field conditions. *Sci. Hortic.* **2012,** 138, 50–54. DOI:10.1016/j.scienta.2012.02.004

Alipour, M.; Abdollahi, H.; Abdossi, V.; Ghasemi, A.; Adli, M.; Mohammadi, M. Evaluation of vegetative and reproductive characteristics and distinctness of some quince (*Cydonia oblonga* Mill.) genotypes from different regions of Iran. *Seed Plant Imp. J.* **2014,** 30, 507–529. DOI:10.22092/spij.2017.111226

Aliyev, J.A. *Azerbaijan: Country Report to the FAO International Technical Conference on Plant Genetic Resources*; FAO International Technical Conference, Leipzig, Germany, **1996.**

Amirahmadi, Z.; Abdollahi, H.; Ayyari, M. Variations in flavonoid compounds of the leaves and fruits of quince (*Cydonia oblonga* Mill.) genotypes from northern regions of Iran. *Iran. J. Hortic. Sci.* **2017,** 48, 329–337. DOI:10.22059/ijhs.2017.134753.859

Anonymous. *Norton Priory, Museum & Gardens-The Walled Garden*; Norton Priory Publication, UK, **2014.**

Antonelli, M. The regenerative ability of quince BA29 *in vitro. Adv. Hortic. Sci.* **1995,**9, 3–6.

Avicenna, A.A. *The Canon of Medicine;* New York, AMS Press, USA, **1973.**

Aygun, A.; Dumanoglu, H. Shoot organogenesis from leaf discs in some quince (*Cydonia oblonga* Mill.) genotypes. *Tarim Bilimleri Dergisi* **2006,**13, 54–61. DOI:10.1501/0001437

Azarabadi, S.; Abdollahi, H.; Torabi, M.; Salehi, Z.; Nasiri, J. ROS generation, oxidative burst and dynamic expression profiles of ROS-scavenging enzymes of superoxide dismutase (*SOD*), catalase (*CAT*) and ascorbate peroxidase (*APX*) in response to *Erwinia amylovora* in pear (*Pyrus communis* L). *Eur. J. Plant Pathol.* **2016,** 147, 279–294. DOI 10.1007/s10658-016-1000-0

Babić, M.; Babić, L.; Pavkov, I.; Radojčin M. Changes in physical properties throughout osmotic drying of quince (*Cydonia oblonga* Mill.). *J. Process. Ene. Agri.* **2008,** 12, 101–107.

Babić, M.; Babić, L.; Radojčin, M.; Pavkov, I. The quince (*Cydonia oblonga* Mill.) firmness changing during osmotic drying. *J. Process. Ene. Agri.* **2007,** 11, 82–85.

Baker, B.S.; Bhatia, S. K. Factors effecting adventitious shoot regeneration from leaf explants of quince (*Cydonia oblonga*). *Plant Cell Tissue Org. Cul.* **1993,**35, 273–277. DOI:10.1007/BF00037281

Baninasab, B.; Mohammadi, S.; Khoshgoftarmanesh, A. H.; Ghasemi, A. Responses of quince (*Cydonia oblonga* Mill.), pear and *Crataegus* rootstocks to Fe-deficiency stress in soilless Culture.

J. Sci. Technol. Greenh. Cul. **2015**, 5, 127–137.

Baxter, P. *Growing Fruit in Australia, for Profit or Pleasure. 5th edition*; Pan Macmillan, Melbourne, Australia, **1998**.

Bayazit, S.; Imrak, B.; Küden, A.; Kemal Güngör, M. RAPD analysis of genetic relatedness among selected quince (*Cydonia oblonga* Mill.) accessions from different parts of Turkey. *HortScience (Prague)* **2011**,38, 134–141

Bell, L.R.; Leitão, J.M. *Cydonia*. In *Wild Crop Relatives: Genomic and Breeding Resources, Temperate Fruits*; Kole, C., Ed.; Springer-Verlag Press. Heidelberg, Germany, **2011**; p 1.

Bellini, E.; Giordani, E. Conservation of under-utilized fruit tree species in Europe. *Acta Hortic.* **2000**,522, 165–174. DOI:10.17660/ActaHortic.2000.522.18

Bellini, E.; Giordani, E. The Online European Minor Fruit Tree Species Database. EMFTS Database. Ed. [Online] 1999, 1 www.unifi.it/project/ueresgen29/netdbase/db1.htm (accessed Mar 7, **2009**).

Benedek, P.; Szabó, T.; Nyéki, J. New results on the bee pollination of quince (*Cydonia oblonga* Mill.). *Acta Hortic.* **2001**, 561, 243–248. DOI:10.17660/ActaHortic.2001.561.35

Benzarti, S.; Hamdi, H.; Lahmayer, I.; Toumi, W.; Kerkeni, A.; Belkadhi, K.; Sebei, H. Total phenolic compounds and antioxidant potential of quince (*Cydonia oblonga* Miller) leaf methanol extract. *Int. J. Innov. Appl. Stu.* **2015**, 13, 518–526.

Beyaz, A.; Ozturk, R.; Acar, A. I.; Turker, U. Determination of enzymatic browning on quinces (*Cydonia oblonga*) with image analysis. *J. Agri. Mach. Sci.* **2011**, 7, 411–414.

Birişik, N.; Baloğlu, S. Books of Abstracts, 21st International Conference on Virus and other Graft Transmissible Diseases of Fruit Crops, Neustadt, Germany, July 5–10, 2009; International Council for the study of virus and other graft transmissible diseases

of fruit crops (ICVF). *International Society for Horticultural Science* **2010**, 427, 257–262.

Bobev S.; Angelov L.; Govedarov G.; Postman J. Field susceptibility of quince hybrids to fire blight in Bulgaria. *APS Annual Meeting Report* **2009**,99, S13.

Bonn, GW.; van der Zwet, T. *Distribution and Economic Importance of Fire Blight*. In *Fire Blight: the Disease and its Causative Agent, Erwinia amylovora*; Vanneste J.L., Ed.; CAB International, Wallingford, UK, **2000**; p 37. DOI:10.1079/9780851992945.0037

Bunnag, S.; Dolcet-Sanjuan, R.; Mok, D.W.S.; Mok, M.C. Responses of two somaclonal variants of quince (*Cydonia oblonga*) to iron deficiency in the greenhouse and field. *J. Am. Soc. Hortic. Sci.* **1996**, 121, 1054–1058.

Çelik, H.; Katkat, A.V. Some parameters in relation to iron nutrition status of peach orchards. *J. Biol. Env. Sci.* **2007**, 1, 111–115.

Çetin, M.; Soylu, A. Studies on the fertilization biology of standard quince cultivars (*Cydonia oblonga* Mill.). *Bahçe* **2006**, 35, 83–95 (in Turkish).

Chaplin, M.H.; Westwood, M.N. Relationship of nutritional factors to fruit set. *J. Plant Nut.* **1980**, 2, 477–505. DOI:10.1080/01904168009362791

Chartier-Hollis, J.M. The induction and maintenance of caulogenesis from undifferentiated callus of quince (*Cydonia oblonga*). *Acta Hortic.* **1993**,336, 321–326. DOI:10.17660/ActaHortic.1993.336.42

Chong, C.; Chan, W.W.; Taper, C.D. Sorbitol and carbohydrate content in apple skin. *J. Hortic. Sci.* **1972**, 47, 209–213. DOI:10.1080/00221589.1972.11514459

Cinelli, F.; Loreti, F.; Muleo, R. Regeneration and selection of quince BA29 (*Cydonia oblonga* Mill.) somaclones tolerant to lime-induced chlorosis. *Acta Hortic.* **2004**, 658, 573–579. DOI:10.17660/ActaHortic.2004.658.87

Costa, G.; Ramina, A. *Temperate Fruit Species*. In *Horticulture: Plants for People and Places*; Dixon, G.R.; Aldous, D.E., Eds.; Springer Press, Heidelberg, **2014**; Volume 1: Production Horticulture; p. 97.

Croitoru D.C.; Baciu, A.A. Preliminary results obtained in the experimental area the agro productivity study of some quince varieties in the fruit growing basin Târgu Jiu. *J. Hortic. Fores. Biotech.* **2013**, 17, 250–254.

Curtis, J.; Tallis, N. *Forgotten Empire: The World of Ancient Persia*; The British Museum Press, London, UK, **2005**.

D'Onofrio, C.; Morini, S. Development of adventitious shoots from *in vitro* grown *Cydonia oblonga* leaves as influenced by different cytokinins and treatment duration. *Biol. Plant.* **2005**, 49, 17–21. DOI:10.1007/s10535-005-7021-8

D'Onofrio, C.; Morini, S. Increasing NaCl and $CaCl_2$ concentrations in the growth medium of quince leaves: I. Effects on somatic embryo and root regeneration. *In Vitro Cell. Dev. Biol.-Plant* **2002a**, 38, 366–372. DOI:10.1079/IVP2002308

D'Onofrio, C.; Morini, S. Increasing NaCl and $CaCl_2$ concentrations in the growth medium of quince leaves: II. effects on shoot regeneration. *In Vitro Cell. Dev. Biol.-Plant* **2002b**,38, 373–377. DOI:10.1079/IVP2002309

D'Onofrio, C.; Morini, S. Somatic embryo, adventitious root and shoot regeneration in *in vitro* grown quince leaves as influenced by treatments of different length with growth regulators. *Sci. Hortic.* **2006**,107, 194–199. DOI:10.1016/j.scienta.2005.05.016

D'Onofrio, C.; Morini, S.; Bellocchi, G. Effect of light quality on somatic embryogenesis of quince leaves. *Plant Cell Tissue Org. Cul.* **1998**, 53, 91–98. DOI:10.1023/A:1006059615088

Dadpour, M.R. Grigorian, W.; Nazemieh A.; Valizadeh, M. Application of epi-illumination light microscopy for study of floral ontogeny in fruit trees. *Int. J. Bot.* **2008**, 4, 49–55. DOI:10.3923/ijb.2008.49.55

Dalkiliç, Z.; Osman Mestav, H. *In vitro* pollen quantity, viability and germination tests in quince. *Afr. J. Biotech.* **2011**, 10, 16516–16520. DOI:10.5897/AJB11.2637

Daryaee, T. List of fruits and nuts in the Zoroastrian tradition: an Irano-Hellenic classification. *Name-ye Iran-e Bastan* **2006**-07, 6, 75–84 (in Persian).

Davatchi, A.; Esmaïli, M. The quince moth *Euzophera bigella* Zell. (Lep. Phyticidae) in Iran. *Entomol. Phytopath. Appl.* **1970**, 29, 67–79.

Déri, H.; Szabó L. G.; Bubán T.; Orosz-Kovács Z.; Szabó T.; Bukovics, P. Floral nectar production and composition in quince cultivars and its apicultural significance. *Acta Bot. Hun.* **2006**, 48, 279–290. DOI:10.1556/ABot.48.2006.3-4.4

Dolcet-Sanjuan, R.; Mok, D.W.S.; Mok, M.C. Plantlet regeneration from cultured leaves of *Cydonia oblonga* L. (quince). *Plant Cell Rep.* **1991**, 10, 240–242. DOI:10.1007/BF00232566

Dumanoglu, H.; Tuna-Gunes, N.; Aygun, A.; San, B.; Akpinar, A.E.; Bakir, M. Analysis of clonal variations in cultivated quince (*Cydonia oblonga* "Kalecik") based on fruit characteristics and SSR markers. *New Zealand J. Crop Hortic. Sci.* **2009**, 37, 113120. DOI:10.1080/01140670909510256

Duron, M.; Decourtye, L.; Druart, P. *Quince* (*Cydonia oblonga* Mill.). In *Biotechnology in Agriculture and Forestry*; Bajai, Y.P.S., Ed.; Springer, Berlin, Heidelberg, Germany, **1989**; Vol. 5: Trees II; p. 42.

Ebert, G.; Kirkby, E.A. *Fertilizing for High Yield and Quality: Pome and Stone Fruits of the Temperate Zone*; IPI Bulletin No. 19, International Potash Institute, Horgen, Switzerland. **2009**. DOI:10.3235/978-3-9523243-6-3

Encyclopedia of Life. *Cydonia oblonga* (Quince). www.eol.org/

pages/637321?category (accessed Jan 3, **2018**).

EPPO. Pathogen-tested material of *Malus, Pyrus* and *Cydonia*. *EPPO Bull.* **1999**, 29, 239–252. DOI:10.1111/j.1365-2338.1999.tb00828.x

Érsek, T.; Belbahri, L.; Nagy, Z.Á.; Bakonyi, J.; Crovadore, J.; Lefort, F. Medlar decline caused by *Phytophthora cactorum* in Hungary. *Plant Pathol.* **2008**, 57, 775. DOI:10.1111/j.1365-3059.2008.01857.x

Esumi, T.; Tao, R.; Yonemori, K. Comparison of early inflorescence development between Japanese pear (*Pyrus pyrifolia* Nakai) and quince (*Cydonia oblonga* Mill.). *J. Jap. Soc. Hortic. Sci.* **2007a**, 76, 210–216. DOI:10.2503/jjshs.76.210

Esumi, T.; Tao, R.; Yonemori, K. Relationship between floral development and transcription levels of *LEAFY* and *TERMINAL FLOWER 1* homologs in Japanese pear (*Pyrus pyrifolia* Nakai) and quince (*Cydonia oblonga* Mill.). *J. Jap. Soc. Hortic. Sci.* **2007b**, 76, 294–304. DOI:10.2503/jjshs.76.294

Evrenosoğlu Y.; Misirli A. Investigations on the pollen morphology of some fruit species. *Turk. J. Agri. For.* **2009**, 33, 181–190. DOI:10.3906/tar-0801-47

Fikret Balta M.; Muradoglu, F. Fruit set of Turkish quince cv. "Ekmek" (*Cydonia oblonga* Mill.) under open-pollination conditions. *Bio Sci. Res. Bull. Biol. Sci.* **2006**, 22, 1–2.

Fisichella, M.; Morini, S. Somatic embryo and root regeneration from quince leaves cultured in ventilated vessels or under different oxygen and carbon dioxide levels. *In Vitro Cell. Dev. Biol.-Plant* **2003**,39, 402–408. DOI:10.1079/IVP2003429

Fisichella, M.; Silvi, E.; Morini, S. Regeneration of somatic embryos and roots from quince leaves cultured on media with different macroelement composition. *Plant Cell, Tissue Org. Cul.* **2000**,63, 101–107. DOI:10.1023/A:1006407803660

Food and Agriculture Organization of the United Nations. FAOSTAT. www.fao.org/faostat (accessed Nov 14, **2017**).

Francini, A.; Sebastiani, L. Phenolic compounds in apple (*Malus×domestica* Borkh.): Compounds characterization and stability during postharvest and after processing. *Antioxidants (Basel)* **2013**,2, 181–193. DOI:10.3390/antiox2030181

Frantskevich, N.A. Wild relatives of crop plants and their conservation in the basin of the river Ai-Dere (Kara-Kala region of the Turkmen SSR). *Byulleten'-vsesoyuznogo ordena Lenina I ordena Druzhby Narodov Instituta Rastenievodstva Imeni N I Vavilov* **1978**,81, 86–91 (in Russian).

Ganopoulos, I.; Merkouropoulos, G.; Pantazis, S.; Tsipouridis, C.; Tsaftaris, A. Assessing molecular and morpho-agronomical diversity and identification of ISSR markers associated with fruit traits in quince (*Cydonia oblonga*). *Gen. Mol. Res.* **2011**,10, 2729–2746. DOI:10.4238/2011.November.4.7

Gerami, E.; Hassanzadeh, N.; Abdollahi, H.; Ghasemi, A.; Heydari, A. Evaluation of some bacterial antagonists for biological control of fire blight disease. *J. Plant Pathol.* **2013**, 95, 127–134. DOI:10.4454/JPP.V95I1.026

Ghahremani, Z.; Alipour, M.; Ahmadi, S.; Abdollahi, H.; Mohamadi, M.; Ghasemi, A.A.; Adli, M. Selecting effective indices for evaluation of fire blight resistance in quince germplasm under orchard settings. *Acta Hortic.* **2014**, 1056, 247–251. DOI:10.17660/ActaHortic.2014.1056.41

Ghasemi, A.; Nassiri, J.; Yahyaabadi, M. Study of the relative tolerance of quince (*Cydonia oblonga* Mill.) rootstocks to different bicarbonate concentrations. *Seed Plant Pro. J.* **2010**, 26, 137–151. DOI:10.22092/sppj.2017.110400

Ghozati, E.; Abdollahi, H.; Piri, S. Comparison of total phenolic contents of the leaves and fruits of quince genotypes and its effects on the fire blight resistance.

Seed Plant Imp. J. **2016**, 32, 331–345. DOI:10.22092/spij.2016.113062

Giorgota, A.; Preda, S.; Isac, M.; Tulvinschi, M. Development of a micropropagation protocol for the Romanian quince (*Cydonia oblonga*) cultivar "Aurii" and rootstocks "BN70" and 'A TYPE'. *Acta Hortic.* **2009**, 839, 105–110. DOI:10.17660/ActaHortic.2009.839.11

Guiné, R.P.F.; Barroca, M.J. Books of Abstracts, 6th Central European Congress on Food, Novi Sad, Serbia, May 23–26, 2012; International Union of Food Science and Technology: Paris, France, **2012**; p. 666.

Hagin, J.; Tucker, B. *Fertilization of Dryland and Irrigated Soils*; Springer-Verlag Berlin Heidelberg, Germany, **1982**. DOI:10.1007/978-3-642-68327-5

Hajebrahimi, S.; Banihashemi, Z. Host range of *Phytophthora parsiana*: a new high temperature pathogen of woody plants. *Phytopathol. Mediterr.* **2011**,50, 159–165. DOI:10.14601/Phytopathol_Mediterr-3055

Hanke, M. V.; Flachowsky, H.; Peil, A.; Hättasch, C. No flower no fruit – genetic potential to trigger flowering in fruit trees. *Gene Geno. Genom.* **2007**,1, 1–20.

Hegedûs, A.; Stefanovits-Bányai, É.; Szabó, Z.; Nyéki, J.; Pedryc, A.; Halász, J. Books of Abstracts, 1st Symposium on Horticulture in Europe, Vienna, Austria, February 17–20, 2008; International Society for Horticultural Science: Leuven, Belgium, **2008**; p 280.

Hodgson, I. Common quince: a fruit of distinction. *The Garden* December 2011, **2011**, 64–66.

Holevas, C.D.; Biris, D.A. *Bitter Pit-Like Symptoms in Quinces: Effect of Calcium and Boron Sprays on the Control of the Disorder.* In *Mineral Nutrition of Fruit Trees*; Atkinson, D.; Jackson, J.E.; Sharples, R.O.; Waller, W.M., Eds.; Butterworths, Boston, USA, **1980**; p. 319.

Horst, R.K. *Field Manual of Diseases on Fruits and Vegetables*; Springer, Netherlands, **2013**.

Hrustić, J.; Grahovac, M.; Mihajlović, M.; Delibašić, M.; Ivanović, M.; Nikolić, M.; Tanović, M.; Molecular detection of *Monilinia fructigena* as causal agent of brown rot on quince. *Pestic. Phytomed. (Belgrade)* **2012**,27, 15–24. DOI:10.2298/PIF1201015H

Hummer, K.E.; Janick, J. *Rosaceae: Taxonomy, Economic Importance, Genomics.* In *Genetics and Genomics of Rosaceae, Plant Genetics and Genomics: Crops and Models*; Folta, K.M.; Gardiner, S.E., Eds.; Springer, New York, **2009**; p. 1.

Hummer, K.E.; Pomper, K.W.; Postman J.; Graham, C.J.; Stover, E.; Mercure, E.W.; Aradhya, M.; Crisosto, C.H.; Ferguson, L.; Thompson, M.M.; Byers, P.; Zee, F. *Emerging fruit crops.* In *Fruit Breeding, Handbook of Plant Breeding*; Badenes, M.L.; Byrne, D., Eds.; Springer, Boston, MA, **2012**; Vol. 8; p. 97.

Jones, A.L.; Aldwinckle, H.S. *Compendium of Apple and Pear Diseases*; American Phytopathological Society Press, St. Paul, USA, **1990**.

Kader, A.A. *Postharvest Technology of Horticultural Crops*; University of California, Division of Agriculture and Natural Resources, Oakland, California, USA. Publication No. 3311, **1992**.

Khandan, A.; Abdollahi, H.; Hajnajari, H. *National Guidelines for Distinction, Uniformity and Stability Examination in Quince (Cydonia oblonga* Mill.); Seed and Plant Certification and Registration Institute Publication, Karaj, Iran, **2011**.

Khatamsaz, H. *Iran's Flora, Rosaceae*; Research Institute of Forests and Rangelands, Tehran, Iran, **1992**. (in Persian).

Khoramdel Azad, M.; Nasiri, J.; Abdollahi, H. Genetic diversity of selected Iranian quinces using SSRs from apples and pears.

Biochem. Gen. **2013**, 51, 426–442. DOI: 10.1007/s10528-013-9575-z

Khosravinezhad, F.; Abdollahi, H.; Kashefi, B.; Hassani, M.; Salehi, Z. Study on *in vitro* propagation of some promising quince (*Cydonia oblonga*) cultivars. *Iran. J. Hortic. Sci.* **2016**, 47, 135–144. DOI:10.22059/ijhs.2016.58219

Khoubnasabjafari, M.; Jouyban, A. A review of phytochemistry and bioactivity of quince (*Cydonia oblonga* Mill.). *J. Med. Plants Res.* **2011**, 5, 3577–3594.

Kimura, T.; Yamamoto, T.; Ban, Y. *Identification of Quince Varieties Using SSR Markers Developed from Pear and Apple*; International Union for the Protection of New Varieties of Plants, Geneva. Switzerland, 2005.

Koc, B.; Eren, I.; Kaymak Ertekin, F. Modelling bulk density, porosity and shrinkage of quince during drying: The effect of drying method. *J. Food Eng.* **2008**, 85, 340–349. DOI:10.1016/j.jfoodeng.2007.07.030

Kuden, A.; Tumer, M.A.; Gungor, M.K.; Imrak, B. Pomological traits of some selected quince types. *Acta Hortic.* **2009**, 818, 73–76. DOI:10.17660/ActaHortic.2009.818.9

Kuzucu, C. F.; Sakaldas, M. The effects of different harvest times and packaging types on fruit quality of *Cydonia oblonga* cv. "Eşme". *J. Fac. Agri. Harran Uni.* **2008**, 12, 33–39.

Leblay, C.; Chevreau, E.; Raboin, L.M. Adventitious shoot regeneration from *in vitro* leaves of several pear cultivar (*Pyrus communis* L.). *Plant Cell, Tissue Org. Cul.* **1991**, 25, 99–105. DOI:10.1007/BF00042180

Lim, T.K. *Edible Medicinal and Non-Medicinal Plants*, Vol. 4, Fruits; Springer, Heidelberg, Germany, **2012**. DOI:10.1007/978-94-007-4053-2_45

Maggioni, L.; Fischer M.; Lateur M.; Lamont E. J.; Lipman E. *Report of a Working Group on Malus/Pyrus*. International Plant Genetic Resources Institute, Rome, Italy. **2002**.

Marino, G.; Molendini, L. *In vitro* leaf-shoot regeneration and somaclone selection for sodium chloride tolerance in quince and pear. *J. Hortic. Sci. Biotech.* **2005**, 80, 561–570. DOI:10.1080/14620316.2005.11511978

Martínez-Valero, R.; Melgarejo, P.; Salazar, D.M.; Martínez, R, Martínez, J.J.; Hernández, F.C. A. Phenological stages of the quince tree (*Cydonia oblonga*). *Ann. Appl. Biol.* **2001**, 139, 189–192. DOI:10.1111/j.1744-7348.2001.tb00395.x

Mazarei, M.; Zakeri, Z.; Hassanzade, N. The status of fire blight disease on pome fruits in west Azarbaijan province and Ghazvin in 1991–1992. *Iran. J. Plant Pathol.* **1994**, 30, 7–9.

McGee, H. *On Food and Cooking, the Science and Lore of the Kitchen*; Scribner Publisher, New York, USA, **2004**.

McGinnis, L. Quest for quince: expanding the NCGR collection. *Agricultural Research* January 2007, **2007**, 20–21.

McMorland Hunter, J.; Dunster, S. *Quinces: Growing & Cooking: The English Kitchen*; Prospect Books Publisher, UK, **2014**.

Meech, W. W. *Quince Culture. An illustrated Handbook for the Propagation and Cultivation of the Quince*, Orange Judd company, New York, USA, **1888**.

Mehrabipour, S.; Abdollahi, H.; Adli, M. Response of some quince (*Cydonia oblonga* Mill.) genotypes from Guilan and Khorasan provinces to fire blight disease. *Seed Plant Imp. J.* **2012**, 28, 67–84. DOI:10.22092/spij.2017.111092

Mehrabipour, S.; Abdollahi, H.; Hassanzadeh, N.; Ghasemi, A. A. The role of some quince stock (*Cydonia oblonga*) genotypes in susceptibility to fire blight disease. *Appl. Entomol. Phytopathol.* **2010**, 78, 25–42.

Mengel, K. Iron availability in plant tissues-iron chlorosis on calcareous soils. *Plant*

Soil **1994**, 165, 275–283. DOI:10.1007/BF00008070

Milić, D.; Vukoje, V.; Sredojević, Z. Production characteristics and economic aspects of quince production. *J. Process. Ene. Agri.* **2010**, 14, 36–39.

Mirabdulbaghi, M.; Abdollahi, H. Foliar nutrient response in some Iranian quince genotypes. *J. Hort. For.* **2014**,6, 92–98. DOI:105897/JHF2014.0363

Mitcham, E.J.; Elkins, R.B. *Pear Production and Handling Manual*; University of California, Agricultural and Natural Resource Communication Service, California, USA, **2007**.

Mohammadi, M.; Nadi, S.; Abdollahi, H. Tolerance to the late spring frost in some quince (*Cydonia oblonga* Mill.) genotypes in Karaj climate. *Seed Plant Imp. J.* **2017**, 32, 461–477. DOI:10.22092/spij.2017.113083

Moradi, S.; Koushesh Saba, M.; Mozafari, A. A.; Abdollahi, H. Antioxidant bioactive compounds changes in fruit of quince genotypes over cold storage. *J. Food Sci.* **2016**, 81, H1833–H1839. DOI:10.1111/1750-3841.13359

Moradi, S.; Koushesh Saba, M.; Mozafari, A. A.; Abdollahi, H. Physical and biochemical changes of some Iranian quince (*Cydonia oblonga* Mill.) genotypes during cold storage. *J. Agri. Sci. Technol.* **2017**, 19, 377–388.

Morini, S.; D'Onofrio, C.; Bellocchi, G.; Fisichella, M. Effect of 2,4-D and light quality on callus production and differentiation from *in vitro* cultured quince leaves. *Plant Cell, Tissue Org. Cul.* **2000**, 63, 47–55. DOI:10.1023/A:1006456919590

Morini, S.; D'Onofrio, C.; Fisichella, M.; Loreti, F. Effect of high and low temperature on the leaf regenerating capacity of quince BA29 rootstock. *Acta Hortic.* **2004**, 658, 591–597. DOI:10.17660/ActaHortic.2004.658.89

Mosharaf L.; Ghasemi A. The effect of harvesting time on prolonging the storage time of Isfahan quince cultivar. *J. Crop Pro. Process.* **2004**, 8, 181–190.

Muleo, R.; Fisichella, M.; Iacona, C.; Viti, R.; Cinelli, F. Different responses induced by bicarbonate and iron deficiency on microshoots of quince and pear. *Acta Hortic.* **2002**, 596, 677–681. DOI:10.17660/ActaHortic.2002.596.117

Murashige, T.; Skoog, F. A revised medium for rapid growth and bio assays with tobacco tissue cultures. *Physiol. Plant.* **1962**, 15, 473–497. DOI:10.1111/j.1399-3054.1962.tb08052.x

Nagy-Déri, H. Morphological investigations on anthers and pollen grains of some quince cultivars. *Acta Biolo. Szegediensis* **2011**, 55, 231–235.

Nanos, G.D.; Mpezou, A.; Georgoudaki, T. Effects of 1-MCP and storage temperature on quince fruit quality. *Acta Hortic.* **2015**, 1079, 453–458. DOI:10.17660/ActaHortic.2015.1079.59

NCGR-Corvallis. National Clonal Germplasm Repository of U.S. National Plant Germplasm System. Taxon: *Cydonia oblonga* Mill. https://npgsweb.ars-grin.gov (accessed Oct 24, **2017**).

Németh, G. Benzyladenine stimulated rooting in fruit tree rootstocks cultured *in vitro*. *Zeitschrift für Pflanzenphysiologie* **1979**, 95, 396–389. DOI:10.1016/S0044-328X(79)80209-3

Nikkhah, S.; Ganji Moghadam, E. Effects of warm solutions of thiabendazole and packaging on qualitative and quantitative characteristics of Quince cv. Gorton. *Pajouhesh- va- Sazandegi in Agron. Hort.* **2009**, 21, 61–67 (in Persian).

Ogawa, J.M.; English, H. *Diseases of Temperate Zone Tree Fruit and Nut Crops*; University of California, Division of Agriculture and Natural Resources, Oakland, Publication 3345. **1991**.

Oliveira, A. P.; Pereira, J. A.; Andrade, P. B.; Valentão, P.; Seabra, R. M.; Silva, B. M.

Phenolic profile of *Cydonia oblonga* Miller leaves. *J. Agri. Food Chem.* **2007**,55, 7926–7930. DOI:10.1021/jf0711237

Özbek, S.; Ozel Meyvecilik. C.U.Z.F. *Yayinlari*; Adana. **1978**. (in Turkish).

Paunoviç, S.; Rankoviç, M. Relationship between quince fruit deformation virus and some pome fruit viruses. *Acta Hortic.* **1998**,472, 125–134. DOI:10.17660/ActaHortic.1998.472.12

Phillips, H. *Pomarium Britanicum: An Historical and Botanical Account of Fruits Known in Great Britain*; Innes Printer, Oxford, UK. **1820**.

Pinar, H.; Kaymak, S.; Ozongun, S.; Uzun, A.; Unlu, M.; Bircan, M.; Ercisli, S.; Orhan, E. Morphological and molecular characterization of major quince cultivars from Turkey. *Not. Bot. Hort. Agri.* **2016**, 44, 72–76. DOI:10.15835/nbha44110228

Postman, J. *Cydonia oblonga*: The unappreciated quince. *Arnoldia* **2009**, 67, 2–9.

Postman, J. D. Quince (*Cydonia oblonga* Mill.) center of origin provides sources of disease resistance. *Acta Hortic.* **2012**, 948, 229–234. DOI:10.17660/ActaHortic.2012.948.26

Postman, J. D. World *Pyrus* collection at USDA genebank in Corvallis, Oregon. *Acta Hortic.* **2008**, 800, 527–534. DOI:10.17660/ActaHortic.2008.800.69

Psallidas, P.G.; Tsiantos, J. *Chemical Control of Fire Blight*. In *Fire Blight: the Disease and its Causative Agent, Erwinia amylovora*; Vanneste, J.L., Ed.; CAB International, Wallingford, UK, **2000**; p 199. DOI:10.1079/9780851992945.0199

Quoirin, M.; Lepoivre, P. Improved media for *in vitro* culture of *Prunus* sp. *Acta Hortic.* **1977**, 78, 437–442. DOI:10.17660/ActaHortic.1977.78.54

Radović, A.; Nikolić, D.; Milatović, D.; Durović, D.; Dordević, B. Fruit characteristics of quince cultivars in the region of Belgrade. *J. Pom.* **2015**, 49, 15–19. (in Serbian).

Rahemi, M.; Akbari, H. Effects of heat treatments and packing type on quality and storage period of quince fruit. *Iran. J. Hortic. Sci. Technol.* **2004**, 4, 83–94.

Rajabi, G. R.; Dastgheib-Beheshti, N. Supplementary studies on quince moth in Iran. *Appl. Ent. Phytopath.* **1979**, 47, 53–70.

Ramos, A. M.; Ibarz, A. Thixotropy of orange concentrate and quince puree. *J. Tex. Stu.* **1998**, 29, 313–324. DOI:10.1111/j.1745-4603.1998.tb00173.x

Robinson, R. *Plant Sciences, Vol. 2*; Macmillan Reference, New York, USA, **2001**.

Rodríguez-Guisado, I.; Hernández, F.; Melgarejo, P.; Legua, P.; Martínez, R.; Martínez, J.J. Chemical, morphological and organoleptical characterization of five Spanish quince tree clones (*Cydonia oblonga* Miller). *Sci. Hortic.* **2009**,122, 491–496. DOI:10.1016/j.scienta.2009.06.004

Römheld, V.; Nikolic, M. *Iron*. In *Handbook of Plant Nutrition*; Barker, A.V.; Pilbeam, D.J., Eds.; CRC Taylor & Francis Group, Boca Raton, FL, USA, **2007**; p 329.

Rop, O.; Balík J.; Řezníček, V.; Juríková, T.; Škardová, P.; Salaš, P.; Sochor, J.; Mlček, J.; Kramářová, D.: Chemical characteristics of fruits of some selected quince (*Cydonia oblonga* Mill.) cultivars. *Czech J. Food Sci.* **2011**, 29, 65–73.

Scaramuzzi, F. Contributo allo studio delle cultivar di cotogno da frutto. *Rivista di Frutticoltura e di Ortofloricoltura Italiana* **1958**, 41, 21–50.

Serra, S. Salt Stress Responses in Pear and Quince: Physiological and Molecular Aspects. PhD Dissertation, Alma Mater Studiorum – Università di Bologna, Bologna, Italy, **2009**.

Sharafi, Y. Investigation on pollen viability and longevity in *Malus pumila* L. *Pyrus communis* L. *Cydonia oblonga* L. *in vitro*. *J. Med. Plants Res.* **2011**, 5, 2232–2236.

Sharma, R.; Joshi, V.K.; Rana, J.C. Nutritional composition and processed products of quince (*Cydonia oblonga* Mill). *Ind. J. Nat. Pro. Res.* **2011**, 2, 354–357.

Silva, B.M.; Andrade, P.B.; Ferreres, F.; Domingues, A.L.; Seabra, R.M.; Ferreira, M.A. Phenolic profile of quince fruit (*Cydonia oblonga* Miller) (pulp and peel). *J. Agri. Food Chem.* **2002a**, 50, 4615–4618. DOI:10.1021/jf0203139

Silva, B.M.; Andrade, P.B.; Mendes, G.C.; Seabra, R.M.; Ferreira, M.A. Study of the organic acids composition of quince (*Cydonia oblonga* Miller) fruit and jam. *J. Agri. Food Chem.* **2002b**,50, 2313–2317. DOI:10.1021/jf011286+

Simidchiev, T. Investigations on the nectar and honey productivity of the quince (*Cydonia vulgaris* Pers./). *Nauch. Trud. Vissh. Selskostop. Inst. Vasil Kolarov (Plovdiv)* **1967**, 12, 241–253 (in Bulgarian).

Simoons, F.J. *Food in China: A Cultural and Historical Inquiry*; CRC Press, Boca Raton, FL, USA, **1990**.

Sola, S. What quince tree will grow in Georgia?. Ehow. Ed. [Online] *2011*, 1 www.ehow.com (accessed Nov 12, **2012**)

SPII. *Special Issues of the First Report on Seed & Plant Improvement Institute Released Cultivars*; Seed & Plant Improvement Institute Press, Karaj, Iran, **2015** (in Persian).

Stancevic, A.; Nikolic, M. Quince breeding in Yugoslavia. *Acta Hortic.* **1992**, 317, 107–110. DOI:10.17660/ActaHortic.1992.317.10

Stojanović, C.; Babić, M.; Babić, L.; Pavkov, I.; Radojčin, M.; Lončarević, V. Osmotic drying of quince (*Cydonia oblonga* Mill.) in sucrose solution. *J. Process. Ene. Agri.* **2010**, 14, 44–48.

Sykes, J.T.A description of some quince cultivars from western Turkey. *Econo. Bot.* **1972**,26, 21–31. DOI:10.1007/BF02862258

Szabó, T.; Nyéki, J.; Soltész, M.; Szabó, Z.; Tóth, T. Time of flowering and fertilisation of quince varieties. *Int. J. Hortic. Sci.* **1999**,5, 9–15.

Talaei, Z.; Abdollahi, H.; Sorkhi, B.; Tatari, M. Determination of self-incompatibility *S*-alleles in native quince (*Cydonia Oblonga* Mill.) germplasm of Iran. **2017**, NCBI Accession No: MF281258.

Tang, Y.; Yu, X.; Mi. M.; Zhao, J.; Wang, J.; Zhang, T. Antioxidative property and antiatherosclerotic effects of the powder processed from *Chaenomeles speciosa* in ApoE$^{-/-}$mice. *J. Food Biochem.* 34, **2010**, 535–548. DOI:10.1111/j.1745-4514.2009.00297.x

Tatari, M.; Abdollahi, H.; Mousavi, A. Effect of pollination on dropping of flowers and fruits in new quince (*Cydonia oblonga* Mill.) cultivar and promising genotypes. *Sci. Hortic.* **2018**, 231, 126–132. DOI:10.1016/j.scienta.2017.10.045

Tatari, M.; Mousavi, A. Impact of harvesting time and length of cold storage period on physiological and quality traits of four quince genotypes (*Cydonia oblonga* Mill.). *J. Hortic. Res.* **2017**, 25, 67–79. DOI:10.1515/johr-2017-0008

Tateo, F.; Bononi, M. Headspace-SPME analysis of volatiles from quince whole fruit. *J. Ess. Oil Res.* **2010**, 22, 416–418. DOI:10.1080/10412905.2010.9700360

The Connecticut Agricultural Experiment Station. The Connecticut Agricultural Experiment Station list of Quince (*Cydonia*) Plant Health Problems. www.ct.gov/caes (accessed Nov 27, **2017**).

Thomidis, T.; Tsipouridis, C.; Isaakidis, A.; Michailides, Z. Documentation of field and postharvest performance for a mature collection of Quince (*Cydonia oblonga*) varieties in Imathia, Greece. *New Zealand J. Crop Hortic. Sci.* **2004**, 32, 243–247. DOI:10.1080/01140671.2004.9514302

Thompson, M.M. Temperate fruit crop germplasm in Syria. *Plant Genet Resource Newsletters of International Board for*

Plant Genetic Resources (IBPGR) **1986**, 68, 29–34.

Torkashvand, M. *In vitro* Establishment and Micropropagation of Superior Quince Genotypes of Iran and Evaluation of Their Genetic Potential by Molecular Markers. MSc Dissertation, Azad University of Tehran, Tehran, Iran, **2014** (in Persian).

Tsuneya, T.; Ishihara, M.; Shiota, H.; Shiga, M. Volatile components of quince fruit (*Cydonia oblonga* Mill.). *Agri. Biol. Chem.* **1983**, 47, 2495–2502. DOI:10.108 0/00021369.1983.10865983

Türk, R.; Memicoglu, M. The effects of different localities and harvest time on the storage period of quince. *Acta Hortic.* **1994**, 368, 840–849. DOI:10.17660/ ActaHortic.1994.368.100

Tukey, H.B. Dwarfed Fruit Trees. Cornell University Press, Ithaca, USA. **1964**.

Tuna-Gunes, N. Changes in ethylene production during preharvest period in quince (*Cydonia vulgaris* L.) and the use of ethylene production to predict harvest maturity. *Europ. J. Hortic. Sci.* **2003**,68, 212–221.

Tuna-Gunes, N. Ripening regulation during storage in quince (*Cydonia oblonga* Mill.) fruit. *Acta Hortic.* **2008**,796, 191–196. DOI:10.17660/ActaHortic.2008.796.24

Tuna-Gunes, N.; Dumanoglu, H. Some fruit attributes of quince (*Cydonia oblonga*) based on genotypes during the pre-harvest period. *New Zealand J. Crop Hortic. Sci.* **2005**, 33, 211–217. DOI:10.1080/011406 71.2005.9514352

Tuna-Gunes, N.; Dumanoglu, H.; Poyrazoglu, E.S. Use of 1-MCP for keeping postharvest quality of "Ekmek" quince fruit. *Acta Hortic.* **2012**, 934, 297–302. DOI:10.17660/ActaHortic.2012.934.37

Tuna-Gunes, N.; Koksal, A.I. Ethylene biosynthesis of quince during storage. *Acta Hortic.* **2005**, 682, 177184. DOI:DOI:10.17660/ ActaHortic.2005.682.17

Tuna-Gunes, N.; Koksal, A.I. Relationships between some fruit characteristics and sensory evaluation in quince (*Cydonia oblonga* Mill.) fruits. *Acta Hortic.* **2007**, 741, 125–132. DOI:10.17660/ ActaHortic.2007.741.14.

Turchi, R. *Collezioni Germoplasma Frutticolo Presenti sul Territorio Regionale, V. Il Germoplasma Toscano delle Specie Legnose da Frutto*; ARSIA, Regoone Toscana, Italy, **1999**. (in Italian).

Umano, K.; Shoji, A.; Hagi, Y.; Shibamoto, T. Volatile constituents of peel of quince fruit, *Cydona oblonga* Mill. *J. Agric. Food Chem.* **1986**, 34, 593–596. DOI:10.1021/ jf00070a003

University of Reading. National fruit collection. www.nationalfruitcollection. org.uk (accessed Jun 23, **2017**).

UNSE. *UNECE Standard Concerning the Marketing and Commercial Quality Control of Quince*; Agricultural Standards Unit, Economic Cooperation and Trade Division, United Nations Economic Commission for Europe, United Nation, Publication No. FFV-62, **2014**.

UPOV. *Guidelines for the Conduct of Tests for Distinctness, Uniformity and Stability of Quince (Cydonia Mill. sensu stricto)*; International Union for the Protection of New Varieties of Plants. Geneva, Switzerland. **2003**.

USDA. *Quince Food Composition. National Nutrient Database of USDA for Food Composition*; United States Department of Agriculture, Agricultural Research Service, USA, **2017**.

van der Zwet, T.; Beer, S. V. *Fire Blight–Its Nature, Prevention and Control: A Practical Guide to Integrated Disease Management*; US Department of Agriculture, Washington DC, USA. **1999**. DOI:10.5962/bhl.title.134796

van der Zwet, T.; Keil, H.L. *Fire Blight: A Bacterial Disease of Rosaceous Plants*; United States Department of Agriculture,

Agricultural Handbook No. 510, USA, **1979**.

Vanneste, J.L. *Erwinia amylovora*. In *Pathogenesis and Host Specificity in Plant Diseases: Histopathological, Biochemical, Genetic and Molecular Bases*; Singh U.S.; Singh, R.P.; Kohmoto, K.; Eds.; Pergamon Press, Oxford, London, UK, **1995**; Vol. 1: Prokaryotes; p. 21.

Vermerris, W.; Nicholson, R. *Phenolic Compound Biochemistry*; Springer Press, Dordrecht, the Netherlands, **2006**.

Vinson, J.A.; Zubik, L.; Bose, P.; Samman, N.; Proch, J.; Dried fruits: excellent *in vitro* and *in vivo* antioxidants. *J. Am. Coll. Nut.* **2005**, 24, 44–50. DOI:10.1080/07315724.2005.10719442

Vinterhalter, B.; Nešković, M. Factors affecting *in vitro* propagation of quince (*Cydonia oblonga* Mill.). *J. Hortic. Sci.* **1992**, 67, 39–43. DOI:10.1080/00221589.1992.11516218

Voicu, A.; Campeanu, G.; Bibicu, M.; Mohora, A.; Negoita, M.; Catana, L.; Catana, M. Achievement of a fortifying product based on apples, quinces and sea buckthorn. *Not. Bot. Hort. Agri. Nap.* **2009**, 37, 224–228. DOI:10.15835/nbha3713125

Vossen, P.M.; Silv, D. *Temperate Tree Fruit and Nut Crops*. In *California Master Gardener Handbook, 2nd ed.*; Pittenger, D., Ed.; University of California, Division of Agricultural and Natural Resource,

Communication Service & Information Technology, California, USA, **2015**; p 459.

Westwood, M.N. *Temperate Zone Pomology: Physiology and Culture*; Timber Press, Portland, Oregon, USA, **1993**.

Wojdylo, A.; Oszmiański, J.; Teleszko, M.; Sokół-Łętowska, A. Composition and quantification of major polyphenolic compounds, antioxidant activity and colour properties of quince and mixed quince jams. *Int. J. Food Sci. Nut.* **2013**, 64, 749–756. DOI:10.3109/09637486.2013.793297

Yamane, H.; Tao, R. Molecular basis of self-(in)compatibility and current status of *S*-genotyping in *Rosaceous* fruit trees. *J. Japan. Soc. Hortic. Sci.* **2009**, 78, 137–157. DOI:10.2503/jjshs1.78.137

Yezhov, V.N.; Smykov, A.V.; Smykov, V.K.; Khokhlov, S.Y.; Zaurov, D.E.; Mehlenbacher, S.A.; Molnar, T.J.; Goffreda, J.C.; Funk, C.R. Genetic resources of temperate and subtropical fruit and nut species at the Nikita Botanical Gardens. *HortSci.* **2005**, 40, 5–9.

Yolaçtı, S. Düsük ve Yüksek Sıcaklıkların Bazı Meyve Ağaçlarında *in Vitro* Polen Gelisimi Üzerine Etkileri. MSc Dissertation. Fırat Üniv. Fen Bil. Ens. Elazığ, Turkey, **2006** (in Turkish).

Zeller, O. Entwicklungsgeschichte und morphogenese der blütenknospen von der quitte (*Cydonia oblonga* Mill.). *Angewandte Botanik* **1960**, 34, 110–120.

CHAPTER 4

PEACH

MONIKA GUPTA[1*], RACHNA ARORA[1], and DEBASHIS MANDAL[2]

[1]*Department of Fruit Science, Punjab Agricultural University, Ludhiana, India*

[2]*Department of Horticulture, Aromatic & Medicinal Plants, Mizoram University, Aizawl 796004, Mizoram, India*

Corresponding author. E-mail: monika-fzr@pau.edu

ABSTRACT

Peach (*Prunus persica* (L.) Batsch), native to China, is an important deciduous crop of temperate regions. Globally, production-wise, it ranks 3rd, just behind apples and pears. The fruits are a rich source of minerals, vitamins, and antioxidants. Thriving well in deep, fertile, well-drained, sandy loam to clay loam soils, peaches need about 600–900 chilling hours for successful cultivation; however, some low chill cultivars can be successfully grown under subtropical conditions. The cultivars of peach are categorized into two major groups, namely, clingstone and freestone based on the adherence of pulp to the stone. The flesh color and texture are the other predominant factors distinguishing peach

cultivars. Flowers in peach develop in the leaf axils after one-year-old growth. Pollination occurs through insects. The fruit development in peach follows a double-sigmoid curve. The fruit is drupe, with mesocarp being its edible portion, and follows a double-sigmoid growth curve pattern. Fruits are closely spaced on the branches. Both flower and fruit thinning are essential to improve size, yield, and quality. Fruits being climacteric in nature are harvested when physiologically mature. For yellow-fleshed varieties, the change of the fruit color from green to light green, and for white-fleshed varieties, the change from green to whitish green or creamy white or cream are the indicators of harvest maturity. Although there is the availability of diverse genetic material having wider adaptability, there is utmost need to strengthen the breeding program. The future thrust areas for peach improvement should stress upon improved cultivars resistant to spring frost, insect-pest and diseases, extended harvest season, improved fruit size, weight, color, flavor and aroma, lengthening of shelf life, regulating tree size, and broadening the adaptive range.

4.1 GENERAL INTRODUCTION

Peach is a deciduous tree and can be grown in a dry, continental, or temperate climate. Peaches are classified as the member of the genus *Prunus* and species *persica*. Other members belonging to the genus *Prunus* are nectarines, Japanese plum, European plum, apricot, almond, sour cherry, and sweet cherry. The classification of peach was first of all done by Carolus Linneaeus in 1753. In 1801, August Johann Georg Batsch botanically named peach as *Prunus persica*.

Peach is native to the region of Northwest China, where it was first domesticated and cultivated. Romans acquired peaches from China to Persia in the 1st or 2nd century BC. Then it got spread throughout the Europe. Then it was introduced in Mexico, California, and Arizona. From Florida in 1565, it spread into the states of Louisiana, Arkansas, Georgia, the Carolinas, New Jersey, and New York by the 1800s.

Peaches are consumed fresh as well as processed products such as canned peach, dried, frozen preserves, juice and beverages; pickles, jams, nectar, and marmalade are also prepared. Peaches are not popular only for their nutritional value and human health benefits but are also representative of many cultural traditions in China. Peach kernels are a common constituent of the traditional Chinese medicine to regulate blood pressure, counter-inflammation, and cure allergies. In Korea, the peach is considered as the fruit of happiness and prosperity. The blossoming flowers of peach indicate the onset of spring season for the Vietnamese. In Vietnam, Bonsai trees of peach are used for decoration purposes to welcome New Year. Tresidder claims, "The artists of Renaissance symbolically used peach to represent heart, and a leaf attached to the fruit as the symbol for tongue, thereby implying speaking truth from one's heart." Ripe peaches were used as a symbol of good health.

4.2 AREA AND PRODUCTION

Globally, peach occupies the third position in terms of production among all temperate fruit trees, just after apples (89.33 million tons) and pears (27.34 million tons). This total peach production is estimated at over 24.98 million tons worldwide. The production has increased about three times since 1966, from 5.8 to 27.34 million tons, as compared with 2016 production mainly due to the accelerating production in China. Asia contributes 56.5% share in the total world production of peaches, followed by Europe (25%), America (14%), and Africa (4.5%). The five largest producer countries of peaches are China, accounting for 57.9% of the world production, followed by Spain (6.12%), Italy (5.71%), the USA (3.7%), and Greece (3.39%) (FAOSTAT 2016). Peach with 16.4 lakh ha ranked second globally in terms of the area of production after apples (52.9 lakh ha). Over the last 50 years, the worldwide area under peaches has been increased one and half times, from 6.3 lakh ha to 16.4 lakh ha. China (51.1%) occupies the largest area for peaches, followed by Spain (5.3%) and Italy (4.2%).

4.3 MARKETING AND TRADE

The peach and nectarine production forecast a gain from 458,000 metric tons to 21.2 million tons. This is mainly due to the increase in the production of peaches in China and the European Union. Peach production may decrease in America due to unfavorable climatic conditions. The overall 21.57 lakh tons of peaches was exported globally with an export value of 21.7 lakh US$ in 2016. According to the World Export Scenario of peaches, the major five exporters are China (36%), Spain (20.3%), Italy (17.9%), USA (4.6%), and Greece (4.6%). Germany, Russian Federation, France, United Kingdom, and Poland are the major importers of peaches. Greece and China have emerged as the major markets for the production of processed and value-added products of peaches. Peaches are also processed in Australia, Argentina, USA, Spain, South Africa, Chile, and Italy (FAOSTAT 2016, USDA/PSD 2017).

4.4 COMPOSITION AND USES

Peaches being delectable in nature are also a rich source of vitamins such as vitamin C; minerals such as potassium, phosphorous, magnesium, fluoride, and iron; and antioxidants such as *lutein, zeaxanthin*, and *ß-cryptoxanthin* (Table 4.1). Peaches are having 85% of water content, 7.5%–8.5% sugars, 0.6%–1.2% proteins, 0.3% fat content, and 1.2%–1.4% fiber content. The main organic acid present in peaches is malic acid. Peaches are low in calories. The pulp of the peach fruit contains prunasin, a glycoside, while amygdalin is present in seeds. Chlorogenic acid, epicatechin, catechin, and cyanidine are the major phenols present in the mesocarp of peach fruits. These phenolic compounds are responsible for the development of color and flavor in peaches. Different peach cultivars vary in their chemical compositions due to their genetic attributes, cultivation practices, sun exposure, and maturation time, etc. (Lima et al., 2013). The cultivar with the highest total sugar concentration was "Maciel," followed by "Diamante" and "T. Beauty." The lowest total sugar levels were found in the cultivars "C.1122" and "C.843." The varieties richest in carotenoids were "Maciel," "Diamante," "C.845," "C.843", "C.1050," and "Sensação." Total phenols were higher in white or less yellow fruits desirable for fresh as well as processing purposes. The aromatic compounds such as (E)-2-nonenal (tallowy and fatty) and 2-methylhexyl propanoate (fruity, green, ripe, and sweet) were detected only in "Royal Glory"; (Z)-2-heptenal (fatty) and (E)-2-octenal (green and fatty) were found in "Maria Marta" and "Norman"; 3(2H)-benzofuranone, octyl acetate (orange) and (E)-2-hexenyl hexanoate (green, fruity, and cheesy) were found in "Royal Glory" and "Redhaven"; and cisgeraniol (rose-like odor) was found only in "Redhaven" (Kralz et al., 2014).

Peaches protect the human body from aging, infectious agents, essential for maintaining the night vision, healthy mucus membranes and skin, and to cleanse the kidneys and the bladder. Peaches are helpful in lowering the blood pressure and cholesterol levels of the human body. The boiled bark and leaf extract were used as sedatives for relieving gastric irritations and cough. The infusion of its flowers was recommended to treat jaundice and as a purgative. Peaches are widely used in the cosmetics industries also. Peach kernel oil is used in biofertilizers, cattle feed, and pharmaceutical preparations.

TABLE 4.1 Nutritive Value of Peaches Per 100 g Pulp

Nutrient	Nutritive Value
Energy	39 kcal
Carbohydrates	9.54 g
Protein	0.91 g
Total fat	0.25 g
Cholesterol	0 mg
Dietary fiber	1.5 g
Folates	4 µg
Niacin	0.806 mg
Pantothenic acid	0.153 mg
Pyridoxine	0.025 mg
Riboflavin	0.031 mg
Thiamin	0.024 mg
Vitamin A	326 IU
Vitamin C	6.6 mg
Vitamin E	0.73 mg
Vitamin K	2.6 µg
Sodium	0 mg
Potassium	190 mg
Calcium	6 mg
Copper	0.068 mg

TABLE 4.1 *(Continued)*

Nutrient	Nutritive Value
Iron	0.25 mg
Magnesium	9 mg
Manganese	0.61 mg
Phosphorus	11 mg
Zinc	0.17 mg
Carotene-ß	162 µg
Crypto-xanthin-ß	67 µg
Lutein–zeaxanthin	91 µg

Source: USDA National Nutrient Database.

4.5 ORIGIN AND DISTRIBUTION

Peaches were originated in China between the Tarim Basin and the north slopes of the Kunlun Mountains and date back to 2000 BC. The primary centers of peach diversity are recorded to be mountainous areas of Tibet and Southwest China. However, the secondary centers of peach diversity are Iran, Central Asia, Caucasus, Crimea, Italy, Moldavia, California, and Spain. In Japan and India, peaches appeared about 4700–4400 BC during the Jomon period and 1700 BC during the Harappan period. By 300 BC, the peach cultivation spread from China through Persia to Greece. According to the historical background, Alexander the Great introduced peaches to Europe. Spanish introduced the peaches to America in the 16th century and to England and France in the 17th century. However, its commercial cultivation started after 19th century in America. Peaches and nectarines are cultivated in 71 countries. Among these, peaches are grown commercially in China, Spain, Italy, the United States, and Greece. Globally, China contributes more than 50% of its share in peach production among all the countries. Georgia, California, and South Carolina are the leading peach producing states of the United States. In India, peaches are commercially cultivated in northwestern states of Jammu and Kashmir, Himachal Pradesh, and in Uttar Pradesh hills. The peach cultivars having low-chilling requirement can also be grown in parts of Punjab, Haryana, etc.

4.6 BOTANY AND TAXONOMY

Peaches are deciduous in nature. The trees can grow up to a height of 13–33 feet having straight and smooth trunk, low headed and glabrous twigs, with a reddish–greenish bark in the first year that turns dark grey–silver later on. The leaves are lanceolate (10–20 cm long) with serrated margins and pinnate venation. The flowers are pink with five petals, solitary, simple, perfect, and normally sessile. The flowers bloom early than the leaf emergence after the completion of winter dormancy. The fruit buds are produced laterally on one-year-old shoots as well as spurs. Botanically, the peach fruit is drupe, developed from a single ovary, also known as a stone fruit. The outer fuzzy skin of the fruit is called exocarp, the edible portion is mesocarp, and the hard pit is endocarp. However, the fuzzless peach is known as nectarine. The fruits are of variable color (velvety skin with yellow or red color, flesh is of yellow or white color), they are round or elongated, and weigh 180–230 g per fruit. The peach cultivars are categorized into freestone and clingstone based on the adherence of pulp to the stone. The flesh of freestone cultivars separates easily from the seed, while that of the clingstone cultivars does not separate from the flesh. The white-fleshed peach cultivars are sweet with little acidity; however, the yellow fleshed cultivars are acidic blended with sweetness. Peach fruit contains single seed, known as the kernel. The kernel is large, redbrown and tastes sweet as well as bitter depending upon the variety.

Ground color, flesh color, and over color (% age) of the fruit rate of ethylene generation, total soluble solids and pH were lesser in the lower positions of canopy than the upper position. Shady conditions adversely affect the fruit size and fresh weight (Luchsinger, 2002). The life span of peach trees is 12–15 years commercially; however, they may survive up to 20–30 years depending upon the proper management technologies adopted by the growers.

4.6.1 TAXONOMICAL DETAILS

Peach belongs to the family Rosaceae, subfamily Prunoideae, genus *Prunus*, subgenus *Amygdalus* and species *persica*. *Prunus persica* (L.) Batsch is a diploid species with the chromosome number of $2n = 16$. All the commercial cultivars relate to *Prunus persica* (L.) Batsch. The wild species of peach are *P. davidiana*, *P. fergamensis*, *P. kansuensis*, and *P mira*. *P. davidiana* is used as a rootstock, and it tolerates the drought conditions; however, this rootstock is highly susceptible to the nematodes. *P. fergamensis* has variable fruit types and resistant to powdery mildew. The flowers of *P. kansuensis* are resistant to frost, but fruit quality is very poor. *P mira*, also used as a seedling rootstock, is considered as ancestor of *Prunus persica*.

4.7 VARIETIES AND CULTIVARS

Globally, peaches can be distinguished by their color, namely, yellow, white, or red flesh, with yellow being the most common. The flesh texture may be melting or nonmelting. The fruit shape is round; however, the flat-shaped cultivars are known as "Peento." Peaches may be freestone or clingstone. They may have the fuzz. Peaches having low acid content at maturity are called "Honey" peaches having high sugar content preferred in Asia. Peaches with high acid content are liked in the US markets. The important cultivars of leading peach producer countries are enlisted below.

1. *China*: Feicheng Tao, Shenzhou Mitao, YidouMitao, Fenghuayulu, and Baihuashuimi. Shumitao peaches are mostly grown in southern parts of China, popularly known as "Honey Peach."
2. *United States:* Dixired, Surecrop, Redhaven, Norman, Belle of Georgia, Elberta, Scarlet Prince, July Prince, Red Globe, J. H. Hale, Alexander, and Dixigem.
3. *Italy:* Gilda Rossa, Iris Rosso, Maria Angela, Maria Bianca, Maria Delizia, Maria Rosa, Regina Bianca, Regina di Londa, Greta, and White Crest.
4. *Spain:* Subirana, Donutnice, UFO4, Platifun, Sweet Dream, O'Henry, Summer Sweet, Garcicia, Nectarlight, Big Top, African Bonnigold, and Catherina.
5. *France:* Francoise, Lucie, Primerose, Tendresse, Benedicte, Douceur.
6. *India:* For cold regions: July Elberta, Elberta, Peshwari, Quetta, Burbank, and Stark Earliglo. Low-chilling cultivars of the subtropical belt are: Flordasum, Flordared,

Shan-e-Punjab, Sharbati, and Sunred (nectarine).

The characteristics of important peach cultivars grown worldwide are as follows:

1. Elberta: It is a large fruited freestone variety. The fruits are smooth skinned, pale yellow with red splash. The flesh is yellow, firm, juicy, sweet, and of excellent quality.
2. J. H. Hale: It is the self-unfruitful cultivar of peach. The fruits are large sized, round, and freestone. Skin and flesh both are yellow colored, firm, and of excellent quality.
3. Alexander: It is a large fruited, round freestone variety. Skin color is greenish red and flesh color is greenish white, juicy, and sweet with distinct aroma.
4. Dixigem: It is a medium-sized freestone variety. Its skin is yellow colored with red blush, and flesh is light yellow colored.
5. Babcock: It has medium-sized freestone fruits. Skin is of light pink color with red blush and flesh is white in color and very sweet.
6. Redhaven: Fruits are medium-sized freestone and the flesh is bright red and firm.
7. Flordasun: The fruits are medium-sized freestone, their

skin is yellow with red blush, and flesh is yellow colored.

8. Sharbati: Fruits are medium, clingstone, and skin is greenish yellow with red tinge. Flesh is white colored, juicy, and aromatic.

9. Candor: It has medium-sized fruits, roundish-oblong shape with slightly pointed tip, skin is bright yellow with dark red blush, and flesh is bright yellow, firm, and more acidic, freestone.

10. Baifeng: Fruit size is medium, flesh is white and self-melting, mid season variety, rich in vitamin C content, clingstone.

11. Veteran: Fruit size is medium to large, semifreestone to freestone, the color of flesh as well as skin is yellow, mid maturing.

12. Early Amber: Fruits are medium, semifreestone, skin is bright red, having firm yellow flesh and early maturing.

13. Maygold: Medium-sized fruits, skin yellow with red blush, clingstone, early maturing, and flesh is firm and melting in nature.

14. Augustprince: It is a stone-free late season peach cultivar having yellow fleshed, developed by the cross of Sunprince and BY87P943 at the United States Department of Agriculture (USDA). Large fruits with less pubescence, 7–8 cm in diameter with 70%–80% bright red and attractive yellow ground color, and moderately resistant to bacterial spot.

15. Burpeach twenty one: Hybrid peach having yellow fleshed firm fruits, developed by the cross of A25.045 and O.P. at California, USA. Fruits are 80%–90% red with yellow ground color, flesh firm, sweet, and mildly acidic.

16. Burpeach twenty two: It is an early peach hybrid between unnamed peach X Tropic Beauty, fruit is having 70%–80% red blush over yellow ground color, and firm flesh with mild acidity.

17. Candy Princess: This is a yellow-fleshed hybrid peach developed by the cross of Super Bright and an unnamed Peach cultivar in the USA. Fruits are globose to oblate, dark red blending with reddish orange background over yellow ground color, and flesh firm, crisp with sweet flavor and mild acidity.

18. Carolina Gold: It is a freestone late season peach hybrid with a parentage of Biscoe and NC-C5S-067, developed at North Carolina State University, USA. Fruits are having red blush over golden yellow ground color and large in size

with excellent quality, and flesh is resistant to oxidative browning. Trees are moderately resistant to bacterial spot.

19. Challenger: It is a hybrid early season yellow-fleshed freestone peach suitable for local market in North Carolina, USA, developed by the cross of Redhaven and bulk pollen of cross of Reliance and Biscoe. Fruits are medium sized with excellent quality, and trees are highly resistant to bacterial spot.

20. China Pearl: White-fleshed hybrid peach developed by the cross of Contender and PI 134401 (introduced from China), developed at North Carolina State University, USA. Fruits are large, round, freestone, and low in acid, while trees are moderately susceptible to bacterial spot.

21. Crimson Princess: It is a large fruited peach hybrid with a cross between Ruby Diamond and Crimson Lady at California. Fruits are dark red over yellowish pink background, having firm and sweet flesh, melting and clingstone.

22. Crimson Rocket: It is a yellow-fleshed freestone peach developed by USDA, having melting flesh, 80% red blush on yellow ground skin color suitable for dessert use as having good sugar acid blending, with flesh usually remaining firm until ripening.

23. Diamond Candy: It is a freestone hybrid peach of California developed by the cross of Diamond Ray with an unnamed peach. Fruits are symmetrical, uniform, and large with dark red over reddish orange ground skin color having firm, sweet flesh with mild, subacid flavor.

24. Early Augustprince: It is a stone-free hybrid peach developed by the cross of Sunprince and BY87P943 by USDA. Fruits are larger and having 70%–80% bright red with yellow ground color skin, less pubescence, and moderately resistant to bacterial spot.

25. Galactica: It is saucer-shaped freestone hybrid peach developed by the cross of NCN-4 with Hangchow (Saucer peach of China) at North Carolina State University, USA. Fruits are having 80% red blush and white flesh with excellent quality but susceptible to bacterial spot.

26. Galaxy: It is a saucer-shaped clingstone peach of USDA with having 50% red blush on large-sized fruit with white, firm, melting flesh; sweet but low in acid.

27. Goodwin: It is a clingstone peach developed by University of California, Davis (UC Davis), USA; it is suitable for canning having golden yellow less pubescent skin with bright yellow nonmelting flesh.

28. GP-27: Firm yellow-fleshed late season freestone peach of California having large fruit with 25%–75% dark red to orange red covering over yellow–orange ground color skin.

29. Gulfcrest: It is a USDA-released mid-chill clingstone peach of large size with 95% blush over yellow ground color and yellow-colored nonmelting flesh.

30. Gulfcrimson: It is a nonmelting yellow-fleshed mid-chill clingstone peach developed by USDA having large attractive fruit with 80% blush over yellow ground color.

31. Gulfking: USDA-released mid-chill clingstone peach with round fruit having red over yellow–orange ground color and yellow flesh.

32. Flat Wonderful™ (H28-52-96270): It is a saucer-shaped clingstone peach developed by Rutgers University, USA, having medium-sized fruits with mottled red over color and orange red under color skin with yellow-colored sweet flesh.

33. Intrepid: It is mid-season freestone developed by North Carolina State University, USA, suitable for local markets. Fruits are pubescent, yellow fleshed, and have excellent flavor and quality.

34. Ivory Duchess: It is a Californian clingstone peach having large uniform fruits with deep red mottled over dark pink background skin color and white, crisp, subacidic firm flesh.

35. Late Ross: It is nonmelting yellow-fleshed clingstone peach developed by UC Davis, USA; suitable for canning.

36. Lilleland: It is a mid-late season clingstone peach developed by UC Davis, USA, having nonmelting yellow flesh, suitable for canning.

Finn and Clark (2008) described about many new peach cultivars; based on the information, some important peach varieties are listed in Table 4.2.

Peach   257

TABLE 4.2   Some Newly Released Peach Varieties

Variety Name	Type	Fruit Character	Plant Character	Released/ Originated From
Beaumont™ (MSUP8706)	Hybrid: SH424 × Fayette	Uniform, round to oblong fruits, freestone, medium dark blush over yellow ground color, yellow-colored flesh having good firmness, sweet, and moderate acidity	Medium spreading, flowers nonshowy and self-fertile	Michigan State University, USA.
Messina™ (NJ352)	Hybrid: D90-9 nectarine × NJ318	Large, freestone fruits, bright orange–red over yellow orange ground color, yellow fleshed with sweet flavor and moderate acidity	Late season, vigorous tree having nonshowy self-fertile flowers	Rutgers University, USA.
Vitoria™ (NJ353)	Hybrid: Biscoe × Fairtime	Fruits are large, freestone, red over with yellow orange under color, yellow fleshed and sweet flavor with moderately acidic	Moderately vigorous tree having showy self-fertile flowers with leaf glands reniform	Rutgers University, USA.
P.F. 11 Peach	Parentage unknown	Fruits are round, freestone, 85% red over yellow ground color, excellent flavor, yellow-colored firm flesh	Mid-season variety. Spreading type tree with medium vigor, flower nonshowy pink in color. Resistant to brown rot and bacterial spot	Coloma, Michigan, USA.
P.F. 19-007	Parentage unknown	Spherical in shape, having 90% red cover over yellow surface, flesh is extremely firm, yellow, freestone, nonmelting, and free from fiber	Spreading type, medium vigor tree having pink nonshowy flowers resistant to brown rot and bacterial spot	Coloma, Michigan, USA.
P.F. 22-007	Parentage unknown	Spherical in shape, mid-late variety having 80% red cover over yellow surface fruit skin, firm, nonmelting, freestone, yellow flesh, free from fiber	Spreading type tree having medium vigor, resistant to bacterial spot	Coloma, Michigan, USA.

258 Temperate Fruits: Production, Processing, and Marketing

TABLE 4.2 *(Continued)*

Variety Name	Type	Fruit Character	Plant Character	Released/ Originated From
P.F. 5D Big	Parentage unknown	Very early maturing semicling peach having spherical large fruits with 85% red covers upon yellow ground color, yellow-colored nonmelting fiberless flesh	Moderately spreading tree with medium vigor, resistant to brown rot and bacterial spot diseases	Coloma, Michigan, USA.
P.F. 7A Freestone	Parentage unknown	Early maturing, spherical fruits, with 90% red covers over yellow ground color having firm nonmelting and freestone flesh	Trees are moderately upright, vigorous, medium density; flowers are showy, pink, and plants are resistant to brown rot and bacterial spot	Coloma, Michigan, USA.
P.F. Lucky 12	Parentage unknown	Round, freestone fruits having 95% dark red on yellow ground color skin, yellow-colored firm flesh, free from fiber	Mid-season medium vigor, spreading tree having resistant to brown rot and bacterial leaf and fruit spot	Coloma, Michigan, USA.
Peach Tree	Hybrid	Fruits are early, saucer shaped, 60% red color over yellow grounded skin, white, and firm flesh	Upright vigorous tree having large red–pink flowers	Modesto, California.
Plawhite 5	Seedling selection	Medium size, slightly flat fruit having 80%–85% red cover over yellow green ground color skin, medium firm, and melting, sweet fleshed clingstone peach	Very early season variety having upright to upright spreading canopy with showy flowers	Cartaya, Spain

Peach 259

TABLE 4.2 *(Continued)*

Variety Name	Type	Fruit Character	Plant Character	Released/ Originated From
S5848	Parentage unknown	Broad oblate fruits having gray–orange covers 90% of surface over yellow under color skin, fine texture, yellow, soft, very juicy freestone flesh	Trees are moderately vigorous and spreading type with red–purple flowers	Angers, France.
Snow Duchess	Hybrid, Snow Princess × Crimson Lady	Round, large, very deep red with moderate pink background color fruit having freestone, firm, sweet but subacidic flesh, white in color	Upright, medium vigor, dense tree having showy, large self-fertile purplish pink flowers	Le Grand, California, USA.
Spring Princess	Hybrid	Very large fruits with deep red mottled on reddish orange color skin having melting, clingstone, yellow flesh	Upright, vigorous tree with self-fertile, pale purplish pink flowers	Le Grand, California, USA.
Sugarpeach II	Hybrid Coral Princess × local nectarine	Round, large, very deep red mottled over medium red orange skin color, firm, clingstone, and yellow-fleshed peach	Large spreading vigorous trees having large purplish pink flowers	Le Grand, California, USA.
Super Zee	Hybrid	Fruits are early, medium-sized, globose, skin is red over yellow with firm yellow-fleshed having good flavor and clingstone	Large, upright, vigorous and medium dense trees with self-fertile reddish pink flowers	Modesto, California, USA.
Sweet Henry	Hybrid	Round and large-sized fruit with firm, yellow-fleshed clingtsone peach having red over yellow color skin	Tree is large, vigorous and upright with reddish pink large flowers	Modesto, California, USA.
Sweet-N-UP	Hybrid	It is a freestone peach with red blushed on yellow ground color skin, yellow-colored melting flesh	Trees are with nonshowy self-fertile flowers	USDA-ARS.

260 Temperate Fruits: Production, Processing, and Marketing

TABLE 4.2 *(Continued)*

Variety Name	Type	Fruit Character	Plant Character	Released/ Originated From
TexKing	Hybrid: Goldprince × TX3290-2	Fruits are large, red blushed over yellow color with firm yellow clingstone flesh	Trees have medium-chilling requirement, semispreading with showy self-fertile flowers and moderately resistant to bacterial spot	Texas A&M University, USA
TexPrince	Hybrid: Flordaking × P60-12	Round-shaped fruit with 60%–80% reddish blush over yellow ground color having yellow-colored melting flesh	Trees are with medium-chilling requirement, nonshowy self-fertile flowers and moderately resistant to bacterial spot	Texas A&M University, USA
Thai Tiger TXW1193-1	Unknown	Low-chilling peach having round with flat tipped fruits with 60%–80% red blush over yellow ground color skin and melting, clingstone yellow flesh	Trees are with showy self-fertile flowers	Texas A&M University, USA
Thai Tiger TXW1490-1 Thai Tiger TXW1491-1 Thai Tiger TXW1C4	Tropic Beauty × Selfed	Low-chilling yellow-fleshed clingstone peach having 60%80% reddish blushed over yellow ground color skin	Trees are with low-chilling requirement and showy self-fertile flowers	Texas A&M University, USA
Tropic Peach One (also known as Tropic Prince)	Tropic Beauty × Selfed	It is a low-chilling yellow-fleshed clingstone peach having 40%–80% reddish blush over yellow ground color skin and round-shaped fruit	Semispreading trees having showy self-fertile flowers	Texas A&M University, USA

Peach   261

TABLE 4.2 *(Continued)*

Variety Name	Type	Fruit Character	Plant Character	Released/ Originated From
TruGold	Haploid seedling of P-2-1 (Redhaven × Veefreeze), doubled with colchicine treatment, seed propagated	Medium to large, attractive, round, freestone, yellow-fleshed peach	Trees are with nonshowy self-fertile flowers	Washington State University and USDA-ARS.
UF Beauty UF Blaze UF Sun	Hybrid: 90-50CN × UF Gold	All yellow-fleshed clingstone, nonmelting peach; differ in fruit size, color, and chilling requirement. UF Beauty is having red skin with dark red stripes, whereas UF Blaze is bright red on deep yellow ground color, and UF Sun has 50% red skin with dark stripes. UF Sun having the lowest chilling requirement of 100–150 hours, followed by UF Beauty 200 hours and UF Blaze 300 hours	Trees are semispreading and vigorous with showy self-fertile flowers and highly resistant to bacterial spot	University of Florida, USA.
Vitall	Hybrid	Medium-sized clingstone peach having firm, yellow-fleshed, nonmelting flesh, suitable for canning	Trees are with nonshowy self-fertile flowers and moderately resistant to *Cytospora* canker; however, susceptible to bacterial spot	University of Guelph, Ontario, Canada.

TABLE 4.2 *(Continued)*

Variety Name	Type	Fruit Character	Plant Character	Released/ Originated From
Vista Snow	Hybrid	Medium-sized globose fruit having 70% red color over yellow ground skin, and flesh is firm, white, and clingstone	Trees are large, upright, and vigorous with showy red pink flowers	Modesto, California, USA.
Vollie	Hybrid: Redskin × Kalhaven	Fruits are of beautiful color and having reddish around pit, yellow flesh, and freestone	Trees are with nonshowy self-fertile flowers and moderately resistant to Cytospora canker, but susceptible to bacterial spot	University of Guelph, Ontario, Canada.

4.7.1 NECTARINES

The fuzzless peaches are known as nectarines. These are the genetic mutants of peaches having smaller size and more red color on the fruit surface and characteristic aroma. Nectarines may be categorized as yellow- and white-fleshed cultivars and freestone and clingstone. The yellow-fleshed cultivars are more acidic and having less sugar content; however, the white-fleshed nectarines are less acidic with a higher sugar content. The freestone nectarines are consumed fresh, whereas clingstones are used for cooking and processing.

The characteristics of important nectarine cultivars grown worldwide are as follows:

1. *Nectared:* Large and round fruits, flesh is yellow, firm, and of good taste, freestone, mid-maturing.

2. *Sunred:* It has small- to medium-sized fruits, and skin color is bright red with yellow firm flesh, semifreestone, early maturing.

3. *Sunrise:* Medium-sized fruits, skin is thick, having mahogany red; flesh is yellow, firm, and mildly acidic.

4. *Sunripe:* Excellent fruit quality, but the color of the fruits is dull red.

5. *Sunlite:* It has medium-sized fruits, late maturing cultivar.

Finn and Clark (2008) described about many new nectarine cultivars; based on the information, some important nectarine varieties are listed in Table 4.3.

Peach   263

TABLE 4.3   Some Newly Released Nectarine Varieties

Variety Name	Type	Fruit Character	Plant Character	Released/ Originated From
Brunectwentyone	Hybrid	Large, uniform, round fruited clingstone nectarine with 50%–60% red cover over yellow–orange ground color, and flesh is firm, yellow colored, juicy, and nonmelting	Trees are medium large, dense, upright but moderately vigorous with showy light flowers	Fowler, California, USA.
Burnectnineteen	Hybrid: Crimson Baby × Unnamed Peach	Uniform, round fruits having 85%–95% red cover over yellow–orange skin with firm, juicy, clingstone, yellow-colored flesh	Early variety and trees are medium large, dense, moderately vigorous, and upright with showy light-pink flowers	Fowler, California, USA.
Burnectwenty	Hybrid	Fruits are uniform, large, and round with 45%–55% red cover over yellow–orange-colored skin, and flesh is yellow, firm, and clingstone	Late-season variety with medium large, dense, moderately vigorous, upright trees having showy light-pink flowers	Fowler, California, USA.
Burnectwentythree	Hybrid: Burpeachone × Zhang Yu Pan	Fruits are saucer shaped, uniform, but occasional lobbing with 70%–90% dark red cover over light yellow ground color skin, and white, firm, nonmelting flesh, and clingstone	Medium dense, upright, moderately vigorous trees with showy light-pink flowers	Fowler, California, USA.
Grand Bright	Hybrid: Ruby Diamond × Unnamed nectarine	Fruits are large, globose, dark red over reddish orange gorund color having very firm, yellow color, clingstone flesh	Large, vigorous, upright trees with self-fertile purple–pink flowers	Le Grand, California, USA.

264 Temperate Fruits: Production, Processing, and Marketing

TABLE 4.3 *(Continued)*

Variety Name	Type	Fruit Character	Plant Character	Released/ Originated From
Late Bright	Hybrid: September Red × Unnamed Peach	Fruits are very large, globose, deep red streaked over reddish orange and yellow ground color skin, and flesh is firm and yellow in color. It is a late-season clingstone nectarine	Trees are large, vigorous, and upright with showy purple–pink flowers	Le Grand, California, USA.
May Pearl	Hybrid: June Pearl × Rose Diamond	Medium-sized globose fruits with dark red mottled skin, and flesh is white, firm, and melting. It is a clingstone nectarine	Trees are medium sized, vigorous, upright, and having showy self-fertile purplish pink flowers	Le Grand, California, USA.
May Pearl II	Hybrid: Unnamed White-fleshed peach × Rose Bright	Fruits are medium-sized, globose, dark red over reddish orange background color skin and clingstone, with firm, white color flesh	Medium sized, dense, vigorous, spreading trees with showy self-fertile, purplish pink flowers	Le Grand, California, USA.
Nectagala	Hybrid: Zaitabo × Maillarbelle	Fruits are round, large and with 90% dark red blush over orange–red ground color having very firm, yellow, clingstone flesh	Trees are medium large, dense, vigorous, and upright with showy pale-pink flowers	Elne, France.
Nectalady	Hybrid	Large, uniform, round-shaped fruit with red cover (90%–100%) over orange–red ground color and very firm, yellow flesh. It is a late-season clingstone nectarine	Trees are medium large, dense, vigorous, and semispreading with showy pale-pink flowers	Elne, France.

Peach 265

TABLE 4.3 *(Continued)*

Variety Name	Type	Fruit Character	Plant Character	Released/ Originated From
Nectarmagie Nectaperle	Hybrid Nectamagie: Maillarnecta × O.P. Nectaperle: Maillarosette × Zaitabo	Both are early season variety with very firm white flesh and clingstone nectarines. Bright purple red (90%−100%) cover over pink−red ground color	Semispreading, medium dense, large, vigorous tree with showy pale-pink flowers; however, it is nonshowy flowers in Nectaperle	Elne, France.
Nectapink	Hybrid: Zaitabo × Maillarbelle	Fruits are large, uniform, round to ovate, 80%−90% dark red cover over orange−red ground color having very firm, semisweet, yellow flesh. It is a mid-season clingstone nectarine	Trees are medium large, dense, upright, vigorous with showy pale-pink flowers	Elne, France.
Nectaprima	Hybrid: Zaitabo × Armking	Fruits are large, round, 90%−100% dark red skin with very firm, yellow-colored flesh. It is an early season clingstone nectarine	Trees are medium large, dense, semispreading, and vigorous with showy pale-pink flowers	Elne, France.
Nectareine	Hybrid: Zaitabo × Andano	Fruits are large, round to oblate, uniform with 80%−90% dark red over orange−red ground color skin having very firm yellow flesh. It is a mid-season clingstone nectarine	Trees are dense, semi-upright, medium large with showy pale-pink flowers	Elne, France.

266 Temperate Fruits: Production, Processing, and Marketing

TABLE 4.3 *(Continued)*

Variety Name	Type	Fruit Character	Plant Character	Released/ Originated From
Nectariane	Hybrid: Andano × Zaitabo	Fruits are large, round to oblong, 80%–90% dark red color over orange–red ground color skin and very firm yellow flesh. It is an early season clingstone nectarine	Trees are medium large, dense, semi-upright, and vigorous with nonshowy flowers	Elne, France.
Nectaroyal	Hybrid	Large, round to oblate, uniform fruits with 80%–90% dark red over bright orange–red ground color having very firm yellow flesh. It is a semilate-season clingstone nectarine	Trees are medium large, dense, semi-upright, and moderately vigorous with showy pink flowers	Elne, France.
Pacific Sweet	Hybrid	Large, round to oblong uniform fruits having very deep red over reddish yellow background skin color and firm, crisp subacidic, yellow-colored flesh. It is a clingstone nectarine	Trees are large, vigorous, spreading with self-fertile purplish pink flowers	Le Grand, California, USA.
Polar Light	Hybrid	It is an early season clingstone nectarine that requires low chilling. Fruits are large, globose, red with yellow ground color skin, and flesh is firm, white with good sugar and acidic balance	Trees are large, vigorous, upright, medium dense with showy reddish pink flowers	Modesto, California, USA.

Peach   267

TABLE 4.3 *(Continued)*

Variety Name	Type	Fruit Character	Plant Character	Released/ Originated From
Red Bright	Hybrid: Ruby Diamond × unnamed peach	Fruits are large, uniform, globose with deep red over orange–yellow ground cover and having firm yellow-colored flesh. It is a clingstone nectarine	Trees are medium, dense, vigorous, upright with showy purplish pink flowers	Le Grand, California, USA.
Sauzee King	Hybrid	Medium-sized early season fruit with white, firm fleshed, and red skin over yellow ground color and clingstone	Trees are large, upright, vigorous, with showy reddish pink flowers	Modesto, California, USA.
Sugar Pearl	Hybrid: Summer Bright × Unnamed White Nectarine	Fruits are large, clingstone, very deep red colored with crisp, firm, white flesh	Medium-sized tree with medium vigor, spreading having nonshowy self-fertile flowers	Le Grand, California, USA.
Sugarine I	Hybrid: Bright Pearl × Spring Bright	Fruits are medium, round, very deep red in color with firm yellow-fleshed and clingstone nectarine	It is an early season variety having medium, spreading, vigorous tree with showy self-fertile, purplish pink flowers	Le Grand, California, USA.
Sunectwentyone	Hybrid	Fruits are round to elongated; skin is red over yellow ground color. Flesh is sweet, very firm, and yellow in color	It is an early season variety having vigorous, dense, spreading, upright tree with nonshowy self-fertile flowers	Bakersfield, California, USA.
Viking Pearl	Hybrid	Large globose clingstone nectarine having red to pink color fruits with crisp, firm, sweet, and white-colored flesh	Trees are medium sized, spreading with nonshowy flowers	Le Grand, California, USA.

4.8 BREEDING AND CROP IMPROVEMENT

Old cultivars of peaches were not very attractive, unsuitable for handling and shipping, having low shelf life, and did not satisfy commercial requirements. By keeping all these points in mind, the major peach-breeding objectives planned later aimed at developing improved cultivars resistant to spring frost, insect-pest and diseases, extended harvest season, improved fruit size, weight, color, flavor and aroma, lengthening of shelf life, controlling tree size, and broadening the adaptive range. Peach breeding started in the late 1890s and early 1900s in California and New Jersey in the United States. Michigan started its breeding program in 1924, and the Redhaven cultivar was developed, which dominated the eastern USA for decades (Iezzoni, 1987). Peach breeding has also been undertaken in Argentina, Australia, Brazil, Canada, China, France, Italy, Mexico, etc. (Okie et al., 1985; Yoshida, 1988; Wang and Lu, 1992). The 20th century has been observed as the "Golden Age of Peach Breeding" (Sansavini et al., 2006). Among the major yellow-fleshed varietal releases, the USA and Europe contributed about 55% and 30%, respectively. France, Japan, South Korea, and China had its edge in developing a number of white-fleshed cultivars.

Some of the breeding programs have been initiated with the aim of developing hardiness of dormant buds in Canada, Michigan, and New York (Layne, 1982; Callahan et al., 1991) or creating low-chill cultivars for subtropical regions as in Florida, Texas (Sherman and Rodriguez-Alcazar, 1987), and Georgia (Krewer et al., 1997) in the USA. Regal (a self-fertile hybrid of Harvester and Surecrop), Gala (open pollinated seedlings of Harvester), and Glory (La Premiera × L61-3-5) are some of the promising hybrids developed by the USA. Regal is known for tolerating winter injury and bacterial spot disease, whereas Gala is an early mid-season cultivar. The fruits of Glory are yellow fleshed, round, and freestone.

4.9 SOIL AND CLIMATE

Peaches flourish well in sandy loam to clay loam soils. The soil must be deep, fertile, and having good water-drainage capacity. The vegetative growth of peach orchards is highly affected under abiotic stress conditions such as water logging and high pH soils. The ideal pH of the soil for peach orchards should be 5.8–6.8. It can grow well in soils with an electrical conductivity less than 0.5 mmho/cm, calcium carbonate below 5%, and lime under 10%. The ideal soils up to more than 1 m of their

Peach

depth should be rich in organic matter. The compact soils having less than 10%–20% pores should be avoided for peach cultivation. In the USA, the south-facing sloppy lands (4%–8% slope) are considered ideal for peach cultivation as they increase the air drainage during the periods of spring frost. Deep valleys are avoided for the establishment of peach orchards due to water-logging conditions in such areas.

Peaches require temperate climatic conditions for their successful cultivation. They require a chilling period of 600–900 hours at a temperature of 7.2 °C (45 °F) or below. The peach cultivars having low-chilling requirement (250–300 hours) develop good color and ripen at low temperature in winters and high temperature (40 °C) in summers. Peaches cannot be grown successfully in areas having temperatures below minus 12°C (10 °F) due to the wood damage. The areas subjected to frost after mid-April should be avoided for peach cultivation. Peaches require clear and warm weather during fruiting season. Very hot conditions develop astringency in peaches; however, cool and wet weather favors the disease development. Peaches require full sunshine for an enhanced yield and better quality, whereas partial shade results in yield reduction.

4.10 PROPAGATION AND ROOTSTOCK

In peaches, T-budding is the most commonly used method for scion propagation. T-budding should be done in autumn months when the mother trees are in a dormant condition. In the Northern Hemisphere, spring budding as well as June budding can also be performed successfully. The buds should be taken from virus free mother plants. Care should be taken to choose the buds that are situated in the central part of the shoot due to their better shape. The budding height should be 10–15 cm from the ground level. In spring, the scion should be shortened above the graft union immediately before bud break. These budded plants should be kept in nursery when they attain a height of 1.2 to 1.4 m. June budding should be preferred in areas with mild spring. The plants attain 60–80 cm height by October and will become ready for sale within a single year. Besides this, peach can also be propagated through air layering (Castro and Silveira, 2003) and herbaceous, semihardwood cuttings, or woody cuttings of scion peach tree cultivars (Mindello et al., 2008). Hardwood cuttings may be treated with indole-3-butyric acid (IBA) as root promoting hormone (Dutra et al., 2002). The stenting technique, that is, simultaneous grafting of the rooted cuttings may be used for the rapid multiplication of peach, but the success rate is very low. However, there is advancement in breeding practices for peach scion and rootstocks with the application of micropropagation (Gomez et al., 2005). Peach rootstocks are generally propagated sexually by using the seeds after breaking the seed dormancy by removing the seed test and GA_3 application. Axillary buds taken from healthy shoots (Felek et al., 2017) are used as explants for the in vitro propagation of peach cv. Garnem.6-Benzyladeninepourine, with IBA and gibberellic acid may be used for shoot initiation and shoot proliferation. IBA is the preferable auxin used for in vitro rooting of the peach rootstock GF 677 (Ahmad et al., 2004).

Rootstocks play an important role to enhance the productivity and to improve the efficiency by affecting the tree survival, tree vigor, precocity, fruit size, fruit quality, incidence toward various insect-pests and diseases. Peach seedlings are the most promising rootstock for peach, but these are not successful in heavy clayey soils as well as poorly drained soils where pH is above 7.5. Peach seedlings are also susceptible to nematode attack. Nemaguard and Okinawa seedling rootstocks are reported to be resistant to nematodes. Besides peach seedlings, other species such as *Prunus armeniaca* (resistant to root-knot nematodes), *P.*

tomentosa (dwarfing), and *P salicina* (semi vigorous) are also used as peach rootstocks under varied conditions. Interstems (grafting a very small portion of third genotype between the rootstock and scion) are used to induce the dwarfing effect, cold hardiness and resistance to insect pest, and diseases in peach trees. The "Nemaguard" rootstock exhibits maximum vigor, followed by "P30-135" and "K146-43." Peach trees where K146-43 is used as an interstock and "Nemaguard" as a rootstock have the vigor in between these two rootstocks (Tombesi et al., 2010).

4.11 LAYOUT AND PLANTING

An orchard layout is essential for providing the maximum number of trees per unit area. Adequate space, adequate sunlight, cold air drainage, and water drainage are desired for successful establishment and longer tree survival of peach trees. Before the plantation in the orchard, soil-sampling tests for pH, nutrients, and pathogens should be done. Based on the soil-testing report, amend the soil by incorporating phosphorus, potassium, and lime before planting the seedlings. Peach trees should be planted during late winter or early spring when they are in the dormant condition. One-year-old seedlings are preferred for the plantation in a new orchard. Light and frequent irrigation should be done for the proper establishment of seedlings.

The plant density should be 6 m × 6 m, accommodating 275 plants/ha in the square system of planting. The yield is considerably less in low-density orchards; however, intercultural operations are easy to manage in such orchards (Perry and Fernandez, 1993). High-density orchards accommodate approximately 1250 plants/ha. This leads to minimize the land wastage by maximizing the productivity per unit area. But now a days, the concept of meadow orcharding is becoming popular due to the short life span of peach trees. These meadow orchards are of ultrahigh density, accommodating about 100,000 plants/ha. The biggest advantage of this system is of high yield due to more plants per unit area. The fruiting zone also remains near the ground during the initial years of planting. But only early maturing cultivars regenerate sufficient flower buds for annual production. However, the late-bearing cultivars may bear biennially in this system. Fruits remain very small in size. Under ultrahigh-density planting, the microclimate of the tree changes, leading to invite many insect pests and diseases (Annandale et al., 2004). Constant fertigation is required to get high annual production as the trees get deprived off nutrients very quickly. Besides this, proper weed management should be

required either by hand hoeing or through herbicide spray.

4.12 IRRIGATION

Irrigation plays an important role to promote the vegetative growth and enhance the yield and productivity of peach trees. Adequate irrigation to peach trees results in less winter injury and disease incidence and helps maintain the fruit size and uniformity in fruit ripening. Excessive irrigation causes the root rot, limiting the root growth and development. Over irrigation also results in nutrient deficiencies such as iron chlorosis by causing the leaching of nutrients from the root-zone area. However, water stress adversely affects the vegetative growth as well as fruit size (Teskey and Shoemaker, 1982) by disturbing the photosynthetic activity of the plants. According to Childers (1983), fully grown peach trees require 373 ha mm of water per season through conventional irrigation. Microirrigation through the drip system allows the precise application of water, thereby maximizing the water use efficiency. Peach orchards require 1.5 inches water per week through the drip irrigation system (90% efficiency). However, through the overhead system (60% efficiency) about 2.3 inches of irrigation water needs to be applied every week from the pit-hardening stage till maturity

(Frecon, 2002). For commercial peach cultivation, the irrigation water should be clean and free from the accumulation of the toxic salts. Irrigation water is considered adequate if it has a sodium absorption ratio below 3 or total salts below 1000 ppm. In peaches, irrigation may be used as surface irrigation, microirrigation, and drip irrigation, depending upon the climate, soil type, topography, and water availability. In the flood system, 2–4 irrigations per month, in microsprays and sprinklers, irrigation at weekly intervals, and in the drip system daily irrigation are desired (Bryla et al., 2005). However, irrigation methods affect the vegetative growth and production of peach trees. During the initial three years of growth, trees irrigated by drip irrigation acquire large size and produce higher yields than furrow or microirrigated trees (Bryla et al., 2003).

4.13 NUTRIENT MANAGEMENT

The productivity and vigor of the peach trees are improved greatly with adequate nutrient supply. Fruiting is an exhaustive process for the fruit trees. It removes a large amount of nutrients from the soil, which need to be supplied annually to prevent the adverse effects of nutrient deficiency. The nutrient supply must be optimum to encourage sufficient vegetative growth and timely fruit set. The deficiency of any of essential nutrients may affect the tree health and ultimately the productivity. On the other hand, excessive nutrients might result in disturbing the balance between vegetative growth and fruiting and also lead to the contamination of the environment.

The most reliable diagnostic tool used for assessing the nutrient needs is the leaf-nutrient analysis. The leaf sampling in peaches is done from the middle of shoots from the current season's growth in the month of mid-May to mid-July (50–110 days after fruit set) as this period is considered consistent in terms of nutrient concentration in the plant. Macronutrients (%) namely, N (2.6–3.5), P (0.14–0.40), K (2–3), Ca (1.5–3), Mg (0.3–0.8), and S (0.14–0.40); and micronutrients (ppm) such as Zn (20–50), B (30–70), Fe (80–250), Mn (40–200), and Cu (5–16) have been referred to as optimum leaf nutrient standards for peach orchards (Layne and Bassi, 2008). The nutrient level in flowers is positively correlated with the leaf-nutrient level at 60–120 days after full bloom (Sanz and Montanes, 1995). Dormant shoots or roots are also used to predict the tree-nutrient status at early stages (Johnson et al., 2006).

The method of application, time, and dose of nutrients are important to make the application most effective and optimize the uptake and absorption. The application of nutrients depends upon many factors such as the age of peach trees, their vigor, soil-nutrient status, and composition and climatic conditions of the region. Fully grown peach trees require approximately 40 kg farm yard manure, 500 g N, 250 g P_2O_5, and 700 g K_2O per tree in Himachal Pradesh, whereas in Uttar Pradesh, the dose of fertilizers is about 300 g N, 500 g P_2O_5, and 300 g K_2O per tree (Chadha, 2002).

4.13.1 NITROGEN

The optimum nitrogen level needs to be maintained in peach orchards with annual application according to the age of the tree. However, over- and under fertilization should be avoided. The deficiency symptoms exhibited by trees due to N deficiency are yellowing of leaves, red color on leaves and twigs, stunted tree growth, and smaller sized fruits of poor quality. The excessive doses cause more vegetative growth, increased pruning costs, lesser fruit set, and poor fruit quality due to dense canopy. In India, nitrogen is applied as a combination of organic and inorganic supplements. The new approach being followed in many parts of the world to avoid nitrogen pollution in the environment is foliar application in combination with soil application. Foliar application

is effective if applied to peach trees before leaf senescence (Furuya and Umemiya, 2002). This way, nitrogen (generally applied at a concentration of 1% or less) is mobilized out of the leaf and is stored in the tree for utilization in emerging new growth in spring. Decrease in fruit size has been observed in the case of the foliar application of urea; however, half-soil application in combination with foliar urea showed results at par with the soil application only (Johnson et al., 2001).

The timing and method of nitrogen application are important to increase the effectiveness and reduce the wastage. The soil application of nitrogen is often divided into two splits: before bloom and during vegetative season. Multiple applications through fertigation are the most effective. Band application is another method preferred over broadcasting.

4.13.2 PHOSPHORUS

In some parts of the world, phosphorus is applied only at the time of planting. However, in the case of young plantations grown in sandy soils, the deficiency is commonly observed. The growth of plants is reduced. Symptoms do not appear on leaves initially, but in the case of severe deficiency, purplish color starts appearing on leaves. The fruits do not develop properly and have poor quality. Healthy trees had phosphorus concentration in the range of 0.14%–0.25% (Robinson et al., 1997). Phosphorus in the form of super phosphate is readily absorbed by the trees. The fertilizer should be applied in the tree basin and thoroughly mixed by hoeing. The application should cover the whole area below the tree canopy.

4.13.3 POTASSIUM

Potassium plays an important role in the activation and regulation of enzyme systems inside the plants. Deficiency of potassium is responsible for disturbing the plant water relations, opening of stomata, thereby affecting the process of photosynthesis (Mengel and Kirkby, 2001). Therefore, its deficiency is the major cause of concern in orchards. The deficiency symptoms observed in peach orchards are yellowing of leaves, necrosis of leaf margins, and scorched appearance. Shoot growth is restricted, and the number of flower buds is greatly reduced. Leaf K values between 2% and 3% indicate optimum values of potassium fertilizers in the orchard. The correction of K deficiency is accomplished with soil applications of various formulations of K-containing fertilizers. Generally, a single application every 3–4 years is sufficient (Weir and Cresswell, 1993).

4.13.4 MAGNESIUM

It's deficiency not commonly observed in peach orchards. These appear in acidic soils where very high rainfall is there. The symptoms start with pale green color on leaf tips and some marginal necrosis on leaves. In the case of severe deficiency, shoot growth is restricted, flower bud formation is inhibited, and ultimately the yield is reduced. For the correction of deficiency, the soil or foliar application has proved effective. The deficiency can be corrected by the foliar application of magnesium sulfate @ 6.8 kg/acre at the petal fall stage (Walsh, 2005).

4.13.5 ZINC

Zinc plays an important role in the formation of auxins. It is a precursor of many proteins. Therefore, auxin formation is restricted in its deficiency leading to stunted growth. The leaves show interveinal chlorosis and become small and pale. The internodes of leaves get shortened and the whorl of leaves on tip of branches gives a rosette appearance. The deficiency can be corrected with the foliar application of $ZnSO_4$. A dose of 10–25 ppm can be applied as foliar application. Soil application is effective, but its cost is more as reported by many workers (Mann et al., 1986; Arce et al., 1992).

4.13.6 BORON

The peach orchards generally do not exhibit B deficiency; however, toxicity may be observed due to the high mobility of the nutrient (Gupta et al., 1985). The North Indian soils are not deficient in boron and applications are therefore not needed. Even though B can be leached from the soil fairly easily because of high mobility, even then this nutrient can be stored in relatively large amounts in peach trees (Dye et al., 1984), thus carrying on the problem from one year to the next. The toxicity symptoms may appear as necrosis in leaf margins and near the mid-rib. Developing fruits are malformed and may show cracking. The concentrations causing damage might be above 300 mg/kg dry weight (Nable et al., 1997). However, in the case of boron deficiency, two foliar sprays of boron (0.34 kg/acre) can be applied at full bloom and the petal fall stage (Walsh, 2005).

4.13.7 IRON

Iron deficiency is a major problem in peach orchards, particularly in calcareous soils. The major symptoms start in younger leaves showing interveinal chlorosis, giving a netting appearance. Severe deficiency results in reduced flowering and fruit production. The soil application of chelated Fe (50–250 g) as well as foliar application (0.5%–1%) can be used as a correction measure on spring flush, summer flush, and late summer flush, if needed (Chadha, 2002).

4.13.8 MANGANESE

The deficiency is commonly observed in peach orchards. Light-green small-sized spots appear on the marginal area of leaves. In the case of severity, the flowering and fruit set are reduced. The application of 20 ppm $MnSO_4$ as foliar spray has been found most effective (Layne and Bassi, 2008).

4.14 TRAINING AND PRUNING

The growth habit of different peach cultivars ranges from weeping to very upright. Training and pruning of peach trees are essential to allow proper light penetration, air circulation, and even the distribution of fruits around the tree canopy. By following correct techniques of training and pruning, the productivity of the orchards can be increased. Training systems provide the tree with a strong framework and shape, and it is the deciding factor for the fruit-load-bearing capacity of the tree. These direct the growth of the tree in such a way that various cultural operations such as spraying and harvesting can be performed at a reduced cost. Regular pruning of the desired parts ensures new growth and fruit bearing all around the tree annually and increases the productivity of the orchard.

4.14.1 SYSTEMS OF TRAINING

The success of a training system in an area depends on the amount of light intercepted, leaf-nutrient content, and ultimately the yield.

4.14.1.1 OPEN CENTER SYSTEM

This system is most commonly adopted in countries such as Italy and USA. In this system, three primary scaffold branches are retained on the tree with two secondary scaffolds on each (Layne and Bassi, 2008). Tree height is restricted up to 3–4 m. In Italy, the branches in this system are trained to grow more horizontal to get abundant sunlight. This system is known there as open vase and is most popular among all systems. The standard-type trees are easily trained to open center or vase shape. The spraying and harvesting operations are relatively less difficult. However, the trees have a weaker framework than the modified leader system.

4.14.1.2 Y SYSTEM/TATURA TRELLIS

It is the most suitable system for high-density planting. The system was developed for mechanical harvesting of peaches. The growth of the central leader is restricted near the base and two primary scaffolds are retained and trained to grow in the shape of alphabet "Y." The plants in this system are able to intercept maximum sunlight, are higher yielding, and of better quality than the traditional planting system (Anonymous, 2018). This system is most suitable for high-density plantation from 1000 to 2000 plants/ha.

4.14.1.3 PILLAR SYSTEM

In this system, the trees are kept very upright with narrow crotch angles. The multiple leader trees are more productive and require less pruning than central leaders. The trees are allowed to grow naturally with selective pruning only.

4.14.1.4 CENTRAL LEADER SYSTEM

In this system, the tree is allowed to grow more naturally. Only the side branches are well distributed and allowed to grow in all directions. This system provides a strong framework and well-distributed branches with strong crotch angles. Though the pruning cost is less, this system of training is not suitable for peach orchards as these grow very tall and difficult to maintain. The trees become too shady. After a few years, the bearing wood is lost in the inner canopy. That is why this system is seldom followed in peach orchards.

4.14.1.5 MODIFIED LEADER SYSTEM/PALMETTE SYSTEM

One-year-old plants are headed back to a height of about 90 cm at the time of planting. The side shoots are removed with one or two buds remaining on them. Sprouting of lateral branches occurs during the coming spring and summer season. In coming dormant/autumn season, 4–5 healthy and well-spaced branches are selected above 45 cm height from ground level and the rest are removed. No branch is allowed to grow up to 45 cm. The main central branch is called the "leader." When the next growing season approaches, there is the development of new branches on the selected primary shoots. On the main leader, 3–4 well-placed second-tier branches are selected. The remaining undesirable branches are removed. Heading back of the leader should be done near to some outgrowing lateral branch. This practice helps to restrict the vertical growth of the leader branch. The system has the advantages of the

central leader as the root system, and the framework is strong. At the same time, the canopy growth is restricted (Bal, 1998).

4.14.1.6 PRUNING

In peach, fruit is borne on one-year-old branches. The secondary shoots are rarely formed from the adventitious buds. The plants require more exposure to light for growth. Another thing to be kept in mind is the sensitivity of cut ends of the tree to bacterial canker infection which starts from a wound or cut end of the branch. Therefore, time and method of pruning should be critically decided. Pruning in peaches involves heading back, thinning out, and bench cutting. Heading back is done on one-year-old branches by shortening them. Thinning out is also performed in some one-year-old shoots expected to bear fruit, keeping at least one viable bud known as the renewable bud. Old, undesirable, and unproductive shoots should be headed back. Thinning out of 40% one-year-old shoots is practiced to manage tree canopy, improve fruit size, and enhance the fruit quality. The most suitable time for peach pruning is during the dormant season. In high-density planting, as maintained in European Mediterranean countries in large numbers and being preferred for planting in other places of the world now, the control of tree size and canopy is very important. The restricted canopy spread can only ensure the good fruit quality; otherwise, the reciprocal shading will adversely affect the fruit quality and yield. Older trees need vigorous pruning. Cutting back of the trees to good outward growing laterals helps in renewal. The bearing will be less the following year as the fruiting wood is lost. But the bearing will be greatly improved in coming years. Hand pruning has proved to be very expensive for the fruit growers; therefore, mechanical pruning options had to be relied upon to get quick results in no time. However, only topping is being practiced in peach trees mechanically and selective manual pruning still needs to be performed. Heavy fruit set and significant reduction in fruit size have been observed as a result of mechanical pruning. In such cases, extensive fruit thinning becomes a necessity.

Development of new rootstocks that can restrict tree vigor and the use of growth regulators for better flowering and fruiting is the future area of research.

4.15 INTERCROPPING AND INTERCULTURE

Young plants are planted in the orchard may take about 2 to 3 years to come into bearing. Intercrops are grown usually in prebearing period at the normal planting distance for one year only in the area between trees and rows lying vacant because of the limited spread of roots and canopy of the fruit trees in orchards. These intercrops can be vegetables and pulses and are uprooted at appropriate time when primary fruit trees start yielding commercially. So, it is important to have the proper knowledge of an intercrop suitable for an orchard and things to be taken care of, for its cultivation in an orchard without affecting the fruit trees adversely. The intercrops are fed separately, in addition to the fruit crops so that they do not compete with the orchard plants. However, for high and ultra high density, which are being preferred for plantation, the intercropping is completely avoided. The peach tree itself is used as a filler tree in orchards, such as mango, litchi, and pear which come into bearing very late.

Weeds compete for nutrition, space, and water in the orchards. These also become hosts for a lot of disease, causing insects. These need to be managed preferably before they flower or seed is set in them, as some flowering weeds attract pollinators and disrupt the pollination process of peach trees. Interculture helps in better soil aeration and aids in the destruction of weeds, insect larvae, and disease spores. Without interculture, the fruits will fall prey to insect pests and diseases, and size and quality will remain inferior. Only shallow interculture is recommended as the deep cultivation may damage the fibrous roots present in the upper layers of the soil. Crop residue may be used as mulch in the orchard to check weed incidence. The cover crops can be sown in the rainy season after harvesting of fruit, and then incorporated by ploughed up in winter. Polythene mulches are also being used in orchards and have proved to be very effective against weeds; however, these are not economical to use. There are some chemical herbicides used to control weed population in peach orchards, such as Diuron (pre-emergence) and 2,4-D(amine) and glyphosate (post emergence). The repeated use of herbicides encourages the development of resistance and should be avoided. High-density plantation aids in better weed management due to a lesser portion of orchard land lying vacant. Some of organic products used as herbicides such as vinegar, clove oil, cinnamon oil, acetic, and citric acid (Granatstein et al., 2006) appear to be effective when used during early stages of weed growth. Moreover, these are very costly and control only to a limited extent.

4.16 FLOWERING AND FRUIT SET

Flowers in peach develop in the leaf axils of one-year-old growth. These are pink in color, perfect, solitary or paired, perigynous. The terminal bud appearing on a shoot is vegetative. Flower buds in peach are borne in such a way that a leaf bud appears in between two floral buds. There is some chilling requirement below 45 °F necessary to break the bud dormancy. This requirement varies greatly from cultivar to cultivar. While selecting a cultivar for an agro climatic zone, the chilling requirement is the first and foremost criterion to know its adaptability to the place. The warmer area can fulfil lesser chilling hour requirement; hence, low-chilling cultivars are grown there and vice versa. Once the chilling hours are met, flowers are produced in early spring before the appearance of leaves. Flower bud differentiation occurs from the time of initiation of bud to the opening of flower. At the time of flower opening, the sperm cell and egg cell become functional. The flowers are 2.5–3 cm in diameter with five petals. The receptacle is cup shaped. Anthesis is the final phase in flowering. At anther dehiscence, the pollens are shed and come in contact with the stigma, and the pollen tube grows through the style and reaches ovary. The union of gametes is called fertilization. Following bloom, the cup-like receptacle dries and is called the shuck. It then splits and falls off as the fruit grows.

Pollination in peaches occurs through insects through homogamy. In most of the peach varieties, the fruit set is through self-pollination. Some of the peach cultivars such as J. H. Hale and June Elberta are self-sterile. The time taken to reach harvest maturity from fruit set varies from 10 to 11 months, depending upon the cultivar. Hormonal stimulus from the developing embryo prevents the fruit from dropping and causes the ovary to enlarge.

4.17 FRUIT GROWTH, DEVELOPMENT, AND RIPENING

The fruit is yellow or white fleshed, with fuzz (peaches) or without (nectarines). The fruit development in peach can be divided into three stages:

1. Stage I: It lasts for approximately 30–40 days after full bloom in *Alberta*. In this stage, the size of the pericarp and seed increases very fast due to rapid cell division, whereas embryo develops very slowly.
2. Stage II: It starts 40 days after full bloom and continues for four weeks in cv. *Alberta*. In this stage, the development of fruit size is relatively slow. This is the pit-hardening stage, also known as the lag phase. The pit of the fruit hardens, and the embryo development continues. Lignification of endocarp is an important feature of this stage.
3. Stage III: It is also known as final swell and starts 65 to 70 days after bloom in Alberta and lasts six weeks thereafter. In it, the pericarp increases rapidly in size due to cell enlargement in the mesocarp. The fruit completes its growth by accumulating starch, proteins, and lipids. The fruit growth largely depends upon the rate of net water accumulation as the larger quantity of water transported through the phloem increases the fruit size rapidly. Stage III contributes maximum to the increase in volume. About 66% increase in size takes place at this stage. The fruit development in peach follows a double-sigmoid curve. The fruit is drupe, and the edible portion is called the mesocarp. The outer fuzzy skin is called the exocarp and the stony pit is called endocarp. Nectarines have smaller size than peaches. Lack of fuzziness in nectarines makes them more susceptible to mechanical injuries. White-fleshed peaches are softer than yellow-fleshed peaches and have poor keeping quality.

4.18 FRUIT RETENTION AND FRUIT DROP

Up to 90% flowers may set fruits in peach (Szabo and Nyeki, 1999); therefore, fruits are closely spaced on the branches, and there is a competition between fruits for carbohydrates and available assimilates. Fruit thinning is essential to manage this competition (Costa and Vizzotto, 2000). Both flower and fruit drop occur in response to physiological conditions and environmental too. Many flowers and fruitlets may drop shortly after bloom if their ovules were not fertilized. The drop of young fruits, also known as June drop, occurs commonly in April and May. Even then, a bulk of fruit remains on the trees.

Peach trees need flower and fruit thinning; otherwise, the fruit size does not develop properly, and the quality deteriorates. The fruit thinning is practiced in peach before the pit-hardening stage of the fruit. The earlier the thinning performed, the more the final fruit size. Fruit thinning is a practice commercially performed by hand. However, manual thinning is very expensive. Mechanical hand-thinning attachments and high-pressure water spray systems to perform the thinning operations have been tried for blossom thinning but with mixed results. Limb shakers have been successfully tried for fruit thinning in peaches (Rosa et al.,

2008). Mechanical Darwin String thinner at bloom and drum shaker at the flowering stage have proved to be the most beneficial economically in peach orchards (Miller et al., 2011). About 60 days after full bloom, GA_3 applied @ 25 mg/l to the basal part of the shoots has been found to reduce flowering and increase the fruit yield (Stern and Ben-Arie, 2009).

The multiple application strategy seems to be a promising approach involving chemical and/or cultural techniques to reduce the crop load after through monitoring during dormancy, flowering, and fruit set (Costa et al., 2005). New formulations such as photosynthesis inhibitor metamitron (Brevis), ABA, and 1-aminocyclopropane-1-carboxylic acid are emerging as promising chemical thinners (Greene and Costa, 2013). The fruits are retained at closer spacing if the crop is to be raised for processing purposes as the un-thinned trees are very high yielding.

Girdling is found to have positive influence on size development of the fruit and even reducing the fruit drop. There is a significant increase in fruit size with girdling in a single year and not repeated in subsequent years as there are reports of the overall yield reduction in subsequent years in the case of repeated girdling (Taylor, 2004). Early tree girdling has been found to reduce stem water potential and shoot growth, resulting in vigor reduction (Tombesi et al., 2014).

4.19 HARVESTING AND YIELD

Peach fruits are climacteric in nature, and their harvesting time influences the fruit quality and marketability. If harvested at the immature stage, the fruits do not ripen properly, have inadequate development of flavor, and low consumer acceptance. Harvesting at the overmature stage results in quick softening of the flesh and shortened postharvest life. Mature fruit has physiologically completed its development and is capable of ripening on its own. Therefore, the maturity indices are the criteria used to assess and schedule the optimum time of harvesting. Outer skin color of the fruit is the most commonly used maturity index used to estimate the optimum time of harvest. For yellow-fleshed varieties, the change of fruit color from green to light green and for white fleshed, the change from green to whitish green or creamy white or cream are the indicators of harvest maturity. Firmness of the flesh is another reliable maturity index. Average time taken by the fruit to reach harvest maturity from fruit set is about 70 days. It may vary from cultivar to cultivar and the geographical location of the place. All fruits do not mature simultaneously on the tree. Therefore, 3–4 pickings are required at the 3–4 day interval.

The fruits are hand picked using ladders and collected in baskets or crates. The pickers should wash their hands on completing the picking a lot to avoid the spread of disease spores. Cushioning material is spread inside the container to avoid bruises while handling. The collecting sacs and containers are sanitized properly. The diseased and damaged fruits should be kept separately and should not be kept in the field as it may act as a source of fungal infection for future crops. Such fruits should be disposed off immediately by separate handling staff. Harvested fruits should not be exposed to direct sunlight or heat. The fruits should be precooled and graded before packaging and transportation to packhouses. Physically damaged fruits are suitable for processing or animal feed only, while small, spotted, or misshapen fruits are sold in low value markets. The average yield varies from 70 to 120 kg/tree in subtropical conditions (Bal, 1998).

4.20 PROTECTED/HIGH-TECH CULTIVATION

For peach growing in greenhouses, the planting systems to be adopted should be such as to restrict the growth of the trees. High-density plantation in peaches has encouraged dwarfing of the trees and growing them in protected structures such as greenhouse. However, the planting densities above 1500 trees/ha have resulted in a significant reduction in fruit quality and yield over a period of several years (Caruso et al., 1999). Reduction in tree vigor allows the reproductive phase to come early and hence the early fruiting. The trees bear precociously than open-field conditions, with an earliness of 13–20 days as compared to open-field conditions and higher mean yield (Furukawa et al., 1990), while solar lean to houses and high tunnels are capable of advancing the harvest by 90 and 30 days, respectively (Layne et al., 2013). Y or Tatura training systems have proved to be more suitable for greenhouse growing. Many low-chilling cultivars ripen quite early in the season, extend the season of peach fruit availability, and increase the income of the orchardists under greenhouse cultivation (Gao et al., 2004). Moreover, the plants are protected from strong winds, hails, frost, and the pesticide application is greatly reduced. China is the largest producer of peaches and nectarines in the world. Three types of structures are mainly installed: solar heated or lean to type east–west-oriented or north–south-oriented high-tunnel structures (Layne et al., 2013).

The grafted trees procured from nursery during the dormant season are planted in the greenhouse at 1 m × 1.5 m, 1.2 m × 1.5 m, or 1.5 m × 2 m with best fertilization and irrigation practices followed to encourage growth for the first year (Layne et al., 2013). Later in the first year, the nitrogen supply is restricted, and PPP 333 can be sprayed on new shoots to promote reproductive growth, and winter pruning is performed in the dormant season (Layne et al., 2013). The use of dwarfing rootstocks, summer pruning, autumn girdling, and application of growth regulators have been performed in greenhouse peaches to get desired results. Greenhouse temperature is the most critical element for obtaining the crop of good quality. High temperature during bloom and fruit set is harmful causing massive fruit drop; hence, it should not be allowed to go beyond 23 °C during that period, and temperature regulation should be done by manual or automatic ventilation (Erez et al., 1998). Evaporative cooling is required to prevent chilling negation, which may occur inside the greenhouse due to high temperature in areas where day temperature inside may rise above 19 °C during November. The micro sprinklers aid in reducing the bud temperature by providing evaporative cooling (Erez et al., 1998). The pest and disease management is easier inside the greenhouse than open conditions.

The propagation of peach plants has been successfully tried in the hydroponics system. The time taken by peach plantlets to reach the transplantable stage is significantly reduced than open-field conditions in the hydroponics system (Souza et al., 2011). However, the initial set-up costs are quite high. The costs are compensated later on due to rapid crop growth, reduced insect-pest infection, more productivity, and better fruit quality of the hydroponically raised plants (Souza et al., 2011). The "sundrop" apricots were found to be slow in rooting and establishment in the aeroponics system and more than 50% seedlings could not survive, while in the hydroponics system, the seedlings treated with BAP cytokinin performed slightly better (Arzani, 2000).

4.21 PACKAGING AND TRANSPORT

Generally, specific packaging is designed for specific fruits. Packaging should be able to hold the weight of the fruit and protecting it at the same time from compression bruising, vibration injuries, impact bruising etc. For distant markets, the hard-ripe fruits are harvested. The containers must be designed in such a way that they have enough stacking strength. In a properly designed package, the fruit can be cooled down quickly due to the provision of pack ventilation holes (maximum up to 5%) for air circulation. For retail markets, the fruits are generally packed in corrugated fiberboard cartons or smooth light-weight containers or plastic trays to handle with convenience and to avoid abrasion. The retail packs are generally attractive looking with eye-catching appeal and properly labeled, with the name of fruit, variety, grade, name of the producer, and contact information. For wholesale marketing and transportation, the fruits are packed in a large packaging assembly called pallet bins. The palletization helps in safe stacking and handling of the produce to distant markets. For stone fruits to be transported to distant markets, a complete cold chain system exists right from packing to customer shelf. A moisture barrier is needed to reduce weight loss and dehydration of the fruit. A thin plastic film placed inside the carton under and over the fruit serves the purpose of minimizing the loss. Films are to be used only in the case of the complete cold-chain system as if the fruit is allowed to warm in between, the respiration rate of the fruit will increase, resulting in moisture losses and rapid development of fungal rots.

4.22 POSTHARVEST HANDLING AND STORAGE

Fruits can be successfully stored at 0–0.3 °C and 85%–90% relative humidity for two weeks for freestone cultivars and four weeks for cling-stones. They must be handled gently during and after harvest. Handling during loading and unloading needs to be gentle. The damage sometimes is not visible on the fruit at the time of mishandling; however, it deteriorates the keeping quality later on and reduces the value of product. Bulk containers have to be packed to transport to distant places. Vibration during the transport and the pressing of the fruit against each other causes abrasion and bruises. Careful packaging is needed to avoid the friction of the fruits, and ventilation holes in packaging must not be blocked. Sometimes packing has to be carried out in the field itself. In such cases, there should be dry, clean, and shady area for keeping the fruit before packing. Postharvest handling also includes different packhouse operations such as precooling, sorting, sizing, washing, grading, packing, and temperature management during transportation until storage. The storage life of peaches varies from 7 to 50 days depending on the cultivar. Cultivars with short- and long storage life must not be mixed or stored at the same place. Quick precooling of the fruit, preferably with the forced air-cooling method quickly brings the fruits to −1 to 0 °C and RH of 90%–95%. Fruits can be cooled in hydro coolers up to around 5–10 °C in the field itself if early packaging is needed Sizing and grading are done to fetch better prices of the produce. Larger sizes are preferred in international markets fetching highest prices. Sizing rings are used for this purpose, either manual or machine rollers. The sizing codes assigned in terms of fruit diameter and circumference are: AAAA (≥90 mm and 28 cm), AAA (80–90 mm and 25–28 cm), AA (73–80 mm and 23–25 cm), A (67–73 mm and 21–23 cm), B (61–67 mm and 19–21 cm), C (56–61 mm and 17.5–19 cm), and D (51–56 mm and 16–17.5 cm) (Maloney, 2006).

The conveyer system of sorting is used to handle a larger bulk of fruits. The packhouse area should be clean with workers maintaining good hygiene. Precooling, proper handling, and hygiene play a major role in extending the storage life of the fruit. Temperature is another critical factor to maximize the storage life. Pellet stacks should be properly ventilated in between with no obstruction to airflow. Controlled atmosphere (CA) storage of the fruits is sometimes performed than normal storage to retain fruit firmness and ground color. CA storage is done at 1%–2% O_2 + 3.5% CO_2 (Maloney, 2006); however, it is not common for peaches.

Generally, the bulk fruits of peach can ripen without external application of ethylene. Only the temperature needs to be taken care of. The ripening at lower temperatures (<2

°C–3 °C) is much slower than 20 °C. During ripening, the cell-wall-degrading enzymes become active. The respiring fruit utilizes oxygen and metabolizes sugars into CO_2, chemical energy, and heat energy. If the respiration is slowed down due to temperature reduction or increasing the dose of CO_2 concentration (<5%), the softening enzymatic reactions are retarded, and sugar metabolism is greatly reduced. Temperature adjustment is needed as per the days until the produce becomes available to the consumers. The harvested peach fruit is physiologically active, and it becomes senescent after attaining ripening. The freestone varieties ripen quickly than clingstones. However, improper storage conditions result in undesirable flavors, flesh discoloration, and internal browning. Physiological cause of internal breakdown of the fruit is low temperature chilling injury (CI). Peach cultivars are more susceptible to CI than nectarines; therefore, the susceptible cultivars need to be stored above 5 °C for extending the postharvest life (Cristoto and Kader, 2000).

4.23 PROCESSING AND VALUE ADDITION

Apart from relishing raw peaches, there is huge demand for its processed products, facilitating the year-round availability of the fruit.

Fruits have very short shelf life, and their preservation allows the prevention of microbial growth, reduces waste, and maintains the sensory and nutritional quality of the fruit. The preservation includes the following.

Dehydration in which the water content of the fruit is removed using warm air. It is the oldest technique of fruit preservation. It helps the preservation of the fruit by reducing the risk of spoilage by mould development. Oven drying is the simple and most popular technique. Sun drying is another technique that uses natural sunlight to dry the produce. The fruit is exposed to sun in thin layers. The technique, however, has many drawbacks, as the drying is nonuniform, more time is required, and poor sanitation conditions and the energy input cannot be controlled. The fruit slices can be dehydrated by immersing them in hypertonic solution, for example, concentrated sugar syrup so that dehydration occurs through osmosis, also known as osmotic dehydration. Electric dehydrators are also used for dehydrating the fruits and the best dehydrating agents which maintain the nutritive value of the fruit and provide more uniformity to the food product.

The fruit can be successfully preserved through canning the processed fruits either through hermetic sealing (heated sealing in airtight containers) or pressure canning. Peach jams, jellies,

Peach

squashes, and spreads are quite popular products that contribute to the value addition of the fruit.

4.24 DISEASES, PESTS, AND PHYSIOLOGICAL DISORDERS

4.24.1 FUNGAL DISEASES

Peach is attacked by a lot of fungal diseases that affect the overall productivity of the fruit, quality, and ultimately the economic yield of the orchard. Most of these disease incidences occur at preharvest stages of the fruit at blooming, on leaves, branches, fruits, trunks, and even roots. These diseases are caused by heterotrophic spore producing microorganisms. Out of these, the pathogens infecting roots are soil borne. With increased travel and trade, the chances of pathogen movement and disease spread have increased, which should be restricted or taken care of. The new bio-control practices and bio-technological tools seem promising for strengthening the future management strategies.

4.24.1.1 BROWN ROT

Most common fungal disease worldwide is brown rot, caused by *Monilinia fructicola* and *M. laxa*. The infection may start on developing fruits and causes brown rotting in mature fruits, resulting in fruit decay. On twigs and branches, it causes blossom and twig blight. The previous year fruits having brown rot inoculum left on the floor or trees are the primary sources of spread of spores. The secondary inoculum for infection may be infected shoots and twigs. High atmospheric humidity is most conducive for the pathogen to spread fast.

An integrated management strategy is effective against the spread of infection and control of the disease. The infected fruit and twigs should be removed from the orchards. Alternate host, such as wild peach should not be allowed to grow nearby. Fungicide application is effective at the full bloom stage; however, the sanitation of orchard floor is of prime importance.

4.24.1.2 SHOT HOLE

The disease equally affects peaches, nectarines and apricots. It is caused by *Wilsonomyces carpophilus* (Lev.). Symptoms appear on twigs, leaves, buds as well as fruits. On twigs, leaves and fruits, symptoms appear as slightly sunken purplish spots of size 2–10 mm which turn brown later on. Over wintering fungus on buds acts as primary source of inoculum. Leaf leisons abscise in dry and hot spell to form shot hole appearance. However, in wet cool environment, the leisons contain spores and do not abscise. The infected leaves may fall occasionally. Fungicide application after leaf drop, just after emergence of leaves and at fruit set stage is effective in protecting against twig and bud infection.

4.24.1.3 SCAB

Scab is the most common fungal disease of *Prunus* species and is caused by *Venturiacarpophila*. The infection may occur on twigs, leaves and fruits but is more prominent on fruits. On leaves, symptoms appear as irregular lesions which are slightly darker in color than the surrounding healthy tissue. Water soaked lesions can be seen on fruits which later on become slightly raised brown in color with yellowish hallow with the formation of conidiospores in the center of lesions. The appearance of scab on leaves is similar to bacterial spot however the latter can be well differentiated by small irregular spots and girdling of limbs in severe infection.

The disease can be managed by application of fungicides at fruit set stage and 2 weeks later. The tree canopy should be managed to allow proper ventilation and sunlight penetration.

4.24.1.4 POWDERY MILDEW

It mainly occurs in semi arid regions of the world. The main causal organism is *Podoshaera pannosa*

and *P. leucotricha*. Nurseries and small trees are more infected with it. Leaves show colonies of powdery masses which results in stunted growth, chlorosis and defoliation. Fruits are more susceptible, especially before pit hardening. The disease, like other fungal diseases, flourishes more in high relative humidity conditions. White cottony fungus grows on fruit surface in the form of circular spots and in severe cases, it may cover the entire fruit. Fruits are sometimes deformed and depressed.The fruits having blemishes caused by powdery mildew become unmarketable. The disease is managed with 4–5 applications of fungicides from fruit set till pit hardening stage of fruit development.

4.24.1.5 PHYTOPHTHORA ROOT ROT AND CROWN ROT

The disease, caused by *Phytophthora sp*, is the most common in nurseries and orchards worldwide. The infection occurs through roots at all stages of development of the tree. The fungi are mainly soil borne, and the conditions conducive to the disease are poor soil drainage and heavy rainfall areas. Terminal growth of the trees is impaired, and the trees show die-back symptoms. Killing of the bark and sometimes girdling of the trunk occurs. Infected roots show decay and disintegration of the tissue. Rotting of crown areas of the tree and even development of cankers are found in some cases.

The disease can be managed by careful irrigation management and proper drainage. The rootstocks resistant to *Phytophthora* can be used in high rainfall areas or poor drainage soils. Trees should be planted on raised beds or ridges. Fungicides can be used to protect the plants from root or crown rot.

4.24.2 MINOR DISEASES

4.24.2.1 VERTICILLIUM WILT

The disease is caused by *Verticillium dahlia*. The disease is of minor importance; however, can do serious damage in case the orchard is planted in areas where wilt susceptible crops are generally grown year over year, for example, cotton, tomato, chilly, and pepper. The symptoms appear as sudden wilting and falling of leaves on one or more branches of the tree. The whole young tree can even be killed under low-fertility conditions.

The disease incidence can be successfully managed by avoiding the orchard plantation in the infected soils. This can be done by avoiding plantation in such areas where susceptible crops were grown for many years. Soil fertility and tree nutrition should be maintained.

4.24.3 BACTERIAL DISEASES

Bacterial diseases of peach include bacterial canker, crown gall, and bacterial leaf spot.

4.24.3. BACTERIAL CANKER AND GUMMOSIS

The most common bacterial disease caused by *Pseudomonas morseprunosum* and *P syringae*. The tree trunk, branches, shoots, spurs, blossoms, buds, leaves, and fruits all are affected with the disease. The small, discolored, irregular lesions appear on leaves, branches, and fruits. Girdling of limbs may even occur in the case of severe infection. Gum may exude out of the cracked barks, exposing the tissue beneath. Old, weaker trees are more susceptible than young and healthy ones. Humid and cold conditions are most suitable for the development and spread of the disease. For the management of the disease, alleviating predisposing factors have to be kept in mind (Cao et al., 2005). The site should be suitable for peach growing. The tree health should be maintained with proper nutrition. The pruning should be delayed to the end of dormancy to minimize the tree stress. For chemical management, the wounds on trunk, branches, and twigs need to be cleared before the commencement of rains and mashobra paint (225 g Lanolin, 12 g Stearic acid, 150 g

Marpholin, and 25 g Streptocycline in 5.5 l of water) can be applied on cut ends (Anonymous, 2018).

4.24.3.2 BACTERIAL LEAF SPOT

It is the most common disease in orchards established in light and sandy soils and in humid warm conditions. The disease affects leaves, fruits, and twigs. Greyish lesions appear on leaves which later on become necrotic. The whole leaves having lesions become chlorotic and abscise. The bacterial colonies can be observed under the microscope to differentiate between bacterial lesions or those caused by pesticides or fungi, particularly that of scab. The scab lesions are more circular in shape and more restricted to the outer surface of the fruit. The canker develops on the terminal buds extending downward to the leaves and branches; however, it is confined to the bark only.

The management can be effective with the use of resistant varieties; however, these are limited. Preventive chemical control measures can be tried by the use of Cu-containing material; however, Cu toxicity has to be avoided.

4.24.3.3 CROWN GALL

The disease is caused by *Agrobacterium tumefaciens* and is most commonly found in peach nurseries. The crown portion of the roots develops swollen tumor-like tissues known as galls, which may be of 1–3 cm diameter or even more. The growth of young trees may get restricted after the development of galls. Any type of injury to roots while handling the nursery plants or interculture operations favors the development of galls in the crown-root portion.

The treatment of nursery plants with a bio-control strain *A. radiobacter* before planting in the field is an effective preventive bio-control measure. The planting material should be free from the disease, and the ones having the infection should not be disposed off. The nursery source should be reliable one. Injury to the roots should be avoided after plantation to avoid the disease infection.

4.24.4 VIRAL DISEASES

Viral diseases of peaches are spread by aphid vectors generally and are spread by the use of infected plant material for propagation.

4.24.4.1 PLUM POX

Plum pox is caused by the *Plum Pox virus* and has resulted in huge economic losses worldwide, particularly in North and South America

(Dal Zotto et al., 2006) and in the Kazakhstan and Chinese region of Asia (Navratil et al., 2005). The virus infects almost all cultivated as well as wild species of *Prunus*. Yellowing and discoloration of leaves occurs with vein clearing and yellowish lines. The petals are discolored and necrotic areas appear on fruits that become misshapen and premature drop may occur. The infection generally spreads by the aphids and use of the infected propagation material. For efficient control measures, the use of the infected plant material for propagation should be strictly avoided. The prophylactic measures should be adopted for disease eradication, which mainly include the reliable plant material. Some resistant plant material should be developed through breeding programs and biotechnological tools.

4.24.5 INSECT PESTS

The insect pests of the peach fruit can reduce the tree vigor, profitability, as well as quality. Major insects are aphids, beetles, borers, caterpillars, and fruit flies which cause serious economic losses to the orchard.

4.24.5.1 PEACH LEAF CURL APHID

This aphid (*Brachycaudus helichrysi*) is a polyphagus pest. It sucks the sap from young leaves, resulting in curling up and yellowing. The vegetative buds get dried up affecting the growth. In the case of severe infestation, the fruit set is also hindered, and the developing fruits fall prematurely. For biological control, the parasitic wasp *Oomyzusscaposus* has been found to parasitize the aphid up to 40%–50% (Singh and Kaur, 2015). Insecticides for the chemical control of sucking pests should be applied as spray application immediately after fruit set and repeated after 15 days (Bal, 1998). The infestation may result in severe economic losses if not managed in time.

4.24.5.2 PEACH BLACK APHID

It (*Pterochlorus persicae*) is a black-colored aphid, somewhat bigger than leaf curl aphid. These gather in large numbers on bark, limbs, tender shoots, and branches of the tree and suck sap. The fruit set is greatly reduced because of the draining out of sap from branches, and even developing fruits may fall prematurely. Similar control measures to the leaf curl aphid can be taken.

4.24.6 PHYSIOLOGICAL DISORDERS

4.24.6.1 PIT SPLITTING

Pit is split or torn apart along the dorso ventral suture. The symptoms are not visible until the fruit is cut open. Fruits are sometimes flattened at one end. The causes of pit splitting may be excessive fertilization and/or excessive irrigation. To prevent it, excessive nitrogen application, especially close to the harvest time, should be avoided. Misshapen fruits should be discarded at the time of packaging.

4.24.6.2 INTERNAL BREAKDOWN/CHILLING INJURY

This disorder occurs during the prolonged storage of the fruit and is the main limiting factor in storage and shipping of stone fruits. The flesh of the fruit shows browning, lacks flavor and juiciness, and fails to ripen properly. The pit cavity appears to be black, and red pigmentation accumulates in the flesh (known as bleeding). Symptoms appear when fruits are taken out of the cold store and kept at room temperature. Early season cultivars are less susceptible to CI than late-season cultivars. The best way to eliminate the problem is temperature management. Storage of the fruit below 0 °C but above freezing point is beneficial to avoid CI symptoms and extending the market life.

4.24.6.3 SKIN BURNING/ STAINING/INKING/ DISCOLORATION

Affected fruits show burns on the fruit skin in the form of discolored black or brown spots or stripes. These symptoms are also called inking symptoms and are confined to the fruit skin only. Excessive storage and inappropriate cooling practices during storage are the reason behind this. Physical damage during harvesting/washing/grading or moisture losses form the fruit surface are some other reasons. Sometimes, the fungicide application close to harvest when comes in contact with bruised fruits may cause inking. To prevent skin burning, physical damage to the fruit during harvesting/postharvest handling should be avoided. Forced air cooling may cause skin burning; therefore, room cooling should be preferred. Excessive moisture loss from the fruit should be avoided.

KEYWORDS

- peach
- *Prunus persica*
- flowering
- fruit set
- nutrition
- maturity
- ripening
- shelf life

REFERENCES

Ahmad, T., Ur-Rahman, H., Laghari, M.H. Effect of different auxin on *in vitro* rooting of peach rootstock GF677. *Sarhad Journal of Agriculture*, 20 (3): 373–375, 2004.

Annandale, J.G., Jovanovic, N.Z.G.S., Campbell, N.D., Sautoy, L.P. Two dimensional solar radiation interception model for hedgerow fruit trees. *Agricultural and Forest Meteorology*, 121: 207–225, 2004.

Anonymous. Package of practices for cultivation of fruits, Punjab Agricultural University, Ludhiana, 2018.

Arce, J.P., Storey, J.B., Lyons, C.G. Effectiveness of three different zinc fertilizers and two methods of application for the control of "Little-Leaf" in peach trees in south Texas. *Communications in Soil Science and Plant Analyses* 23: 1945–1962, 1992.

Arzani, K. The response of "Sundrop" apricot seedlings to soilless culture medium and exogenous cytokinin application. *Acta Horticulturae*, 516: 67–73, 2000.

Bal, J.S. *Fruit Growing*. Kalyani Publishers, Ludhiana, 1998.

Bryla, D.R., Trout, T.J., Ayars, J.E., Johnson, R.S. Growth and production of young peach trees irrigated by furrow, micro spray, surface drip or subsurface drip systems. *HortScience*, 38: 1112–1116, 2003.

Bryla, D.R., Dickson, E., Shenk, R. Influence of irrigation method and scheduling on patterns of soil and tree water status and its relation to yield and fruit quality in peach. *HortScience*, 40: 2118–2124, 2005.

Callahan, A., Scorza, R., Morgens, P., Mante, S., Cordts, J., Cohen, R. Breeding for cold hardiness searching for genes to improve fruit quality in cold-hardy peach germplasm. *HortScience*, 26: 522–526, 1991.

Cao, T., Duncan, R.A., McKenry, M.V., Shachel, K.A., DeJong, T.M., Kirkpatrick, B.C. Interaction between nitrogen-fertilized peach trees and expression of *syrB*, a gene involved in syringomycin production in *Pseudomonas syringae* pv. *syringae*. *Phytopathology*, 95: 581–586, 2005.

Caruso, T., Giovannini, D., Marra, F.P., Sottile, F. Planting density, above ground dry matter partitioning and fruit quality in greenhouse grown "Flordaprince" peach (L. Batsch) trees trained to free standing Tatura. *Journal of Horticultural Science and Biotechnology*, 74 (5): 547–552, 1999.

Castro, L.A.S., Silveira C.A.P. Vegetative propagation of peach by air layering technique. *Revista Brasileira de Fruticultura, Jaboticabal*, 25: 368–370, 2003.

Chadha, K.L. *Handbook of Horticulture*. Directorate ICAR, Pusa, New Delhi, 2002.

Childers, N.F. *Modern Fruit Science*. Hort Pub Gainesville, Florida, 1983.

Costa, G., Vizzotto, G. Fruit thinning of peach trees. *Plant Growth Regulation*, 31: 113–119, 2000.

Costa, G., Dal Cin, V., Ramina, A. Practical, physiological and molecular aspects of fruit abscission. *Acta Horticulturae*. 727: 301–310, 2005.

Cristoto, C.H., Kader, A. *Peach: Postharvest Quality Maintenance Guidelines*. University of California, Davis, CA, 2000.

Reuter, D., Robinson, J.B. (eds.). *Plant Analysis: An Interpretation Manual*, 2nd edn. CSIRO Publishing, Collingwood, Australia, pp. 349–382, 1997.

Dal Zotto, A., Ortego, J.M., Raigón, J.M., Caloggero, S., Rossini, M., Duchase, D.A. First report in Argentina of plum pox virus causing Sharka disease in *Prunus. Plant Disease Note*, 90: 523, 2006.

Dutra, L.F., Kersten, E., Fachinello, J.C. Cutting time, indolebutyric acid and tryptophan in rooting of peach tree cuttings. *Scientia Agrícola*, 59: 327–333, 2002.

Dye, M.H., Buchanan, L., Dorofaeff, F.D., Beecroft, F.G. Boron toxicity in peach and nectarine trees in Otago. *New Zealand Journal of Experimental Agriculture*, 12: 303–313, 1984.

Erez, A., Yablowitz, Z., Korcinski, R. Greenhouse peach growing. *Acta Horticulturae*, 465: 593–600, 1998. http://dx.doi.org/10.1590/S1413-70542011000200013

FAOSTAT. http://www.fao.org/faostat/en/#data 2016.

Felek, W., Mekibib, F., Admassu, B. Micropropagation of peach, *Prunus persica* (L.) Batsch. cv. Garnem. *African Journal of Biotechnology*, 16: 490–498, 2017.

Finn, C.E., Clark, J.R. Register of new fruit and nut cultivars list 44. *HortScience*, 43 (5): 1321–1343, 2008.

Frecon, J.L. *Best Management Practices for Irrigating Peach Trees*. Rutgers Cooperative Extension, New Jersey Agricultural Experiment Station, Rutgers, 2002.

Furukawa, Y., Kataoka, T., Shimomura, M. Productivity of high density peach orchard using of free rootstock (*Prunus persica* Thunb). In: *Proceedings of 23rd International Horticultural Congress*, 4169 Frienze, Italy, 1990.

Furuya, S., Umemiya, Y. The influence of chemical forms on foliar-applied nitrogen absorption for peach trees. *Acta Horticulturae*, 594: 97–103, 2002.

Gao, H., Wang, S., Wang, J. 2004. Fruit protected cultivation in China. *Acta Horticulturae*, 633: 5966, 2004.

Gomez, P.M., Perez, S.R., Rubiio, M., Dicenta, F., Gradziel, T., Sozzi, G.O. Application of recent biotechnologies to *Prunus* tree crop genetic improvement. *Ciencia e investigación agraria: revista latinoamericana de ciencias de la agricultura*, 32 (2): 73–96, 2005.

Granatstein, D., Kirby, E., Brockington, M., Hogue, G., Mullinix, K. Effectiveness of weed control strategies for organic orchards in Central Washington. Year 2 Report, 2006.

Greene, D., Costa, G. Fruit thinning in pome and stone fruit: State of art. Proceedings of EUFRIN Thinning Working Group Symposia. *Acta Horticulturae*, 998: 93–102, 2013.

Gupta, U.C., Jame, Y.W., Campbell, C.A., Leyshon, A.J., Nicholaichuk, W. Boron toxicity and deficiency: a review. *Canadian Journal of Soil Science*, 65: 381–409, 1985.

Iezzoni, A.F. The "Redhaven" peach. *Fruit Varieties Journal*, 41: 50–52, 1987.

Johnson, R.S., Andris, H., Day, K., Beede, B. Using dormant shoots to determine the nutritional status of peach trees. *Acta Horticulturae*, 721: 285–290, 2006.

Johnson, R.S., Rosecrance, R., Weinbaum, S., Andris, H., Wang, J. Can we approach complete dependence on foliar-applied urea nitrogen in an early-maturing peach? *Journal of the American Society for Horticultural Science*, 126: 364–370, 2001.

Kralj, M.B., Jug, T., Komel, E., Fazt, N., Jarni, K., Zivkovic, J., Mujic I. Aromatic compound in different peach cultivars and effect of preservatives on the final aroma of cooked fruits. *Hemijska Industrija*, 68: 767–779, 2014.

Krewer, G., Beckman, T.G., Sherman, W.B. Moderate chilling peach and nectarine breeding and evaluation program at Attapulgus, Georgia. *Acta Horticulturae*, 465: 171–175, 1997.

Layne, D.R., Bassi, D. *The Peach: Botany, Production and Uses*. CABI, Boston, MA, 2008.

Layne, D.R., Wang, Z., Niu, L. Protected cultivation of peach and nectarine in China-industry observations and assessments. *Journal of the American Pomological Society*, 67 (1): 18–28, 2013.

Layne, R.E.C. Cold hardiness of peaches and nectarines following a test winter. *Fruit Varieties Journal* 36: 90–98, 1982.

Lima, A.J.B., Alvarenga, A.A., Malta, M.R., Gebert, D., Lima, E.B. Chemical evaluation and effect of bagging new peach varieties introduced in southern Minas Gerais-Brazil. *Food Science and Technology*, 33: 434–440, 2013.

Luchsinger, L., Ortin, P., Reginato, G., Infante, R. Influence of canopy fruit position on the maturity and quality of "ANGELUS" peaches. *Acta Horticulturae*, 592: 515–521, 2002.

Maloney, B. Post harvest handling guide for Moldovan peaches and nectarines. *Agribusiness Development Project*, USAID, 2006.

Mann, M.S., Sidhu, B.S., Chahil, B.S., Mann, S.S. Effect of different rates of zinc applied to soil and foliage of peach (*Prunuspersica* Batsch) on zinc concentration in leaves, fruit yield and quality. In: Chadha, T.R., Bhutani, V.P., Kaul, J.L. (eds.), *Advances in Research on Temperate Fruits*. Dr Y.S. Parmar University of Horticulture and Forestry, Solan, India, pp. 189–191, 1986.

Mengel, K., Kirkby, E.A. *Principles of Plant Nutrition*. 5th ed., Kluwer Academic Publishers, Dordrecht, 2001.

Miller, S.S., Schupp, J.R., Baugher, T.A., Wolford, S.D. Performance of mechanical thinners for bloom or green fruit thinning in peaches. *HortScience*, 46 (1): 43–51, 2011.

Mindêllo, N.U.R., Telles, C.A., Biasi, L.A. Adventitius rooting of peach cultivars semihardwood cuttings. *Scientia Agrária*, 9: 565–568, 2008.

Nable, R.O., Banuelos, G.S., Paull, J.G. Plant and soil. *Boron Toxicity*, Chapter 12, Kluwer Academic Publishers, the Netherlands, 193: 181–198, 1997.

Navrátil, M., Safarova, D., Karesova, R., Petrzik, K. First incidence of plum pox virus on apricot trees in China. *Plant Disease Note*, 89: 338, 2005.

Okie, W.R., Ramming, D.W., Scorza, R. Peach, nectarine and other stone fruit breeding by the USDA in the last two decades. *HortScience*, 20: 633–641, 1985.

Perry, J.M.R.L., Lauri, P.É., Fernandez, T.R. Apple rootstock studies in Michigan. *Compact Fruit Tree*, 26: 97–99, 1993.

Robinson, J.B., Treeby, M.T., Stephenson, R.A. Fruits, vines and nuts. *Computers and Electronics in Agriculture*, 61: 213–221, 2008.

Sansavini, S., Gamberini, A., Bassi, D. Peach breeding, genetics and new cultivar trends. *Acta Horticulturae*, 713: 23–48, 2006.

Sanz, M., Montañes, L. Flower analysis as a new approach to diagnosing the nutritional status of the peach tree. *Journal of Plant Nutrition*, 18: 1667–1675, 1995.

Sherman, W.B., Rodriguez-Alcazar, J. Breeding of low chill peach and nectarines for mild winters. *HortScience*, 22: 1233–1236, 1987.

Singh, S., Kaur, G. Natural control of peach leaf curl aphid, *Brachycaudushelichrysi* (Kaltenbach) (Hemiptera; Aphididae) by *Oomyzusscaposus* (Hymenoptera: Eulophidae) on peach. *Journal of Biological Control*, 29 (3): 167–168, 2015.

Smith, S.A. Peaches: Value added food products. In: Smith, S.A. (ed.), *Consumer Food Safety Specialist*, Washington State University, Washington, DC, 2017.

Souza, A.G., Chalfun, N.N.J., Faquin, V., Souza, A.A. Production of peach grafts under hydroponics conditions. *Ciência e Agrotecnologia*, 35 (2): Print version, ISSN 1413-7054. http://dx.doi.org/10.1590/S1413-70542011000200013

Stern, R.A., Ben-Arie, R. GA_3 inhibits flowering, reduces hand thinning and increases fruit size in peach and nectarine. *Journal of Horticultural Science and Biotechnology*, 84 (2): 119–124, 2009

Szabò, Z., Nyéki, J. Self-pollination in peach. *International Journal of Horticultural Science*, 5: 76–78, 1999.

Taylor, K.C. Cable-tie girdling of peach trees approximates standard girdling results. *The Journal of the American Pomological Society*, 58 (4): 210–214, 2004.

Teskey, B.J.E., Shoemaker, J.S. *Tree Fruit Production*, A VI Pub, Co., Inc., Westport, pp. 187–291, 1982.

Tombesi, S., Kevin, R.D., Johnson, R.S., Phene, R., Dejong, T.M. Vigour reduction in girdled peach trees is related to lower midday stem water potentials. *Functional Plant Biology.* http://dx.doi.org/10.1071/FP14089. 2014.

Tombesi, S., Johnson, R.S., Day, K.R., DeJong, T.M. Interactions between rootstock, inter-stem and scion xylem vessel characteristics of peach trees growing on rootstocks with contrasting size-controlling characteristics. AoB PLANTS, 2010: plq013. doi:10.1093/aobpla/plq013, 2010.

USDA/PSD. http://apps.fas.usda.gov/psdonline/psd Home.asp, 2017.

Walsh, C. Nutrient management. *Environmental Science and Technology.* University of Maryland, 2005.

Wang, Z., Lu, Z. Advances of fruit breeding in China. *HortScience*, 27: 729–732, 1992.

Weir, R.G., Cresswell, G.C. *Plant Nutrient Disorders 1. Temperate and Subtropical Fruit and Nut Crops.* Inkarta Press, Melbourne, Australia, 1993.

Yoshida, M. Peach cultivars, breeding and statistics in Japan. In: Childers, N.F., Sherman, W.B. (eds.), *The Peach.* Horticultural Publ., Gainesville, FL, 1988.

CHAPTER 5

PLUM

LOBSANG WANGCHU[1*], THEJANGULIE ANGAMI[1], and DEBASHIS MANDAL[2]

[1]*Department of Fruit Science, College of Horticulture and Forestry, CAU, Pasighat 791102, Arunachal Pradesh, India*

[2]*Department of Horticulture, Aromatic and Medicinal Plants, Mizoram University, Aizawl 796004, Mizoram, India*

Corresponding author. E-mail: lobsang1974@gmail.com

ABSTRACT

Plum holds a conspicuous and prominent position in the global scenario. Fresh plum fruits are delicious, nutritious, and have wide varieties of sizes and colors prized both for its exquisite fresh fruit flavor and in the fruit-processing industry. Plums are by far the most diverse of all the *Prunus* species and could be the most diverse of all deciduous fruit crop species. The diversity of plums is also reflected in their names. There are plums, prunes, gage plums, egg plums, and Mirabelle; and the wild plums such as cherry plums, bullaces, damsons, and sloes. The fruit constitutes an important source of minerals, vitamins, sugars, and organic acid in addition to protein, fat, and carbohydrate. Native species of plums exist in nearly every temperate zone in the world where there is sufficient chilling to break dormancy. With diverse genetic material, plums are the ideal species to play a central role as a fresh fruit for local or regional markets. Adapted cultivars have wider adaptability and can be found or bred for in any temperate region of the world.

5.1 INTRODUCTION

Plum is considered to be one of the important deciduous stone fruits of temperate regions and known to be the first temperate fruit species to entice human interest (Faust and Suranyi, 1999). In plums, the seed is covered by the lignified endocarp which is called stone. The mesocarp forms the fruit flesh which is encircled by the exocarp, the skin. Plums are generally consumed fresh for dessert with a small quantity used for the preparation of beverages. Plums have been known since yore for their beneficial properties which led to the increasing interest due to many health advantages and benefits to the human body owing to the presence of good content of phytonutrients, which are regarded a rich source of natural antioxidants. The plum was known to Greeks and Romans. Romans contributed to their cultivation and spread them throughout Europe. Plums having a higher content of sugar along with harder and firm flesh are usually dried with stones in them without fermentation; these are referred to as Prunes and are considered a well-relished nutritious product. In Central Europe, dried plum fruits contribute, to a large extent, to the carbohydrate and vitamin supply of the rural population during winter times. Out of more than 2000 varieties of plums, very few are of commercial significance (Somogai, 2005), which are grown in temperate regions, among which China, Romania, and the USA are some top leading nations in terms of production (Blazek, 2007). Although the cultivation of plums under Indian condition is not very old; however, because of their great acceptance to the consumer either as fresh or their various products, they have gained popularity and are occupying an important position as a commercial fruit crop. The important plum-producing states in India are Jammu & Kashmir, Himachal Pradesh, hilly regions of Uttarakhand, plains of Punjab, UP, and with Northeastern states (at smaller scales). The Nilgiri Hills in the south also offer immense scope for area expansion of its cultivation.

5.2 PRODUCTION LEVEL AND AREAS

5.2.1 INTERNATIONAL

China, Romania, USA are some leading nations in plum production. Top 10 major plum producers in the world in 2018 as per the estimated data from the Food and Agriculture Organization of the United Nations (FAOSTAT, 2018) are given in Table 5.1

TABLE 5.1 World Plum Production Estimates

Position	Countries	Production (in MT)
1	People's Republic of China	6,788,107
2	Romania	842,132
3	Serbia	430,199
4	United States of America	368,206

TABLE 5.1 *(Continued)*

Position	Countries	Production (in MT)
5	Iran	313,103
6	Turkey	296,878
7	India	251,389
8	Chile	229,951
9	Morocco	205,222
10	Ukraine	198,070

5.3 MARKETING AND TRADE

World plum production increased from 8.1 million tons in 1999 to 12.6 million tons in 2018 (FAOSTAT, 2018). In Europe, about 90% of the plums produced are European plums, whereas in Asia, 82% are Japanese plums. In North America, 53% of the produced plums are European, 38% Japanese, and nearly 9% *Prunus americana*, *Prunus simonii*, *Prunus nigra*, and *Prunus P. munsonia* (Suranyi and Erdos, 1998). In Germany, about 50,000 tons of European plums are sold by producer markets annually. In total about 300,000 tons of plums are harvested including fruits produced for direct selling, brandy production, and in home gardens. In Asia, China is dominant in plum production. Huge area enlargement in China may be the reason for the strong increase in plum production. China alone contributed 6.7 million tons of the world plum production in 2018 (FAOSTAT, 2018). Countries with the highest production of *Prunus domestica* are Germany, Romania, Bulgaria, and Serbia. In North America, about 90% of the market fruits are produced in California with most of the production used for drying, with "French Prune" being the important cultivar. The value of production of dried prunes was 1246 billion US$, as per the average production of the years 1995–2004. The decrease of production was caused by a decrease of about 25% price for dried prunes during the last 10 years (USDA-NASS, 2005). The United States is a net exporter of plums. The top market for the US plum exports in 2004 was Canada (56%), followed by Taiwan (11%), Mexico (8%), and Hong Kong with 7% (AgMRC, 2006). In Germany, half of the plum production is used in bakeries, 30% for fresh consumption, and the rest for brandy production. Also, over the few past decades, the plum production in Bulgaria, Romania, and Serbia has decreased due to problems with the plum pox virus (PPV), which causes the Sharka disease. The South African plum production has increased from 60,000 tons in 2009–2010 to nearly 80,000 tons in 2013–2014, giving a clear report that the biggest share of plum production is destined for the export market, while the remainder is consumed as fresh produce, and with the least production share dedicated for processing. The most cultivated plum varieties in South Africa include Laetitia, Songold, and Sapphire (SA Fruit Trade Flow, 2014).

5.4 NUTRITIONAL COMPOSITION AND HEALTH BENEFITS

Plum fruit, being a good source of sugar, contains small amount of starch at the immature phase, which increases later during ripening. The important acids present in plum are quinic acid and malic acid, imparting the acidic taste of fruits. The presence of vitamin C is of nutritional significance in plums. The presence of phenolic substances, namely, caffeic acid, catechin, chlorogenic acid, rutin, and phenolic acid impart astringent characteristics of the fruit. The major anthocyanins in plum are the cyanidin-3-rutinoside and peonidin-3-rutinoside. Amygdalin, a glucoside, exists in the seed of the fruit. Major amino acids present in the fruit are asparagine, aspartic acid, glutamine, glutamic acid, serine, valine, leucine, and proline. Flavonoids and carotenoids are also present in the plum fruit. The plum fruit also contains a good amount of minerals such as K, Na, Ca, Mg, Zn, and Fe. Prunes contain low sodium, higher level of dietary fiber, iron, beta carotene, and potassium. Prune juice can be utilized as humectants and a natural laxative.

The proximate proportion of nutrients in the plum fruit as per Bhutani and Joshi (1995) revealed that the principal edible portion of plum pulp contains moisture ranging from 86 to 88 g per 100 g of edible portion. Total sugar (6.7%–9.9%), titratable acidity (16.3–30.5 meq $H^+/100$ g), dietary fiber (1.3%–2.4%), malic acid (0.14–2.54 meq $H^+/100$ g), citric acid (0.03–0.04 meq $H^+/100$ g), quinic acid (0.12–0.41 meq $H^+/100$ g), protein (0.4%–0.8%), energy (125–187 kJ), vitamin C (4–11 mg per 100 g), thiamin (0.02–0.05 mg per 100 g), carotene (0.26–0.78 mg per 100 g), flavanones (46.3–57 mg per 100 g), anthocyanin (926 mg per 100 g), pectin (0.81%–0.98%), K (120–190 mg per 100 g), Na (0–3 mg per 100 g), Ca (6–8 mg per 100 g), Mg (4–7 mg per 100 g), Fe (0.1–0.4 mg per 100 g), and Zn (0.1 mg per 100 g).

There are many health benefits of the plum fruit. It is known to control and regulate the action of the digestive system, which aids in relieving constipation. Vitamin C in the fruit aids in developing resistance in the body against infectious agents and scavenges free radicals. Carotenoid, a health-promoting phytochemical, acts against aging and disease caused by free radicals and reactive oxygen species. Plums contain a good amount of potassium, a main component of the human body and cell fluids, which aids in regulating blood pressure. The fruit also possesses moderate sources of vitamin B complex groups that help the body metabolize proteins, carbohydrates, and fats. A certain level of vitamin K is also present in the plum fruit, which plays an important role in blood-clotting factors and aids in diminishing Alzheimer's disease in the elderly (Birwal et al., 2017).

5.5 ORIGIN AND DISTRIBUTION

There are about 42 species of plum but the cultivated plums are derived from two species namely *Prunus domestica* (European plum) and *Prunus salicina* (Japanese plum) (Chadha, 2019). Five centers of origin for plums had been identified by Watkins (1976). These include Europe for *Prunus domestica* (European plum), Western Asia for *Prunus insititia* (Damson plum), Western and Central Asia for *Prunus cerasifera* (*P. cerasifera*) (cherry plum), China for *Prunus salicina* (Japanese plum), and North America for *Prunus americana* (American plum). Plum species evolved from the North American center of origin have been placed under the category of *Prunocerasus*, whereas the Asian and European plums have been placed under the category of *Euprunus* (Knight, 1969).

Native places of other species have been identified as Italy, Greece, and Yugoslavia for the Italian plum (*Prunus cocomilia* Ten.), Europe and Asia for the Blackthorn sloe (*Prunus spinosa* L.), North China for the Apricot plum (*Prunus simonii* Carr.), Manchuria, along the Ussuri River for the Manchurian plum (*Prunus ussuriensis* Kov. and Kost.), the Northeastern United States for the Allegheny plum or sloe (*Prunus alleghaniensis* Porter), the Southern United States for the Sand plum (*Prunus angustifolia* Marsh.), etc. Wild strains of *Prunus salicina* are reported in Hubei and the Yannan region (Okie and Weinberger, 1996). *Prunus domestica* is assumed to be relatively young species, and its wild state is unknown. Luther Burbank reported the Caucasus Mountains near the Caspian Sea as the origin for *Prunus domestica* and its ancestors (Malcolm, 2006). Crane and Lawrence (1956) suggested that Asia Minor as the place where natural hybrids between *P. cerasifera* and *Prunus spinosa* originated first, and the dissemination of seeds of such hybrids from Iran and Asia Minor might have been the progenitors of *Prunus domestica* in Europe.

5.6 BOTANY AND TAXONOMY

A plum or gage is a stone fruit that belongs to the family Rosaceae of the genus *Prunus*, and subfamily *Prunoidae*, having a basic chromosome compliment of $n = 8$. The European plum (*Prunus domestica*) with a hexaploid genome ($2n = 6x = 48$) and the Japanese plum (*Prunus salicina*) with a diploid chromosome set ($2n = 2x = 16$). Plums are categorized as *Euprunus* and *Prunocerasus* and subgenus *Prunophora*. In the section *Prunocerasus*, the North American species bearing small fruits such as *Prunus americana*, *Prunus angustifolia*, *Prunus hortulana*, *Prunus munsoniana*, and *Prunus maritima* with diploid nature are subsumed. Meanwhile, the European group that contains the plum species in Europe and Asia, namely, *P. cerasifera*, *Prunus spinosa*, and *Prunus salicina* belongs to the *Euprunus* section (Neumüller, 2011), ranging from diploid to hexaploid.

The important cultivated plums are of two types: the European plum (*Prunus domestica*) and the Japanese plum (*Prunus salicina*). Other species are of interest to plum breeders for imparting winter hardiness and other desirable characteristics to the existing cultivars or for use as rootstocks (Table 5.2).

TABLE 5.2 Important Plum Species

Species	Desirable Characteristics in Breeding
Prunus alleghaniensis Porter	Fruit, ornamental
Prunus Americana Marsh	Winter hardiness, fruit
Prunus angustifolia Marsh	Earliness, disease resistance
Prunus cerasifera Ehrh.	Rootstock, fruit
Prunus cocomilia Ten.	Fruit
Prunus domestica L.	Fruit productivity
Prunus hortulana Bailey	Fruit
Prunus insititia L.	Fruit
Prunus maritima Marsh	Ornamental, fruit
Prunus maxicana Wats	Fruit
Prunus munsoniana Wight and Hedre	Fruit, disease resistance
Prunus nigra Ait	Winter hardiness, fruit
Prunus rivularis Scheele	Resistance to biotic factors like diseases
Prunus salicina Lindl.	Firmness and size of the fruit
Prunus simoni Carr.	Firmness and size of the fruit
Prunus spinose L.	Fruit, domestica parent
Prunus subcordata Benth.	Fruit, dwarfing rootstock
Prunus umbellata Ell.	Fruit

The subgenus *Prunophora* is distinguished from other subgenera (peaches and cherries) in the shoots having a terminal bud and solitary side buds (not clustered), 1–5 flowers are grouped together on short stems, and the fruit having a groove running down one side, and a smooth stone or pit. The trees of plum are moderately rigorous bushy trees attaining a height of 4–8 m or more. In the European plum, the tree bark is smooth, leaves are coarsely serrated, thick with glossy dark green above and pale green underneath with considerable pubescent. The flowers are borne largely on spurs. The Japanese plum has a rough bark even when young. Leaves are of medium size, sharply pointed, and almost free from pubescence. Flowers are abundantly produced in the cluster of 3 in a bud on many budded short or compact spurs on last season's shoot. The Japanese plum can be readily distinguished from the European plum by its rough bark and more persistent fruiting spurs throughout the tree.

5.7 CULTIVARS

The plum species and varieties can be grouped under the following types:

1. European plum

Prunus domestica is the most important commercial plum in Europe, but it is also grown in other continents of cooler regions and has been divided into the following subgroups considering its characters. The European plum is originated in Eastern Europe or Western Asia (Cambrink, 1993) and has been in cultivation in the European continent for over the past 2000 years. *Prunus domestica*, being hexaploid, originated by means of hybridization of diploid (*P. cerasifera*) and tetraploid (*Prunus spinosa*). The European plum exhibits different unlimited range of color variations such as yellow, green, red, and blue skin colors.

Prunes: This is commercially the important group. Generally, the fruits are smaller in size with a higher sugar content (up to 30 °Brix) and their shape varies from oval to elongated. Hence, the fruits can be well used for drying without removing the pith (Neumüller, 2011). During cooking or baking, prunes keep their texture very well and loose less saps than plum. The color is usually purple and young shoots of prune trees are not pubescent. Important cultivars are Agen (French), Stanley, Sugar, Imperial Epineuse, Italian, German, Giant, and Tragedy. Typical prunes also include "Prune d'Agen" and "German Prune," which is the most spread prune in Europe called "Hausz-wetsche" in Germany, "Pozegaca" in Yugoslavia, "Besztercei" in Hungary, "Casalinga" in Italy, "Quetsche Commune" in France, "Vinete romanesti" in Romania, and "Kustandilska" in Bulgaria (Neumüller, 2011).

Reine Claude (Green Gage): Reine Claude has round fruits in different colors, ranging from green

(Green Gage) to yellow (Oullins) and purple (Graf Althans). The flesh is sweet, juicy with aroma, and of high quality. The fruits are mostly used for fresh consumption and sometimes for brandy production. Its cultivars include Reine Claude, Bavay, Jefferson, Washington, Imperial Gage, and Hand.

Yellow egg: Yellow egg is a smaller group of plums meant for processing, especially canning. The fruit is large, long, and oval, exhibiting yellow flesh with yellow skin. Its cultivars include Yellow Egg, Red Magnum, Bonum, and Golden Drop.

Lombard: Lombard is a group of large oval red or pink plums somewhat of fair quality. Its cultivars include Bradshaw, Pond, Victoria, and Lombard.

Imperatrice: Imperatrice includes all blue plums. Important varieties include President, Grand Duke, Diamond, and Tragedy.

Some European species like *Prunus insititia* includes the small-sized fruits, namely, Damsons, Mirabelle, and Bullaces. The plants of this species are of dwarf stature and compact with smaller leaves and flowers. Damsons are excellent for preserves and jams but are not consumed fresh, whereas Mirabelle has excellent flavor.

2. Japanese or Oriental plum

It is an important plum species (*Prunus salicina*) for developing high-quality cultivars. It is originated from China. Japanese plums have the tendency to bloom earlier and thus affected by frost. Cultivars of Japanese plums vary habit. They exhibit spready and upright growth habits. Important cultivars are Santa Rosa, Beauty, Meriposa, Kelsey, etc. Satluj Purple is a suitable cultivar introduced in the plains of Punjab for commercial purpose (Dhatt et al., 1992).

3. American Plum

Most American plums are self-unfruitful; hence, pollinizers are required to be placed in the orchards (Bhutani and Joshi, 1995). Some species are described briefly as follows.

Prunus americana **Marsh:** The color of this fruit peel is yellow or orange along with golden yellow flesh. Leading cultivars are Desoto, Hawkeye, Wyant, Weaver, and Terry.

Prunus hortulana **Bailey:** This species is resistant to brown rot and is used in processing industries for jams and marmalades. Important cultivars include Wayland and Golden Beauty.

Prunus munsoniana **Wight and Hedrick:** This species is also resistant to brown rot and spring frost. The important variety includes Wild Goose.

Prunus besseyi **Bailey:** This dwarfing rootstock species is aimed to utilized for hybridization purpose.

Prunus subcordata Benth: This species originated in South–Central and Northeast California. The fruits of this species are commercially used in processing industries for the preparation of preserves and jellies.

Some important cultivars of plums and their characteristic attributes are briefly described in Tables 5.3 and 5.4

5.8 BREEDING AND CROP IMPROVEMENT

5.8.1 INTRASPECIFIC HYBRIDIZATION

5.8.1.1 FLOWERING TIME

Japanese plum cultivars bloom quite early in the season. The blooming period relies not only on the species but also on the variety. The blooming time depends also on the region. The time difference between the full bloom of early and late blooming genotypes is more prolonged in warmer regions than in cooler regions. The blooming time of the individual flower also depends on the position of the flower bud on the tree. The Japanese plums generally set more flower buds on long shoots. The flowering period of the Japanese plum is generally shorter than that of the European plum. New varieties of the European plum usually set flower buds on long shoots, which generally delays in blooming, thereby ensuring a better fruit set even in bad weather conditions. The length of the blooming period is genetically determined, but largely influenced by the environment as well.

5.8.1.2 FERTILITY

Fertility is expressed as the percentage of number of fruits developed out of a known number of flowers, and it is genetically determined. The degree of self-fertility is a result of different external and internal factors, flower quality as well as, temperature. Temperature influences the speed of the pollen tube growth and the aging of the ovule. Tests for partial self-fertility are made by comparing the fruit set after cross-pollination to that after self-pollination. In the European plum, some self-fertile cultivars are German prune, Stanley, Hanita, Katinka, etc.; partially self-fertile cultivars include Italian prune, Tophit, Ersinger, etc.; and self-sterile genotypes are Green Gage, President, Valor, Opal, etc. On average, the Japanese plums have a higher flower set and bigger fruits than that of the European plum cultivars, and a fruit set of 5%–10% is sufficient for a good yield.

5.8.1.3 INTERSTERILITY

Cross-incompatibility prevents fertilization between combinations

TABLE 5.3 Chief Characteristics of Some European Plum Cultivars and Prunes

Cultivars	Maturity Time	Fruit Shape	Stone Adhesion	Fruit Size	Skin Color	Flesh Color	Fruit Quality	Remarks
Queenston	Early	Oval	Free	Medium	Blue	Yellow	Fair, acceptable market quality	—
California Blue	Early	Oval	Nearly free	Large	Blue	Yellow	Fair	—
Lombard	Early	Oval	Free	Large	Red	Yellow	—	Good for canning
Early Italian	Early	Oval	Free	Medium	Blue	Yellow	Fair	Ripen better in cool conditions
Stanley	Early	—	—	Medium	Blue	Amber	—	Good pollinizer
Grand Duke	Mid-season	Oval	Cling	Large	Blue	Yellow	Fair	—
Damson	Mid-season	Roundish	Free	Small	Dark blue	Yellow	Poor flavor	Hardy cultivar used for jam
Italian (Fellenberg)	Mid-season	Oval	Free	Medium	Dark blue	Greenish yellow	Firm, very good quality	Standard prune and heavy bloom
Reine Claude (Green Gage)	Late	Oval	Free	Medium	Yellowish green	Golden yellow	Very good quality	Fresh, canning
President	Late	Oval	Free	Large	Purplish	Deep yellow	Juicy, good	Fresh
Vision	Late	Oval	Free	Large	Blue	Yellow	Eating quality	—
Victoria	Late	Oval	Free	Medium	Red	Yellow	—	Frost resistant
Albion	Late	Oval	Cling	Large	Purplish black	Golden yellow	Juicy, good quality	—

TABLE 5.4 Important Characteristics of Some Japanese Cultivars

Cultivars	Maturity Time	Fruit Shape	Stone Adhesion	Fruit Size	Skin Color	Flesh Color	Fruit Quality	Remarks
Early Golden	Very early	Round	Free	Medium	Reddish blue	Yellow	–	Vigorous, biennial bearer
Beauty	Very early	Heart	Cling	Medium	Greenish yellow to bright crimson	Amber with scarlet	Good	No need for pollinizers
AU-Amber	Early	–	Cling	Small	Dark red purple	Yellow	High	Suitable for storage
Kelsey	Mid-season	Heart	–	Large	Greenish yellow	Greenish yellow	Juicy fruit, good	Good for shipping
AU-Producer	Mid-season	–	Free	Small	Dark red purple	Dark red	Good	Fresh
Santa Rosa	Late and early strains	Oblong conic	–	Large	Purplish crimson	Amber	Very good	Prolific bearer
AU-Roadside	Late	–	Semicling	Medium	Dark red	Dark red	Good	–
Burbank	Uneven	Round	Cling	Medium	Greenish yellow to bright crimson	Amber with scarlet	Flesh firm	Requires 2–3 pickings
Mariposa	–	Heart	–	Large	Greenish yellow, mottled with red	Red	Fair	Requires cross-pollination
Wickson	Early	Heart	–	Large	Greenish yellow to red	Yellow	–	Need pollinizer

of plum varieties. Among the European plum cultivars, a low frequency of intersterility was found, which may be due to the hexaploidy of the species. In the Japanese plum, intersterility occurs more often. In several studies, intersterility with the "Italian Prune" was observed. However, this incompatibility is unclear.

5.8.1.4 STERILITY

A low fruit set may be the result of morphological sterility based on short style, small stigmata, or underdeveloped ovary. This phenomenon is more frequent in the Japanese plum than that in the European plum. On young plum trees, more sterile pistils are found than on older trees. The low fertility of seedlings in the first or second year of flowering is based on the ontogenesis of the plant, which is a sign of the juvenility of the plant. Suranyi (1994) explored the flower anomalies of plums and found that the traits of sterility are inherent along with seasonal effects.

5.8.1.5 POLLINATION

Pollination is the transfer of the pollen to the stigma. The prime time for pollination is the first two days of the opening of the flower. In both the plum types, the results of cross-pollination rely on the female parent than on the quality of the pollinator.

Good pollinators within the European plum are cultivars such as "Stanley" and "Cacanska lepotica." Good pollinators produce about 50,000 pollen grains per flower. In "Stanley" and "Italian Prune" cultivars, more than 70,000 were found (Hartmann and Stosser, 1994). In crossing experiments, the quality of the pollinator is important and must be considered. The quality of the pollen depends on the deposition of starch. The highest content is found just before the opening of the flower; high starch content in the pollen grain correlates with the speed of the pollen tube growth (Lorenz, 2000).

5.8.2 INTERSPECIFIC HYBRIDIZATION

Interspecific hybridization is of prime importance in plum breeding. Between different species of the genus *Prunus* it is commonly used for the improvement of rootstocks. For instance, the rootstock "Marianna" is an interspecific hybrid between *P. cerasifera* and *Prunus munsoniana*. For scion breeding, interspecific hybrids at present are comparatively low. Plumcots, a hybrid between *Prunus salicina* and *Prunus armeniaca*, are of commercial interest. Interspecific hybridizations aids in transferring important traits, from one species to another. For example, the cold hardiness of *Prunus spinosa*, *P. cerasifera*, and *Prunus americana* can be transferred to *Prunus salicina* and *Prunus domestica*. The superior fruit quality of *Prunus domestica*, manifested in terms of organic acids, sugars, and aromatic compounds, makes it a potential partner for crossing for improving the inferior fruit quality of other *Prunus* species. The European plum is one *Prunus* species with genotype resistant to the PPV, mediated by hypersensitive response. The Japanese plum genotypes often excel other species in its fruit size, good transport, and storage ability of the fruits. Thus, hybrids between the European and Japanese plums will be promising in improving the pomological value of both species.

5.8.3 MUTATION INDUCTION AND GENETIC ENGINEERING

At the current state of knowledge, genetically modified plum varieties may not be necessary for the plum production. Classical breeding methods are far from being the limit of the improvement of plum genotypes. Very few efforts were made on plum in inducing mutations (Srinivasan et al., 2005). As mutations are often unstable in somatic tissue resulting in chimeric plants, it is preferred to induce mutagenesis during gametes development and utilize for breeding purposes. Induction of mutagenesis in the European plum using x-rays to obtain spur types of some plum

cultivars has been reported (Cociu et al., 1997). Despite the large efforts, genetic transformation and regeneration in plums have been successful in single cases, a part of the coat protein gene of the PPV was transferred to the genome of seedlings of the *Prunus domestica* genotype "B69158" (Scorza et al., 1994). One of these genetically modified seedlings, the clone "C5," showed a level of resistance to PPV. However, this kind of resistance is not advantageous for the known quantitative resistance in existing cultivars as the genetically modified plants can get infected with the virus and can serve as a host of PPV.

5.9 CLIMATE AND SOIL

Plums are usually grown under rainfed conditions. Areas receiving 100–125 cm rainfall, well distributed throughout the growing season are suitable. The distinctive feature of a plum is that it has a definite requirement of winter chilling for fruitfulness, failing which it remains vegetative. Dormancy is such a developmental phase in a tree which is necessary to allow the trees to survive unfavorable low winter temperatures. Plum trees require sufficient low temperatures during winter months (winter chilling) to enter the dormant phase or rest period. The European plum can be grown in areas that experience chilling for 800–1500 hours below 7 °C, whereas the Japanese plum requires 100–800 hours. The Japanese plum can be grown stretching from the subtropical plain up to an elevation of 1525 above mean sea level (msl). The European plum thrives well in hills from 1525 to 2745 above msl. The blossoms of the Japanese plums are sensitive to spring frost not because they are less hardy than other plums but because they bloom very early in the spring. Areas free from spring frost are conducive to plum cultivation.

The European plums perform better on clay soils, whereas the Japanese plums prefer lighter soils (Teskey and Shoemaker, 1978). Nevertheless, deep, well-drained loamy soils are best for both types of plums with a depth of at least 600 mm with soil pH between 5.5 and 6.5 (Department of Agriculture, Forestry and Fisheries [DAFF], 2010). Plums planted on gentle slopes provide good protection to soil against erosion and stabilize the slope. Waterlogged, poorly drained soil with excessive salts should be avoided. It is necessary to carry out the soil profile pits for inspecting the soil for determining the best method to be used in preparing the soil.

5.10 PROPAGATION AND ROOTSTOCK

Plums are mostly grown on plum rootstocks though seedling of peach, Japanese apricot, and almond. In India, shield budding is a common method in which wild apricot is used as a rootstock. Clone GF-43 (*Prunus domestica*), GF-655/2, Montizo 20, and Montpol 21 (*Prunus institia*), and *Prunus* species selections are used as rootstocks for plums. Myrobalan B, Myrobalan 29C, Marianna GF 8/1, Marianna 2624, Marianna 8-6 (Madidon), Marianna 9-52, Pixy, St. Julian A, Ishtara/Feriana, GF-31, and GF-677 are clonal selections for plums and prunes (Rehalia and Tomar, 2004). T-budding or chip budding is commonly followed in autumn or spring on seedling rootstock or clonal rootstock. These can be propagated by leafy, softwood, and hardwood cutting under intermittent mist (Bhutani and Joshi, 1995). The propagation by cleft or tongue grafting is also performed when the stock and the scion are still in dormant. For raising the rootstock seedlings, the seeds are kept under alternate layers of moist sand for the duration of 1–3 months for stratification, depending upon species at temperature 3 °C–5 °C to break the rest period before they germinate. It was reported that the late Mr D.T. Fish recommended the use of Green Gages as stocks for dwarf plums. The common Green Gage is raised from seeds as it is of moderate growth, which

Plum

in turn will have a dwarfing influence upon any plum worked upon it (Sanders, 1998).

In vitro propagation techniques for the multiplication of planting materials have also been developed; one such protocol for surface sterilization and micropropagation of plum cv. Stanley was developed using axillary bud explants. During sterilization protocol establishment, the highest significant survival value was recorded when explants disinfected with 2% sodium hypochlorite for 15 min and 0.1% mercuric chloride for 7 min. The best shoot response and proliferation were obtained on full strength MS media supplemented with 0.5 mg/l 6-Benzylaminopurine, benzyl adenine in combination with 0.1 mg/l of indole-3-butyric acid (IBA), whereas the best rooting response was observed on half-strength MS media supplemented with 1.0 mg/l IBA. Therefore, these concentrations are recommended for the in vitro propagation of sufficient, true to type, and disease-free plants of *Prunus domestica* L. cv. Stanley (Wollela, 2017).

5.10.1 ROOTSTOCKS AND THEIR CHARACTERISTICS

5.10.1.1 MYROBALAN

Myrobalan (*P. cerasifera*) is the most widely used rootstock, particularly for the European plums (Hartmann and Kester, 1972), though it is compatible with the Japanese plum cultivars also. It is susceptible to oak root fungus (*Armillariella mellea*) but resistant to crown rot and drought.

Myrobalan 29C: It is a vigorous stock which may delay fruit ripening. It is not suitable under heavy soils, though it can withstand water logging due to high rainfall or heavy irrigations. It is immune to the root-knot nematodes (*Meloidogyne incognita* and *Meloidogyne javanica*) but susceptible to the root-lesion nematode (*Pratylenchus vulnus*).

Myrobalan B: It is regarded as a standard clonal rootstock for vigorous types of plum trees; it delays ripening and is a good co-absorber (Okie, 1987).

Clonal selections from Myrobalan seedlings have also been made at some other research centers of Europe.

Myrobalan GF 31: The rootstock yields a vigorous tree and can be multiplied easily through cuttings. It results in early bearing and suitable even dry stony soils.

Myrobalan 2-7: A vigorous, drought-tolerant rootstock; exhibits good absorption of K.

Myrobalan 5-Q: The rootstock has the tendency to delay ripening.

5.10.1.2 MARIANNA

Like the Myrobalan this is a vigorous rootstock, but it is characterized by its resistance to nematodes, crown gall, and the oak root fungus, and is easy to propagate through cuttings. It originated in Marianna, Texas, USA in the 1870s. Botanically, it is a cross of Myrobalan (*P. cerasifera*) with *Prunus munsoniana* and is a hexaploid ($2n = 48$). Few clonal selections produce from this stock are as follows:

1. **Marianna 2624:** It produces semivigorous trees resistant to water logging and root-knot nematodes. It is, however, susceptible to bacterial canker and is unsuitable for heavy soils. Nitrogen is efficiently utilized by this stock.
2. **Marianna 4001:** The rootstock yields a vigorous tree having drought tolerant and good anchorage (Okie, 1987).
3. **Marianna GF 8-1:** Though the trees produced on this stock are vigorous, they are precocious. It is resistant to waterlogging, viruses, and root-knot nematode. Okie (1987) reported that the rootstock is suitable for both light and heavy soils and is also tolerant to high-pH soil.
4. **Buck Plum** (Morrison's plum stock): A vigorous rootstock compatible with most European and Japanese plums. It is resistant to crown gall.

5.10.1.3 PRUNUS DOMESTICA

In Europe, selections from *Prunus domestica* have been widely used. Following are some of the distinguishing features of these selections.

1. **Ackermann's** (Marunke): A selection made at the Ackermann–Torgace Nursery in Thuringen, Germany, where it was found wild (Tukey, 1964). It produces a semivigorous tree, hastens fruit ripening, in spite of low productivity.
2. **Brompton:** It is of English origin and is easy to propagate through hardwood cuttings. It produces a semivigorous tree and may delay fruit ripening.
3. **Brussel:** Developed in Holland, it has been widely used in Europe. It is easily propagated by layering and makes a large tree.
4. **Common Plum:** This stock has been used for long in England. It can be easily propagated through both layers and hardwood cuttings. It is not compatible with all the plum cultivars, but when compatible, it is of value as a semidwarfing plum rootstock.

Plum

5.10.1.4 PRUNUS INSITITIA SELECTION

St. Julien stocks: St. Juliens has been regarded as the dwarfing stock for plums for long. They are propagated from seed as well as by layers and cuttings. Many clonal selections have been made and the more important ones are enumerated below.

1. **St. Julien A:** It is of value as a semidwarfing rootstock for the plum, which results in early bearing and exhibits good absorption of Ca.
2. **St. Julien K:** This is an ultradwarfing plum rootstock.
3. **St. Julien GF 655-2:** It produces a semidwarfing tree with a very good productivity rating (Okie, 1987). The rootstock is found susceptible to high-pH soil but resistant to bacterial canker and viruses.
4. **St. Julien Hybrid 1:** The rootstock is selected from a cross between *Prunus insititia* and *Prunus domestica*, giving rise to a vigorous tree.
5. **St. Julien Hybrid 2:** Also selected from a cross between *Prunus insititia* and *Prunus Domestica,* it is a semidwarfing rootstock. It is unsuitable for light stony soils.
6. **St. Julien W 61:** A promising semivigorous rootstock that can be vegetatively propagated

under mist. Trees on it yield high with large fruits.

5.10.1.5 AMERICAN PLUM ROOTSTOCKS

***Prunus americana*:** This rootstock has been reported (Tukey, 1964) useful as a moderately dwarfing rootstock. It is shown to have a good compatibility with a vast range of European prune cultivars, damsons, Japanese plum, and the Native American types.

Beach Plum (*Prunus maritima*): It has been classified as a very dwarfing stock (Okie, 1987), with poor anchorage and low productivity. It is propagated through seed.

Nanking Cherry (*Prunus tomentosa*): A dwarfing rootstock with drought and frost resistant. It can easily be propagated by hardwood cuttings.

5.10.1.6 ROMANIAN ROOTSTOCKS

Rosior Varatic (*Prunus domestica*): It is a self-sterile seed-propagated stock. Seedlings are more uniform than those of Myrobalan (*P. cerasifera*). Trees are efficient producers and well adapted to heavy, calcareous soils.

Corocodus 163 (*P. cerasifera*): It is quite vigorous and is resistant to low winter temperature. It can be propagated easily by hardwood cuttings (Parnia et al., 1988).

5.10.1.7 OTHER SPECIES AND HYBRIDS

Almond (*Prunus amygdatus*) and peach (*Prunus persica*) have also been used as rootstocks for the plum. Clonal selections of almond × peach hybrids have been made. Some of the characteristic features of these rootstocks are detailed below.

1. **Bitter Almond** (*Prunus amygdatus*): It produces a semivigorous tree.
2. **Peach** (*Prunus persica*): It is a semivigorous rootstock of plum.
3. **GF 677:** A selection from *Prunus amygdalus* × *Prunus persica*. It is vigorous and resistant to drought but susceptible to root-knot nematode and waterlogging.
4. **GF 557:** A selection from *Prunus amygdatus* × *Prunus persica* that produces a vigorous tree, resistant to drought, root-knot nematode. It is, however, very sensitive to waterlogging.

5.11 LAYOUT AND PLANTING

Certain points are to be taken into consideration before orchard establishment, namely, slope, plant spacing and planting density, orchard design, date and depth of planting, and pollinizers. Suitable orchard planning should be done to obtain optimum land use and to assure cost-effective orchard management. A suitable irrigation system should be installed before the orchard establishment. In low-lying areas, for instance, the bottom of a valley cold pockets usually occur. Cold air gets accumulate in the lower lying areas of an orchard, leading to frost damage since cold air is heavier than warm air (DAFF, 2010). In flat and valley areas, a square system of planting is recommended, whereas in sloppy areas, a contour or terrace system is recommended. The best time to plant 1–2-year saplings are during the dormant stage. In order to avoid winter injury, spring planting is recommended except in regions where winter cold is not severe.

Larger square pits of 1 m in width × 1 m in length × 80 cm in depth should be dug out a month earlier to accommodate the root system of the grafted saplings, followed by planting (DAFF, 2010). In order to avoid collar rot and scion rooting, care should be taken that the bud union remains approximately 15–20 cm above the soil surface, making sure that the soils around the plants are pressed hard to prevent air pocket followed by irrigation. However, care should also be taken not to compact the soil in the planting hole as it would have a negative effect on the soil aeration, which will lead to poor root development. A small mound to be made at the base with loose soils to act as mulch and avoid water stagnation; further, the saplings should be headed back or trim the central leading branch approximately 30 cm above the bud union. Generally, the 6×6 m^2 planting distance is recommended. A closer spacing of 5.4×5.4 m^2 may be adopted in sandy soils; however, with the availability of dwarfing rootstocks as many as 800–1200 trees per ha have been reported (Sansavini, 1990). Planting of different types of hedgerow along with tree species spaced 2.4–3 m in rows; 4.8×6.6 m^2 apart are being under observations (Bhutani and Joshi, 1995). Pollinizers are required for better pollination in plum as most of the plum cultivars are self-unfruitful (Arora and Singh, 1990). Santa Rosa and Beauty, partially self-fruitful cultivars, are benefitted from pollinizers. For a better fruit set and good harvest, the pollinizer's variety should be planted near every third tree of every third row.

5.12 ORCHARD FLOOR MANAGEMENT

5.12.1 NUTRIENT MANAGEMENT

It has been reported that a hectare bearing plum orchard removes 30.1–39 kg N, 4.3–6.5 kg P_2O_5, and 24.2–38 kg K_2O annually, whereas nonbearing trees remove only 1.2 kg N, 0.4 kg P_2O_5, and 1.1 kg K_2O annually per hectare. It is obvious that fertilizer requirement ranges from region to region and depends on many factors such as climate, plant species, plant age, rootstock, soil fertility, soil type, soil moisture, and soil pH. Under Meghalaya conditions, for bearing plum trees of 10–13 years age, a dose of about 100 g N, 200–205 g P, and 20–100 g K per tree per year is considered optimum for the maximum yield with good physico-chemical attributes. In plain areas of Punjab, the application of 200 g N + 75 g P + 200 g K per tree with a basal dose of farm yard manure (FYM) @ 35 kg per tree applied in December together with the entire dose of P and K, and half of N one month before flowering and another half month later resulted in the maximum yield per tree. In Himachal Pradesh, it is recommended to apply 10 kg of well-decomposed FYM, 50 g N, 25 g P, and 60 g K for every year age of plant up to 10 years. N is applied in two split doses: once in spring before flowering and the other one month later. The deficiency of zinc is often observed in plums growing in sandy soils during the summer month. It can be corrected by foliar sprays containing 3 kg zinc sulfate in 500 l of water. Boron deficiency generally leads to misshapen fruits and the development of corky spots, resulting in fruit cracking. This deficiency can be managed by spraying 0.1% boric acid in the month of June. Thakur and Thakur (2014) reported that plum trees when treated with 75% NPK + biofertilizers (60 g each/tree basin) + green manuring (Sunhemp @ 25 g seeds/tree basin exhibited the highest annual shoot growth, tree height, tree volume, fruit set, and fruit yield.

Leaf-nutrient analysis has proved to be the most important guide to determine the deficiencies and excesses in fruit tree nutrition. Plum requires adequate amount of all major and micro-elements. The most suitable time for sampling of leaves for foliar diagnosis has been determined during the second half of June or July. Leece (1975) computed the data on the optimum level of nutrient elements in leaves of plums on a dry weight basis are N (2.4%–3%), P (0.14%–0.25%), K (1.6%–3%), Ca (1.5%–3%), Mg (0.30–0.80), Na (0.02%), Cl (0.3%), Fe (100–250 ppm), Cu (6–16 ppm), Mn (40–160 ppm), Zn (20–50 ppm), B (25–60 ppm).

5.12.2 IRRIGATION

Plum is mostly grown under rainfed condition in India. The hilly regions receiving more than 120–150 cm rain fall distributed during the growing season in addition to some rainfall in winter periods may be considered conducive to plum cultivation. The critical irrigation requirement period is May and June during the rapid fruit development period. Irrigation is highly essential in young plants to overcome moisture stress. Bearing plants can endure the moisture stress condition better because of their better root system. Irrigation to bearing plum orchard increases yield by more than 50% with improve fruit quality. Irrigation at 50% field capacity at a 12-day interval in May and an 8–9 day interval in June is recommended for a higher yield of quality fruits. Few major factors should be taken into account on deciding the volume and frequency of irrigation. The total water requirement of the entire season varies from 15 to 36 inches, which depends on factors such as soil type, topography, size, age and canopy arrangement of the tree, types of mulching, and the irrigation system. Irrigation can be done by furrow, portable pipe, or basin flooding. The drip irrigation system is one of the good options as compared to flood or furrow irrigation because of less water requirement per year.

5.12.3 WEED CONTROL AND INTERCULTURE OPERATIONS

Weed control is one of the crucial orchard floor management in plum orchards. Hand weeding, a conventional method, is found costly and often slanderous to the surface-feeder roots. Herbicides such as glyphosate, atrazine, simazine, diuron, and oxyfluorfen are found effective for controlling weeds in the plum orchard. Besides controlling weeds, weed management practice has a significant influence on growth, productivity, leaf-nutrient status, and fruit quality (Bhutani and Joshi, 1995). Weeds should be kept short within the work row. However, a strip of not more than 1 m on both sides of the tree should be kept free of weeds. It is suggested to avoid the use of herbicides during the initial years of the tree because of the possible injury to unestablished trees; rather, weeds can be checked by mulching and tillage. In the second and succeeding years, herbicides can be applied to control weeds. Besides weed control, the use of different mulching materials regulates soil temperature and conserves better soil moisture and nutrient uptake by the plants, which results in the improvement of fruit quality in terms of increased TSS and reduced acidity especially under black polythene mulch (Chauhan and Shylla, 2006). Cultivation of leguminous crops as cover crops is generally advised for young and bearing plums. Studies revealed that plum trees perform better under sod or sod mulch (Teskey and Shoemaker,

1978). Shylla and Chauhan (2004) reported that intercropping produced larger and heavier fruits than other soil management practices, though it recorded a 15.95% less yield. In rainfed areas, generally cover crops are not recommended as it will compete with the main fruit trees for nutrients and moisture. However, where irrigation facilities exist, leguminous cover crops can be grown to prevent soil erosion.

5.13 TRAINING AND PRUNING

A young plum tree is pruned or trained to provide the structural strength and maximum fruiting area and fruitfulness at an early age. Plum trees are generally trained as a modified central leader system or an open center system. Japanese varieties with spready nature are commonly trained as open center, and varieties with upright growth habit like Santa Rosa and Stanley are trained as a modified leader system. The trees are generally obtained from the nursery as one-year-old whips that are headed back to a height of about 60 cm. Heading back at a height lower than this usually results in trees branching very close to the ground. In the case of older branched trees obtained from the nursery not more than 3-5 appropriately placed existing branches may be selected to develop as scaffold branches, and if the trees sought to be trained as a modified leader system, then keep the central growth dominant over other scaffold branches. Primary branches should be spaced 15–20 cm apart along the trunk. Secondary branches are to be selected during the second, third, and fourth dormant seasons. At the end of the fourth year of growth and pruning, 7–9 well-spaced secondary branches are obtained which do not hinder light penetration in the center of the tree, thereby stimulating fruiting there. Small branches are thinned out lightly to facilitate sunlight, spraying and other cultural operations and to develop a healthy spur system (Teskey and Shoemaker, 1978). In bearing trees, pruning is directed toward increasing the size of the fruits, preventing the breakage of branches due to heavy crop, and promoting continuous new growth. The Japanese varieties bear fruits both on spurs and on one-year-old wood and have the tendency to overbear; therefore, a heavier pruning is recommended than that in the European plum. The life-span of the plum spurs is 5–8 years, while it may be necessary to prune for some spur renewal each season, removal of most of the new growth (75%–85%) is recommended. The extent of pruning requires to be so regulated so as to induce an annual extension growth of 25–50 cm in the young bearing trees of both the types. An annual extension growth

of 15 cm is considered adequate for the older trees of the European plum. The older trees of the Japanese plums require 25–30 cm of annual extension growth for proper fruiting. Naturally spreading varieties, such as Burbank, the straight-out growing branches should be cut back. In most European varieties, little pruning forms a well-shaped trees (Childers, 1983). The other systems of training also adopted in the European plum growing areas are hedgerows with central leader, pyramid, summer hedged, palmettes, etc., which show promise for mechanically handling of trees (Kar, 2000). Mika and Buler (2011) reported a new training system i.e. central leader, spindle, which was applied to obtain trees suitable for mechanical harvesting. The leader was not headed after planting, and summer training treatments were performed. From the third year onward, renewal pruning was done after fruit harvesting. This new training and pruning system resulted in very rapid tree growth, much young wood, fruit bud formation on young wood, and early bearing. Trees appeared to be suitable for hand and mechanical harvesting within four years from planting.

5.14 FLOWERING, FRUIT GROWTH, AND DEVELOPMENT

After the floral initiation, the development of flowers takes place rapidly within the bud. All the flower parts are distinctly formed by the time the tree goes into winter dormancy. A slow development takes place in winter followed by rapid development in spring, leading to the opening of flowers and shedding of flowers. Consequently, the duration of flowering of a specific tree can vary from one week to several weeks (Wertheim and Schmidt, 2005). Peach and Plum blossoms are similar in form and structure. However, plums have smaller, white, and longer flowers stalks. The flowers are primarily borne in umbrella-shaped clusters, consisting of 2–3 single individual flower. These clusters in turn are borne on short spurs. Plum flowers of almost all cultivars are self-sterile. Therefore, cross-pollination is necessary to assure a good fruit set. The major pollinating agents are honeybees. The stigma of a flower is receptive for six days. Hence, pollination should take place during this period for a good fruit set. It is recommended to place at least six beehives per hectare in an orchard (DAFF, 2010).

The double-sigmoid growth pattern is a characteristic of a stone fruit such as plum. In this growth curve, four distinct stages (S1–S4) could be established. S1 is the first exponential growth phase and characterized by cell division and elongation, S2 shows little or no fruit growth but the endocarp hardens to form a solid stone, S3 is the second exponential growth phase due to cell enlargement, and in S4, the fruit growth rate decreases and fruit ripening occurs (Valero et al., 2013). Prior to fertilization, one of the ovules of plum fruit get aborted. During the subsequent development of the fruit, the ovary's outer wall gives three prominent layers of the pit. The fruit skin is formed through two hypodermal layers in the ovary which consists of the cuticle, the epidermis, and few layers of collenchymatous cells. The flesh possesses the vascular bundles of the fruit that are secured in the lignified tissue of the pit. During the process of growth and maturation of fruits, various changes occur, including physical and biochemical changes (Muste, 2008). The fruit mass and size of the plum fruit increase during growth and development (Vlaic et al., 2014). The sugar content escalates throughout the growth and development. Vlaic et al. (2014) reported an increased in TSS contents throughout the growth and fruit development from 6.4° to 21.3 °Brix for Stanley variety and 8.55°–23.95 °Brix for the variety Tuleu Gras. Similar results were found by Ciobanu et al. (2011) when studying Stanley and Tuleu Gras varieties. Developing fruits are acidic due to the accumulation of many organic acids, namely, malic, citric, tartaric, quinic, oxalic, fumaric, and succinic acid, although mature fruits do not taste acidic because of the large amounts of accumulated sugars and the decrease of total acidity that usually occur during ripening (Valero et al., 2013). It is generally well known that during fruit ripening, the phenolic concentration lowers, whereas the flavonoids concentration grows (Manach et al., 2004). Volatiles that determine the flavor are also produced. Wax develops on the fruit skin. The respiration rate is high during growth because the plum is a typical climacteric fruit. As maturity approaches, it decreases to a preclimacteric minimum and increases irreversibly to a maximum during ripening (Cambrink, 1993). At this stage, the fruit becomes soft and sweet with a characteristic flavor. During ripening, the chlorophyll content decreases (Vlaic et al., 2018) and carotenoids content increases, and the pectic substances in the cell walls change from an insoluble to a soluble form, resulting in the softening of the fruit (Bhutani and Joshi, 1995).

5.15 CROP AND QUALITY REGULATION

Generally, plums tend to bear heavy crops and bear under-sized fruits of low quality, and such fruits may not fetch good price in the market. Therefore, thinning is necessary to reduce limb breakage, to increase fruit size and uniformity in fruit color, and to stimulate floral initiation for the next season crop. Generally, thinning after pit hardening stage does not produce maximum sized fruits. Thinning is generally done by hand, mechanically, or by using chemicals from full bloom to four weeks after petals fall. The advantages of chemical thinning over hand or mechanical thinning are reduced thinning costs, better fruit size, and improved fruit quality (Chauhan and Awasthy, 1993). Naphthalene acetic acid (10–50 ppm), 3-Cyclopiazonic (100–300 ppm), 2,4,5-T (2–25 ppm), Sevin (1000–2500 ppm), Dinitro-ortho-cresol (1000–2500 ppm), and Paclobutrazol (1000–2500 ppm) have been recommended for fruit thinning in plums between full bloom and pit hardening (Rehalia and Tomar, 2004). The mechanical thinning treatment reduces the fruit set, the number of fruits per tree, and yield, but increases fruit size and fruit weight, promotes higher vegetative growth, and avoids biannual bearing. On the other hand, ethephon at 250 μL L^{-1} reduces the fruit set and increases the fruit size of the Katinka plum (Pavanello et al., 2018). The main point to keep in mind while thinning is to remove all the small undersized, misshaped, insect and disease-attacked fruits. For obtaining well-colored fruits of good size, the retained fruits should have spacing of 5–8 cm apart depending on varieties (Bal, 1997).

5.16 MATURITY, HARVESTING, AND YIELD

Vegetatively propagated plum generally comes to bearing 3–5 years after planting and remains in productive life up to 25–35 years, depending on care and management practices. The optimum maturity of plums at which fruits are to be harvested will depend upon the important consideration chiefly whether the fruits are to be consumed in the local markets, canned, or dried. Those to be consumed fresh locally can be harvested at a more mature stage. For a distant market, fruits are harvested at 50% color development and remain firm. For canning, the fruits should be ripe and firm with maximum sugar content. For determining the appropriate time of harvest of fruits, various indexes are used. These are as follows.

1. Color development: Every variety of plum has its distinct skin color at maturity, and when the full color develops, it is considered ready for harvesting.
2. Flesh firmness: Depending on the firmness of fruits at which they are harvested for distant market without precooling ranges from 4.1 to 1.9 kg.
3. Days after full bloom for maturity of a variety is more or less fixed for a particular zone.
4. Total soluble solids of fruits should reach the peak value at maturity. The sweetness due to sugar content is one of the most important quality indicators of the fruit. The sugar content increases toward fruit maturity due to total soluble solids, resulting to a rich taste may lead to better consumer preferences (Crisosto et al., 2004). Attaining the threshold value of 12.5% TSS in the European plum has been reported to be a good standard for maturity.

Plum fruits do not ripen uniformly at a time. Hence, they have to be harvested in 2–3 pickings. Plum fruits being perishable in nature must be handled with care. The baskets used for picking should be lined with soft materials on the inner surface. Immediately after harvesting, fruits should be kept preferably under the shade. For the distant market, fruits should be harvested early in the day with immediate transfer to the packing shed. For canning purpose, fruits are handled with less care.

Yield will greatly depend on the age of the tree as well as on the management practices adopted. On average, a fully grown plum tree yields 40–75 kg fruits. The mean performance of different plum varieties for fruit yield in the climatic conditions of Kashmir, India, as stated by Kumar et al. (2018) is given in Table 5.5.

TABLE 5.5 Performance of Different Plum Varieties for Fruit Yield

Varieties	Mean Fruit Yield (kg/tree)
Meriposa	57.91
Au-cherry	46.68
Tarkol	46.18
Beauty	44.50
Monarch	43.40
Green Gage	34.02
Methley	32.40
Black Amber	28.70
Au-Rosa	23.00
Frontier	22.40
Kanto-05	20.99
Santa Rosa	20.79
Red Beauty	19.22
Grand Duke	15.58
President Plum	15.09

5.17 GRADING AND PACKAGING

Grading of fruits into different sizes is an important aspect, and it forms an integral part of better marketing. Fruits picked at an appropriate stage of maturity combined with size grading command better market than ungraded fruits. Three standards or sizes are recommended under the "AGMARK" system of grading plum, and the box sizes used for marketing are given in Table 5.6.

Fruits are packed in the wooden boxes of 37 cm × 16.5 cm × 16.5 cm size. The interior of the wooden box all around is wrapped with a paper to prevent the fruits from coming in contact with the rough surface of the box. The bottom of the box is spread with wooden shaves or some soft packaging material, and the fruits are placed in layers. After each layer, a sheet of paper is kept that acts as a cushion for the next fruit layer. The top layer of fruits is finally covered with a sheet of paper and some soft materials, and then the lid of the box is nailed. Packaging materials include wood, cardboard boxes, and plastic film. The plastic film is recent technology that is reported to be an effective means in reducing moisture and nutrient loss from fresh fruits.

TABLE 5.6 Grading and Packing Standards for Plum

Grade	Fruit Size Diameter (mm)	Box size (cm)	No. of Layers	No. of Fruits Per Layer
Special	42 and above	37 × 16.5 × 16.5	3	28–32
Grade 1	36–42	37 × 16.5 × 16.5	4	38–43
Grade 11	Below 36	37 × 16.5 × 16.5	4	50–56

5.18 STORAGE

Although plums are highly perishable and its shelf life is short, its storability depends on the stage of maturity of fruits. Picking at the correct stage of maturity is one of the best ways of ensuring optimum storage. The main objective is to harvest the fruit and place it in optimum storage conditions for the maximum storage life when it is neither undermature nor overmature and at the same time to retain the ability to ripen at full quality. Plums are generally stored at 0 °C with 85%–90% relative humidity. Under these conditions, fruits can be stored for the duration of 2–4 weeks, depending upon species, variety, stage of maturity of fruits, etc. Fruit storability extends up to 12 days at room temperature when fruits are dipped in 4% calcium chloride for 2 min. The promising influence of 1-methylcyclopropene (1-MCP) on delaying the ripening process of the "Angeleno" plum fruit under the air and controlled atmosphere was reported where the application of 1-MCP before air storage was found the best means to control the ripening process for the short- and medium-storage periods, that is, 40–60 days (Menniti et al., 2006). The controlled atmosphere storage has been practiced overseas by maintaining 2%–3% O_2 and 2%–8% CO_2 where fruits can be retained for 2–3 months. Steffens et al. (2014) recommended the CA condition for storing "Laetitia" plums (a Japanese plum) is 2 kPa O_2 + 2 kPa CO_2, since it allows for a slower apparent ripening of the fruit (lower evolution of the skin red color) and low intensity of internal breakdown. The modified atmosphere packaging technique was useful to delay the ripening process of plum cultivars through a delay in the changes in color and the losses of firmness and acidity, and in turn an extension of shelf-life could be achieved (Díaz-Mula et al., 2010).

5.19 PROCESSING AND VALUE ADDITION

Plum processing usually depends on drying of the fresh plum, canning, and preparation of jams, jellies, beverages, wines, and brandies. Earlier, natural sun drying was quite common; however, today dehydration of plum is generally done.

1. **Jams and jellies:** Mixed fruit jam and jellies are commonly prepared from plum fruits. Late varieties have a better acid–sugar balance and a high dry matter content. Hence, they are more suited for the purpose than early ripening (Rein et al., 1988).

2. **Prune juice:** In general, prune juice is a water extract of dried

prunes. Salunkhe and Bolin (1972) developed a process based on the use of enzyme to rapidly hydrolyze the pectin for easy filtration. For juice extraction, fruits are steamed for 8–10 min and passed through a pulper to get rid of the peel and seed, and then they are treated with enzyme for 6–12 hours. The juice is filtered adjusted to 22.5°–23 °Brix, flash pasteurized to 88 °C, and filled in hot cans at 82 C–85 °C. The addition of pectinolytic enzyme increases the juice yield, causing a slight change in total soluble solids, pH, and acidity, and a decrease in viscosity improving the color without affecting the flavor (Joshi et al., 1991). The presence of anthocyanin in prune juices would be a clear indication of adulteration (Raynal and Moutounet, 1989).

3. **Dried plums:** About 75% of the world's supply of dried prunes are produced in California and in the Pacific Northweast (Bhutani and Joshi, 1995). Although sun drying was very common earlier, today they are mostly dehydrated to prevent spoilage during rain. The French prune and Imperial sugar cultivars are mostly selected for dehydration. Dried plums or prunes are prepared for drying by cleaning with airblast and water sprays, dipping in cold or hot water, followed by spreading in a single layer on trays, and then dehydrated to 18% moisture in a forced-air tunnel dehydrator.

4. **Wine:** Plum fruit pulp is let undergo the fermentation process to give quality appealing and acceptable wine. The pulp is fermented to produce the wine. As per Vyas and Joshi (1982), for the preparation of wine, the fruit should be fully ripe, diluted with water in a 1:1 ratio, further adding 0.3% pectinol, 150 ppm SO_2, and enough sugar to raise the TSS to 24 °Brix. The addition of pectinolytic enzyme before fermentation facilitates pressing increases the juice yield and hastens wine clarification.

5. **Vermouth:** The word Vermouth is derived from the German word "Wermut" ("Wer" means man and "Mut" means courage). Vermouth is as an aromatized fortified wine (alcohol 15%–21%) flavored with a mixture of spices and herbs. The Vermouth is of sweet or Italian

type and dry or French type. In Italian type, alcohol content varies from 15% to 17% with 12% to 15% sugar, whereas French have 18% alcohol with 4% reducing sugar. Vermouth is quite popular in European countries and the USA besides their commercial production in Russia and Poland (Amerine et al., 1980).

6. **Plum brandy:** Brandies derived from fruits other than grapes are referred to as fruit brandies. The drink is known by the name Slivovitz, a native drink of Romania, Hungary, and Serbia. Besides overproduction, a significant amount of collage fruit usually having defects in fruits made the growers realize the fact to exploit the fruit for the preparation of their native drink Slivovitz from plum.

As per Sharma et al. (2011), the fruit of plum is most suitable for the manufacture of dehydrated fruit-based powder, chutney mix, ready-to-serve drink mix, and dried pickle. Consequently, the developed technology has a scope for commercial exploration at the industry level for manufacturing shelf-stable products of the plum fruit for their efficient and profitable utilization, thereby ensuring the reduction in postharvest losses and better returns to the growers.

5.20 DISEASES AND PESTS

5.20.1 DISEASES

1. **Bacterial Canker** (*Pseudomonas syringae*): The oozing of gums on fruits and bark or outer sapwood is common. Cool, wet period of early spring, frosty weather during bloom, and early growth favor the incidence of the bacterial canker. The disease may be controlled by fall and winter spray of copper fungicides such as copper oxychloride (0.3%).

2. **Crown Gall** (*Agrobacterium tumefaciens*): The organisms live in the roots of plants and in the soil, and infection takes place through wounds. Galls stimulated by the presence of bacteria result from hyperplasia and hypertrophy of cells. Using gall-free nursery stock and gall-resistant rootstocks can avoid the infection.

3. **Brown rot** (*Monilina fructicola*): Both blossom and fruit are affected. Symptoms include browning and blossom death. Tiny brown spots enlarge to a large, soft, watery, and brown area on fruits. The disease extends up to shoots causing wilting and cankering on young branches.

4. **Sharka or plum pox**: Aphids are the agents that transmit Sharka or PPV and also by budding and grafting material. Chlorotic leaves and shallow depressions on fruits are common symptoms. Following proper orchard sanitation, removing, and burning the infected trees can manage and check the spread of disease. Aphids can be controlled by the application of organophosphorus insecticides.

5. **Oak Root Fungus:** The disease is caused by *Armillarella mellea* and is responsible for large annual losses in most cultivars of tree fruits. Fig, persimmon, the black walnut (*Juglans hindsii*), and French pear are the only fruit tree species that are resistant to the fungus (Thomas et al., 1953). The fungus infects the roots and the lower part of the trunk. Myrobalan 29 and Mariana 2624 show more resistance to the fungus than other plum rootstocks, but even these can be severely injured or killed by the disease. The disease can be controlled by fumigating the soil after the harvest of fruits with carbon bisulfide. The fumigant should not be applied closer than 2–2.7 m from any living tree as the material is toxic to living roots.

6. **Crown rot** (*Phytophthora* spp.): Primarily a disease of the young trees, its onset begins during extended periods of high soil moisture. Cankers develop on the bark below the ground; infected trees appear pale after leafing. If the cankers girdle the trunk, the tree will eventually die. There is no satisfactory control measure once the disease appears. However, it can be prevented by planting the trees at the same depth at which they grew in the nursery or even deeper (graft union should be kept out 30–45 cm above the soil level). Ensuring proper drainage in the tree basins is crucial for preventing the outbreak of the disease.

5.20.2 PESTS

1. **San Jose scale** (*Quadraspidiotus perniciosus*): The adults feed on sap from limbs, twigs, and fruits. Small grayish specks appear on the bark covered with a gray layer of overlapping scales that appear as if they were sprayed with wood ash. Dead leaves adhering to fruit spurs during the dormant season indicate

the presence of the San Jose scale. Spraying of universal oil during the spring @ 10 litres in 490 litres of water during the dormant season can give effective control. Insecticides such as chlorpyriphos (0.4%) and dimethoate (0.03%) can be sprayed to kill the crawlers during May.

2. **Stem borer** (*Anarsia lineatella*): Borers attack the fruits and twig terminals. To control, holes to be cleared with a flexible wire and should be plugged by inserting cotton soaked in petrol or kerosene, followed by covering the hole with clay mud thoroughly.

3. **Plum fruit moth** (*Laspeyresia pomonella*): The larvae penetrate and enter inside the fruit and tunnel deeply into the fruit tissue. The entrance is generally surrounded by grass and droplets of gum. Carbaryl (0.1%) is applied about a month before the anticipated date of harvest.

4. **Peach twig borer** (*Anarsia lineatella*): A common pest, the borer attacks plum fruits and twig terminals. The cultivars ripening during the hatching period require utmost protection, since ripening fruits are very vulnerable. Egg hatch dates may vary with the prevailing season, but usually one brood occurs in mid-May and another in early July. Properly timed sprays can control the insects at these times.

5. **Mites:** Three species of mites, namely, *Tetranychus urticae* (the two spotted mite), *T. pacificus* (the Pacific mite), and *Panonychus ulmi* (the European red mite) have been reported to cause damage to plum trees. These feed on leaves, and when present in large numbers can cause leaf browning and defoliation. Dormant spray treatments are effective against the European red mite, but sometimes foliar acaricide applications are necessary.

6. **Nematode:** Nematode namely *Meloidogyne incognita* infestation leads to malfunctioning of the roots, as a result of which the above ground parts of plants show stunting and yellowing of leaves. Proper rootstock selection can reduce the damage caused by root knot nematode. Marianna 2624, Myrobalan 29C, and Nemaguard are resistant rootstock. Carbofuran granules are also applied (100–300 g/tree) after the harvest of the crop.

KEYWORDS

- **Plum**
- **Rosaceae**
- **temperate**
- ***Prunus domestica***
- ***Prunus salicina***
- **prunes**
- **Green Gage**
- **Santa Rosa**
- **Slivovitz**

REFERENCES

AgMRC. 2006. *Commodity Profile: Plums, Fresh Market.* Agricultural Issues Center, University of California. pp.1–6.

Amerine, M.A., Kunkee, K.E., Ough, C.S., Singleton, V.L. and Webb, A.D. 1980. *The Technology of Wine Making*, 4th ed., AVI Publishing, Westport, CT.

Anon. 2007a. *Common European Plum or Prunus domestica.* http://www. britannica. com/eb/topic-128272/common-European-plum. 29/01/2007.

Anon. 2007b. Plum history. *Fruit Patch Sales.* http://www.fruitpatch.net/products/plums.htm.

Arora, R.L. and Singh, R. 1990. Pollination requirement in subtropical plum (*Prunus salicina*). *Indian Journal of Horticulture.* 34: 143.

Bal, J.J. 1997. Japanese Plum. In: *Fruit Growing.* Kalyani Publishers. Ludhiana. p. 381.

Bhutani, V.P. and Joshi, V.K. 1995. Plum. In: *Handbook of Fruit Science and Technology: Production, Composition, Storage and Processing* (D.K. Salunkhe and S.S. Kadam, eds.), Marcel Dekker, New York. p. 206.

Birwal, P., Deshmukh, G., Saurabh, S.P. and Pragati, S. 2017. Plums: A brief introduction. *Journal of Food Nutrition and Population Health.* 1:1.

Blazek, J. 2007. A survey of the genetic resources used in plum breeding. In 8th International Symposium on Plum and Prune Genetics. *Breeding and Pomology.* 734: 31–45.

Cambrink, J.C. 1993. Plums and related fruits. In: *Encyclopedia of Food Science, Food Technology and Nutrition* (R. MaCre, R.K. Robinson, and M.J. Sadler, eds.), Academic Press, London, p. 3630.

Chadha, L.K. 2019. Plum. In: *Handbook of Horticulture* (vol.1), Indian Council of Agricultural Research. Directorate of Knowledge Management in Agriculture, New Delhi, pp. 328–335.

Chauhan, J.S. and Shylla, B. 2006. Influence of orchard floor management practices on soil moisture and temperature in a plum orchard grown under mid hill zone condition. In: *Temperate Horticulture: Current Scenario* (D.K. Kishore, S.K. Sharma, and K.K. Pramanick, eds.), New India Publishing Agency, New Delhi, pp. 267–272.

Chauhan, P.S. and Awasthy, R.P. 1993. Fruit thinning in temperate fruits. In: *Advances in Horticulture: Fruit Crops* (K.L. Chadha and O.P. Pareek, eds.), Malhotra Publishing House, New Delhi, Vol. 3, p. 1243.

Childers, N.F. 1983. *Modern Fruit Science.* Horticultural Publication, Gainesville, FL, p. 242.

Ciobanu, A., Cichi, M., Călinescu, M. and Iancu, D. 2011. Research regarding the dynamic accumulation of sugar at some plum tree varieties. *Annals of the University of Craiova-Agriculture, Montanology, Cadastre Series.* 41: 20–25.

Cociu, V., Botu, I., Rinoiu, N., Pase, I. and Moduran, I. 1997. *Prunul.* Editura Copphys., Bukarest.

Cociu, V., Botu, I., Rinoiu, N., Pase, I. and Moduran, I. 1997. *Prunul.* Editura

Copphys., Bukarest. In: *Breeding Plantation Tree Crops: Temperate Species.* (S.M. Jain and P.M. Priyadarshan, eds.), Springer, pp. 161–231.

Crane, M.B. and Lawrence, W.J.C. 1956. *The Genetics of Garden Plants*, 4th ed., MacMillan Company, London.

Crisosto, C.H., Garner, D., Crisosto, G.M. and Bowerman, E. 2004. Increasing "Blackamber" Plum. *Postharvest Biology and Technology.* 34: 237–244.

Department of Agriculture, Forestry and Fisheries (DAFF). 2010. *Plums-Production Guidelines.* DAFF, Republic of South Africa. p. 35.

Dhatt, A.S., Chanana, Y.R., Nijar, G.S., Minhas, D.P.S., Bhindra, A., Dhillon, D.S and Uppal, D.K. 1992. Studies on the performance of a new table plum suitable for northern plains. *Indian Journal of Horticulture.* 49(2): 172–174.

Díaz-Mula, H.M., Martínez–Romero, D., Castillo, S., Serranoa, M. and Valero, D. 2010. Modified atmosphere packaging of yellow and purple plum cultivars. Effect on organoleptic quality. *Postharvest Biology and Technology.* 61: 103–109.

FAOSTAT, 2018. www.faostat.fao.org.

Hartman, H.T. and Kester, D.E. 1972. *Plant Propagation,* Prentice Hall of India Pvt. Ltd., p. 702.

Faust, M. and Suranyi, D. 1999. Origin and dissemination of plum. *Horticultural Reviews.* 23: 179–231.

Hartmann, W. and Stosser, R. 1994. Die Befruchtungsbiologie bei einigen neueren Sorten von Pflaumen und Zwetschen (Prunus domestica) [Fertilization Biology of some new Plum Varieties (*Prunus domestica*)]. *Erwerbsobstbau.* 36: 37–41.

Joshi, V.K., Chauhan, S.K. and Lal, B.B. 1991. Extraction of juices from peaches, plums and apricot by petinolytic treatment. *Journal of Food Science and Technology.* 28: 64.

Kar, P.L. 2000. Plum. In: *A Textbook on Pomology: Temperate Fruits* (T.K.

Chattopadhyay, ed.), Kalyani Publishers, Ludhiana, Vol. 1V, pp. 72–97.

Knight, R.L. 1969. Prunus. *Tech. Commun. Common. Bull. Hort. Plantation Crops.* 31: 649.

Kumar, D., Srivastava, K.K. and Singh, S.R. 2018. Morphological and horticultural diversity of plum varieties evaluated under Kashmir conditions. *Tropical Plant Research.* 5(1): 77–82.

Leece, D.R. 1975. Diagnostic leaf analysis for stone fruit. 4. Plum. *Australian Journal of Experimental Agriculture and Animal Husbandry.* 15: 112.

Lorenz, J. 2000. Influence of the fruit curtain on pollen quality and strong accumulation in the reproductive blood parts in plum (*Prunus* x *domestica* L.). Dissertation, University of Hohenheim, Grauer Verlag, Stuttgart.

Malcolm, P. 2006. History of plum trees and their hybrids. www.thephantomwriters. com.

Manach, C., Scalbert, A., Morand, C., Remesy, C. and Jimenez, L. 2004. Polyphenols: food sources and bioavailability. *The American Journal of Clinical Nutrition.* 79: 727–747.

Menniti, A.M., Donati, I. and Gregori, R. 2006. Responses of 1-MCP application in plums stored under air and controlled atmospheres. *Postharvest Biology and Technology.* 39: 243–246.

Mika, A. and Buler, Z. 2011. Intensive plum orchard with summer training and pruning. *Advances in Horticultural Science.* 25(3): 193–198.

Muste, S. 2008. *Vegetable Raw Materials in the Food Industry*, Academic Press Publishing House, Cluj-Napoca.

Neumüller, M. 2011. Fundamental and applied aspects of plum (*Prunus domestica*) Breeding. *Fruit, Vegetable and Cereal Science and Biotechnology.* 5(1): 139–156.

Norton, J.D., Boypau, G.E., Smith, D.A. and Abrahams, B.R. 1991. Santa Rosa plum. *Horticultural Science.* 26: 213.

Okie, W.R. 1987. Plum rootstocks. In: *Rootstocks for Fruit Crops* (R.C. Rom and R.F. Carlson, eds.), John Wiley and Sons, New York, pp. 321–360.

Okie, W.R. and Weinberger, J.H. 1996. Plums. In: *The Encyclopedia of Fruits and Nuts* (J. Janick and R.E. Paull, eds.), CAB International, UK. pp. 694–704.

Pavanello, A.P., Zoth, M. and Ayub, R.A. 2018. Manage of crop load to improve fruit quality in plums. *Revista Brasileira Fruticultura.* 40(4): e-721.

Raynal, J. and Moutounet, M .1989. Intervention of phenolic compounds in plum technology. Mechanism of anthocyanin degradation. *Journal of Agricultural and Food Chemistry.* 37: 1051.

Rehalia, A.S. and Tomar, C.S. 2004. Plum and Prune. In: *Recent Trends in Horticulture in the Himalayas—Integrated Development under the Mission Mode* (K.K. Jindal and R.C. Sharma, eds.), Indus Publishing, New Delhi. p. 128.

Rein, S.J., Otto, K. and Jacob, H. 1988. On the baking quality of important plum and damson varieties. *Erwerbsobstbau.* 30(7): 180.

Salunkhe, D.K. and Bolin, H.R. 1972. Dehydrated protein fortified with juice. *Food Product Development.* 6: 84.

Sanders, T.W. 1998. The Plum. In: *Fruit and Its Cultivation.* Biotech Books, Tri Nagar, Delhi. p. 160.

Sansavini, S. 1990. The fruit industry in Italy. *Chronica Horticulturae.* 30: 3.

Scorza, R., Ravelonandro, M., Callahan, A.M., Cordts, J.M., Fuchs, M., Dunez, J. and Gonsalves D. 1994. Transgenic plums (*Prunus domestica* L.) express the plum pox virus coat protein gene. *Plant Cell Reports.* 14: 18–22.

Sharma, K.D., Sharma, R. and Attri, S. 2011. Instant value added products from dehydrated peach, plum and apricot fruits.

Indian Journal of Natural Products and Resources. 2(4): 409–420.

Shylla, B. and Chauhan, J.S. 2004. Influence of orchard floor management practices on cropping and quality of Santa Rosa plum grown under mid hill conditions. *Acta Horticulturae.* 662: 213–216.

Somogai, L.P. 2005. *Plums and Prunes Processing of Fruits Science and Technology*, 2nd ed., CRC Press, Boca Raton, FL, Vol. 21, pp. 513–530.

South African Fruit Trade Flow. 2014. https://www.namc.co.za/wp–content/ uploads/2017/10/South–African–Fruit– Trade–Flow–September–2014–Issue–15. pdf.

Steffens, C.A., Do Amarante, C.V.T., Alves, E.O. and Brackmann, A. 2014. Fruit quality preservation of "Laetitia" plums under controlled atmosphere storage. *Annals of the Brazilian Academy of Sciences.* 86(1): 485–494.

Srinivasan, C., Padilla, I. M. G. and Scorza, R. 2005. *Prunus sp.* almond, apricot, cherry, nectarine, peach and plum. In: *Biotechnology of Fruit and Nut Crops* (R.E. Litz, ed.), CABI Publishing, Wallingford, UK, pp. 512–542.

Suranyi, D. 1994. Ontogenetic characteristics in flowers of some plum cultivars. *Acta Horticulture* (ISHS). 359: 278–286.

Suranyi, D. and Erdos, Z. 1998. Szilva. In: *Gyumolcsfajta ismeretes hasznalat* (M. Soltesz, ed.), Mezogazda, Budapest, pp. 258–287.

Teskey, B.J.E. and Shoemaker, J. 1978. *Tree Fruit Production.* 3rd ed. AVI, Westport, CT. p. 358.

Thakur, N. and Thakur, B.S. 2014. Studies on the effect of integrated nutrient management on growth and yield of plum cv. Santa Rosa. *The Asian Journal of Horticulture.* 9(1): 112–115.

Thomas, H.E.; Wilhum , S. and Maclean, N.A. 1953. *Plant Diseases. The Yearbook of Agriculture*, USDA, Washington, D.C., pp. 702–703.

Plum

Tukey, H.B. 1964. *Dwarfed Fruit Trees*, MacMillan Co., New York. p. 562.

USDA-NASS. 2005. *California Prune (Dried Plum) Forecast*. California Statistical Office.

Valero, D. and Serrano, M. 2013. Growth and ripening stage at harvest modulates postharvest quality and bioactive compounds with antioxidant activity. *Stewart Postharvest Review*. 3: 5.

Vlaic, R.A., Muresan, A.E., Muresan, V., Scrob, S.A., Moldovan, O.P., Mitre, V. and Muste, S. 2014. Physico–chemical changes during growth and development of three plum varieties. *Bulletin UASVM Food Science and Technology*. 71(12): 131–135.

Vlaic, R.A., Muresan, V., Muresan, A.E., Muresan, C.C., Paucean, A., Mitre, V., Chis, S.M. and Muste, S. 2018. The changes of polyphenols, flavonoids, anthocyanins and chlorophyll content in plum peels during growth phases: From fructification to ripening. *Notulae*

Botanicae Horti Agrobotanici. 46(1): 148–155.

Vyas, K.K. and Joshi, V.K. 1982. Plum wine making: Standardization of a methodology. *Indian Food Packer*. 36(5): 80.

Watkins, R. 1976. In: *Evolution of Crop Plants* (N.W. Simmonds, ed.), Longmans, London. pp. 243–246.

Weinberger, J.H. 1975. *Plums. Advances in Fruit Breeding* (J. Janick and J.N. Moore eds.). Purdue University Press, Lafayette, IN, USA, p. 336.

Wertheim, S.J. and Schmidt, H. 2005. Flowering, pollination and fruit set. In: *Fundamentals of Temperate Zone Tree Fruit Production* (Tromp, Webster and Wertheim, eds.), Backhuys Publishers, Leiden, p. 400.

Wollela, E.K. 2017. Surface sterilization and in vitro propagation of *Prunus domestica* L. cv. Stanley using axillary buds as explants. *Journal of Biotech Research*. 8: 18–26.

CHAPTER 6

SWEET CHERRY

BERTA GONÇALVES[1,2*], ALFREDO AIRES[1], IVO OLIVEIRA[1],
SÍLVIA AFONSO[1], MARIA CRISTINA MORAIS[1], SOFIA CORREIA[1],
SANDRA MARTINS[1], and ANA PAULA SILVA[1,3]

[1]*Centre for the Research and Technology for Agro-Environmental and Biological Sciences, CITAB, Universidade de Trás-os-Montes e Alto Douro, UTAD, Quinta de Prados 5000-801, Vila Real, Portugal*

[2]*Department of Biology and Environment, Escola das Ciências da Vida e Ambiente, Universidade de Trás-os-Montes e Alto Douro, UTAD, Quinta de Prados 5000-801 Vila Real, Portugal*

[3]*Department of Agronomy, Universidade de Trás-os-Montes e Alto Douro, UTAD, Quinta de Prados 5001-801, Vila Real, Portugal*

Corresponding author. E-mail: bertag@utad.pt

ABSTRACT

Sweet cherry is a nonclimacteric fruit belonging to the genus *Prunus*, mainly grown in countries with temperate climate. It is one of the most import fresh fruits and is highly appreciated for its sweetness, color, and nutritional value. In addition, it is mostly consumed as a fresh fruit, but it can be eaten as dried, marmalade, jam, fruit juice, in pastry and yogurts. Over the past years, many studies have focused on the use of innovative cropping techniques to increase cherry quality. Although advances have been made in research, consistent scientific information about sweet cherry production and its constraints is still limited to some countries, producer's organizations, universities, and technical schools. Therefore, the main goal of this chapter is to provide the reader with a broad view of the key factors that play a significant role in sweet cherry production, presenting important and uptdated information about this issue. The chapter has been divided into: General Introduction; Area and Production; Marketing and Trade; Composition and Uses; Origin and Distribution; Botany and Taxonomy; Varieties and Cultivars; Breeding and Crop Improvement; Soil and Climate; Propagation and Rootstock; Layout and Planting; Irrigation; Nutrient Management; Training and Pruning; Intercropping and Interculture; Flowering and Fruit Set; Fruit Growth; Development and Ripening; Fruit Retention and Fruit Drop; Harvesting and Yield; Protected/High-tech Cultivation; Packaging and Transport; Postharvest Handling and Storage; Processing and Value Addition; and Disease, Pest, and Physiological Disorders.

6.1 GENERAL INTRODUCTION

Sweet cherry (*Prunus avium* L.) is the domesticated form of the wild cherry. Sweet cherries have been recorded as a food source since 400–500 BC (Kolesnikova, 1975), although the first written testimony to cherry, from Theophrastus, dates back to 300 BC. In Europe, cherry cultivation began with the Romans (Zohary and Hopf, 2000). Further dissemination of cherry trees throughout the Mediterranean Basin, and even to England, following the expansion of the Roman Empire, resulted in the development of different ecotypes, landraces, and varieties, that resulted in the recent cultivars (Iezzoni et al., 1991). It is a species indigenous to Northern Iran, Ukraine, and other countries south of the Caucasus Mountains, and distributed in European regions extending from southern Sweden to Greece, Italy, and Spain (De Candolle, 1883). Some researchers suggest that birds contribute to the dissemination of this species, from Northern India to the plains of Southern Europe (Webster, 1996). In the 17th century, with human migration from Europe to the Americas, cherry made its way into the new world (Scorza and Hammerschlag, 1992). It reached the Pacific coast in the late 19th century, where a local nursery owner developed one of the major cultivars, Bing (Faust and Surányi, 1997). Nowadays, sweet cherries rank among the most popular table fruits worldwide, as reflected by their production, the area covered by cherry orchards, and by the economic value of the production. The quality of cherry is not only linked to physical characteristics, such as size, color, and firmness of the pulp, but also to chemical traits, such as flavor, sweetness or sourness (Dever et al., 1996). It is generally agreed that cherries are considered tasty snacks with good nutritious value, full of antioxidants, giving them great health benefits.

6.2 AREA AND PRODUCTION

The latest Food and Agriculture Organization Corporate Statistical Database (FAOSTAT) data (2016) indicate a worldwide production of 2,317,956 tons and the area covered by cherry orchards of 439,692 ha, with an average yield of around 5 tons/ha. The largest producer of cherries is Turkey (Table 6.1), with 599,650 tons, followed by the USA (288,480 tons) and Iran (220,393 tons). The fourth country in the list, Chile, recorded about half (123,224 tons) of the production of Iran, and subsequent countries are below the 100,000 tons threshold: Uzbekistan, Italy, and Spain, with around 95,000 tons, Romania and Greece with 73,834 tons and 71,858 tons, respectively, with Syria closing the top 10 of the most important production countries, with 69,153 tons. Over the last 50 years (1967–2016), production has almost doubled, from 1,252,503 tons to the already referred 2,317,956 tons (Figure 6.1A).

TABLE 6.1   Top 10 Countries for Production, Harvested Area, and Yield (FAOSTAT, 2018)

Top 10 Countries					
Production (tons)		Area (ha)		Yield (tons/ha)	
Turkey	599,650	Turkey	84,746	Slovenia	23.2
USA	288,480	USA	37,110	Suriname	20.5
Iran	220,393	Syria	35,004	Switzerland	13.9
Chile	123,224	Italy	29,970	Romania	12.1
Uzbekistan	95,267	Iran	28,397	Belarus	10.3
Italy	94,888	Spain	25,252	Canada	9.19
Spain	94,138	Chile	24,498	Albania	9.18
Romania	73,834	Russia Federation	18,154	Uzbekistan	8.81
Greece	71,858	Greece	13,678	Israel	8.52
Syria	69,153	Uzbekistan	10,808	Armenia	8.43

Until 1991, worldwide production of cherries was always about 1,500,000 tons or lower. From that year onward, a trend is visible, with a steady increase of values, although with some recorded decreases that are most likely linked to the influence of climate conditions, but also linked to some reductions in the production area. Data from 2017 harvest indicate a decrease in the world production of 3%, linked to weather damage to orchards in the European Union and Turkey (United States Department of Agriculture, 2017). The USA production data are likely to reflect an increase linked to favorable climate conditions recorded, namely those observed in the Pacific Northwest and California. In Europe, these values decrease, caused by considerable losses provoked by frosts that occurred in half of the major productive countries, including Turkey, where reduced fruit set caused due to summer hail combined with substantial rain at harvest in several regions was recorded. The 10 major values of area devoted to cherry production are almost exclusively represented by those countries with higher production (Table 6.1), except Russia, which replaces Romania. Turkey also leads in this specific parameter, with 84,746 ha, followed by USA (37,110 ha) and Syria (35,004 ha). With less than 30,000 ha are Italy, Iran, Spain, and Chile (29,970, 28,397, 25,252, and 24,498 ha, respectively). The last three countries in this top 10 are Russia (18,154 ha), Greece (13,678 ha), and Uzbekistan (10,808 ha). The area covered with cherry orchards remained more or less constant from 1967 to 1984 (Figure 6.1A). Data from 1985 reveal a considerable increase, almost the double (169,294 to 290,471 ha), and from that point on, it mimics the trend recorded for the production, with a steady increase, to reach the values of 439,692 ha. The values of yield follow an inverse trend of those recorded for production and area. From 1967 to 1985, a decrease of yield is observed, from 12.5 tons/ha to 5.3 tons/ha (Figure 6.1B). Minor changes are found from that year on

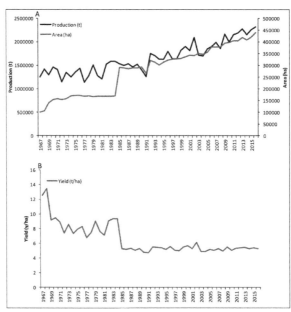

FIGURE 6.1 Production (tons) and the area occupied by sweet cherry (ha) (A) and yield (tons/ha) (B) evolution in the last 50 years (FAOSTAT, 2018).

that area linked not only to changes in the area of orchards, but also to values of production influenced by climatic conditions. Among the top 10 most cherry producing countries, Romania is the one that presents a higher value of yield, reaching 12 tons/ha, followed by Uzbekistan with 8.8 tons/ha, the only two countries of the top 10 producers and with area covered with cherry orchards that are in the top 10 of yield (Table 6.1). Interesting is the fact that the top three countries in this indicator are Slovenia (23.2 tons/ha), Suriname (20.5 tons/ha) and Switzerland (13.9 tons/ha), which, combined, only have 814 ha of orchards that represent 0.24% of the total area occupied by sweet cherry orchards.

6.3 MARKETING AND TRADE

The Food and Agriculture Organization (FAO) updated data (FAOSTAT, 2018) regarding trade balances, from 2013, indicate a total quantity of imports of 370,135 tons, being the leading import country Russia, responsible for as much as 19.7% of the global quantity. Major trade partners for imports of the Russian Federation are Turkey and Greece, from where 46% of the imported cherry are originated. Other major importing countries are China (both mainland and Hong Kong) (mainly from Chile), Germany (that relies on Austria for imports), Canada (preferentially imports from the USA), Kazakhstan (from Kyrgyzstan), Austria (main partner is Turkey), and United Kingdom (imports coming mainly from Spain). These seven countries are responsible for 70% of the total importation of cherry worldwide (Table 6.2). Importation of cherries is valued, from data of 2013, in about 1600 million US dollars (USD), worldwide, with China responsible for 30% of this value (497 million USD). Regarding exports, the worldwide value is of 360,811 tons, mainly due to the trades made by USA (69,795 tons, corresponding to 19% of the total, mainly for Canada and Republic of Korea), with several other countries of the top 10 producers, also making the top exporting ranking, namely, Chile (for China and the USA), Turkey (for Germany and Russia, particularly), Greece (key partners are Bulgaria and Russia), or Spain (to the United Kingdom and Germany), which together represent over 40% of the total. The total amount of exports is valued in about 1560 million USD, with most of them being contributed by the USA (27%), followed by Chile (24%) and Turkey (10%).

338 Temperate Fruits: Production, Processing, and Marketing

TABLE 6.2 Major Exporting Countries of Cherries and Their Main Importers (FAOSTAT, 2018)

Countries	Total Exports (tons)	Importing Country	Quantity (tons)
United States of America	69,795	Canada	24,601
		Republic of Korea	12,411
		China	7628
Chile	53,684	China	34,806
		United States of America	8212
		Brazil	2667
Turkey	53,467	Germany	16,562
		Russian Federation	13,712
		Bulgaria	5598
Greece	23,570	Russian Federation	8578
		Bulgaria	4880
		The Netherlands	2588
Spain	21,923	United Kingdom	5173
		Germany	3270
		Italy	3040

Prices at the producers, recorded for 2016 (FAOSTAT, 2018), varied from the 17331 USD/tons, in Japan, to 172 USD/tons, in Peru. In the major countries of production, prices were of 1168 USD/tons, 2500 USD/tons, and 1301 USD/tons, in Turkey, USA, and Iran, respectively. Producers' price index (PPI), which records how the average annual change over time in the selling prices received by farmers, and taking the 2000–2006 period as the 100 index show an increase of prices in Turkey (PPI of 179), USA (data of 2014, PPI of 124), and Iran, where the increase is quite surprising (PPI of 764).

6.4 COMPOSITION AND USES

Sweet cherry production is essentially for fresh market (McCune et al., 2010), although as much as 40% of its production can be either processed or discarded. Those that are processed can be converted into frozen products, canned, brined, dried, used for juice or wine production, or transformed into jams and jellies (McLellan and Padilla-Zakour, 2004). The chemical composition of sweet cherries has been discussed and studied many times, as it is linked to consumers' acceptance, from an organoleptic point of view, but also concerning health-related properties (Zheng et al., 2016). Furthermore, the presence or absence of some compounds can be of interest for processing purposes. For instance, levels of sugars can influence the final alcohol content and sweetness of cherry wine, a product that can also be affected by organic acid content of cherries (Sun et al., 2012). Sweet cherry fruits present the greater part of their nutrients and bioactive compounds in the fleshy fruit and skin. The average centesimal composition of sweet cherry (Table 6.3), as reported by USDA-ARS (2018), is very similar to other reports (Bastos et al., 2015), but can change slightly among cultivars (Wills et al., 1983; Pacifico et al., 2014).

TABLE 6.3 Composition (g/100 g Fresh Weight [FW]) of Sweet Cherry in the USDA Database

Water	82.25
Carbohydrate	16.01
Protein	1.06
Total lipid	0.20
Ash	0.48
Energy	263 kJ/63 kcal

Source: USDA-ARS (2018).

Cherries are among the most important fruits, in what concerns the amounts of nutrients and phytochemicals (McCune et al., 2010). Although the centesimal composition does not vary much, when comparing cultivars or locations of orchards, the amounts of other nutrients and bioactive compounds, such as polyphenols, of which cherries have significant levels, can change influenced by pre- and postharvest factors, such as cultivar, ripeness stages, pre- and postharvest treatments, climatic conditions, and geographical origin (Poll et al., 2003; Gonçalves et al., 2007; Faniadis et al., 2010; Cao et al., 2015). The content of carbohydrates of sweet cherries is mainly composed of glucose, fructose, sucrose, maltose, and sorbitol, with the first two accounting for about 90% of total sugars (Usenik et al., 2008). However, sugar content depends on the cultivar, agronomic factors, environmental conditions, and ripening stages (Ballistreri et al., 2013; Serrano et al., 2005; Usenik et al., 2015). The cherry glycemic index is low (Brand-Miller and Foster-Powell, 1999), and it is a source of vitamins (Schmitz-Eiberger and Blanke, 2012) and minerals (Yığıt et al., 2009). It presents reduced levels of fat, and it is cholesterol-free (McCune et al., 2010). Some of the major fruit characteristics responsible for the consumers' acceptance of cherries are total soluble solids (TSS) and the titratable acidity (TA). Early works (e.g., Kappel et al., 1996) indicate that sweet cherries should have a TSS over than 15 °Brix, which, in fact, occurs for the large majority of cultivars (Table 6.4). The content of sugars tends to increase as ripening, as it also occurs for acidity (Serrano et al., 2005; Giné-Bordonaba et al., 2017), but the former increases at a higher rate. Although TA plays a key role in consumer acceptance, within a given cherry ripening stage, its measurement is less relevant than TSS because TA changes less than TSS (Crisosto et al., 2003).

TABLE 6.4 pH, Total Soluble Solids (TSS), and Titratable Acidity (TA) of Sweet Cherry Cultivars

Cultivar	pH	TSS (°Brix)	TA (% Malic Acid)	TSS/TA
0-900 Ziraat	3.8	16.6	0.8	20.9
13S-10-10	3.4	18.6	1.0	18.3
13S-20-30	3.8	13.6	0.6	21.6
13S-43-48	4.0	13.5	0.5	24.8
13S-51-24	3.8	16.4	0.8	21.0
Ambrunes	4.4	18.0	0.5	34.0
Belge	3.7	17.8	0.9	18.9
Bing	3.7	16.7	0.8	20.2
Black Gold[a]	4.1	14.6	0.6	N.A.
Black Pearl	N.A.	16.9	0.6	27.7
Black Star	3.7	22.7	1.3	17.0
Blaze Star	4.0	20.3	1.0	20.2
Brazilian Grown Cherry	4.1	18.7	N.A.	34.1
Brooks	N.A.	N.A.	0.9	N.A.
Burlat	4.2	18.4	0.8	22.2
Comp Stella	3.9	19.7	0.8	23.3
Cristalina	N.A.	N.A.	0.8	N.A.
Dalbasti	3.7	15.2	0.7	21.0
Di Giardino[b]	3.7	16.3	0.6	29.4
Di Nello[b]	4.0	12.9	0.4	32.8
Donnantonio	4.4	20.7	0.7	27.6
Ducinola Nera	4.2	22.5	1.0	22.2
Durona di Cesena	3.7	13.7	0.7	18.8
Early Star	4.0	15.2	1.0	15.0
Ferrovia	4.0	19.7	0.9	22.1
Gabbaladri	4.3	21.5	0.7	31.5

Sweet Cherry 341

TABLE 6.4 *(Continued)*

Cultivar	pH	TSS (°Brix)	TA (% Malic Acid)	TSS/TA
Genovese	4.2	18.4	0.8	23.7
Giorgia	3.9	18.9	1.0	18.2
Grace Star	3.9	17.2	0.9	18.3
Lambert	3.7	17.2	0.9	19.9
Lapins	3.9	16.4	0.6	25.9
Maiolina Grappolo	4.3	16.6	0.7	24.5
Marchiana[b]	3.5	16.1	0.6	25.2
Maredda	4.2	19.4	0.8	24.1
Marvin-Niram	N.A.	N.A.	1.2	N.A.
Merton Late	3.7	16.1	0.7	22.2
Minnulara	4.8	21.8	0.6	37.2
Moreau	4.2	17.6	0.8	22.2
Napoleona Grappolo	4.6	15.2	0.6	27.4
Napoleona Verifica	4.1	16.4	0.9	17.6
Napoleona Forestiera	4.3	17.3	0.9	20.1
Newstar	N.A.	N.A.	1.2	N.A.
No. 57	N.A.	N.A.	1.0	N.A.
NY-6479	N.A.	N.A.	1.3	N.A.
Pico Colourado	4.2	20.0	0.6	39.0
Pico Negro	4.3	15.0	0.6	25.0
Prime Giant	N.A.	N.A.	1.0	N.A.
Puntalazzese	4.1	22.3	1.3	16.7
Rita	14.8	N.A.	N.A.	39.0
Salmo	3.8	24.5	0.8	29.0
Samba[c]	17.1	N.A.	N.A.	23.0
Santina	N.A.	16.5	0.6	26.7
Somerset	N.A.	N.A.	0.9	N.A.
Sonata	N.A.	N.A.	1.3	N.A.
Starks Gold	3.7	14.1	0.8	18.6
Summit	4.0	19.6	0.7	28.6
Sunburst	3.9	19.7	0.9	21.7
Sweet Earl	3.9	13.5	0.8	17.8
Sweetheart	3.9	20.8	0.9	22.2

342 Temperate Fruits: Production, Processing, and Marketing

TABLE 6.4 *(Continued)*

Cultivar	pH	TSS (°Brix)	TA (% Malic Acid)	TSS/TA
Toscana	4.2	19.5	0.9	20.9
Van	3.5	17.9	0.9	19.9
Vista	3.6	15.2	0.8	18.7
Zio Peppino	4.1	13.7	0.7	19.1

Source: Chockchaisawasdee et al. (2016), [a]Budak (2017), and [b]Souza et al. (2014).

N.A., not available.

The total acidity, another important component of fruit organoleptic quality, is largely dependent on the presence of malic and citric acids, the major organic acids found in ripe fruits (Ballistreri et al., 2013; Serradilla et al., 2011), although shikimic, fumaric, ascorbic and oxalic acids can also be found (Bastos et al., 2015). The content of sugars and organic acids is dependent on several factors, including cultivar, as clearly described by Usenik et al. (2008) (Table 6.5).

TABLE 6.5 Mean Sugars (g/kg FW) and Organic Acids (mg/kg FW) Content of Different Sweet Cherry Cultivars

Cultivar	Glucose	Fructose	Sorbitol	Sucrose	Malic Acid	Citric Acid	Shikimic Acid	Fumaric Acid
Badascony	106.7	91.2	21.7	8.9	4.7	0.3	7	3
Burlat	68.9	57.6	6.7	8.6	3.6	0.1	7.2	7.6
Early Van Compact	123.0	97.1	26.7	9	5.8	0.3	17.7	2
Fercer	84.4	64.2	18.9	9.4	8.1	0.5	26.7	6.6
Fernier	105	92	22	11.8	6.6	0.3	10	1.4
Ferprime	62.2	51.5	4.5	5.8	3.6	0.1	6.6	3.9
Ferrador	99.6	88	17.8	9.7	5.3	0.3	7.1	1.3
Ferrovia[a]	N.A.	51	N.A.	N.A.	6.4	N.A.	N.A.	N.A.
Lala Star	118	101.5	23.3	12.5	5.8	0.2	11.4	2.8
Lapins	93.7	79.9	14.4	7.6	3.5	0.1	8.4	1.7
Noire de Meched	101	84.4	20.2	9.7	6.6	0.4	15.2	5.1
Sciazza[a]	N.A.	64	N.A.	N.A.	7.7	N.A.	N.A.	N.A.
Sylvia	61.8	47.6	11	3.6	4	0.1	9.5	1

Sweet Cherry

TABLE 6.5 *(Continued)*

Cultivar	Glucose	Fructose	Sorbitol	Sucrose	Malic Acid	Citric Acid	Shikimic Acid	Fumaric Acid
Vesseaux	85.1	71.2	18.1	9.7	5.4	0.1	15.5	1.9
Vigred	74.6	61.1	12.3	8.1	6.5	0.3	9.9	2.1
USDA data[b]	65.9	53.7	N.A.	1.5	N.A.	N.A.	N.A.	N.A.
Unidentified[c]	60.2	54.7	16.2	N.A.	7.2	0.7	N.A.	3.7

Source: Usenik et al. (2008), [a]Esti et al. (2002), [b]USDA-ARS (2018), [c]Bastos et al. (2015), N.A., not available.

Cherries are a good source of minerals, namely in what concerns potassium (260 mg/100 g edible portion) (McCune et al., 2010). Other minerals present in cherry composition are calcium (14 mg/100 g), magnesium (10 mg/100 g), phosphorous (20 mg/100 g), sodium (USDA-ARS, 2018), zinc (0.69 mg/100 g), and iron (1.16 mg/100 g) (Souza et al., 2014). Besides minerals, cherry is also an interesting fruit, when looking for vitamin intakes. In fact, it contains both hydrosoluble (B and C) and liposoluble (A, E, and K) vitamins (Mirto et al., 2018). Major vitamin present is vitamin C (7–50 mg per 100 g edible portion), followed by vitamin E (0.1 mg per 100 g edible portion) and vitamin K (2 µg per 100 g edible portion) (McCune et al., 2010), whereas the vitamin A content is around 3 mg retinol activity equivalent (RAE) per 100 g edible portion (Serradilla et al., 2016).

Several other compounds present in cherries are linked to their health-promoting characteristics (Tables 6.6 and 6.7). They are an exceptional source of numerous phenolic compounds, namely hydroxycinnamate, flavonols, procyanidins (Gao and Mazza, 1995; Liu et al., 2011; Usenik et al., 2010), anthocyanins, namely 3-*O*-glucoside and 3-*O*-rutinoside of cyanidin (Gonçalves et al., 2007; Kelebek and Selli, 2011), hydroxycinnamic acid derivatives (neochlorogenic, *p*-coumaroylquinic, and chlorogenic acids) (Usenik et al., 2010; Liu et al., 2011) flavonols (quercetin-3-glucoside, quercetin-3-rutinoside, and kaempferol-3-rutinoside), and flavan-3-ols (catechin and epicatechin) (Jakobek et al., 2009). Some of them are clearly related to the skin, as this specific part of the fruits can be of a different color (dark or clear red, or even white) and linked to the anthocyanin content (Chaovanalikit and Wrolstad, 2004;

Kayesh et al., 2013). Sweet cherries are also an excellent source of tannins (Tomás-Barberán and Espín, 2001), with content ranging from 32 to 75 mg 100 g^{-1} of FW (Prvulović et al., 2012). Phenolics range from 60 mg/100 g in "Brooks," to 687 mg/100g in "Santina," while anthocyanins varied from 0.6 mg/100 g in "Starks Gold," a cultivar of white coloration, to 98.1 mg/100 g in "Santina" (Table 6.7).

TABLE 6.6 Two Major Anthocyanin Content (mg Cyanidin 3-glucoside Equivalents (CGE/100 g FW and Two Major Phenolic Content (mg/100 g FW) of Several Sweet Cherry Cultivars

Cultivar	Anthocyanin Content (mg CGE/100 g FW)		Phenolic Content (mg/100 g FW)	
	Cyanidin 3-glucoside	Cyanidin 3-rutinoside	Neochlorogenic Acid	p-Coumaroylquinic Acid
Black Star	8.35[a]	68.28	47.52	6.01
Blaze Star	4.73[a]	51.87	23.22	26.62
Burlat	34.84	46.92	9.18	11.28
Donnantonio	1.40	29.64	9.59	5.08
Ducignola Nera	3.21	35.67	71.45	7.55
Early Star	13.42	32.11	11.69	15.97
Ferrovia	2.08	22.88	26.99	6.39
Gabbaladri	0.40	5.69	45.33	1.97
Genovese	1.71	22.68	19.50	1.62
Giorgia	3.28[a]	25.20	25.27	3.35
Grace Star	1.41[a]	22.57	29.59	3.48
Maiolina Grappolo	9.67	26.17	32.51	2.65
Maredda	23	67.82	74.09	6.78
Minnulara	1.04	25.99	22.56	1.01
Moreau	10.97	18.61	13.39	15.81
Napoleona Forestiera	6.48	6.52	10.47	15.53
Napoleona Grappolo	0.94	12.41	11.45	9.45
Napoleona Verifica	0.80	8.95	11.60	12.61
Puntalazzese	8.16	48.26	34.66	0.95
Sunburst	1.46	21.84	16	9.25

Sweet Cherry

TABLE 6.6 *(Continued)*

Cultivar	Anthocyanin Content (mg CGE/100 g FW)		Phenolic Content (mg/100 g FW)	
	Cyanidin 3-glucoside	Cyanidin 3-rutinoside	Neochlorogenic Acid	*p*-Coumaroylquinic Acid
Sweet Early	5.54	24.34	6.27	8.40
Sweetheart	3.83[a]	22.42	9.88	6.44
Toscana	2.85	25.51	35.30	3.13
Zio Peppino	0.55	6.40	7.91	12.34

Source: Ballistreri et al. (2013).
[a]Second major anthocyanin is peonidin 3-rutinoside.

TABLE 6.7 Total Anthocyanin and Total Phenolic Contents of Sweet Cherry Cultivars

Cultivar	Total Anthocyanins (mg/100 g)	Total Phenolics (mg gallic acid equivalents/100 g)
0-900 Ziraat	24.6	69.7
Ambrunes	N.A.	100
Belge	29.1	115.4
Black Gold[c]	30	92
Black pearl	69.8	447
Black Star	85.6	145.7
Blaze Star	62.4	138.8
Brazilian Grown Cherry[a]	26.7	314.5
Brooks[b]	10	60
Burlat	84.6	122
Cristalina	65	150
Cristalina[b]	79	155
Dalbasti	20.4	67.5
Donnantonio	32.8	103.8
Ducinola Nera	40.1	149.4
Durona di Cesena	17.4	78.9
Early Star	47.2	87
Ferrovia	27	97.4
Gabbaladri	6.2	105.3
Genovese	25.1	105
Giorgia	32.5	115

346 Temperate Fruits: Production, Processing, and Marketing

TABLE 6.7 *(Continued)*

Cultivar	Total Anthocyanins (mg/100 g)	Total Phenolics (mg gallic acid equivalents/100 g)
Grace Star	27.2	114.6
Hartland[c]	76	147
Hedelfingen[c]	40	96
Maiolina Grappolo	37.9	146
Maredda	94.2	162.2
Marvin-Niram	45	80
Merton Late	26.2	79.3
Minnulara	27.9	124.1
Moreau	31.5	154.5
Napoleona Grappolo	14.3	85
Napoleona Verifica	10.5	89.8
Napoleona Forestiera	13.3	123.6
Newstar[b]	20	75
Newstar	25	75
No. 57	20	70
NY-6479	15	80
Pico Colourado	N.A.	50
Pico Negro	N.A.	150
Prime Giant	15	80
Puntalazzese	58.3	156
Regina[c]	41	104
Salmo	N.A.	N.A.
Santina	98.1	687.4
Somerset	25	80
Sonata	130	100
Starks Gold	0.6	64.4
Sunburst	24.7	102.8
Sweet Early	70.6	110.8
Sweetheart	28.1	89.6
Toscana	29	131.7
Vista	14.8	61.1
Zio Peppino	7.6	86.1

Source: Chockchaisawasdee et al. (2016), [a]Souza et al. (2014), [b]Serrano et al. (2009), and [c]Kim et al. (2005).

N.A., not available.

Sweet cherries are also present in their composition carotenoids and the most important are β-carotene (38 μg per 100 g FW) and lutein/zeaxanthin (85 μg per 100 g FW) (Tomás-Barberán et al., 2013). They are precursors of vitamin A; and β-carotene, β-cryptoxanthin, α-carotene, lycopene, and lutein are present in cherry at levels of 0.02 mg/g dry weight (Leong and Oey, 2012), while zeaxantine was also identified (Demir, 2013).

Several pieces of evidence are available regarding the effects of cherries in health. The intake of cherries can be useful to prevent cancer, associated with the content of anthocyanins, that has been proved to significantly reduce the number and size of cecal tumors in rats (Kang et al., 2003), and other protective effects have been recorded in cancer cell lines (Chen et al., 2005). These chemopreventive activities are connected to their capacity to stimulate the expression of Phase II detoxification enzymes, to inhibit mutagenesis caused by environmental toxins and carcinogens, to induce cellular apoptosis and differentiation, to decrease cellular proliferation by modulating diverse signal transduction pathways (Wang and Stoner, 2008). Sweet cherry has also been associated with beneficial effects on preventing cardiovascular diseases. Reduction of oxidant stress in cardiac tissue (Xu et al., 2004), decreased vascular inflammation, and indirect reduction of foam cell formation, a precursor for the development of atherosclerotic plaque (Cannon, 1998), have been found with the exposure of endothelial cells to cyanidin-3-glycoside. Other evidence obtained with rat experiments showed the reduced incidence of irregular and rapid heart rates and less cardiac damage from heart attacks (Bak et al., 2006). Although obtained with extracts of tart cherry, it can be expected to get similar results with sweet cherry samples. A study by Kelley et al. (2006) reported the link between a decrease of levels of C-reactive protein (CRP) and nitric oxide (NO) in healthy subjects after ingestion of sweet cherries (280 g/day) for 28 days. The importance of this study lies in the fact that high levels of serum CRP being one of the most important indicators of inflammation and a significant risk factor for cardiovascular disease (Ridker, 2001).

The reduction of clinical diabetes can also be attributed to cherry and its components. In fact, the anthocyanins and quercetin present in sweet cherries can reduce oxidative stress (McCune and Johns, 2003) that is associated with the complications of diabetes (Cunningham, 1998), reducing its onset (Montonen et al., 2004). Several other studies show the effect of anthocyanins and anthocyanidins in insulin production, either in cell cultures (Jayaprakasam et al., 2005; Ghosh and

Konishi, 2007) or using rat models and cherry anthocyanins (Jayaprakasam et al., 2006; Adisakwattana et al., 2011), 3-O-β-d-glucoside specifically (Tsuda et al., 2003), or even diet enriched with tart cherry (Seymour et al., 2007). As previously mentioned, the low glycemic index of sweet cherries should also be pointed out, when compared with many other fruits (Foster-Powell et al., 2002), which makes them an appropriate fruit-based snack for individuals with diabetes.

Another important health effect that can be accomplished by sweet cherries is their effect on the human inflammatory process. Low-grade inflammation is a potential risk for a variety of other chronic diseases, such as cancer, cardiovascular disease, obesity, and arthritis (Nicklas et al., 2004). One of the key steps to reduce inflammatory processed is the inhibition of the cyclooxygenase enzymes (COX) responsible for inflammatory response. The work by Seeram et al. (2003) showed the cyanidin and malvidin effects in inhibiting the expression of COX, with a higher activity than polyphenols found in green tea, due to the chemical structure of sweet cherries that exhibits a hydroxyl group in the B ring of the compound. Furthermore, results from the same workgroup showed even a greater inhibition of COX-2 than ibuprofen and naproxen (Seeram et al., 2001). The beneficial effects of cherries on

inflammatory processes were also proved by the reduction of arthritis in male Sprague Dawley mice (He et al., 2006). The daily intake of 280 g or approximately 2.5 cups of sweet cherries for four weeks reduced the levels of inflammatory markers in humans (Kelley et al., 2006). Eating sweet cherries can also reduce gout symptoms, as showed by the reduction of serum urate, a marker for this disease (Jacob et al., 2003). Other works present good possibilities of indirectly reducing Alzheimer's disease (Yoshimura et al., 2003; Heo et al., 2004), or even symptoms of jet lag (McCune et al., 2010).

6.5 ORIGIN AND DISTRIBUTION

Sweet cherry culture is located in most of the world's countries with temperate climate, in the latitudes between 35 °N and 55 °S. In the Northern Hemisphere's temperate zones, cultivars mature from the end of April (in southern growing regions) to June–July (main season), while the harvesting season finishes in late August in Norway. In the Southern Hemisphere, the majority of cultivars are harvested in December and January, coinciding with the lucrative markets of North America and Western Europe, as well as Southeast and East Asia (Bujdosó et al., 2017). As shown in Section 6.2, the latest FAOSTAT data (2016) indicate a worldwide production of 2,317,956 tons, led by Turkey, with 599,650 tons, and followed by the USA (288,480 tons) and Iran (220,393 tons). Other important countries are Chile, Uzbekistan, Italy, Spain, Romania, Greece, and Syria. In Turkey, the most important cherry-producing region is the Aegean Region, followed by the Mediterranean and Eastern Marmara regions (Table 6.8). The cherry is generally grown in the temperate climate zones of these regions, with Isparta province being the most important one (55,657 tons), followed by Konya (55,426 tons) and Manisa (46,648 tons). The USA is the second major producer country as its production is located in several states (Table 6.9), mainly in Washington, California, and Oregon (National Agricultural Statistics Service, 2017). In Iran, the third major producer, sweet cherry is mainly grown in the Isfahan, Alborz, Tehran, and Khorasan regions (Bujdosó et al., 2017).

TABLE 6.8 Sweet Cherry Production Regions in Turkey

Provinces	Production (2018) (tons)
Isparta	55,657
Konya	55,426
Manisa	46,648
Izmir	46,574
Afyonkarahisar	40,387

Source: http://www.tuik.gov.tr.

TABLE 6.9 Sweet Cherry Production Regions in the USA

Region	Production (2015) (tons)
California	60,100
Idaho	1810
Michigan	13,430
Montana	740
New York	770
Oregon	38,700
Utah	230
Washington	222,650

Source: National Agricultural Statistics Service (2017).

6.6 BOTANY AND TAXONOMY

The sweet cherry ($2n = 2x = 16$) is a deciduous tree of large stature, able to reach nearly 20 m in height, with attractive peeling bark and allogamous species, generally self-incompatible (Rodrigues et al., 2008). It belongs to the family Rosaceae, subfamily Prunoideae, genus *Prunus*, and subgenus *Cerasus* divided into eight subsections according to the form of the calyx, pubescence of style, and indentation of mature leaves (Ingram, 1948).

The growth habit of sweet cherry trees typically conforms to the architectural model of Rauh (Hallé et al., 1978), with all axes—trunk and lateral branches—both orthotropic (upright growth at least initially before the possible secondary re-orientation of the branch) and monopodial (vegetative extension of one apical meristem). Spurs are located in the distal two-thirds or half of the annual growth of long shoots, and fruiting is lateral on the preformed parts of all the axes, whether long or short shoots (Lauri, 1992), leaving bare wood the following years.

To the characterization of sweet cherry trees, a list of descriptors is available (IPGRI, 1985). This must be the foundation of every description made of new or unknown varieties of sweet cherry. The most commonly used are summarized in Table 6.10.

TABLE 6.10 List of Descriptors and Scales for Sweet Cherry

Descriptor	Unit of Measure/Scores
Tree vigor	Very weak (1), weak (3), medium (5), strong (7), very strong (9)
Tree habit	Upright (1), semi-upright (2), spreading (3), drooping (4)
Yield size	Very small (1), small (3), medium (5), large (7), very large (9)
Yield weight	g
Fruit shape	Cordate (1), reniform (2), oblate (3), circular (4), elliptic (5)
Color of skin	Yellow (1), yellow with blush (2), orange red (3), light red (4), red (5), brown red (6), dark red (7), blackish (8)

TABLE 6.10 *(Continued)*

Descriptor	Unit of Measure/Scores
Length of stem	Very short (1), short (3), medium (5), long (7), very long (9)
Fruit flesh color	Cream white (1), yellow (2), pink (3), red (4), dark red (5)
Juice color	Colorless (1), cream yellow (2), pink (3), red (4), purple (5
Fruit firmness	Soft (3), medium (5), firm (7), very firm (9)
Fruit juiciness	Weak (1), medium (5), strong (7)
Sweet–acid balance	Very sweet (1), sweetish (3), balanced (5), acidulated (7), very acid (9)
Stone shape	Medium elliptic (1), broad elliptic (2), circular (3)
Taste quality	Bad (3), fair (5), good (7)
Beginning of flowering	Very early (1), early (3), medium (5), late (7), very late (9)
Date of full flowering	Early (3), mid-season (5), late (7), extremely late (9)
Time of ripening	Very early (1), early (3), medium (5), late (7), very late (9)
Harvest maturity	Early (3), mid-season (5), late (7), extremely late (9)

Source: IPGRI (1985) and Höfer and Giovannini (2017).

Several other traits are also important to evaluate, namely length and width of the leaf blade, petiole length, the basal and apical angles of the blade, and leaf form. In flowers, their open diameter,

Sweet Cherry   351

petal length, and width, as well as pistil length and number of stamens should also be evaluated (Sánchez et al., 2008). Flowers are white, presenting elongated pedicels, born in racemose clusters of 2–5 flowers on short spurs with multiple buds at tips, being the distal one vegetative and continues spur growth. The ovary is perigynous with distinct hypanthium. All these characteristics can vary, depending on the cultivar or the rootstock. The effect of cultivars can be seen in Table 6.11, for tree vigor and habit. The works by Rodrigues et al. (2008) and Sánchez et al. (2008) are also good examples of the differences found among cultivars, regarding leaves, flowers, and fruit traits.

TABLE 6.11 Tree Habit in Some Sweet Cherry Cultivars Grown in Portugal and Spain

Cultivar	Tree Vigor (Reference 100 for Burlat)	Tree Habit
Precoce Bernard[a]	93	Spreading
Burlat[a]	100	Upright
Francesa de Alenquer[a]	52	Upright
Lisboeta[a]	46	Upright
Tardif de Vignole[a]	68	Upright
Saco (Montalegre)[a]	60	Upright
Saco (Cova da Beira)[a]	52	Drooping
Maringa[a]	80	Spreading

TABLE 6.11 *(Continued)*

Cultivar	Tree Vigor (Reference 100 for Burlat)	Tree Habit
Ambrunés Especial[b]	N.A.	Spreading–drooping
Aragonesa[b]	N.A.	Spreading–drooping
Burlat[b]	N.A.	Upright
California[b]	N.A.	Medium spreading
Costalera[b]	N.A.	Very upright
De Valero[b]	N.A.	Spreading–drooping

Source: [a]Rodrigues et al. (2008) and [b]Sánchez et al. (2008).

6.7 VARIETIES AND CULTIVARS

Over 100 cultivars of sweet cherry are commercially cultivated in the world, each one with distinctive characteristics. Major worldwide cultivars of importance are "Alex," "Attika," "Bing," "Brooks," "Burlat," "Chelan," "Hedelfingen," "Kordia," "Lambert," "Lapins," "Rainier," "Regina," "Skeena" (Figure 6.2), "Sonata," "Staccato" (Figure 6.3), "Stella," "Sweetheart" (Figure 6.4), "Techlovan," "Tulare," and "Van" (Schuster, 2012). A brief description of the most common cherry cultivars is present in Table 6.12.

Considering the major producing countries, Turkish production is mainly based on cultivar "0900

Ziraat" grafted on *Prunus mahaleb* and "Mazzard" (*Prunus avium*) rootstocks. Fruits of this cultivar are of high quality, firm, with skin of mahogany color. About 90% of Turkey's exports are of this cultivar, even though it presents some drawbacks like it is not a self-fertile cultivar. Therefore, it needs pollinators and has low fruit set, which lead to low productivity (Mert and Soylu, 2007). For pollination purposes, growers in Turkey use cultivars like "Lambert," "Merton," "Late," "Bigarreau," "Gaucher," and "Starks Gold." Other important cultivars are "Stark Gold," "Regina," and "Lambert" (Bujdosó et al., 2017; Demirsoy et al., 2017). In the USA, the used cultivars are different if cherries are for fresh consumption or processing. For fresh consumption, the most important cultivar is "Bing" produced either in California, Washington, or Oregon. For California, other important cultivars are "Burlat," "Brooks," "Coral Champagne," "Chelan,"

"Early Garnet," "Garnet," "Rainier," "Royal Rainier," and "Tulare." In other growing regions, such as Washington and Oregon, cultivars that spread out the harvest season are also used, namely "Chelan," "Santina," "Tieton," "Early Robin," "Benton," "Rainier," "Attika," "Lapins," "Skeena," "Regina," and "Sumtare." Cultivars for processing, usually grown in Michigan, are mostly "Emperor Francis," "Gold," "Napoleon," "Sam," and "Ulster" grafted on "Mazzard" and *Prunus mahaleb*. Although existent, the areas devoted to cherry for fresh consumption are reduced, relying on cultivars such as "Attika," "Benton," "Cavalier," "Ulster," "Summit," "Hudson," and "Regina." In Iran, the third most important producer, some of the major cultivars are local ones, such as "Sorati Lavasan," "Zarde Daneshkade," "Shishei," "Siahe Mashhad," although cultivars "Bing," "Lambert," and "Van" are also found.

TABLE 6.12 Varietal Description of the Most Popular Sweet Cherry Cultivars

Cultivar	Origin	Pollination	Susceptibility	Growth	Fruit
Bing	Oregon, USA	Self-incompatible	Low resistance to rain-cracking and double fruiting	Moderately vigorous; upright growth habit	Large and firm fruit; purple–red skin color
Burlat	France	Self-incompatible	Resistant to bacterial canker; moderately resistanceto rain-cracking	Moderately vigorous; upright growth habit	Large and very firm fruit; dark red or purple skin color

Sweet Cherry   353

TABLE 6.12   *(Continued)*

Cultivar	Origin	Pollination	Susceptibility	Growth	Fruit
Lapins	Summerland, Canada	Self-compatible	Low resistance to rain-cracking	Upright growth habit; very productive	Large and very firm fruit; dark red skin color
Regina	Germany	Self- incompatible	High resistance to rain-cracking; moderately resistant to powdery mildew	Upright growth habit; low productivity	Very large and firm fruit; dark red skin color
Santina	Summerland, Canada	Self-compatible	Low resistance to rain-cracking; moderately tolerant to splitting	Semi-open growth habit; vigorous	Medium-large and firm fruit; dark red skin color
Skeena	Canada	Self-compatible	Resistant to rain-cracking	Spreading growth habit; very productive	Very large and firm fruit; wine-red to mahogany skin color
Staccato	Summerland, Canada	Self-compatible	Moderately susceptible to rain-cracking; very sensitive to powdery mildew	Spreading growth habi	Large and firm fruit; dark red skin color
Summit	Summerland, Canada	Self-compatible	Moderately susceptible to rain-cracking	Spreading growth habit; vigorous	Very large fruit with medium or firm consistency; dark red to black skin color
Sweetheart	Summerland, Canada	Self-compatible	Moderately sensitive to rain-cracking; very susceptible to powdery mildew	Open growth habit	Moderately large and very firm; red skin color
Van	Summerland, Canada	Self-incompatible	Moderately susceptible to rain-cracking	Vigorous growth	Medium size and firm fruit; dark red skin color

Source: Fogle et al. (1973), Kappel et al. (1998), Kappel et al. (2006), Long et al. (2007), Quero-Garcia et al. (2017), and Liu et al. (2018).

FIGURE 6.2 Fruits of "Skeena."

FIGURE 6.3 Fruits of "Staccato."

FIGURE 6.4 Fruits of "Sweetheart."

6.8 BREEDING AND CROP IMPROVEMENT

Breeding objectives in sweet cherry are focused, as for many other horticultural commodities, in fruit quality, namely those related to size, firmness, and flavor. Other main objectives are self-fertility, extended harvest season (Sansavini and Lugli, 2005), adaptability to mechanical harvest, with recent efforts devoted to precocity, productivity, and resistance to rain-induced cracking, and to diseases and pests (Kappel et al., 2012).

Since fruit quality receives much of the attention, and, as size being one of the first attributes recognized by consumers are therefore key to successful commercialization, both for producers and retailers. According to Kappel et al. (1996), optimum weith of the fruit is around 11–12 g, but it can be affected by the leaf area per fruit (Roper and Loescher, 1987) or crop load (Proebsting, 1990). Although it is known that the number of cells present in the fruit mesocarp, the major contributor to fruit size, is genetically controlled (Olmstead et al., 2007). Firmness has also been addressed, and nowadays a new cultivar has significantly higher firmness than standard cultivars (Looney et al., 1996). Both parameters have been successfully changed, as recent fruit size gains have been achieved with several cultivars, while firmness has also increased from Bing, a standard cultivar with a fruit firmness of 170 g/mm, to "Sweetheart" that presents a firmness of 299 g/mm (Kappel, 2005). However, it appears that these characteristics can be negatively correlated with some genetic backgrounds, being as highly heritable but also very polygenic, with a complex genetic determinism that must be further studied (Campoy et al., 2015).

Self-fertility has been an important goal of breeding, and its development was a major step stone for the increase in sweet cherry productivity around the world. Before this, growers had to plant an appropriate number of suitable pollinizing (cross-compatible) cultivar(s) to ensure acceptable cropping. The first self-fertile commercial cultivar was Stella, released in 1968 by the Canada Department of Agriculture Research Station at Summerland, British Columbia (Lapins, 1971). Nowadays, almost all commercially available self-fertile cultivars can trace back to Stella (Sansavini and Lugli, 2005). To reduce this lack of diversity on original self-compatibility, other self-compatible landraces, such as Cristobalina or Kronio, are now being used as parents in several breeding programs (Quero-García et al., 2017).

Sweet cherries are a highly seasonal fruit with a rather short harvest period, which limits their commercialization opportunities. Obtained cultivars that extent this period, either by anticipated or delayed maturation, will be able to reach a higher price, and has also been one of the goals of breeding (Sansavini and Lugli, 2005). For early ripening cultivars, breeding has been using as parents "Burlat," followed by the recently released "Rita" and "Primulat" (Sansavini and Lugli, 2005).

Harvesting is one of the most critical moments of cherry growing, due to the lack of labor, or to its increasing costs. Mechanical harvesting has already been used in cherries for processing, but fruits that are to be placed in the fresh market are of higher quality standards when reaching the consumers. To be suitable for mechanical harvesting, the fruit must easily detach from the stem. Some cultivars have already been developed to allow this harvesting, namely "Ambrunés," "Fermina," "Linda," "Sumste" (Bargioni, 1996), "Vittoria" (Bargioni, 1970), and "Cristalina" (Kappel et al., 1998) that possess a dry abscission zone between the fruit and stem, where a hardened scar tissue must exist, to prevent juice loss, oxidation, and pathogen attacks (Sansavini and Lugli, 2008).

Precocity in cherries has in many other cultures, allows growers to achieve a quicker return of the investment. For this specific trait, the cultivar "Sweetheart" can be used as a standard as it is considered a quite precocious cultivar (Kappel et al., 2012). Besides cultivar, the effect of rootstock cannot be overlooked, as different cultivars may change their precocity when grafted on different rootstocks (Kappel et al., 2012).

A very important factor that may greatly influence the economical sustainability of sweet cherry cultivation is rain induced cracking. This phenomenon, which appears to be physiological, is determined by environmental factors, more than genetic (Jayasankar and Kappel, 2011). Breeding for this specific characteristic is even harder as data are lacking on understanding the mechanisms of cracking, and by the need of a clear selection tool, as the cracking index does not show a consistent relation between its results and field observations. Furthermore, a suggested relationship between firmness and cracking susceptibility, as pointed out by Brown et al. (1996), has not been found in other studies (Kappel et al., 2000).

Another major objective of breeding is to increase the tolerance to pests and diseases, to reduce losses and costs, and to allow growers to use integrated pest management and organic production. Major problems arise from diseases such as powdery mildew (*Podosphaera* spp.), brown rot (*Monilinia* spp.), bacterial canker (*Pseudomonas* spp.), while the most important pests are the cherry fruit fly (*Rhagoletis* spp.), black cherry aphid (*Myzus cerasi*), and cherry slug (*Caliroa cerasi*). The bacterial canker is the main problem in cooler and wetter areas of production, and,

by the early 1980s, some cultivars were available and labeled as resistant to this disease (Kappel et al., 2012). However, new and more virulent strains were able to overcome such resistance, and, apparently, no total resistance to canker is to be achieved, although breeding must go on and include tolerant cultivars as parents, such as "Colney," "Hertford," "Merton Glory," or "Vittoria" (Bargioni, 1996). Control of the brown rot is not easy, with damages depending on climatic conditions and virulence of local races (Budan et al., 2005). Although no resistance has ever been found, some cultivars such as "Regina," "Early Korvik," "Melitopolska Chorna," and "Valerij Chkalov" have good levels of tolerance to this disease, although not high enough for any to be used as parents (Kappel and Sholberg, 2008). Some wild species of cherries are resistant to leaf spot (*P. sargentii, P. serrulata var. spontanea, P. subhirtellapendula rosea, P. insisa, P. canescens, P. kurilensis, P. nipponica*, and *P. maackii*) and must be used in breeding programs (Schuster and Tobutt, 2004). Resistance to this pathogen has been described for cultivars "Moreau," "Venus," and "PMR-1," linked, in this selection, to a single gene, *Pmr1* (Olmstead et al., 2002).

For the pests, no resistance has been found. However, for *M. cerasi*, clones from *Prunus canescens*,

Prunus incisa, Prunus kurilensis, and Prunus nipponica showed resistance to colonization, and crosses between these clones and cultivar Napoleon resulted in hybrids with some tolerance to colonization (Bargioni, 1996).

Recently, as a major breakthrough for cherry breeding was achieved, by the publication of the sweet cherry genome (Shirasawa et al., 2017). Many fruit quality traits are quantitative trait loci (QTLs), and heritabilities are unknown, and, to date, markers for the traits mentioned above have not been published. QTLs have been identified for fruit weight and firmness (Campoy et al., 2015), size (Rosyara et al., 2013), skin and flesh color (Sooriyapathirana et al., 2010), and fruit-cracking tolerance (Balbontín et al., 2013). Other tools, like marker-assisted selection, may become a significant biotechnological tool, although it is currently limited due to the polygenic nature of fruit quality traits (Salazar et al., 2014). Major advances can also come from the use of DNA information, as clustered regularly interspaced short palindromic repeats (CRISPR) alongside CRISPR-associated protein 9 (Cas9) system (CRISPR/Cas9). A continuous effort to obtain new and improved sweet cherry cultivars is underway all around the world, with major breeding programs taking place summarized in Table 6.13.

TABLE 6.13 Breeding Programs Around the World and Major Objectives

	Country	Breeding Institute	Major Breeding Objectives
Europe	Bulgaria	Fruit Growing Institute	Self-fertility, compact habit, early ripening, and large fruit size
	Czech Republic	Research and Breeding Institute of Pomology Holovousy Ltd	Harvest period, high-quality fruit, resistance to diseases and rain-induced cracking
	France	Institut National de la Recherche Agronomique– EP Innovation	Fruits traits and tree characteristics
	Germany	Julius Kühn-Institut, Institute for Breeding Research on Fruit Crops	Fruits traits and tree characteristics
	Hungary	NARIC Fruitculture Research Institute	Maturity time, fruit quality, large fruit size, low sensitivity cracking, and self-fertility
	Italy	Bologna University, Department of Agricultural Sciences	Self-fertility, early maturit, and consistent high yield
	Romania	Research Institute for Fruit Growing	Early- and late ripening

TABLE 6.13 *(Continued)*

	Country	Breeding Institute	Major Breeding Objectives
	Spain	Centro de Investigaciones Científicas y Tecnológicas de Extremadura	Physicochemical properties of native cultivars
		Murcia Institute of Agri-Food Research and Development	Extending the maturity time, fruit quality and self-fertility
	United Kingdom	East Malling Research	Late ripening cherries with low susceptibility to cracking
	Ukraine	Institute of Horticulture of National Academy of Agrarian Sciences of Ukraine	Fruit weight, productivity, precocity, quality and maturity period
North and South America	Canada	Agriculture and Agri-Food Canada	Fruits traits and tree characteristics
		Politécnica Universidad Católica de Valparaíso	Fruit size and firmness, productivity and chilling requirements
	Chile	INIA–Biofrutales	Harvest period
		Consorcio Tecnológico de la Fruta S.A. and Pontificia Universidad Católica de Chile	High-quality fruit
	USA	Washington State University	Early-, mid-, and late-ripening self-fertile cultivars with a range of skin colors
Asia	China	Institute of Pomology, Dalian Academy of Agricultural Sciences	Productivity, fruit weight, tasting quality, firmness and disease resistance
		Zhengzhou Fruit Research Institute, Chinese Academy of Agricultural Sciences	Quality, high yield, resistance to high temperature and high humidity, and adaptation to soil and climatic conditions
		Institute of Pomology and Forestry, Beijing Academy of Agriculture and Forestry Science	Self-fertility
	Japan	Horticultural Experiment Station, Yamagata Integrated Agricultural Research Center	Early- and late ripening
	Turkey	Atatürk Horticultural Central Research Institute	Fruit quality and self-fertility

Source: Quero-García et al. (2017).

6.9 SOIL AND CLIMATE

Sweet cherries, like other fruit trees, require light (aerated), well-drained soil with pH ranging from 6 to 6.8 and with 1 m, at least, of soil depth (James, 2011). Nonetheless, they can tolerate soils from sandy loam to clay loam as long as they have good drainage, while pH ranging from 5.5 to 8 is considered acceptable (Verma, 2014). Below pH 5.5, the solubility of nutrients, in particular, the manganese and aluminum increase quickly reaching toxic levels, inhibiting root elongation and branching and reducing the vigor of newly planted trees (James, 2011). On the other hand, in soils with pH levels above 8, the growth of cherries can be inadequate due to the limited availability of nutrients such as phosphorous, iron, zinc, and manganese (Melakeberhan et al., 2001). The high sodium content in soils with pH above 8 affects the structure of soil (McCauley et al., 2017) negatively, and thus soils with such pH values should also be avoided. Cherry roots are exceptionally sensitive to the soils with excessive moisture, which may stunt or kill the tree (Verma, 2014). In poorly drained soils, cherry trees often develop fungal and bacterial diseases such as the brown rot bacterial canker (Elmhirst, 2006). Prominent gray streaks and rusty yellow-brownish spots appear in leaves and stems that are indicators of limited and poor aeration (Roper and Frank, 2004). Another important aspect of soil is the salinity level. Cherries, as the majority of fruit trees, are sensitive to salts and salinity levels of 1.9–3.1 dS/m in the soil can decrease the yield by 10%–50% (Kotuby-Amacher et al., 2000), and therefore should be avoided.

Climate components affect the growth and profitability of commercial cherry production directly as cherries are more sensitive to climate variations than other fruit species. They require cold and long winter conditions, with an extended dormancy period of 7 °C or below to ensure a full bud break in spring. Also, most sweet cherry cultivars have chilling requirements of about 1000 hours during the winter (Verma, 2014). The absence of adequate chilling hours may result in a poor bud break, a lengthy bloom period, and a poor overlap between main and pollinizer cultivars, causing an uneven fruit maturity as well as prolonged harvest periods. A cool spring and a mild summer with temperatures above 15 °C during the blossom period ensure maximum pollination and a maximum fruit set (Murray, 2011).

Furthermore, a little rainfall from late spring to summer with low humidity throughout the entire growing season reduces the fruit damage by cracking and reduces disease pressure (Cline et al., 2009; Balbontín et al., 2013). For cherry production, the pattern of rainfall throughout the year is more important than the total annual rainfall (James and Measham, 2011). The two most important and critical periods for cherries are blossom and fruit ripening (Simon, 2006). If the rain continues during the blossoming period, the fruit set is reduced, while persistent rainfall during the fruit-ripening period increases the occurrence of cracking and diseases is relatively higher (Christensen, 1973; James, 2011; James and Measham, 2011). However, their severity will be dependent on the cultivar (Stojanović et al., 2013), and the most resistant ones can be recommended for adverse climatic conditions. Besides, a rain-protective covering system can be used to reduce fruit cracking (Børve et al., 2003; Lang, 2009).

Another important climate parameter is the occurrence of frosts. Cherries are extremely susceptible to frost injury during periods of blossom and fruit set. Szabó et al. (1996) found 82%–100% damaged flowers in sweet cherry cultivars when temperatures decrease below −2.5 °C during full bloom, whereas almost no damage occurred at the same temperature in the balloon stage. Szpadzik et al. (2009) found the death of 13% of flowers during the frost when the temperature was

−4 °C, and more than 90% fruit primordia were suffered significant injuries when temperatures reached −5 °C at the beginning of May. More recently, Matzneller et al. (2016), while studying the vulnerability of a sweet cherry cultivar Summit to spring frosts found diverse levels of cold hardiness and, in general, flower organs became more vulnerable when in more advanced development stages, concluding that late frost occurrence can have more negative impact in the bud and flower development and by consequence in fruit setting than earlier frosts. Salazar-Gutiérrez et al. (2014) observed a similar trend, reporting a progressive vulnerability of the flower buds to low temperatures and cultivars were sometimes sensitive, semisensitive or hardies, dependent on the sampling date. In addition, genotype differences can explain some different behavior of several cultivars to the occurrence of frosts.

Nowadays, the opportunity to have a higher profit with the early harvests has caused the displacement of cherry orchards to more cool and warm regions (Li et al., 2010), creating another type of problems for producers, especially the occurrence of high temperatures, particularly in the summer during the flower bud differentiation. It is believed that during this period, high temperatures (>30 °C) can cause double pistil formation, double fruits and poor fruit set in the following year (Imrak et al., 2014). Therefore, in these situations, shading systems are highly recommended to reduce the high temperatures, inhibiting the double fruit formation. Previously, also Gonçalves et al. (2004a) found that cherry fruit qualities (size, dimension, weight, and soluble solids) were better in a warmer rather than a cooler year, indicating that a warm climate is desirable to achieve better fruit quality. Also, for species with a high chilling requirement like cherry trees, a significant increase in temperature in the future due to climate changing may reduce the chilling accumulation so much that spring phenology becomes delayed leading to production losses (Guo et al., 2014). On the other hand, it is generally accepted that cherries from higher altitude are of better quality than lower elevations. However, this is highly dependent on cultivar genotype. Alburquerque et al. (2008) studying the combined effect of altitude and chilling hours observed a positive relationship between chilling accumulation and altitude of different regions in Spain. They showed that locations above 450 m would be more appropriate for "Ruby," "Somerset," and "Burlat" cultivars, whereas for "New Star" and "Marvin," the limiting altitudes were 550 and 650 m, respectively. Instead, "Cristobalina" can be grown in almost every area of Spain starting from 175 m above sea level. Conversely, Faniadis et al. (2010) found that elevation of orchards from 59 to 456 m of altitude above sea level in Northern Greece had a marked influence on the cherry phytochemical compositional in "Burlat," "Tragana," and "Mpakirtzeika" cultivars. In these cultivars, the levels of antioxidants and antioxidant activity were always higher in orchards placed in the elevated zones. The knowledge of the relationship between altitude and chill accumulation is of great interest when deciding to grow cherry trees, and thus the most important step is to find the appropriate cultivar for each location or region.

6.10 PROPAGATION AND ROOTSTOCK

There are two main ways to propagate a cherry tree: by sexual methods using it seeds or using asexual methods through cuttings or by grafting. Because cherries are usually self-incompatible, resulting in great genetic variability, the seeds are not a good choice for commercial propagation. Their importance resides in professional breeding programs (Hartmann et al., 2011). The propagation of cherries by cutting is most often used in plant nurseries for cloning of rootstock, and gardeners or growers normally use grafting but only when they know exactly what they are getting (Hartmann et al., 2011). Budding or grafting the desired cultivar onto a rootstock has been the chosen method by the nursery, plant growers, and cherry producers.

Another important method used in sweet cherry tree multiplication is the micropropagation based on the stimulation of axillary branching and adventitious shoot formation by the addition of cytokinins to a culture medium (Druart, 2012).

Rootstocks for cherries are produced using plants of certain species or their hybrids and offer several benefits such as: adaptation to unfavorable soil conditions (textural, salts, and drainage), resistance to soilborne diseases or pests and tolerance to climatic conditions (temperature, precipitation, wind, and radiation) and anchorage (Webster, 1980; Usenik et al., 2010; Meland, 2015). *Prunus avium* L., *Prunus cerasus* L., *Prunus canescens* Bois, *Prunus fruticosa* Pall., and Prunus *mahaleb* L. are the major cherry species used in the parentage of rootstocks (Magyar and Hrotkó, 2008; Meland, 2015; Hrotkó and Rozpara, 2017). Other species used as rootstocks or in rootstock-breeding programs include *Prunus × dawyckensis* Sealy, *Prunus incisa* Thunb., *Prunus concinna* Koehne, *Prunus serrula* Franch., *Prunus subhirtella* Miq., *Prunus pseudocerasus* Lindl.,

Prunus tomentosa Thunb., and *Prunus serrulata* Lind (Westwood, 1978; Cummins, 1979; Webster and Schmidt, 1996; Dorić et al., 2014; Hrotkó and Rozpara, 2017).

The selection of rootstocks is possibly the most critical decision when establishing any orchard due to their huge influence on tree growth and vigor, management practices and fruit precocity, size, weight, yield, and nutritional quality. Therefore, it is crucial to select rootstocks suitable to the local conditions, and factors such as graft compatibility, delayed incompatibility, or full incompatibility are critical. In addition, the use of different rootstocks may contribute to the change of phenological stages of scions, tree survival, growing habit, and bearing habits as shown a study of Burak et al. (2008) using a Turkish cultivar "0900 Ziraat," as well as the studies of Larsen et al. (1987) and Whiting et al. (2005) in the USA with the "Bing" cultivar.

Some of the cherry tree rootstocks produce a very good junction with top cultivars, but for some others, the effectiveness of their grafting is very low or even impossible (Webster and Schmidt, 1996). The symptoms of incompatibility include union breaking between scion and rootstock, bad bud uptake, chlorosis in leaves, stunted growth, early reddening, and leaf fall in the autumn, scion

Sweet Cherry 363

or rootstock overgrowth, excessive rootstock suckering and increased early fruiting (Andrews and Marquez, 1993; Gonçalves et al., 2006; Sitarek, 2006; Cantín et al., 2010). Although in some cases the incompatibility symptoms may not appear, particularly when the growth conditions are near the ideal for cherry, the onset of environmental stress may result in the materialization of such symptoms (Hrotkó and Rozpara, 2017). The physiological incompatibility between scions and the best-known cherry rootstocks has been recently reviewed by Gainza et al. (2015) and Hrotkó and Rozpara (2017). Table 6.14 presents some examples of the most common rootstock available in the market and their main characteristics.

TABLE 6.14 Most Common Cherry Rootstocks in the Market and Their Main Characteristics

Rootstock	Main Characteristics	References
Mazzard (*P. avium*)	"Mazzard" is still widely used as standard vigor rootstock for sweet cherry throughout the world; develops a good graft union with almost all sweet cherry cultivars; presents a deep and extensive root system with good anchorage and it does well in a wide range of soils from sandy loam to clay loam; but it is not adequate for poorly drained or wet soils; trees on Mazzard are late to begin bearing, so they are not suitable for modern intensive orchards.	Perry (1987); Webster and Schmidt (1996); Long and Kaiser (2010); Long et al. (2014).
F 12/1 (*P. avium*)	Obtained in the UK in 1944, exhibits physiological compatibility with the majority of commercial cultivars; high vigour; produces a very large standard cherry tree; moderately resistant to frosts; requires fertile soil with good water supply and go badly in heavy with excess of moist; resistant to bacterial canker but high susceptible to crown gall.	Webster and Schmidt (1996); Hrotkó and Rozpara (2017).
Colt (*P. avium* × *P. pseudocerasus*)	Semidwarfing rootstock and smaller than Mazzard; easy to propagate; good anchorage and 80% of vigour; limited adaptation to drought, cold winters temperatures, and lime soils; resistant to *Phytophthora* root rot, bacterial canker and gopher damage but susceptible to crown galls (*Agrobacterium tumefaciens*).	Webster (1980); Long and Kaiser (2010).

TABLE 6.14 *(Continued)*

Rootstock	Main Characteristics	References
Saint Lucie 64 (syn. "SL 64") (*P. mahaleb*)	Obtained in 1954 in Bordeaux, France; present good compatibility with the majority of sweet cherry cultivars; easy to propagate and good productivity and precocity with sweet cherry trees; they go well on deep fertile soils and worst on sandy soils; produce a slightly smaller tree than the "Mazzard" rootstock; in general, it is resistant to diseases and has better cold-hardiness; excellent anchorage.	Edin and Claverie (1987); Claverie, (1996); Edin et al. (1996); Moreno et al. (1996); Long and Kaiser (2010); Drkenda et al. (2012); Hrotkó and Rozpara (2017).
Gisela 12 (*P. cerasus* × *P. canescens*)	A semidwarfing and semivigorous, and adapted to a wide range of soil types as well as hotter climates; good precocity and productivity; good resistance to viral diseases; trees grafted on "Gisela 12" do not produce root suckers, but have a quite poorly developed root system and thus require support.	Gruppe (1985); Lichev and Papachatzis (2009); Long and Kaiser (2010).
Gisela 6 (*P. cerasus* × *P. canescens*)	A semidwarfing, precocious and productive rootstock; it is winter hardy and adapted to a wide range of soil types; fair anchorage and good compatibility; going well in moderate to high-density plantings.	Kappel et al. (2005); Long and Kaiser (2010); Long et al. (2014).
Gisela 5 (*P. cerasus* × *P. canescens*)	It is very precocious with good winter hardiness; needs soils with good fertile and extra irrigation supplementation; fair to the good anchorage is adequate for low-productive cultivars; going well under tunnels and good performance in super high-density plantings.	Gruppe (1985); Long and Kaiser (2010); Sotirov (2015).
Gisela 3 (*P. cerasus* × *P. canescens*)	It is the most dwarfing rootstock in the Gisela series; widely adapted to different soil types and very precocious; promotes a vigorous growth in the two first years but slows once fruiting starts; good growing under tunnels and super-high density plantings.	Gruppe (1985); Long and Kaiser (2010); Sotirov (2015).
Krymsk 5 (*P. fruticosa* × *P. serrulata* var. lannesiana)	Breed in Krymsk, Russia, is considered one of the most interesting dwarfing rootstock; shows good tolerance to frost and bacterial canker and exhibits physiological compatibility with all sweet cherry cultivars; presents good tolerance to heavy soils and good anchorage; good productivity and precocity, but smaller fruit size and early senescence.	Long and Kaiser (2010); Long et al. (2014); Maas et al. (2014); Hrotkó and Rozpara (2017).

Sweet Cherry

TABLE 6.14 *(Continued)*

Rootstock	Main Characteristics	References
Krymsk 6 (*P. cerasus* × (*P. cerasus* × *P. maackii*)	Breed in Krymsk, Russia, this dwarf rootstock grows trees 40% less vigorously; the trees planted in this rootstock come into fruit 2–3 years earlier; the trees develop an extensive root system with good anchorage; like "Krymsk 5," "Krymsk 6" rootstock seem to be adapted to both cold and hot climates as well as heavier soils; present a low susceptibility to diseases.	Long and Kaiser (2010); Long et al. (2014); Hrotkó and Rozpara (2017).
MXM 14 (*P. mahaleb* × *P. avium*)	Adapted to a wide range of soils and has good anchorage; it is also resistant to iron-induced chlorosis caused by calcareous soils; presents a good behavior on lime soils; trees grafted on "MaxMa 14" grow about 30%–40% less than on "Mazzard"; it is considered a semistandard rootstock; this rootstock is moderately susceptible to *Phytophthora*, but more tolerant to bacterial canker and crown gall; nonetheless, it is sensitive to water shortage in the soil and requires irrigation during periods of drought.	Long and Kaiser (2010); Long et al. (2014); Hrotkó and Rozpara (2017).
MXM 60 (Mazzard × Mahaleb (M × M))	Bred in Oregon, USA, is more productive than "Mazzard"; compatible with both sweet and sour cherry cultivars; do well in a wide range of soils and they can be planted on heavier and wet soils; it presents a good and high resistant to bacterial canker, and *Phytophthora* root; they have an excellent anchorage.	Lang (2000); Hrotkó and Rozpara (2017).
Weiroot 154 (*P. cerasus*)	Hybrid of landraces selected in Germany; semidwarfing; good adaptability to clay soils; good compatibility and good fruit size.	Bujdosó et al. (2004).
Edabriz (*P. cerasus*)	Reduces the tree vigour by more than 60%; because of their very thin root system, it should be planted in a very fertile loamy or clay soils; require nutrients and water; the cherry trees on "Edabriz" require support and produce root suckers.	Edin et al. (1996); Hilsendegen (2005); Sitarek and Grzyb (2010).

6.11 LAYOUT AND PLANTING

A good setting is essential for the success of sweet cherries fruit plantings. However, before setting, fundamentals to consider are the market demand and consumer preferences, the soil and climate constraints, the availability of healthy rootstocks and cultivars, among other factors. According to Barritt et al. (1997), there are seven essential components to consider when selecting an existing or designing a new orchard system: rootstock, tree density, tree quality, tree arrangement, support systems, tree training, and tree pruning. Nonetheless, factors such as soil and water quality, the potential for winter damage, spring frosts, pressure from disease infection, and rain cracking (Koumanov and Long, 2017) must be carefully considered to have success in cherry orchards plantings. Also, a good orchard design must be planned, since it may affect the economic strategy and expected a life span of the orchard. Javaid et al. (2017) reported that a good orchard design should always allow an efficient utilization of orchard space, a maximum solar radiation interception and distribution within the orchard canopies to achieve the maximum fruit quality and yield, a minimization of competition between trees for nutrients and moisture (through adequate tree spacing), and a compatibility with different management practices (e.g., pruning, thinning, harvesting, and pest and diseases control).

In a sweet cherry, like in other tree fruit crops, the first step in planting orchards is the preparation of the site to receive the future plants. The preparation of the site includes, among other activities, the ripping of soil as deeply as possible in order to remove any hard layer that might inhibit air, water, and roots penetration in the deep layers; removal of small rocks and old root systems from previous crops; soil chemical properties and their textural and structural analysis; if necessary, addition of lime, phosphorus, and other fertilizer nutrients according to soil test recommendations; and analysis to nematodes presence (Verma, 2014; Javaid et al., 2017; Koumanov and Long, 2017). Cherries are very sensitive to soils with poor drainage. Thus, it is crucial to guarantee that any location selected for cherries has good surface and subsurface drainage. Wet soils also promote the development of potential diseases like *Phytophthora* root rots (Wilcox and Mircetich, 1985) and should be avoided. Therefore, it is crucial to know the main constrains about the site to install a cherry orchard, and the chosen measures should be used in both new and replant sites.

Almost similar to the site preparation is the determination of the size and arrangement of plants in the orchard. In general, the designs mostly used are the squares and rectangular forms followed by the counter of hillsides, if the orchards stay in a harsh land (Long et al., 2015). According to Koumanov and Long (2017), to maximize light interception, wherever possible, the rows should be oriented in a north–south direction, but in hills, the rows should be installed in across the slope, and the cover crop vegetation should be planted to help to prevent erosion and slow water runoff. In general, the spatial arrangement of plants and the distance or spacing between trees will be dependent on rootstock, cultivar, training system, and density adopted. As pointed out by Long et al. (2015), in modern cherry orchard systems with trees planted on dwarfing rootstocks, the plant densities range from 0.5×3.0 m^2 (between trees $\times$ between rows) for super-high-density systems to 1.8×4.5 m^2 for multiple leader systems.

After the preparation of soil, the definition of layout, and plant densities, the next step is the tree planting operation, and for that, planting time is defined. It is widely accepted that the best time to plant trees is immediately after delivery and early in the winter/spring period to minimize any transplant crisis. Later planting might reduce the success of orchards settling due to water stress crisis (Day and Sarchet, 2018). When planting, trees on standard rootstocks must be placed with the graft union a few inches above the soil level and, if necessary, supports must be sited before planting. Basically, there are two major ways of planting trees including sweet cherry trees: manually and mechanically. In both systems, a hole with proper depth is open, manually or mechanically, and

the rootstock is positioned vertically in the center, the roots are spread down in the hole, not bent or excessively trimmed, and the hole is filled with soil. The plants are then watered to improve root/soil, remove excess air, and create a moist environment to start their growth, minimizing the transplant shock. Ideally, the irrigation should be done at the same time of plant setting in order the new trees plants get water immediately.

Independently of the system used to plant trees, the plant always should be planted at correct depth which is dependent on the rootstock type used (Long et al., 2015), and in general it is common to adopt the criteria of planting trees with the bud union at 46 cm above the soil surface (after tree settled and planted). The objective is to avoid the formation of roots in scion.

6.12 IRRIGATION

The cherry tree is a tolerant species to drought; however, supplemental irrigation might be useful to achieve high yield with high-quality fruits, and like in other fruit trees, it is important to irrigate at the proper time with the adequate amount. Factors such as precipitation, temperature, wind, evapotranspiration, radiation, tree age, and development stage of fruits (Engin et al., 2009; Measham et al., 2009; Usenik and Stampar, 2011) must be taken into account when deciding to irrigate. Several studies (Kappel et al., 1996; Marsal et al., 2010; Neilsen et al., 2010, 2014; Livellara et al., 2011; Usenik and Stampar, 2011; Measham et al., 2012) have shown the effects of different watering strategies in cherry quality attributes, particularly firmness, color, and soluble solids. Although several studies have tried to establish the correct amount of water to be applied in cherry orchards aiming high-quality fruits, this is very difficult because it will be dependent on soil properties and climate conditions. For example, Hanson and Proebsting (1996) have established that in regions with low summer precipitation and high vapor pressure deficit (VPD), sweet cherry trees may require 750–1000 mm of irrigation during the growing season. On the other hand, Marsal et al. (2010) in a study conducted with different cultivars and rootstocks concluded that in humid regions 550–700 mm of irrigation is enough to achieve fruits with superior quality. Nonetheless, they have recognized that the rootstocks/cultivar answer to watering regimes would be highly variable, and it is very difficult to establish the water amount per year necessary to reach high-quality fruits. Moreover, this amount will be highly variable with local climate conditions.

In addition to the levels of water, the water quality must also be considered because the cherry tree is sensitive to salts (Grattan, 2016), and salinity can be responsible for loses in yield. Kotuby-Amacher et al. (2000) recorded a loss in cherry yield between 10% and 50% when salinity in water and soil varied from 1.3 to 2.5 dS m^{-1} and from 1.9 to 3.1 dS m^{-1}, respectively.

6.13 NUTRIENT MANAGEMENT

Nutrient management in cherries has been studied extensively over the past decades by several authors (Hanson; 1991; Brown et al., 1995; Baghdadi and Sadowski, 1998; Neilsen et al., 2004, 2007; Wójcik and Wójcik, 2006; Azarenko et al., 2008; Lang et al., 2011; Ouzounis and Lang, 2011; Wójcik and Morgaś, 2015), and based on these studies, it was bpossible to find different nutrient recommendations. In general, it is consensual that any fertilization regime is highly dependent on orchard location, soil type, climatic conditions, tree ages, and rootstock/cultivar genotypes, among other factors. Therefore, it is impossible to recommend a general formulation for all cherry orchards and like in other fruit trees, nutrient requirements per unit area in cherries should always be calculated based on the soil fertility and nutrient removal capacity of leaves and fruits. Thus, accurate analysis of the soil fertility, before and after plant setting, and an analysis of the levels of nutrients in leaves and fruits should be done before any foliar nutrient application. Nevertheless, based on the literature (Hanson and Proebsting, 1996; Nielsen et al., 2017), it is possible to find some boundaries of optimal nutrient levels in which sweet cherry can have good vegetative growth and optimal fructification. Some nutrient concentration ranges considered normal, and only when nutrients are below this range, we must consider any nutrient application (soil or foliar) (Table 6.15).

TABLE 6.15 Nutrient Concentration Range in Leaves and Fruits of Sweet Cherry

Nutrient	Range	
	Leaves	Fruits
Nitrogen	1.9%–3%	110–190 mg/100 g
Phosphorous	0.16%–0.40%	15–25 mg/100 g
Potassium	1.3%–3%	120–220 mg/100 g
Calcium	1%–3%	7–14 mg/100 g
Magnesium	0.3%–0.6%	7–12 mg/100 g
Boron	25–60 ppm	0.2–0.5 mg/100 g

Source: Hanson and Proebsting (1996) and Neilsen et al. (2017).

Also, the needs of nutrients are highly variable with the tree physiological stage, density and training systems (James, 2011) which turns almost impossible to define an accurate nutrient doses along the cherry tree cycle without any analysis to soil, leaves, and fruits. However, some authors have estimated some recommendations for the best cherry growth. For orchards with a density of 300–700 trees/ha, the Haifa's group (Anonym, 2018) for recommended a total annual dose of 100 kg/ha of nitrogen, 55 kg/ha of phosphorus, 130 kg/ha of potassium and 15 kg/ha of magnesium. Recently, Neilsen et al. (2016) for the "Skeena" cultivar recommend 27.5 kg/ha of nitrogen, 33.3 kg/ha phosphorous, 33.3 kg/ha potassium and 0.28 kg/ha of boron, on a year basis. Tough, despite these recommendations, the prudence is the key to avoid any excessive nutrient application and any fertilization regime adopted should be supported by an accurate analysis of the soil, leaves or fruits.

6.14 TRAINING AND PRUNING

Besides the soil type, rootstock-cultivar combination, tree density, and spacing, training and pruning systems are critical to achieving high-fruit quality success. Like in other tree fruits, cherry trees are pruned for a number of reasons including:

1. to establish the tree architecture
2. to limit the tree size

3. to increase the sunlight interception;
4. to promote air circulation;
5. to control their vegetative growth, and
6. to increase the quality of fruit production through the right balance between vegetative growth, flower production, and fruit set.

According to Verma (2014), pruning allows a good tree size and shape, good light distribution in the canopy, and good balance between vegetative and reproductive growth. Pruning is a dwarfing process and consists in the removal of tree parts such as shoots, spurs, leaves, roots, or the apical part, to control the size, its structure, and its vegetative growth. In general, pruning is most frequently done during winter (commonly referred to as dormant pruning) or may be done in summer (also known as summer pruning), but always when the environment is conducive to a proper wound healing and disease pressure is low. Although pruning is fundamental to correct or maintain tree structure and to get maximum returns from plants, it is crucial to avoid any excessive large cuts because they can result in excessive, disproportionate stimulation of sprouts near the cut, leading to an excessive fruitfulness with low fruit size.

Almost simultaneously to pruning, training is the first step in any plant production. In this practice, tree growth is directed to the desired shape developing a strong structure of support branches. The major goal of tree training is to control tree growth and minimize cutting procedures. Different innovative ideas for cherry training have been developed in the last decades (Robinson and Domínguez, 2014; Lang, 2015; Long et al., 2015; Law and Lang, 2016; Ayala and Lang, 2017). Based on these compilations, the most common and actual training systems used in cherry production worldwide are (Figure 6.5):

1. the Kym Green Bush (KGB), based on a free-standing tree with a multiple temporary vertical fruiting units;
2. the Spanish Bush (SB), which is very similar to KGB, but fruits are produced only in small laterals that are renewed regularly;
3. the Steep Leader (SL), in which only a tree of four vertical leaders emerge from based and the fruits are produced on lateral branches (each leader mimics a one-sided spindle tree);
4. the Super Slender Axe (SSA) in which the fruit grows on axillary flower buds produced in a one spindle axe;

5. the Vogel Central Leader (VCL), which is based on a pyramidal tree shape like a christmas tree;
6. the Tall Spindle Axe (TSA) is derived from VCL, in which the tree canopy develops a central trunk leader with four of five vigorous lateral branches;
7. the Upright Fruiting Offshoots (UFO), which is a fully trellised system creating a narrow fruiting wall and like the KGB system, produces fruit in renewable vertical fruiting units, which arise from a low horizontal trunk; and
8. the "Y" Trellis (UFO-Y), which is a derivation of UFO and consists in a dual-plane "Y" canopy architecture (UFO-Y) that is developed and managed similarly to single-plane UFO vertical trellis.

According to the authors cited above, both KGB and SL systems are freestanding, require no trellis, but they need a spacing of 1.9–2.4 m between trees and 4.3–5.5 m between rows. In these systems, the trees achieve a moderate tree vigor, and they can be used easily with full size or semidwarfing rootstocks, and they produce semipedestrian orchards that can be harvested from small ladders. In the SL training system, the trees are quite vigorous, and the plant spacing between trees needs to be 3–4.9 m and 4.3–5.5 m between rows, depending however on the rootstock/cultivar combination. The SSA system is more adequate for dwarfing rootstocks, and the TSA is a training system suitable for dwarfing to semi-vigorous rootstocks, or vigorous rootstocks used on weak soils, with the distance between plants being 1.5–2.4 m between trees and 3.4–4.3 m between rows. Instead, with the VCL training system, the plant spacing could be 1.8–2.7 m in the row and 3.9–4.9 m between trees and is more adequate for dwarfing or semidwarfing rootstocks because with a vigorous rootstock, without any manipulation, the trees will be very tall, which is undesirable. Finally, both UFO and UFO-Y, a narrow, precocious, and easy to prune and harvest fruiting walls, are produced and offer the advantage of greater yields due to higher light interception, as well as the potential for mechanical or mechanical-assisted harvesting. Nonetheless, although these two systems are easy to prune and to maintain at maturity, the establishment of the trellized canopy is more time-consuming and costlier than other systems (Long et al., 2015).

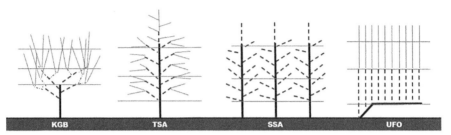

FIGURE 6.5 Examples of some cherry training systems.
Source: Adapted from Lang (2015).

6.15 INTERCROPPING AND INTERCULTURE

Management of sweet cherry orchards is primarily focused on achieving high yields of good fruit quality through balancing reproductive and vegetative growth (Measham et al., 2012). In this aspect, the conservation of soil is one of the most relevant aspects to be considered when planning a new orchard. As already referred, sweet cherry is very sensitive to soil drainage, and in water-logged soils or the presence of limited soil oxygen, its yield is restricted (Long and Kaiser, 2013) and, in some occasions, these could lead to tree's death (Verma, 2014). This situation can be avoided by implemented management strategies to improve soil health (Nielsen et al., 2017) that conserves moisture and moderate temperature to improve the tree growth and yield of sweet cherry orchards (Bhat, 2015). One such strategy consists in using temporary cover crops, which reduce soil erosion and compaction, aerate the soil, and keep the orchard cool (Long and Kaiser, 2013). Current practices generally involve sowing of many types of plants, with legume or grasses (including cereals and clovers) used most extensively (Fogle et al., 1973; Koumanov and Long, 2017). It will be a good practice to mix species to provide diversity, durability, and reduce the incidence of pest mite populations (Epstein and Landis, 2003). Such plants are preferably used between orchard rows (Figure 6.6) since when grown within the tree row, they can compete with sweet cherry for water and soil nutrients, decreasing its growth and yield (Granatstein and Sánchez, 2009). Moreover, sweet cherry trees intercropped with more demanding crops like pea, red clover or French bean could be advantageous for fruit production and fruit quality (Bhat et al., 2016). However, it will be necessary to pay attention to the use of such legume species since they provide nitrogen to the soil and could imbalance its nutrient pool (Epstein and Landis, 2003). On the other hand, intercropping sweet cherry with other crops such as strawberry, cabbage, oats, and maize did not produce the same results (Bhat et al., 2016).

FIGURE 6.6 Cover crops between tree rows.

Most of the operations realized in a cherry orchard aim to reduce weed competition (Koumanov and Long, 2017). To achieve this, mechanical technique, chemical technique, or a combination of them can be used. Tillage is one such technique widely applied before planting (Long and Kaiser, 2013). The application of herbicides, although it is the current practice, should be used with caution. In fact, the effects of herbicides on the soil after the application may last up to four years with negative impacts on the soil health, and hence plant's growth and yield (Long and Kaiser, 2013). The use of herbicides is currently very much conditioned to the type of production practiced, organic or conventional farming. An alternative to herbicide used in weed control consists in the application of organic mulches (Yin et al., 2007). This management strategy could offer some benefits to the soil (Verma, 2014), enhance fruit yield, and reduce water consumption (Yin et al., 2007), but their use is not consensual (Verma, 2014).

6.16 FLOWERING AND FRUIT SET

In general, sweet cherry starts flowering from early to late spring, creating stunning landscapes plenty of fragrant white or pink flowers. Among stone fruits, sweet cherry presents a rapid transition, about three days, between the onset of flowering and the full flowering (Radičević et al., 2011) (Figure 6.7). The occurrence of flowering in this species is largely dependent on the genotypic characteristics of cultivars (Moghaddam et al., 2013). Based on the flowering time, sweet cherry cultivars can be classified into several groups (Table 6.16) from very early to very late flowering, with a difference of 3–9 days between the earliest and the latest flowering cultivars (Hodun and Hodun, 2002).

FIGURE 6.7 Flowering and fruit setting in sweet cherry.

TABLE 6.16 Time of Flowering in Sweet Cherry Cultivars

Early Flowering	Mid-early Flowering	Mid-late Flowering	Late Flowering
Burlat	Compact Stella	Bing	Drogan's Yellow
Lyons Early	Emperor Francis	Germersdorfer	Inge
Moreau	Merchant	Hedelfingen	Lambert
Napoleon	Van	Kordia	
Primulat	Vista	Lapins	
Rita		Stark Hardy Giant	
Souvenir		Summit	
		Sunburst	
		Vega	

Source: Radičević et al. (2011), Sansavini and Lugli (2005), and Kappel (2008).

The ecological and climatic conditions of the local (Garcia-Montiel et al., 2010; Garcia et al., 2014) as well as the year of observation (Letch et al., 2008; Garcia et al., 2014; Radičević et al., 2015; Wenden et al., 2016) may also influence the timing of onset and duration of flowering. As mentioned, an adequate winter chilling in terms of duration and intensity is one of the most important factors to control flowering time in sweet cherry (Mahmood et al., 2000). However, the increased spring and winter temperatures anticipated in climate change scenarios for many regions of the world (IPCC, 2013) would inevitably affect its phenology, growth, and production. Recent works (Drogoudi et al., 2017 and Meland et al., 2017) have reported that sweet cherry tends to flower earlier than it used to in the past decades, and this could be attributed to the warmer temperatures observed during the monitoring period. The advance in the flowering date is making sweet cherry very prone to frost damage (Matzneller et al., 2016), which can result in flower bud death and, consequently, in production loss (Salazar-Gutiérrez et al., 2014). Moreover, it will have serious implications on its fruit set which is, in turn, directly influenced by the beginning, sequence, flow, duration, abundance of flowering (Akšić

et al., 2017), and by the weather conditions in the winter or during the flowering period (Glowacka and Rospara, 2014). As fruit setting takes place once a flower is pollinated, it is crucial that flowers receive compatible pollen. This is particularly important in sweet cherry cultivars, considering that most of them are self-incompatible and intercompatible, meaning that they require another cultivar for effective pollination and fertilization in order to achieve an abundant fruit set (Wang et al., 2010; Radičević et al., 2011; Glowacka and Rospara, 2014). In this case, the pollinizer must be from a different compatibility group and should flower at the same time (Long et al., 2007). Glowacka and Rospara (2014) in a study with sweet cherry cultivars very common in Poland ("Regina," "Sylvia," "Techlovan," and "Vanda") concluded that for "Vanda" the best pollinator was "Techlovan," for the "Regina" was "Sylvia," and for the "Techlovan" was "Vega." Self-incompatibility in sweet cherry is determined by multi-allelic S-locus (Radičević et al., 2011; Lisek et al., 2015) that encodes the pollen and pistil determinants (Cachi et al., 2014). So, knowledge of S-genotypes of sweet cherry cultivars is important not only for breeding purposes but also for selecting appropriate cultivars for cross-pollination and fruit set (Verma, 2014; Sharma et al., 2015). Although not very common, self-compatibility is also observed in sweet cherry cultivars (Marchese et al., 2007). Among commercial sweet cherry cultivars, "Columbia," "Blaze Star," "Isabella," "Lapins," "Skeena," "Stella," and "Sweetheart" are self-compatible (Lon et al., 2007; Liu et al., 2018). Self-compatibility is considered a desirable trait in cherries cultivars since it has many horticultural advantages (Choi and Anderson, 2001): it is less dependent on bee activity, it can be used as universal pollen donors (Verma 2014), it facilitates orchard management, and it aids regular production (Marchjese et al., 2007).

6.17 FRUIT GROWTH, DEVELOPMENT, AND RIPENING

Cherry is a stone fruit and comprised of three parts (Figure 6.8A): epicarp, fleshy mesocarp, and lignified endocarp (stone or pit), with the associated seed (Tukey and Young, 1939). The stages of fruit growth are characterized by a double-sigmoid pattern consisting of three distinct phases (Flore, 1994; Edin et al., 1997): the initial phase corresponds to the exponential growth of a fresh mesocarp due to cell division, and this process will determine in large part the size of the fruit (Stage I); the lag phase corresponds to stone hardening, and this phase is much longer as is the fruit later (Stage II); and the maturation phase represents the thickening of the fruit (Stage III) (Figure 6.8B). In Stages I and III of the fruit growth, the water deficit should be minimized to obtain bigger fruits (Flore and Layne, 1999).

Irrigation and nutritional aspects should be balanced to avoid the loss of firmness and reduction on soluble solids in the fruits (Crisosto et al., 1995). As referred, trees with adequate vigor and light penetration in the canopy by summer pruning (Patten et al., 1983) and a high leaf area-to-fruit ratio (Patten et al., 1983; Whiting and Lang, 2004) are important procedures to ensure the production of high-quality fruits. The most important conditions that affect fruit weight are the tree location in the orchard, fruit location in the tree, and the fruit-to-leaf area ratio (Drake and Fellman, 1987; Flore and Layne, 1999). The fruit size and sweetness decrease when plant density increases (Eccher and Granelli, 2006). Also, this fruit is highly perishable and greatly affected by environmental conditions, as excess rainfall before harvest. Rainfall and high humidity during harvesting are a serious problem to sweet cherry production in most producing areas of the world, causing significant economic loses (Simon, 2006).

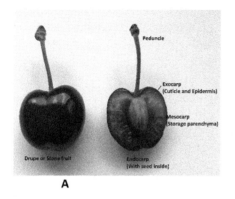 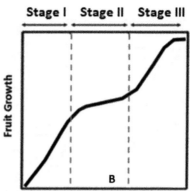

FIGURE 6.8 Illustration of (A) ripe fruit and (B) stages of the fruit growth.

Sweet cherry is a nonclimacteric fruit because during maturation there is no marked peak of ethylene production (Li et al., 1994). Studies proved that fruit respiration, loss of firmness, and change in fruit skin color occur independently of ethylene (Li et al., 1994; Mozetič et al., 2006). Recently, abscisic acid has been studied on nonclimacteric fruit ripening, like sweet cherry, as this acid has been associated with fruit-maturation processes such as softening, and sugar and anthocyanin accumulation (Luo et al., 2014; Ren and Leng, 2014; Shen et al., 2014). During fruit development and maturation, the color changes from green to red. This change is originated by the loss of chlorophylls and the accumulation of anthocyanins in fruit, which represents the most significant indicator of quality (Esti et al., 2002; Gonçalves et al., 2007). Light intensity, temperature, and fruit maturity promote an increase in the compositional constituents, like sugars, total phenolic, and anthocyanin content in sweet cherries (Serrano et al., 2005; Karlidag et al., 2009; Ferretti et al., 2010). Maturation and ripening are physiological factors that influence the cherry texture. Pectinmethylesterase (PME), polygalacturonase (PG), and β-galactosidase (β-Gal) are enzymes responsible for textural changes in fruits. The solubilization of cell walls is caused by the joint activity of PME and PG, leading to an increase in the fruit softening (Remón et al., 2003).

6.18 FRUIT RETENTION AND FRUIT DROP

Fruit drop consists of the detachment of fruits from a branch by the separation of a layer of cells (Celton et al., 2014) through physiological, molecular, and biochemical processes (Costa et al., 2006).

Sweet cherry as other fruit tree species sets an abundance of flowers leading to a production of high numbers of fruit set, more than it can sustain (Celton et al., 2014). Considering the double-sigmoid pattern of fruit growth (Flore, 1994; Edin et al., 1997), it is expected that such a quantity of fruits is beyond ideal. When it happens, there is an intense competition among fruitlets for carbohydrates (Measham et al., 2012) and the tree develops, as an effort to make an efficient use of resources, a physiological drop (Giulia et al., 2013). This drop, very common in early stages of fruit growth, can be viewed as a natural and self-thining mechanism that allows trees to produce high-quality fruits (Nartvaranant, 2016). Hormonal imbalances, namely the production of ethylene (Thomas et al., 1993; Li et al., 2017), environmental factors such as high temperature, rain, and insufficient light (Davarynejad et al., 2014), and genetic factors (Li et al., 2017) are frequently associated with such fruit abscission. Although this premature drop of fruits reduces the number of fruits that can reach maturity, it is not always vigorous to attain the desired fruit quality (Giulia et al. 2013). In these circumstances, a supplementary fruit thinning will be necessary in order to increase fruit retention (Nartvaranant, 2016) and to get fruits of good marketable size and maximum commercial yield (Costa et al. 2006). Manual, mechanical, and chemical techniques of fruit thinning can be applied in sweet cherry orchards (Milić et al., 2015). Among them, manual removal of fruits, although possible, is almost impractical due to the higher cost involved (Whiting and Ophardt, 2005; Whiting et al., 2006). Mechanical fruit thining can be done by a wide variety of equipment (Rosa et al., 2008). Chemical fruit thining consists of chemical sprays of different substances, such as ammonium thiosulfate, vegetable oil emulsion, or gibberellins in the current or previous growing season (Whiting et al., 2006). The efficacy of these methods depends on several factors, mainly climatic conditions (Milić et al., 2015).

Contrary to the early fruit dropping, preharvest fruit drop is a serious problem in many fruit species, including the sweet cherry (Davarynejad et al., 2014) since it causes significant crop losses (Raja et al., 2017). One effective way to control this detrimental process consists of the application of the same plant growth regulators as it occurs in the early fruit drop, that is, gibberellins (Zhang and Whiting, 2013).

6.19 HARVESTING AND YIELD

Knowledge of the ripening process in sweet cherry is essential for the determination of optimal time for harvest, reflecting directly on the sweet cherry quality and customer satisfaction. Fruits should be harvested when fully ripe to achieve good eating quality. Therefore, the harvest may be cautious, protecting the fruit of heat due to the high respiratory rate, and consequently rapid fruit decay (Alique et al., 2003).

Ripening studies of sweet cherry have focused on the identification of optimum maturity stage, taking into account sweet cherry flavor, color of skin, soluble solids, TA, and firmness (Mozetič et al., 2004; Gonçalves et al., 2004a, 2007; Serrano et al., 2009; Serradilla et al., 2010).

Sweet cherries show biochemical and morphological changes during maturation, namely increase in fruit weight, soluble solids, phenolic compounds, anthocyanins, and antioxidant activity that affect and enhance the skin color intensity and decrease in firmness. A recent study established a relationship between sensory evaluations and analytical measurements, to develop guidelines for acceptable sweet cherry fruit firmness (Hampson et al., 2014).

Sweet cherry harvesting is currently performed manually (Figure 6.9) without any mechanization of the process. This type of harvesting has already been used in cherries for processing, but fruits that are to be placed in the fresh market should be of higher quality standards when reaching the consumers.

 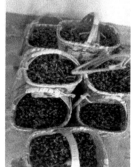

FIGURE 6.9 Manually harvesting of sweet cherries.

6.20 PROTECTED/HIGH-TECH CULTIVATION

Sweet cherry production by year is highly dependent on several factors, including the weather conditions. For example, excessively high humidity and preharvest rain-induced physiological disorders such as fruit cracking (Meland et al., 2014) make trees more susceptible to infections by diseases and pests (Lang, 2009). These problems occurred in many regions of the world where sweet cherry is cultivated and is responsible for substantial economic losses (Balbontin et al. 2013), thereby negatively affecting incomes. Mucha-Pelzer et al. (2006) confirmed that, in Germany, rain and hail during ripening and harvesting season caused yield losses up to 90% in sweet cherry. Nowadays, the big challenge in cherry production is to reduce costs and the labor requirements (Correia et al., 2018) and, at the same time, to improve sweet cherry production combined with high-quality fruit. To overcome these challenges, growers need to adopt new technologies and cultural practices that could reduce or mitigate such problems and strengthen marketing options.

Sweet cherry is one of the fruit trees that is currently being grown under protective environment structures, such as pole and cable covers, high tunnels, or greenhouses (Lang, 2009). These systems are well recommended since this fruit tree has the highest value as well as the highest production risks when compared with the characteristics of other tree fruits of temperate regions (Lang, 2013). The use of rain covers provides a significant risk reduction of fruit cracking in climates with frequent precipitation (Sotiropoulos et al., 2014), acts as a protection from frosts (Lang, 2009), controls fruit ripening (Lang, 2009), benefits plant physiology (Lang, 2014; Sotiropoulos et al., 2014), and improves fruit quality (Børve et al., 2003). Moreover, it also might act against the development of pests and plant diseases (Børve and Stensvand, 2003; Lang, 2009). For example, the incidence of the bacterial canker is substantially reduced in cherry trees produced under high tunnels when compared to those outside the tunnel (Lang, 2009). Also, rain covers can modify crop microclimate, especially air temperature, which constitutes an advantage in sweet cherry production, enhancing earliness and profitability (Børve et al., 2003; Lamont, 2009). Finally, it facilitates field operations and their schedule (Lang, 2014).

The use of rain covers in sweet cherry, especially high tunnels (Figure 6.9), is well disseminated in many European countries (Børve et al., 2003) and American states (Carey et al., 2009), but their implementation is generally very expensive. Therefore, decisions about the design of structure, type of cover (Lang, 2009), and time of covering (Børve et al., 2003), among other factors, are crucial for successful sweet cherry production. In Figure 6.9, there is an example of a high-tunnel structure used in sweet cherry production. Nowadays, there is an increasing demand for intensive sweet cherry production (Koumanov et al., 2018), which is based on smaller trees in denser orchards. As already mentioned, besides the selection of the most appropriate rootstock (Pal et al., 2017) and training system (Radunic et al., 2011), irrigation and fertilization are the key elements in such production system (Koumanov et al., 2018). Concerning irrigation and fertilization, the trees require slow and small applications of water and/or nutrients directly in the root zone, which can be achieved by precise systems such as microirrigation and fertigation (Koumanov et al., 2018).

FIGURE 6.10 High-tunnel structure used in sweet cherry production.

6.21 PACKAGING AND TRANSPORT

Once harvest, sweet cherries are highly perishable, prone to mechanical damage, and their shelf life is limited. Cherries quickly present weight and water loss, changes in sugar–acid balance, bruising of the skin, color changes, softening, surface pitting, and stem browning (Bernalte et al., 2003; Alique et al., 2005). Glossy sweet cherry fruits are considered fresh by consumers, and the glossy appearance deteriorates fast after harvest. Hence, the optimization of storage and transport parameters are factors that can extend fruits' shelf life and thus allows long-distance transportation to the consumer (Remón et al., 2003; Martínez-Romero et al., 2006).

Diverse preservation technologies have been assessed to reduce the deterioration of sweet cherries, to match the increased needs of the current global market. The effects of temperature, relative humidity (RH), and changing the levels of CO_2 and O_2 during controlled atmosphere (CA) storage and in consumer packaging using modified atmosphere packaging (MAP) have been studied (Wani et al., 2014). MAP technology has gained acceptance, mainly in countries with a large cherry production. This technology enables producers to enhance the quality of the cherries during storage and/or transport and retain stem color, flavor and texture of the fruits (Kahlke et al., 2009). Therefore, the fruits may reach the consumers of long-distance markets, with high quality and thus increase consumption. Kader (2001) indicated that the key factors affecting cherry quality during post-harvest storage are temperature and RH. Storage temperature should be constant and affect the postripening behavior, transpiration, senescence, and other physiological mechanisms (Romano et al., 2006). In general, during cold storage, anthocyanins and other phenolic compounds increased in several sweet cherry cultivars (Gonçalves et al., 2004a, 2004b, 2007; Serrano et al., 2009), although the antioxidant activity decreased, which is a chilling injury (CI) symptom (Gonçalves et al., 2004b).

MAP and CA storage have been successfully applied to prolong the shelf life of sweet cherries (Wani et al., 2014). Previous studies with MAP have been reported as effective in delaying physicochemical changes linked to the loss of quality of sweet cherries by increasing the level of CO_2 and decreasing the O_2 content (Petracek et al., 2002; Tian et al., 2004; Wang and Long,

2014). Other authors suggested that cold temperature has more effect on retard decay and senescence of "Bing" sweet cherries than the MAP application (Kupferman and Sanderson, 2001). MAP combined with low-temperature storage has proved to be effective in delaying fruit deterioration while preserving soluble solids, TA, and green color of stems and displaying higher antioxidant capacity, firmness, and brighter color skin (Alique et al., 2003; Remón et al., 2003; Padilla-Zakour et al., 2004; Lara et al., 2015). This treatment promotes an increase of anthocyanins content that can be due to the inhibition of enzyme activity favoring the stability of color (Remón et al., 2000; Conte et al., 2009). However, cherries exhibited a CI symptom, namely the rapid decrease of vitamin C content (Tian et al., 2004).

6.22 POSTHARVEST HANDLING AND STORAGE

During storage, the sweet cherry metabolism promotes changes in phenolic and anthocyanin contents, mainly influenced by cultivar, maturation of fruit stage, and storage conditions (Gonçalves et al., 2004a, 2004b; Mozetič et al., 2006; Serrano et al., 2009). The change of firmness in sweet cherries during storage is due to the increase in the total amount of oxidation enzymes, such as peroxidase (POX), polyphenoloxidase (PPO), pectin-methylperoxidase (PMP), and PG at harvest. These enzymes lead to the breakdown of the cell wall (Remón et al., 2003). Unsuitable storage conditions can lead to the growth and development of bacteria and fungi. These microorganisms prompt fermentation leading to the development of off-flavors due to ethanol and acetaldehyde formation, causing considerable economic losses (Esti et al., 2002). Other study evaluated the effect of hydro-cooling as a precooling treatment, on ripening-related parameters of two sweet cherry cultivars after one week of cold storage (0 °C, 95% RH). Results indicated that hydro-cooling delayed the deterioration and senescence of cherry fruit, maintaining higher quality (Manganaris et al., 2007). A recent study evaluated the effect of postharvest calcium application in hydro-cooling water on several quality attributes of "Sweetheart" and "Lapins" sweet cherry cultivars (Wang et al., 2014). This research revealed that adding $CaCl_2$ in the hydro-cooling treatments increased tissue calcium content, accompanied by reductions in the respiration rate, ascorbic acid degradation, membrane lipid peroxidation, enhanced total phenolic content, and total antioxidant capacity, and resulted in increases of fruit firmness and decreases in TA loss of both cultivars.

The combination of pre- and postharvest treatments has been also recognized as a new strategy for increasing the shelf life of cherries. Preharvest application of hexanal combined with postharvest treatment of hexanal and 1-MCP enhanced the cherry quality after 30 days of storage at 4 °C (Sharma et al., 2010). Treated cherries showed higher brightness, firmness, bioactive compounds content, and antioxidant enzymes activity. Also, new preservation technologies have emerged using MAP applying biodegradable films which should be considered safe for human and environmental friendly (Giacalone and Chiabranco, 2013; Koutsimanis et al., 2015b). Giacalone and Chiabranco (2013) reported that the use of biodegradable films was useful to preserve the quality of cherries by a delay in changes in color and losses of firmness and acidity, and that these polymers could be used in packaging of sweet cherries without negative effects on the final quality. This technology may become a practice in the commercialization of sweet cherries for the direct consumption of high-quality fresh cherries in bio-based packaging (Koutsimanis et al., 2015a).

Environmental friendly solutions for producing and preserving fresh food have been designed, namely the development of new waxes coatings, specifically, edible coatings, has recently been studied to improve food appearance and safety. Edible coatings consist in a thin protective layer that is applied to the skin fruit surface, having the capacity to reduce respiration and transpiration rates and increase storage periods, firmness retention, and decay control (Velickova et al., 2013). Nowadays, the application of edible coatings is one of the most innovative methods to extend the shelf life of fruits, like in sweet cherries (Rojas-Argudo et al., 2005; Martínez-Romero et al., 2006), and other fruits such as tomato (Liu et al., 2007), grapes (Asghari et al., 2013), peach (Ma et al., 2013), and raspberry (Tezotto-Uliana et al., 2014). Indeed, sweet cherry fruits treated with chitosan-coating showed a reduction of water loss, a delay of the skin color changes, TA, and ascorbic acid content during storage (Petriccione et al., 2014). Postharvest chitosan-coated application in cherries also allowed reduce the changes in the content of bioactive compounds.

6.23 PROCESSING AND VALUE ADDITION

Almost all cherry production is for fresh consumption; however, the use of cherries in processing markets has increased considerably in recent years (McLellan and Padilla-Zakour, 2004). A large quantity of sweet cherries are available and can be processed into other cherry products such as frozen, canned, brined, juice, nectars, concentrates, wine, dried forms, marmalade, jams, jellies, compote, fruit source, and puree (McLellan and Padilla-Zakour, 2004; reviewed by Chockchaisawasdee et al., 2016; Jensen, 2017). In addition, chemical and bioactive properties of sweet cherry stems (Bastos et al., 2015), cherry fibers from harvest residues as dietary fiber and functional food ingredients (Basanta et al., 2014), cherry seed as a source of nutritionally important fatty acids (Vidrih et al., 2014), and biodiesel from kernel oil of sweet cherry seed (Demirbas, 2016) have been reported.

Preserving cherries in individually quick freezing technique at −35 °C without the addition of sugar resulted in greater levels of anthocyanins, carotenoids, and vitamin C than the fruits processed by wet-heat or by the freeze-drying technique (Leong and Oey, 2012).

Furthermore, frozen cherry fruits can be included in yogurts and ice creams. For canning cherries, the application of heat at pasteurization and sterilization temperatures is necessary, together with syrup (McLellan and Padilla-Zakour, 2004). However, this process leads to the loss of bioactive compounds, such as anthocyanins and other phenolic compounds during storage (Chaovanalikit and Wrolstad, 2004).

Regarding brined cherry, fruits are packed in a solution containing 1%–1.5% sulfur dioxide (McLellan and Padilla-Zakour, 2004). Nevertheless, the undesirable breakdown of the fruit and the loss of anthocyanins and other phenolic compounds in preserved cherry fruit have been described (Chaovanalikit and Wrolstad, 2004; McLellan and Padilla-Zakour, 2004). Sweet cherry juices are made from fresh fruit or frozen, stored fruit. A typical process for the production of cherry juice consists in a blanching at 85 °C for 3 min in a water bath with the aim to reduce microbial surface and the activity of PPO and POX enzymes (McLellan and Padilla-Zakour, 2004). Moreover, new processing methods have been investigated as a pulsed electric field (PEF) and high hydrostatic pressure with a significant effect on the inactivation of microorganisms (Altuntas et al., 2010).

Functional drinks containing cherry extracts rich in bioactive compounds are highly demanded by consumers (Sun-Waterhouse, 2011). Also, the use of natural dyes has received extensive acceptance by consumers (He and Giusti, 2010). Natural colorants and dyes using anthocyanins from cherries have been isolated, identified, and quantified by HPLC and HPLC-MS (He et al., 2010). Cherry fruit has high sugar content, suitable for fermentation, but with low acidity to give a balanced taste. Taste-active compounds of cherry contribute to assure a satisfactory fermentation, stability, and the desired alcohol content in the finished wine (Niu et al., 2012; Sun et al., 2012), with a significant contribution to characteristic taste or mouthfeel of quality cherry wine. Fully ripe cherries can be dehydrated by traditional drying method that consisting of convective air drying where fruit placed in single layers on trays or shelves are exposed to heat (McLellan and Padilla-Zakour, 2004; Jensen, 2017). Very large diversity and variety of marmalades, jams, compotes, and jellies of sweet cherry can be found in markets. However, heat pasteurization of bulk jam traditionally reduces anthocyanins, phenols, and vitamin C contents (Poiana et al., 2011). Further studies of alternative methods without a long-time cooking and with low-temperature pasteurization need to be investigated.

6.24 DISEASES, PESTS, AND PHYSIOLOGICAL DISORDERS

Sweet cherry like other fruit crop is a challenging one because it is susceptible to the occurrence of many critical damaging insect pests, plant diseases, and physiological disorders (Quero-García et al., 2017), as well as to fruit damages occurring during pre- and postharvest handling (Alique et al., 2005).

These problems are one of the main causes of yield losses and reduced fruit quality when they are not well managed.

Cherry crinkle leaf and deep suture are two of the most common genetic disorders that affect leaves and fruits of sweet cherry. Both disorders cause serious problems in the "Bing" and "Black Tartarian" cultivars and are responsible for substantial yield losses (Southwick and Uyemoto, 1999). In general, leaves appear narrower than normal, and the production of fruits is very low or inexistent, in the presence of a strong cherry crinkle leaf, or with abnormal shapes (Southwick and Uyemoto, 1999). As both disorders are related to climatic conditions and management practices, it is recommended to verify the appearance of symptoms in the beginning of spring.

The occurrence of other physiological disorder such as double fruit (Southwick and Uyemoto, 1999; Engin et al., 2009) also affects the acceptance of fruits by the consumers. The incidence of this disorder is related to warm conditions, and the adoption of management practices that can alter the microclimate could be viewed as a preventive method (Engin et al., 2009).

Cracking of sweet cherry fruit is a physiological disorder characterized by a splitting of the outside layer. Three types of cuticular cracking of cherry fruit can occur: stem end, apical end, and deep side (Christensen, 1972). In spite of many research studies have been carried out, the fundamental mechanism involved in this complex phenomenon has not yet been completely understood. The climate change scenarios indicate an increase in the duration of excess rainfall before harvest (IPCC, 2013) that may lead to greater cherry cracking and, at the same time, could affect the appearance and quality of the fruits (Kays, 1999; Simon, 2006). With the aim to reduce this physiological disorder, several strategies are described, mainly rain protection of sweet cherry fruits with water-resistant covers and the application of minerals or chemicals that delay or reduce water uptake to fruit, increase transpiration of free water from the fruit surface, or improve fruit skin properties (Børve and Meland, 1998; Richardson, 1998; Sekse, 1998; Whiting and Lang, 2004; Whiting and Ophardt, 2005; Usenik et al., 2005, 2009; Lenahan et al., 2006; Kappel and MacDonald, 2007; Thomidis and Exadaktylou, 2013; Wójcik et al., 2013; Sotiropoulos et al., 2014).

Sweet cherry is also a major and minor host of a wide variety of diseases (mainly fungal and bacterial) and insect pests, with a significant impact on its growth, production, and fruit quality. The diseases and

Sweet Cherry

insect pests that regularly occurred in sweet cherry orchards are summarized in Table 6.17.

Control of insect pests and diseases is an exigent task. In fact, the higher consumer's pressure to reduce the usage of broad-spectrum pesticides in agriculture makes the management of pests and diseases in cherry more challenging, particularly in adopting measures within an organic and sustainable production point of view.

Despite this pressure, cherry pests and diseases can be properly managed using a combined use of cultural practices, sanitation measures, resistance, and chemical sprays (Douglas, 2003; Papadopoulos et al., 2017). As referred, appropriate cultural practices include planting cherries in adequate soil and climate conditions, fertilizing with adequate nutrients, and properly correct pruning methods and other practices that minimize eventual tree stresses (James, 2011). Sanitating involves the eradication of parts

of plants infected with pests and diseases in order to remove important sources of inoculum for the next season (Gilligan, 2008). Methods of resistance involve the selection of cultivars with genetic resistance to specific diseases (Johnson, 1983), while the use of chemicals, in some situation, is the only option available, which involves the proper and accurate selection and the adequate timing of its application for specific pests or diseases (Belien et al., 2013; Lux et al., 2016). However, for any measure to be effective, the first step is to know the problem (pest, disease, or physiological disorder) we are facing. Therefore, any inform that provides an overview of the most important cherry problems is necessary and welcome. Tables 6.18 and 6.19 contain some relevant information about symptoms, predisposal factors, damages, and measures used to control the most common diseases and insect pests that occur in sweet cherry.

TABLE 6.17 Most Common Diseases and Insect Pests of Sweet Cherry (The Order/Family of the Causal Organism of Diseases and Insect Pests are Given in Parentheses)

Plant Infection Zone	Diseases	Insect Pests
Rootstock	• *Phytoplasma* Organism (Acholeplasmatales/ Acholeplasmataceae)	
Roots	• *Agrobacterium tumefaciens* (Rhizobiales/Rhizobiaceae) • *Armillaria mellea* (Agaricales/ Marasmiaceae) • *Phytophthora* spp. (Peronosporales/Peronosporaceae)	• *Synanthedon exitiosa* (Lepidoptera/Sesiidae)

TABLE 6.17 *(Continued)*

Plant Infection Zone	Diseases	Insect Pests
Trunk	• *Agrobacterium tumefaciens* (Rhizobiales/Rhizobiaceae)	• *Synanthedon exitiosa* (Lepidoptera/Sesiidae)
Shoots/twigs/spurs	• *Eutypa lata* (Xylariales/ Diatrypaceae) • *Monilinia* sp. (Helotiales/ Sclerotiniaceae) • *Podosphaera clandestine* (Erysiphales/Erysiphaceae) • *Prunus necrotic ringspot virus* (Martellivirales/Bromoviridae) • *Pseudomonas syringae* (Pseudomonadales/ Pseudomonadaceae)	• *Frankliniella occidentalis* (Thysanoptera/ Thripidae) • *Myzus cerasi* (Hemiptera/Aphididae) • *Panonychus ulmi* (Acariformes/ Tetranychidae)
Buds	• *Prunus dwarf virus* (Lepidoptera / Bromoviridae) • *Prunus necrotic ringspot virus* (Lepidoptera/Bromoviridae)	• *Archips argyrospila* (Lepidoptera/ Tortricidae) • *Myzus cerasi* (Hemiptera/Aphididae)
Leaves	• *Blumeriella jaapii (Helotiales/ Dermateaceae)* • *Eutypa lata* (Xylariales/ Diatrypaceae) • *Monilinia* sp. (Helotiales/ Sclerotiniaceae) • *Phytophthora* spp. (Peronosporales/Peronosporaceae) • *Podosphaera clandestine* (Erysiphales/Erysiphaceae) • *Prunus dwarf virus* (Lepidoptera/ Bromoviridae) • *Prunus necrotic ringspot virus* (Lepidoptera/Bromoviridae) • *Pseudomonas syringae* (Pseudomonadales/ Pseudomonadaceae)	• *Archips argyrospila* (Lepidoptera/ Tortricidae) • *Bryobia rubrioculus* (Acariformes/ Tetranychidae) • *Caliroa cerasi* (Hymenoptera/ Tenthredinidae) • *Forficula auricularia* (Dermaptera/ Forficulidae) • *Frankliniella occidentalis* (Thysanoptera/ Thripidae) • *Myzus cerasi* (Hemiptera/Aphididae) • *Orgyia vetusta* (Lepidoptera/Erebidae) • *Panonychus ulmi* (Acariformes/ Tetranychidae) • *Tetranychus urticer* (Acariformes/ Tetranychidae)

Sweet Cherry 389

TABLE 6.17 *(Continued)*

Plant Infection Zone	Diseases	Insect Pests
Flowers	• *Botrytis cinerea* (Helotiales/ Sclerotiniaceae) • *Monilinia* sp. (Helotiales/ Sclerotiniaceae) • *Pseudomonas syringae* (Pseudomonadales/ Pseudomonadaceae)	• *Frankliniella occidentalis* (Thysanoptera/ Thripidae)
Fruits	• *Botrytis cinerea* (Helotiales/ Sclerotiniaceae) • *Monilinia* sp. (Helotiales/ Sclerotiniaceae) • *Podosphaera clandestine* (Erysiphales/Erysiphaceae) • *Prunus necrotic ringspot virus* (Lepidoptera/Bromoviridae) • *Pseudomonas syringae* (Pseudomonadales/ Pseudomonadaceae)	• *Archips argyrospila* (Lepidoptera/ Tortricidae) • *Drosophila suzukii* (Diptera/Drosophilidae) • *Forficula auricularia* (Dermaptera/ Forficulidae) • *Frankliniella occidentalis* (Thysanoptera/ Thripidae) • *Rhagoletis cerasi* (Diptera/Tephritidae) • *Rhagoletis indifferens* (Diptera/Tephritidae) • *Orgyia vetusta* (Lepidoptera/Erebidae)

Source: Gilmore (1960), Blackman and Eastop (1984), Németh (1986), Jones and Sutton (1996), Walton et al. (2010), Çağlayan et al. (2011), Uyemoto et al. (2011), Papadopoulos (2014), James et al. (2017), Puławska et al. (2017), Renick et al. (2008), Hauser et al. (2009), Fleury et al. (2010), Alford (2014), Børve (2017), Harris et al. (2017), and DeFrancesco et al. (2018).

TABLE 6.18 Description of the Main Diseases That Occur in Sweet Cherry

Causal Organism	Commom Name	Symptoms/Damages	Predisposal Factors/ Control measures
Agrobacterium tumefaciens	Crown gall	Soft and spongy galls on roots or trunks; young trees become small, and older trees always develop secondary wood roots. More aggressive in young trees and new plantings than in mature orchards.	• Spread by incorrect nursery practices, and by infested cultivation equipment, as well as by chewing insects. • Application of soil fumigants; use of resistant rootstocks; biological control.

390   Temperate Fruits: Production, Processing, and Marketing

TABLE 6.18 *(Continued)*

Causal Organism	Commom Name	Symptoms/Damages	Predisposal Factors/ Control measures
Pseudomonas syringae	Bacterial canker	Limb dieback with rough cankers, amber-colored gums, leaf spot, blast of young flowers and shoots.	• Dispersal favored by high moisture and low temperatures in spring. • Avoid injury the plant; removal of infected branches; use resistant rootstocks; use of health nursery material; adequate irrigation and fertilization; chemical and biological control.
Prune dwarf virus	Prune dwarf	Leaves narrower than normal with a rough texture.	• Spread by grafting with infected wood, via budding with infected buds and eventual by thrips. • Removal of infected plants; use of thermotherapy (keeping cultures at 38 °C for at least 20 days).
Prunus necrotic ringspot virus	Cherry rugose mosaic	Yellowing, browning, and round holes on leaf tissues; fruits deformed and may ripen later than normal; distorted growth, death of twigs, buds, or young foliage, and stunting of trees.	• Spread by grafting with infected woods or infected buds, or even seeds in nurseries and by Thrips. • Removal of infected plants; use of thermotherapy (keeping cultures at 38 °C for at least 20 days).
Phytoplasma organism	X-disease	Infected trees produce rocky, skinned and pale fruit. On some resistant rootstocks, the symptoms rarely appear but suddenly the tree wilt and collapse above the graft union. The symptoms are more evident at the harvest time.	• Spread by budding or grafting. • Removal of infected plants; use of resistant rootstocks; chemical control.

Sweet Cherry

TABLE 6.18 *(Continued)*

Causal Organism	Commom Name	Symptoms/Damages	Predisposal Factors/ Control measures
Armillaria mellea	Armillaria root rot	White to yellowish mycelia between the bark and the wood; dark brown to black rhizomorphs are often visible on roots surface; premature tree decline and death of the plant.	• Basidiospores spread by the wind. • Removal of infected plants; use of regular cultural practices.
Blumeriella jaapii	Cherry leaf spot	Leaves with round purple to brownish spots on the upperside and white or red spots in the other side; yellowish of leaves that drop prematurely. Defoliation causes loss of vigour, winter-hardiness of buds and wood, reduction of growth and yield.	• Spread by rain and wind; the fungus overwinters in infected leaves on the orchard floor. • Removal of all infected leaves, chemical control.
Monilinia fructicola, M. laxa	Brown rot	Dead of blossoms, spurs, flowers, and leaves; twigs with cankers.	• Moderate temperatures (about 14 °C–25 °C) favor the infection, as well as the rain and humidity during blossom. • Removal of fruit mummies and pruning infect or dead twigs; chemical control.
Phytophthora spp.	Phytophthora root rot	Trees collapse and die soon after the first warm weather of spring; leaves wilt, dry, but remain attached to the tree; reduction of tree growth and early senescence and leaf fall.	• Saturated soil favors infections. • Good soil drainage and more frequent, but shorter irrigations could prevent or reduce the risk of infection; use of resistant rootstocks.

TABLE 6.18 *(Continued)*

Causal Organism	Commom Name	Symptoms/Damages	Predisposal Factors/ Control measures
Podosphaera clandestina	Powdery mildew	White spots on leaves, shoots, or fruits; reduction of fruit quality by flesh degradation. Particularly severe on new plants.	• Spread favored in years of low rainfall, high humidity, and warm temperatures (21 °C–27 °C). • Pruning for good air circulation; removal of infected plants; chemical control.
Botrytis cinerea	Gray mould	Wither of calyx and flower petals; brown lesion when the fruit starts to develop and quickly spreads over the entire fruit.	• Spore dispersal favored by rain and wind; dead parts of the plant, fruit mummies on the orchard floor and other organic matter such as dead or senescent weeds, can be host its spores. • Postharvest applications of calcium chloride and sodium bicarbonate; chemical control.
Eutypa lata	Eutypa dieback	Bark has a dark discoloration with amber-colored gumming and the infected areas in the interior of the wood are discolored brown.	• Preferential infection period during the rainy days; the greatest incidence in the winter, but it can occur in any time of the year, particularly through fresh pruning wounds. • Removal of infected plants; chemical and biological control.

Source: Adapted from Grove (1991), Jones and Sutton (1996), Southwick and Uyemoto (1999), Schuster and Tobutt (2004), Walton et al. (2010), Çağlayan et al. (2011), Uyemoto et al. (2011), Papadopoulos (2014), James et al. (2017), and Puławska et al. (2017).

Sweet Cherry

TABLE 6.19 Description of Principal Insect Key Pests That Occur in Sweet Cherry and Respective Details

Causal Organism	Common Name	Damages	Control Measures
Rhagoletis cerasi	Cherry fruit fly	The damage is provoked by larval feeding in the mesocarp of the infested fruits and usually followed by secondary bacterial and fungal infections. Late fruit infestation is not always easy to detect. Oviposition punctures inflicted immediately before harvest are difficult to detect in the field, and frequently the infestation only becomes visible on grocery shelves or in consumer's houses.	Chemical (contact insecticides or systemic pesticides), sanitary (removal and destruction of infested fruits), biological (natural enemies such as nematodes) control.
Drosophila suzukii	Spotted-wing drosophila	The main damage is on the pulps of fruits. A collapse of fruit tissues starts to begin around the feeding site causing a visible depression on the fruit. In addition, an oviposition scar exposes the fruit to secondary attack by pathogens and other insects, which may cause rotting.	Chemical (insecticides), sanitary (removal and destruction of infested fruits), biological (natural enemies such as parasitoids) control. Use of mixed traps or lamp traps is other ways to control this insect pest.
Myzus cerasi	Black cherry aphid	Curling and distortion of leaves, particularly on young trees. Dense colonies in tree apices. Deformation of shoot growth and formation of open galls.	Chemical (contact or systemic insecticides), cultural (by banding the tree base with glue) and biological (predators such as lady beetles and parasitic wasps) control.
Archips argyrospila	Fruittree leafroller	Larvae feed on leaves and buds, webbing them together to form a protective case. Fruit damage is usually superficial and often occurs when leaves and fruit are affected together.	Biological (predators such as spiders and parasitic wasps), cultural (removal of rolled leaves) and chemical (oil sprays) control.

394 Temperate Fruits: Production, Processing, and Marketing

TABLE 6.19 *(Continued)*

Causal Organism	Common Name	Damages	Control Measures
Caliroa cerasi	Cherry slug	Leaves appear skeletonized and may exhibit only the fine network of veins. Reduction of fruit size.	Biological and chemical control.
Rhagoletis indifferens	Western cherry fruit fly	Pinpoint-sting marks on fruit surface; the appearance of discoloration and gummosis in fruits.	Removal and destruction of fallen and infected fruits; chemical (systemic insecticides) control.
Bryobia rubrioculus	Brown apple mite	Infestations generally confined to a few trees or localized. Chlorosis of leaves which rarely drop.	Biological (predators such as *Galendromus occidentalis* and *Galendromus occidentalis*) control.
Tetranychus urticae	Two-spotted spider mite	Mites usually found on undersides of leaves. Bronze appearance in leaves and a severe infestation exhibit some silken spots and may kill the tree.	Biological (predators such as *Phytoseiulus persimilis)* control.
Forficula auricularia	European Earwig	Irregular holes, ragged leaf edges, or both in leaves of mature trees. However, shoot-tip feeding on young trees may stunt normal growth. The damages on fruit results in narrow and irregular holes depreciating their value.	Sanitary measures (removal of nesting places of the pest), physical removal of insects by vacuuming and biological control.
Panonychus ulmi	European red mite	The damages are a little concern because rarely cause leaf drop. They remove the contents of the leaf cells as they feed, causing leaves to take on a finely mottled appearance.	Biological and chemical control.

Sweet Cherry 395

TABLE 6.19 *(Continued)*

Causal Organism	Common Name	Damages	Control Measures
Synanthedon exitiosa	Peachtree borer	The damages in young trees are severe, once it can girdle and kill. However, older trees can withstand the damage unless there are many larvae, or the tree is attacked several years in a row.	Biological control.
Frankliniella occidentalis	Western flower thrips	Fruit depression, with reddish halos in unripe fruits; brown and rough areas in mature fruits affecting the commercial value of fruits.	Cultural (by avoiding disk the orchard during tree blooming and by using cover crops) and chemical (insecticides) control.
Orgyia vetusta	Western tussock moth	Damages made by larvae are usually insignificant, but they can be serious if the feed is on the surface of fruits, resulting in shallow, dirty, and depressed areas at harvest.	Use of *Bacillus thuringiensis*.

Source: Gilmore (1960), Blackman and Eastop (1984), Németh (1986), Southwick and Uyemoto (1999), Renick et al. (2008), Hauser et al. (2009), Fleury et al. (2010), Walton et al. (2010), Uyemoto et al. (2011), Alford (2014), Papadopoulos (2014), Børve (2017), Harris et al. (2017), Puławska et al. (2017), DeFrancesco et al. (2018).

KEYWORDS

- sweet cherry
- production systems
- management
- cultural practices
- pests
- diseases
- disorders

REFERENCES

Adisakwattana, S.; Yibchok-Anun, S.; Charoenlertkul, P.; Wongsasiripat, N. Cyanidin-3-rutinoside alleviates postprandial hyperglycemia and its synergism with acarbose by inhibition of intestinal α-glucosidase. *J. Clin. Biochem. Nutr.* 2011, 49, 36–41.

Aglar, E.; Yildiz, K.; Long, L.E. The effects of rootstocks and training systems on the early performance of "0900 Ziraat" sweet cherry. *Not. Bot. Horti. Agrobot. Cluj. Napoca.* 2016, 44, 573–578.

Aksic, M.F.; Cerović, R.; Vera, R.; Bakić, I.; Čolić, S.; Meland, M. Vitality and in vitro pollen germination of different "Oblacinska" sour cherry clones. *Genetika.* 2017, 49, 791–800.

Alburquerque, N.; García-Montiel, F.; Carrillo, A.; Burgos, L. Chilling and heat requirements of sweet cherry cultivars and the relationship between altitude and the probability of satisfying the chill requirements. *Environ. Exp. Bot.* 2008a, 64, 162–170.

Alford, D.V. Pests of Fruit Crops: A Colour Handbook, 2nd edn. CRC Press, Boca Raton, FL, 2014.

Alique, R.; Martinez, M.A.; Alonso, J. Influence of the modified atmosphere packaging on shelf life and quality of Navalinda sweet cherry. *Eur. Food Res. Technol.* 2003, 217, 416–420.

Alique, R.; Zamorano, J.P.; Martínez, M.A.; Alonso, J. Effect of heat and cold treatments on respiratory metabolism and shelf-life of sweet cherry, type picota cv. Ambrune. *Postharvest Biol. Tec.* 2005, 35, 153–165.

Altuntas, J.; Evrendilek, G.A.; Sangun, M.K.; Zhang, H.Q. Effects of pulsed electric field processing on the quality and microbial inactivation of sour cherry juice. *Int. J. Food Sci. Technol.* 2010, 45, 899–905.

Andrews, P.K.; Marquez, C.S. Graft incompatibility. *Hortic. Rev.* 1993, 15, 183–232.

Anonym (2018). Fertilization of cherry trees: when and how? http://www.haifa-group.com/portuguese/knowledge_center/recommendations/fruit_trees/fertilization_of_cherry_trees_when_and_how.aspx (accessed April 5, 2018).

Asghari, M.; Ahadi, L.; Riaie, S. Effect of salicylic acid and edible coating based aloe vera gel treatment on storage life and postharvest quality of grape (*Vitis vinifera* L. cv. Gizel Uzum). *Intl. J. Agri. Crop Sci.* 2013, 5, 2890–2898.

Ayala, M.; Lang, G.A. Morphology, cropping physiology and canopy training. In: Quero-García, J., Iezzoni, A., Pulawska, J. and Lang, G. (eds.) Cherries: Botany, Production and Uses. CABI International, Oxfordshire OX10 8DE, UK, 2017, pp. 269–304.

Azarenko, A.N.; Chozinski, A.; Brutcher, L. Nitrogen uptake effciency and partitioning in sweet cherry is influenced by time of application. *Acta Hortic.* 2008, 795, 717–721.

Baghdadi, M.; Sadowski, A. Estimation of nutrient requirements of sour cherry. *Acta Hortic.* 1998, 468, 515–52.

Bak, I.; Lekli, I.; Juhasz, B.; Nagy, N.; Varga, E.; Varadi, J.; Gesztelyi, R.; Szabo, G.; Szendrei, L.; Bacskay, I.; Vecsernyes, M.; Antal, M.; Fesus, L.; Boucher, F.; de Leiris, J.; Tosaki, A. Cardioprotective mechanisms of *Prunus cerasus* (sour cherry) seed extract against ischemia-reperfusioninduced damage in isolated rat hearts. *Am. J. Physiol. Heart. Circ. Physiol.* 2006, 291, 1329–1336.

Balbontín, C.; Ayala, H.; Bastías, R.M.; Tapia, G.; Ellena, M.; Torres, C. Cracking in sweet cherries: A comprehensive review from a physiological, molecular, and genomic perspective. *Chil. J. Agric. Res.* 2013, 73, 66–72.

Ballistreri, G.; Continella, A.; Gentile, A.; Amenta, M.; Fabroni, S.; Rapisarda, P.; Fruit quality and bioactive compounds relevant to human health of sweet cherry (*Prunus avium L.*) cultivars grown in Italy. *Food Chem.* 2013, 140, 630–638.

Bargioni, G. "Vittoria," a new sweet cherry cultivar (in Italian). Revistadella Otro. Italiana, 1970, 63, 312.

Bargioni, G. Sweet Cherry Scions. Characteristics of the principal commercial cultivars, breeding objectives and methods. In: Webster, A.D. and Looney, N.E. (eds.) Cherries: Crop Physiology, Production and Uses. CAB International, Wallingford, UK, 1996, pp. 73–112.

Barritt, B.H.; Dilley, M.A.; Konishi, B.S. Orchard potential depends on solving preplant jigsaw puzzle. *Proc. Wash. Sta. Hort. Assoc.* 1997, 92, 113–116.

Basanta, M.; Escalada Plá, M.F.; Raffo, M.D.; Stortz, C.A.; Rojas, A.M. Cherry fibers isolated from harvest residues as valuable dietary fiber and functional food ingredients. *J. Food Eng.* 2014, 126, 149–155.

Bastos, C.; Barros, L.; Dueñas, M.; Calhelha, R.C.; Queiroz, M.J.R.; Santos-Buelga, C.; Ferreira, I.C. Chemical characterisation and bioactive properties of *Prunus avium* L.: The widely studied fruits and the unexplored stems. *Food Chem.* 2015, 173, 1045–1053.

Beattie, J.; Crozier, A.; Duthie, G.G. Potential health benefits of berries. *Curr. Nutr. Food Sci.* 2005, 1, 71–86.

Belien, T.; Bangels, E.; Vercammen, J.; Bylemans, D. Integrated control of the European cherry fruit fly *Rhagoletis cerasi* in Belgian commercial cherry orchards. In: Proceedings of the EU COST Action FA1104: Sustainable Production of High-quality Cherries for the European Market. COST meeting, Pitesti, Romania, October 2013 (accessed April, 2018).

Bernalte, M.J.; Sabio, E.; Hernández, M.T.; Gervasini, C. Influence of storage delay on quality of Van sweet cherry. *Postharvest Biol. Tec.* 2003, 28, 303–312.

Bhat, R. Response of young "Misri" sweet cherry trees to orchard floor management. *Int. J. Adv. Res.* 2015, 3, 638–648.

Bhat, R.; Wani, W.N.; Sharma, M.K.; Hussain, S. Influence of intercrops on cropping, quality and relative economic yield of sweet cherry cv. Bigarreau Noir Grossa (Misri). *J. Appl. Biotechnol.* 2016, 5, 264–272.

Blackman, R.L.; Eastop, V.F. Aphids on the World's Crops. An Identification and Information Guide. Chichester, UK: John Wiley, 1984.

Børve, J.; Ippolito, A.; Tanovic, B.; Michalecka, M.; Sanzani, S.M.; Poniatowska, A.; Mari, M.; Hrustic, J. Fungal diseases. In: Quero-García, J., Iezzoni, A., Pulawska, J. and Lang, G. (eds.) Cherries: Botany, Production and Uses. CABI International, Oxfordshire OX10 8DE, UK, 2017, pp. 338–364.

Børve, J.; Meland, M. Rain cover protection against cracking for cherry orchards. *Acta Hortic.* 1998, 68, 441–447.

Børve, J.; Skaar, E.; Sekse, L.; Meland, M.; Vangdal, E. Rain protective covering of sweet cherry trees□effects of different covering methods on fruit quality and microclimate. *HortTechnology.* 2003, 13, 143–148.

Børve, J.; Stensvand, A. Use of a plastic rain shield reduces fruit decay and need for fungicides in sweet cherry. *Plant Dis.* 2003, 87, 523–528.

Brand-Miller, J.; Foster-Powell, K. Diets with a low glycemic index: from theory to practice. *Nutr. Today.* 1999, 34, 64–72.

Brown, G.; Wilson, S.; Boucher, W.; Graham, B.; McGlasson, B. Effects of copper-calcium sprays on fruit cracking in sweet cherry. *Sci. Hortic.* 1995, 62, 75–80.

Brown, S.K.; Iezzoni, A.F.; Fogle, H.W. Cherries. In: Janick, J. and Moore, J.N. Fruit breeding. Volume 1: Tree and tropical fruits. John Wiley and Sons. New York, 1996.

Budak, N.H. Bioactive components of *Prunus avium* L. black gold (red cherry) and *Prunus avium* L. stark gold (white cherry) juices, wines and vinegars. *J. Food Sci. Technol.* 2017, 54, 62–70.

Budan, S.; Mutafa, I.; Stoian, I.; Popescu, I. 2005. Screening of 100 sour cherry genotypes for *Monilia laxa* field resistance. *Acta Hortic.* 2005, 667, 145–151.

Bujdosó, G.; Hrotkó, K.; Quero-Garcia, J.; Lezzoni, A.; Puławska, J.; Lang, G. Cherry production. In: Quero-García, J., Iezzoni,

A., Pulawska, J. and Lang, G. (eds.) Cherries: Botany, Production and Uses. CABI International, Oxfordshire OX10 8DE, UK, 2017, pp. 1–13.

Bujdosó, G.; Hrotkó, K.; Stehr, R. Evaluation of sweet and sour cherry cultivars on German dwarfing rootstocks in Hungary. *J. Fruit Ornam. Res.* 2004, 12, 233–244.

Burak, M.; Akçay, M.E.; Yalçinkaya, E.; Türkeli, Y. Effect of some clonal rootstocks on growth and earliness of "0900 Ziraat" sweet cherry. *Acta Hortic.* 2008, 795, 199–202.

Cachi, A.M.; Hedhly, A.; Hormaza, J.I.; Wünsch, A. Pollen tube growth in the self-compatible sweet cherry genotype, "Cristobalina," is slowed down after self-pollination. *Ann. Appl. Biol.* 2014, 164, 73–84.

Çağlayan, K.; Ulubas-Serce, C.; Gazel, M.; Varveri, C. Prune dwarf virus. In: Hadidi, A., Barba, M., Candresse, T. and Jelkmann, W. (eds) Virus and Virus-like Diseases of Pome and Stone Fruits. APS Press, St Paul, Minnesota, 2011, pp. 199–206.

Campoy, J.; Dantec, L.; Barreneche, T.; Dirlewanger, E.; Quero-García, J. New insights into fruit firmness and weight control in sweet cherry. *Plant Mol. Biol. Rep.* 2015, 33, 783–796.

Cannon, R.O. Role of nitric oxide in cardiovascular disease: focus on the endothelium. *Clin. Chem.* 1998, 44, 1809–1819.

Cantin, C.M.; Pinochet, J.; Gogorcena, Y.; Moreno, M.Á. Growth, yield and fruit quality of "Van" and "Stark Hardy Giant" sweet cherry cultivars as influenced by grafting on different rootstocks. *Sci. Hortic.* 2010, 123, 329–335.

Cao, J.; Jiang, Q.; Lin, J.; Li, X.; Sun, C.; Chen, K. Physicochemical characterisation of four cherry species (*Prunus* spp.) grown in China. *Food Chem.* 2015, 173, 855–863.

Carey, E.E.; Jett, L.; Lamont, W.J.; Nennich, T.T.; Orzolek, M.D.; Williams, K.A. Horticultural crop production in high tunnels in the United States: A snapshot. *HortTechnol.* 2009, 19, 37–43.

Celton, J-M.; Dheilly, E.; Guillou, M-C.; Simonneau, F.; Juchaux, M.; Costes, E.; Laurens, F.; Renou, J-P. Additional amphivasal bundles in pedicel pith exacerbate central fruit dominance and induce self-thinning of lateral fruitlets in apple. *Plant Physiol.* 2014, 164, 1930–1951.

Chaovanalikit, A.; Wrolstad, R.E. Total anthocyanins and total phenolics of fresh and processed cherries and their antioxidant properties. *J. Food Sci.* 2004, 69, 67–72.

Chen, P.N; Chu, S.C.; Chiou, H.L.; Chiang, C.L.; Yang, S.F.; Hsieh, Y.S. Cyanidin 3-glucoside and peonidin 3-glucoside inhibit tumor cell growth and induce apoptosis in vitro and suppress tumor growth in vivo. *Nutr. Cancer.* 2005, 53, 232–243.

Chockchaisawasdee, S.; Golding, J.B.; Vuong, Q.V.; Papoutsis, K.; Stathopoulos, C.E. Sweet cherry: Composition, postharvest preservation, processing and trends for its future use. *Trends Food Sci. Technol.* 2016, 55, 72–83.

Choi, C.; Anderson, R.L. Variable fruit set in self-fertile sweet cherry. *Can. J. Plant Sci.* 2001, 81, 753–760.

Christensen, J.V. Cracking in Cherries III. Determination of cracking susceptibility. *Acta Agric. Scand.* 1972, 22, 128–136.

Christensen, J.V. Cracking in cherries VI. Cracking susceptibility in relation the growth rhythm of the fruit. *Acta Agric. Scand.* 1973, 23, 52–54.

Claverie, J. New selections and approaches for the development of cherry rootstocks in France. *Acta Hortic.* 1996, 410, 373–375.

Cline, J.A.; Sekse, L.; Meland, M.; Webster, A.D. Rain-induced fruit cracking of sweet cherries: I. Influence of cultivar and rootstock on fruit water absorption, cracking and quality. *Acta Agric. Scand.* 2009, 45, 213–223.

Conte, A.; Scrocco, C.; Lecce, L.; Mastromatteo, M.; Del Nobile, M.A. Ready-to-eat cherries: Study on different packaging systems. *Innov. Food Sci. Emerg.* 2009, 10, 564–571.

Correia, S.; Schouten, R.; Silva, A.P.; Gonçalves, B. Factors affecting quality and health promoting compounds during growth and postharvest life of sweet cherry (*Prunus avium L.*). *Front. Plant Sci.* 2017, 8, 2166.

Costa, G.; dal cin, V.; Ramina, A. Physiological, molecular and practical aspects of fruit abscission. *Acta Hortic.* 2006, 727, 301–310.

Crisosto, C.; Crisosto, G.; Metheney, P. Consumer acceptance of "Brooks" and "Bing" cherries is mainly dependent on fruit SSC and visual skin colour. *Postharvest Biol. Technol.* 2003, 28, 150–167.

Crisosto, C.H.; Mitchell, F.G.; Johnson, S. Factors in fresh market stone fruit quality. *Postharvest News Info.* 1995, 6, 17–21.

Cummins, J.N. Interspecifc hybrids as rootstocks for cherries. *Fruit Varieties J.* 1979, 33, 85–89.

Cunningham, J.J. Micronutrients as nutriceutical interventions in diabetes mellitus. *J. Am. Coll. Nutr.* 1998, 17, 7–10.

Davarynejad, G.H.; Nyéki, J.; Szabo, T; Lakatos, L; Szabó, Z. Fruit drop pattern of sour cherry cultivars. *Acta Hortic.* 2014, 1020, 185–189.

Day, T.; Sarchet, B. (2018). Growing Fruit Trees in Montana. http://msuextension. org/publications/AgandNaturalResources/ EB0222.pdf (accessed April 3, 2018).

De Candolle, A. 1883. Origin of Cultivated Plants. Deuxiéme Édition. Librairie Germer Baillière et Cie, Paris, p. 377.

De Francesco, J.; Edmunds, B.; Bell, N. Blueberry pests. In: Hollingsworth, C.S., ed. Pacific Northwest Insect Management Handbook. Corvallis, OR: Oregon State University. 2018, pp. J43–J44.

Demir, T. Determination of carotenoid, organic acid and sugar content in some sweet cherry cultivars grown in Sakarya, Turkey. *J. Food Agric. Environ.* 2013, 11, 73–75.

Demirbas, A. Biodiesel from kernel oil of sweet cherry (*Prunus avium L.*) seed. *Energy Sourc. A, Recovery Util. Environ. Effects*, 2016, 38, 2503–2509.

Demirsoy, H.; Demirsoy, L.; Macit, İ.; Akçay, M.E.; Bas, M.; Demirtas, I.; Sarısu, C.; Taner, Y.; Kuden, A. Sweet cherry growing in Turkey: a brief overview. *Acta Hortic.* 2017, 1161, 111–116.

Dever M.; MacDonalds, R.; Cliff, M.; Lane, W. Sensory evaluation of sweet cherry cultivars. *HortScience.* 1996, 31, 150–153.

Dorić, D.; Ognjanov, V.; Ljubojević, M.; Barać, G.; Dulic, J.; Ppanjic, A.; Dugalic, K. Rapid propagation of sweet and sour cherry rootstocks. *Not. Bot. Horti. Agrobot. Cluj.* 2014, 42, 488–494.

Douglas, S.M. (2003). Disease control for home cherry orchards. http://www.ct.gov/ caes/lib/caes/documents/publications/ fact_sheets/plant_pathology_and_ ecology/disease_control_for_home_ cherry_orchards.pdf (accessed April 18, 2018).

Drake, S.R.; Fellman, J.K. Indicators of maturity and storage quality of Rainier sweet cherry. *HortScience.* 1987, 22, 283–285.

Drkenda, P.; Spahić, A.; Spahić, A.; Begić-Akagić, A. Testing of 'Gisela 5' and 'Santa Lucia 64' cherry rootstocks in Bosnia and Herzegovina. *Acta Agric. Slov.* 2012, 99, 129–136.

Drogoudi, P.; Kazantzis, K.; Blanke, M.M. Climate change effects on cherry flowering in Northern Greece. *Acta Hortic.* 2017, 1162, 45–50.

Druart, P. Micropropagation of *Prunus* species relevant to cherry fruit production. In: Protocols for micropropagation of selected economically—important horticultural plants. Methods in Molecular

Biology (Methods and Protocols), Lambardi M., Ozudogru E., Jain S. (eds) Springer Science+Business Media NY, USA, 2012, vol. 994, 490 pp.

Eccher, T.; Granelli, G. Fruit quality and yield of different apple cultivars as affected by tree density. *Acta Hortic.* 2006, 712, 535–540.

Edin, M.; Claverie, J. Porte-greffes du cerisier: Fercahun-Pontavium, Fercadeu-Pontaris, deux nouvelles selections de merisier. *Infos CTIFL* 1987, 28, 1116.

Edin, M.; Garcin, A.; Lichou, J.; Jourdain, J.M. Influence of dwarfing cherry rootstocks on fruit production. *Acta Hortic.* 1996, 410, 239–243.

Edin, M.; Lichou, J.; Saunier, R. Cerise, les Varieties et leur Conduite. Ctifl, 1997, 238.

Elmhirst, J. (2006). Crop profile for sweet cherries in Canada. http://www5.agr.gc.ca/resources/prod/doc/prog/prrp/pdf/cherry_e.pdf (accessed March 23, 2018).

Engin, H.; Sen, F.; Pamuk, G.; Gokbayrak, Z. Investigation of physiological disorders and fruit quality of sweet cherry. *Eur. J. Hort. Sci.* 2009, 74, 118–123.

Epstein, D.; Landis, J. Cherry Orchard Floor Management: Opportunities to Improve Profit and Stewardship. MSU Extension Bulletin E-2890. 2003. 6pp.

Esti, M.; Cinquanta, L.; Sinesio, F.; Moneta, E.; Di Matteo, M. Physicochemical and sensory fruit characteristics of two sweet cherry cultivars after cool storage. *Food Chem.* 2002, 76, 399–405.

FAOSTAT, 2018. Agriculture Data. Available from: http://faostat3.fao.org/faostatgateway/go/to/home/E. (Accessed April 20, 2018).

Faniadis, D.; Drogoudi, P.D.; Vasilakakis, M. Effects of cultivar, orchard elevation, and storage on fruit quality characters of sweet cherry (*Prunus avium* L.). *Sci. Hortic.* 2010, 125, 301–304.

Faust, M.; Surányi, D. Origin and dissemination of cherry. *Horticultural Reviews*, 1997, 19, 263–317.

Ferretti, G.; Bacchetti, T.; Belleggia, A.; Neri, D. Cherry antioxidants: from farm to table. *Molecules.* 2010, 15, 6993–7005.

Fleury, D.; Mauffette, Y.; Methot, S.; Vincent, C. Activity of *Lygus lineolaris* (Heteroptera: Miridae) adults monitored around the periphery and inside a commercial vineyard. *Eur. J. Entomol.* 2010, 107, 527–534.

Flore, J.A. Stone fruit. In: Schaffer, B. and Andersen, P.C. (eds.) Handbook of Environmental Physiology of Fruit Crops. Vol. I: Temperate Crops. CRC Press, Boca Raton, FL, 1994, pp. 233–270.

Flore, J.A.; Layne, D.R. Photoassimilate production and distribution in cherry. *HortScience* 1999, 34, 1015–1019.

Fogle, H.W.; Snyder, J.C.; Baker, H.; Cameron, H.R.; Cochran, L.C.; Schomer, H.A.; Yang, H.Y. Sweet Cherries: Production, Marketing, and Processing. USDA Agri. Handbook 442. Washington D.C. 1973, p. 98.

Foster-Powell, K.; Holt, S.; Brand-Miller, J.C. International table of glycemic index and glycemic load. *Am. J. Clin. Nutr.* 2002, 76, 5–56.

Gainza, F.; Opazo, I.; Muñoz, C. Graft incompatibility in plants: Metabolic changes during formation and establishment of the rootstock/scion union with emphasis on *Prunus* species. *Chil. J. Agric. Res.* 2015, 75, 28–34.

Gao, L.; Mazza, G. Characterization, quantitation, and distribution of anthocyanins and colourless phenolics in sweet cherries. *J. Agric. Food Chem.* 1995, 43, 343–346.

García, F.; Frutos, D.; Lopez, G.; Carrillo, A.; Cos, J. Flowering of sweet cherry (*Prunus avium* L.) cultivars in Cieza, Murcia, Spain. *Acta Hortic.* 2014, 1020, 191–196.

Garcia-Montiel, F.; Serrano, M.; Martinez-Romero, D.; Albuquerque, N. Factors influencing fruit set and quality in different sweet cherry cultivars. *Span. J. Agric. Res.* 2010, 8, 1118–1128.

Ghosh, D.; Konishi, T. Anthocyanins and anthocyanin-rich extracts: Role in diabetes and eye function. *Asia Pac. J. Clin. Nutr.* 2007, 16, 200−208.

Giacalone, G.; Chiabrando, V. Modified atmosphere packaging of sweet cherries with biodegradable films. *Int. Food Res. J.* 2013, 20, 1263−1268.

Gilliagan C.A. Sustainable agriculture and plant diseases: an epidemiological perspective. *Philos. Trans. R. Soc. Lond. B: Biol. Sci.* 2008, 363, 741−759.

Gilmore, J.E. Biology of the black cherry aphid in the Willamette Valley, Oregon. *J. Econ. Entomol.* 1960, 53, 659−661.

Giulia, E.; Alessandro, B.; Mariano, D.; Andrea, B.; Benedetto, R.; Angelo, R. Early induction of apple fruitlet abscission is characterized by an increase of both isoprene emission and abscisic acid content. *Plant Physiol.* 2013, 161, 19521969.

Giné-Bordonaba, J.; Echeverria, G.; Ubach, D.; Aguiló-Aguayo, I.; López, M.; Larrigaudière, C. Biochemical and physiological changes during fruit development and ripening of two sweet cherry varieties with different levels of cracking tolerance. *Plant Physiol. Biochem.* 2017, 111, 216−225.

Glowacka, A.; Rospára, E. Examination of the suitability of different pollinators for four sweet cherry cultivars commonly grown in Poland. *J. Hort. Res.* 2014, 22, 85−91.

Gonçalves, B.; Moutinho-Pereira, J.; Santos, A.; Silva, A.P.; Bacelar, E.; Correia, C., Rosa, E. Scion-rootstock interaction affects the physiology and fruit quality of sweet cherry. *Tree Physiol.* 2006, 26, 93−104.

Gonçalves, B.; Landbo, A.K.; Knudsen, D.; Silva, A.P.; Moutinho-Pereira, J.; Rosa, E. Effect of ripeness and postharvest storage on the phenolic profiles of cherries (*Prunus avium* L.). *J. Agric. Food Chem.* 2004a, 52, 523−530.

Gonçalves, B.; Landbo, A.K.; Let, M.; Silva, A.P.; Rosa, E.; Meyer, A.S. Storage affects the phenolic profiles and antioxidant activities of cherries (*Prunus avium* L.) on human low-density lipoproteins. *J. Sci. Food Agric.* 2004b, 84, 1013−1020.

Gonçalves, B.; Silva, A.; Moutinho-Pereira, J.; Bacelar, E.; Rosa, E.; Meyer, A. Effect of ripeness and postharvest storage on the evolution of colour and anthocyanins in cherries (*Prunus avium* L.). *Food Chem.* 2007, 103, 976−984.

Granatstein, D.; Sánchez, E. Research knowledge and needs for orchard floor management in organic tree fruit systems. *Int. J. Fruit Sci.* 2009, 9, 257−281.

Grattan, S.R. (2016). Drought tip: crop salt tolerance. http://dx.doi.org/10.3733/ucanr.8562 (accessed April 4, 2018).

Grove, G. Powdery mildew of sweet cherry: Influence of temperature and wetness duration on release and germination of ascospores of *Podosphaera clandestina*. *Phytopathology.* 1991, 81, 1271−1275.

Gruppe, W. An overview of the cherry rootstock breeding program at Giessen. *Acta Hortic.* 1985, 169, 189−198.

Guo, L.; Dai, J.; Ranjitkar, S.; Xu, J. Chilling and heat requirements for flowering in temperate fruit trees. *Int. J. Biometeorol.* 2014, 58, 1195−1206.

Hallé, F.; Oldeman, R.; Tomlinson, P.B. 1978. Tropical trees and forest. An architectural analysis. Springer Verlag, New York.

Hampson, C.R.; Stanich, K.; MacKenzie, D.L.; Herbert, L.; Lu, R.; Li, J.; Cliff, M.A. Determining the optimum firmness for sweet cherries using Just-About-Right sensory methodology. *Postharvest Biol. Tec.* 2014, 91, 104−111.

Hanson, E.J. Sour cherry trees respond to foliar boron applications. *HortScience.* 1991, 26, 1142−1145.

Hanson, E.J.; Proebsting, E.L. Cherry nutrient requirements and water relations. In: Webster, A.D., Looney, N.E. (eds) Cherries: Crop Physiology, Production and Uses. CAB International, Wallingford, UK, 1996, pp. 243–257.

Harris, A.L.; Ullah, R.; Fountain, M.T. The evaluation of extraction techniques for *Tetranychus urticae* (Acari: Tetranychidae) from apple (*Malus domestica*) and cherry (*Prunus avium*) leaves. *Exp. Appl. Acarol.* 2017, 72, 367–377.

Hartmann, H.T.; Kester, D.E.; Davies, F.T.Jr.; Geneve, R.L; Geneve. Plant Propagation: Principles and Practices (8th ed.). Upper Saddle River, NJ: Pearson Education, 2011, p. 869.

Hauser, M.; Gaimari, S.; Damus, M. *Drosophila suzukii* new to North America. *Fly Times*, 2009, 43, 12–15.

He, J.; Giusti, M.M. Anthocyanins: Natural colourants with health-promoting properties. *Annu. Rev. Food Sci. Technol.* 2010, 1, 163–187.

He, Y.; Zhou, J.; Wang, Y.; Xiao, C.; Tong, Y.; Tang, J.; Chan, A.; Lu, A. Anti-inflammatory and anti-oxidative effects of cherries on Freund's adjuvant-induced arthritis in rats. *Scand. J. Rheumatol.* 2006, 35, 356–358.

Heo, H.; Kim, D.; Choi, S.; Shin, D.; Lee, C. Potent inhibitory effect of flavonoids in *Scutellaria baicalensis* on amyloid β proteininduced neurotoxicity. *J. Agric. Food Chem.* 2004, 52, 4128–4132.

Hilsendegen, P. Preliminary results of a national German sweet cherry rootstock trial. *Acta Hortic.* 2005, 667, 179–188.

Hodun, G.; Hodun, M. Evaluation of flowering of 80 sweet cherry cultivars and their classification in regard to the season of blooming. Annales Universitis Mariae Curie Sklodowska, Sectio EEE, *Horticultura*, 2002, 10, 189–194.

Höfer M.; Giovannini, D. Phenotypic characterization and evaluation of European cherry collections: A survey to determine the most commonly used descriptors. *J. Hortic. Sci. Res.* 2017, 1, 7–12.

Hrotkó, K.; Rozpara, E. Rootstocks and improvement. In: Quero-García, J., Iezzoni, A., Pulawska, J. and Lang, G. (eds.) Cherries: Botany, Production and Uses. CABI International, Oxfordshire OX10 8DE, UK. 2017, pp. 117–139.

Iezzoni, A.; Schmidt, H.; Albertini, A. Cherries (*Prunus*). Genetic resources of temperate fruit and nut crops. *Acta* Hortic. 1991, 290, 111–176.

Imrak, B.; Sarier, A.; Kuden, A.; Kuden, A.B.; Comlekcioglu, S.; Tutuncu, M. Studies on shading system in sweet cherries (*Prunus avium* L.) to prevent double fruit formation under subtropical climatic conditions. *Acta Hortic.* 2014, 1059, 171–176.

Ingram, C. Ornamental cherries. Ornamental cherries. London. 1948.

IPCC. 2013. Climate Change 2013: The physical science basis exit EPA disclaimer. Contribution of working group I to the fifth assessment report of the Intergovernmental Panel on Climate Change [Stocker, T.F., D. Qin, G.-K. Plattner, M.].

IPGRI, 1985. Cherry descriptors. International Plant Genetic Resources Institute, Rome, Italy. 33 pp.

Jacob, R.; Spinozzi, G.; Simon, V.; Kelley, D. Prior consumption of cherries lowers plasma urate in healthy women. *J. Nutr.* 2003, 133, 1826–1829.

Jakobek, L.; Marijan, S.; Vóca, S.; Šindrak, Z.; Dobričević, N. Flavonol and phenolic acid composition of sweet cherries (cv. Lapins) produced on six different vegetative rootstocks. *Sci. Hortic.* 2009, 123, 23–28.

James, D.; Cieslinska, M.; Pallás, V.; Flores, R.; Candresse, T.; Jelkmann, W. Viruses, viroids, phytoplasmas and genetic disorders of cherry. In: Quero-García, J., Iezzoni, A., Pulawska, J. and Lang, G. (eds.) Cherries: Botany, Production and Uses. CABI International, Oxfordshire OX10 8DE, UK. 2017, pp. 386–419.

James, P. (2011). Australian cherry production guide. https://www.cherrygrowers.org.

au/assets/australian_cherry_production_ guide.pdf (accessed March 22, 2018).

James, P.; Measham, P. Rain and its impacts. In: Australian Cherry Production Guide. Cherry Growers of Australia, Paul James (Eds), Rural Solutions SA, Hobart, Australia, 2011, pp. 196–209.

Javaid, K.; Qureshi, S.N.; Masoodi, L.; Sharma, P.; Fatima, N.; Saleem, I. Orchard designing in fruit crops. *J. Pharmacogn. Phytochem.* 2017, 6, 1081–1091.

Jayaprakasam, B.; Olson, L.; Schutzki, E.; Tai, M.; Nair, M. Amelioration of obesity and glucose intolerance in high-fat-fed C57BL/6 mice by anthocyanins and ursolic acid in Cornelian cherry (*Cornus mas*). *J. Agric. Food Chem.* 2006, 54, 243–248.

Jayaprakasam, B.; Vareed, S.; Olson, L.; Nair, M. Insulin secretion by bioactive anthocyanins and anthocyanidins present in fruits. *J. Agric. Food Chem.* 2005, 53: 28–31.

Jayasankar, S.; Kappel, F. Recent advances in cherry breeding. *Fruit Veg. Cereal Sci. Biotech.* 2011, 5, 63–67.

Jensen, M. 2017. Processing for industrial uses. In: Quero-García, J., Iezzoni, A., Pulawska, J. and Lang, G. (eds.) Cherries: Botany, Production and Uses. CABI International, Oxfordshire OX10 8DE, UK. 2017, pp. 485–505.

Johnson R. Genetic Background of Durable Resistance. In: Lamberti F., Waller J.M., Van der Graaff N.A. (eds) Durable Resistance in Crops. NATO Advanced Science Institutes Series (Series A: Life Sciences), 1983, vol 55. Springer, Boston, MA, USA.

Jones, A.L.; Sutton, T.B. Sour cherry yellows. In: Diseases of Tree Fruits in the East (NCR045). Michigan State University Press, East Lansing, Michigan, 1996, 94 pp.

Kader, A.A. Recent advances and future research needs in postharvest technology of fruits. Bulletin of the International Institute of Refrigeration. LXXXI, 2001, 3–14.

Kahlke, C.J.; Padilla-Zakour, O.I.; Cooley, H.J.; Robinson, T.L. Shelf-life and marketing window extension in sweet cherries by the use of modified atmosphere packaging. *N. Y. Fruit Qual.* 2009, 17, 21–24.

Kang, S.; Serram, N.; Nair, M.; Bourquin, L. Tart cherry anthocyanins inhibit tumor development in ApcMin mice and reduce proliferation of human colon cancer cells. *Cancer Lett.* 2003, 194, 13–19.

Kappel, F. Breeding cherries in the "New World." *Acta Hortic.* 2008, 795, 59–70.

Kappel, F. New sweet cherry cultivars from the Pacific Agr-Food Research Centre (Summerland). *Acta Hortic.* 2005, 667, 53–57.

Kappel, F.; Granger, A.; Hrotkó, K.; Schuster, M. Cherry. In: Badenes, M.L. and Byrne, D.H. (eds) Fruit Breeding, Handbook of Plant and Breeding 8. Springer Science + Business Media, New York, 2012, pp. 459–504.

Kappel, F.; Lane, W.D.; MacDonald, R.; Lapins, K.; Schmid, H. "Santina," "Sumpaca Celeste," and "Sumnue Cristalina" sweet cherries. *HortScience.* 1998, 33, 1087–1089.

Kappel, F.; Lane, W.D.; MacDonald, R.; Lapins, K.; Schmid, H. "Sumste Samba," "Sandra Rose," and "Sumleta Sonata" sweet cherries. *HortScience.* 2000, 35, 152–154.

Kappel, F.; Fischer-Fleming, B.; Hogue, E. Fruit characteristics and sensory attributes of an ideal sweet cherry. *Hort. Sci.* 1996, 31, 443–446.

Kappel, F.; Lang, G.; Anderson, L.; Azarenko, A.; Facteau, T.; Gaus, A.; Southwick, S. NC-140 regional sweet cherry rootstock trial (1998)-results from Western North America. *Acta Hortic.* 2005, 667, 223–232.

Kappel, F.; MacDonald, R.A.; Brownlee, RT. Le cerisier de France 13S2009

(StaccatoMC). *Can. J. Plant Sci.* 2006, 86, 1239–1241.

Kappel, F.; MacDonald, R.A. Early gibberellic acid sprays increase firmness and fruit size of sweetheart sweet cherry. *J. Am. Pomol. Soc.* 2007, 61, 38–43.

Kappel, F.; Sholberg, P. Screening sweet cherry cultivars from the Pacific Agri-Food Research Centre (Summerland) breeding program for resistance to brown rot (*Monilinia fructicola*). *Can. J. Plant Sci.* 2008, 88, 747–752.

Karlidag, H.; Ercisli, S.; Sengul, M.; Tosun, M. Physico-chemical diversity in fruits of wild-growing sweet cherries (*Prunus avium* L.). *Biotechnol. Biotechnol. Equip.* 2009, 23, 1325–1329.

Kayesh, E.; Shangguan, L.; Korir, N.; Sun, X.; Bilkish, N.; Zhang, Y.; Han, J.; Song, C.; Cheng, Z.; Fang, J. Fruit skin colour and the role of anthocyanin. *Acta Physiol. Plant.* 2013, 35, 2879–2890.

Kays, S.J. Preharvest factors affecting appearance. *Postharvest Biol. Tec.* 1999, 15, 233–247.

Kelebek, H.; Selli, S. Evaluation of chemical constituents and antioxidant activity of sweet cherry (*Prunus avium L.*) cultivars. *Int. J. Food Sci. Technol.* 2011, 46, 2530–2537.

Kelley, D.; Rasooly, R.; Jacob, R.; Kader, A.; Mackey, B. Consumption of bing sweet cherries lowers circulation concentrations of inflammation makers in healthy men and women. *J. Nutr.* 2006, 136, 981–986.

Kim, D.; Heo, H.; Kim, Y.; Yang, H.; Lee, C. Sweet and sour cherry phenolics and their protective effects on neuronal cells. *J. Agric. Food Chem.* 2005, 53, 9921–9927.

Kolesnikova, A. Breeding and some biological characteristics of sour cherry in central Russia. 1975, Orel, U.S.S.R.: Priokstoc izdatel'stvo.

Kotuby-Amacher, J.; Koenig, R.; Kitchen, B. (2000). Salinity and Plant Tolerance. Utah State University Extension AG-SO-03, Logan, Utah, USA, https://digitalcommons.usu.edu/cgi/viewcontent.cgi?article=1042andcontext=extension_histall (accessed March 23, 2018).

Koumanov, K. S.; Long, L.E. 2017. Chapter 10: Site preparation and orchard infrastructure. In: Quero-García, J., Iezzoni, A., Pulawska, J. and Lang, G. (eds.) Cherries: Botany, Production and Uses. CABI International, Oxfordshire OX10 8DE, UK. 2017, pp. 223–243

Koumanov, K.S.; Staneva, I.N.; Kornov, G.D.; Germanova, D.R. Intensive sweet cherry production on dwarfing rootstocks revisited. *Sci. Hortic.* 2018, 229, 193–200.

Koutsimanis, G.; Harte, J.; Almenar, E. Development and evaluation of a new packaging system for fresh produce: a case study on fresh cherries under global supply chain conditions. *Food Bioprocess. Tech.* 2015a, 8, 655–669.

Koutsimanis, G.; Harte, J.; Almenar, E. Freshness maintenance of cherries ready for consumption using convenient, microperforated, bio-based packaging. *J. Sci. Food Agric.* 2015b, 95, 972–982.

Kupferman, E.; Sanderson, P. Temperature management and modified atmosphere packing to preserve sweet cherry quality. *Acta Hortic.* 2001, 667, 523–528.

Lamont, W.J. Overview of the use of high tunnels worldwide. *HortTechnology* 2009, 19, 25–29.

Lang, G. How canopy structures can affect sweet cherry productivity (2015). http://www.growingproduce.com/fruits/stone-fruit/how-canopy-structures-can-affect-sweet-cherry-productivity/ (accessed April 7, 2018).

Lang, G.A. Growing sweet cherries under plastic covers and tunnels: Physiological aspects and practical considerations. *Acta Hortic.* 2014, 1020, 303–312.

Lang, G.A. High tunnel tree fruit production: The final frontier? *HortTechnology.* 2009, 19, 50–55.

Lang, G.A. Tree fruit production in high tunnels: Current status and case study of sweet cherries. *Acta Hortic.* 2013, 987, 73–81.

Lang, G.; Valentino, T.; Robinson, T.; Freer, J.; Larsen, H.; Pokharel, R. Differences in mineral nutrient contents of dormant cherry spurs as affected by rootstock, scion, and orchard site. *Acta Hortic.* 2011, 903, 963–971.

Lang, G.A. Growing sweet cherries under plastic covers and tunnels: physiological aspects and practical considerations. *Acta Hortic.* 2009, 1020, 303–312.

Lang, G.A. Precocious, dwarfing, and productive-how will new cherry rootstocks impact the sweet cherry industry? *Hortechnology.* 2000, 10, 719–725.

Lapins, K. Stella, a self-fruitful sweet cherry. *Can. J. Plant Sci.* 1971, 51, 252–253.

Lara, I.; Camats, J.C.; Comabella, E.; Ortiz, A. Eating quality and health-promoting properties of two sweet cherry *(Prunus avium* L.) cultivars stored in passive modified atmosphere. *Food Sci. Technol. Int.* 2015, 21, 1–12.

Larsen, F.; Higgins, S.; Fritts Jr, R. Scion/interstock/rootstock effect on sweet cherry yield, tree size and yield efficiency. *Sci. Hortic.* 1987, 33, 237–247.

Lauri, P. Data on the vegetative context linked to flowering in cherry trees *(Prunus avium). Can. J. Bot.* 1992, 70, 1848–1859.

Law, T.L.; Lang, G.A. Planting angle and meristem management influence sweet cherry canopy development in the "Upright Fruiting Offshoots" training system. *HortScience.* 2016, 51, 1010–1015.

Lenahan, O.M.; Whiting, M.D.; Elfving, D.C. Gibberellic acid inhibits floral bud induction and improve Bing sweet cherry fruit quality. *HortScience.* 2006, 41, 654–659.

Leong, S.; Oey, I. Effects of processing on anthocyanins, carotenoids and vitamin C in summer fruits and vegetables. *Food Chem.* 2012, 133, 1577–1587.

Letch, W.; Malodobry, M.; Dziedzic, E.; Bieniasz, M.; Doniec, S. Biology of sweet cherry flowering. *J. Fruit Ornam. Plant Res.* 2008, 16, 189–199.

Li, B.; Xie, Z.; Zhang, A.; Xu, W.; Zhang, C.; Liu, Q.; Liu, C.; Wang, S. Tree growth characteristics and flower bud differentiation of sweet cherry *(Prunus avium* L.) under different climate conditions in China. *Horticultural Science – UZEI (Czech Republic)* 2010, 37, 6–13.

Li, S.; Andrews, P.K.; Patterson, M.E. Effects of ethephon on the respiration and ethylene evolution of sweet cherry *(Prunus avium* L.) fruit at different development stages. *Postharvest Biol. Technol.* 1994, 4, 235–243.

Li, X.; Kitajima, A.; Habu, T.; Kataoka, K.; Takisawa, R.; Nakazaki, T. Induction and characterization of fruit abscission during early physiological fruit drop in citrus. *The Horticult. J.* 2017, 86, 11–18.

Lichev, V.; Papachatzis, A. Results from the 11-year evaluation of 10 rootstocks of the sweet cherry cultivar "Stella." *Acta Hortic.* 2009, 825, 513–520.

Lisek, A.; Rozpara, E.; Głowacka, A.; Kucharska, D.; Zawadzka M. Identification of S-genotypes of sweet cherry cultivars from Central and Eastern Europe. *Hort. Sci.* 2015, 42, 13–21.

Liu, C.; Qi, X.; Song, L.; Li, Y.; Li, M. Species identification, genetic diversity and population structure of sweet cherry commercial cultivars assessed by SSRs and the gametophytic self-incompatibility locus. *Sci. Hortic.* 2018, 237, 28–35.

Liu, J.; Tian, S.; Meng, X.; Xu, Y. Effects of chitosan on control of postharvest diseases and physiological responses of tomato fruit. *Postharvest Biol. Technol.* 2007, 44, 300–306.

Liu, Y.; Liu, X.; Zhong, F.; Tian, R.; Zhang, K., Zhang, X.; Li, T. Comparative study of phenolic compounds and antioxidant activity in different species of cherries. *J. Food Sci.* 2011, 76.

Livellara, N.; Saavedra, F.; Salgado, E. Plant based indicators for irrigation scheduling in young cherry trees. *Agr. Water Manag.* 2011, 98, 684–690.

Long, L.; Lang, G.; Musacchi, S.; Whiting, M. Cherry Training Systems. Pacifc Northwest Extension, Oregon State University, Corvallis, Oregon, 2015. https://catalog.extension.oregonstate.edu/ sites/catalog/files/project/supplemental/ pnw667/pnw667print_0.pdf (accessed April 3, 2018).

Long, L.; Whiting, M.; Nunez-Elisea, R. Sweet cherry cultivars for the fresh market. Oregon State University. 2007, p. 10.

Long, L.E.; Brewer, L.J.; Kaiser, C. Cherry rootstocks for the modern orchard. In: 57th Annual IFTA Conference, February 2014, in Kelowna, British Columbia, Canada. http://extension. oregonstate.edu/wasco/sites/default/ files/cherryrootstocksmodern-long.pdf (accessed March 29, 2018).

Long, L.E.; Kaiser, C. Sweet cherry rootstocks, 2010. https://catalog.extension. oregonstate.edu/pnw619 (accessed March 28, 2018).

Long, L.E.; Kaiser, C. Sweet cherry orchard establishment in the pacific Northwest: Important considerations for success, 2013. https://catalog.extension.oregonstate.edu/ pnw642 (accessed April 2, 2018).

Looney, N.E.; Webster, A.D.; Kupferman, E. Harvest and handling sweet cherries for the fresh market. In: Webster, W.D. and Looney, N.E (eds.) Cherries: Crop Physiology, Production and Uses. CAB International, Wallingford, UK, 1996, pp. 411–441.

Luo, H.; Dai, S.; Ren, J.; Zhang, C.; Ding, Z.; Li, Z.; Sun, Y.; Ji, K.; Wang, Y.; Li, Q.; Chen, P.; Duan, C.; Wang, Y.; Leng, P. The role of ABA in the maturation and postharvest life of a nonclimacteric sweet cherry fruit. *J. Plant Growth Regul.* 2014, 33, 373–383.

Lux, S.A.; Wnuk, A.; Vogt, H.; Belien, T.; Spornberger, A.; Studnicki, M. Validation of individual-based Markov-like stochastic process model of insect behaviour and a "virtual farm" concept for enhancement of site-specific IPM. *Front. Physiol.* 2016, 7, 363.

Ma, Z.X.; Yang, L.; Yana, H.; Kennedy, J.F.; Meng, X. Chitosan and oligochitosan enhance the resistance of peach fruit to brown rot. *Carbohydr. Polym.* 2013, 94, 272–277.

Maas, F.M.; Balkhoven-Baart, J.; van der Steeg, P.A.H. Evaluation of Krymsk® 5 (VSL-2) and Krymsk® 6 (LC-52) as rootstocks for sweet cherry "Kordia." *Acta Hortic.* 2014, 1058, 531–536.

Magyar, L.; Hrotkó, K. *Prunus cerasus* and *Prunus fruticosa* as interstocks for sweet cherry trees. *Acta Hortic.* 2008, 795, 287–292.

Mahmood, K.; Carew, J.G.; Hadley, P.; Battey, N.H. The effect of chilling and post-chilling temperatures on growth and flowering of sweet cherry (*Prunus avium* L.). *J. Hortic. Sci. Biotechnol.* 2000, 75, 598–601.

Manganaris, G.A.; Ilias, I.F.; Vasilakakis, M.; Mignani, I. The effect of hydrocooling on ripening related quality attributes and cell wall physicochemical properties of sweet cherry fruit (*Prunus avium* L.). *Int. J. Refrig.* 2007, 30, 1386–1392.

Marchese, A.; Boskovic, R.I.; Martínez-García, P.J.; Tobutt, K.R. A new self-compatibility haplotype in the sweet cherry "Kronio," S5′, attributable to a pollen-part mutation in the SFB gene. *J. Exp. Bot.* 2007, 58, 4347–4356.

Marsal, J.; Lopez, G.; del Campo, J.; Mata, M.; Arbones, A.; Girona, J. Postharvest regulated deficit irrigation in "Summit" sweet cherry: Fruit yield and quality in the following season. *Irrig. Sci.* 2010, 28, 181189.

Martínez-Romero, D.; Alburquerque, N.; Valverde, J.M.; Guillén, F.; Castillo, S.;

Valero, D.; Serrano M. Postharvest sweet cherry quality and safety maintenance by aloe vera treatment: A new edible coating. *Postharvest Biol. Technol.* 2006, 39, 93–100.

Matzneller, P.; Götz, K.P.; Chmielewski, F.M. Spring frost vulnerability of sweet cherries under controlled conditions. *Int. J. Biometeorol.* 2016, 60, 123–130.

McCauley, A.; Jones, C.; Olson-Rutz, K. (2017). Soil pH and Organic Matter. http://msuextension.org/publications/AgandNaturalResources/4449/4449-8.pdf (accessed March 23, 2018).

McCune, L.; Johns, T. Symptom-specific antioxidant activity of boreal diabetes treatments. *Pharm. Biol.* 2003, 41, 362–370.

McCune, L.; Kubota, C.; Stendell-Hollis, N.R.; Thomson, C. Cherries and health: A review. *Crit. Rev. Food Sci. Nutr.* 2010, 51, 1–12.

McLellan, M.; Padilla-Zakour, O. Sweet cherry and sour cherry processing. In: Barret, D.M., Somogyi, L.P., and Ramaswamy, H.S. (eds). Processing Fruits: Science and Technology, CRC Press, Boca Raton, FL, 2004, pp. 497–508.

Measham, P.F.; Bound, S.A.; Gracie, A.J.; Wilson, S.J. Crop load manipulation and fruit cracking in sweet cherry (*Prunus avium* L.). *Adv. Hort. Sci.* 2012, 26, 25–31.

Measham, P.F.; Bound, S.A.; Gracie, A.J.; Wilson, S.J. Incidence and type of cracking in sweet cherry (*Prunus avium* L.) are affected by genotype and season. *Crop Pasture Sci.* 2009, 60, 1002–1008.

Melakeberhan, H.; Bird, G.W.; Jones, A.L. Soil pH affects nutrient balance in cherry rootstock leaves. *HortScience.* 2001, 36, 916–917.

Meland, M. Performance of the sweet cherry cultivar "Lapins" on 27 rootstocks growing in a Northern climate. In: Proceedings of the EU COST Action FA1104: Sustainable production of high-quality cherries for the European market. COST meeting, Trebinje, Bosnia and Herzegovina, February 2015.

Meland, M.; Frøynes, O.; Coop, L.; Kaiser, C. Modeling of sweet cherry flowering based on temperature and phenology in a mesic Nordic climate. *Acta Hortic.* 2017, 1162, 19–22.

Meland, M.; Kaiser, C.; Christensen, J.M. Physical and chemical methods to avoid fruit cracking in cherry. *AgroLife Sci. J.* 2014, 3, 177–183.

Mert, C.; Soylu, A. Possible cause of low fruit set in the sweet cherry cultivar 0900 Ziraat. *Can. J. Plant Sci.* 2007, 87, 593–594.

Milić, B.; Keserović, Z.; Dorić, M.; Ognjanov, V.; Magazin, N. Fruit set and quality of self-fertile sweet cherries as affected by chemical flower thinning. *Hort. Sci.* 2015, 42, 119124.

Mirto, A.; Iannuzzi, F.; Carillo, P.; Ciarmiello, L.; Woodrow, P.; Fuggi, A. Metabolic characterization and antioxidant activity in sweet cherry (*Prunus avium* L.) Campania accessions: Metabolic characterization of sweet cherry accessions. *Food Chem.* 2018, 240, 559–566.

Moghaddam, E.G.; Ahmadi Moghaddam, H.A.; Piri, S. Genetic variation of selected Siah Mashhad sweet cherry genotypes grown under Mashhad environmental conditions in Iran. *Crop Breed. J.* 2013, 3, 45–51.

Montonen, J.; Knekt, P.; Jarvinen, R.; Reunanen, A. Dietary antioxidant intake and risk of type 2 diabetes. *Diabetes Care.* 2004, 27, 362–366.

Moreno, M.A.; Montañés, L.; Tabuenca, M.C.; Cambra, R. The performance of Adara as a cherry rootstock. *Sci. Hortic.* 1996, 65, 85–91.

Mozetič, B.; Simčič, M.; Trebše, P. Anthocyanins and hydroxycinnamic acids of lambert compact cherries (*Prunus avium* L.) after cold storage and 1-methylcyclopropene treatment. *Food Chem.* 2006, 97, 302–309.

Mozetič, B.; Trebše, P.; Simčič, M.; Hribar, J. Changes of anthocyanins and hydroxycinnamic acids affecting the skin colour during maturation of sweet cherries (*Prunus avium* L.). *Swiss Soc. Food Sci. Technol.* 2004, 37, 123–128.

Mucha-Pelzer, T.; Müller, S.; Rohr, F.; Mewis, I. Cracking susceptibility of sweet cherries (*Prunus avium* L.) under different conditions. *Commun. Agric. Appl. Biol. Sci.* 2006, 71, 215–223.

Murray, M. Critical temperatures for frost damage on fruit trees, 2011. https://digitalcommons.usu.edu/cgi/viewcontent.cgi?article=1643andcontext=extension_curall (accessed March 23, 2018).

National Agricultural Statistics Service (NASS), Agricultural Statistics Board, United States Department of Agriculture (USDA). 2017. Cherry production.

Nartvaranant, P. Effects of fruit thinning on fruit drop, leaf carbohydrates concentration, fruit carbohydrates concentration, leaf nutrient concentration and fruit quality in Pummelo cultivar Thong Dee. *Songklanakarin J. Sci. Technol.* 2016, 38, 249–255.

Neilsen, D.; Neilsen, G.H.; Forge, T.; Lang, G. Dwarfing rootstocks and training systems affect initial growth, cropping and nutrition in "Skeena" sweet cherry. *Acta Hortic.* 2016, 1130, 199–206.

Neilsen, G.; Kappel, F.; Neilsen, D. Fertigation and crop load affect yield, nutrition and fruit quality of "Lapins" sweet cherry on Gisela 5 rootstock. *HortScience.* 2007, 42, 1456–1462.

Neilsen, G.; Kappel, F.; Neilsen, D. Fertigation method affects performance of "Lapins" sweet cherry on Gisela 5 rootstock. *HortScience.* 2004, 39, 1716–1721.

Neilsen, G.H.; Neilsen, D.; Forge, T. Environmental limiting factors for cherry production. In: Quero-García, J., Iezzoni, A., Pulawska, J. and Lang, G. (eds.) Cherries: Botany, Production and Uses.

CABI International, Oxfordshire OX10 8DE, UK. 2017, pp. 189–222.

Neilsen, G.H.; Neilsen, D.; Kappel, F.; Forge, T. Soil management, fertilization, and irrigation: interaction of irrigation and soil management on sweet cherry productivity and fruit quality at different crop loads that simulate those occurring by environmental extremes. *HortScience.* 2014, 49, 215–220.

Németh, M. Virus, Mycoplasma and Rickettsia Diseases of Fruit Trees. Martinus Nijhoff Publishers, Dordrecht, The Netherlands, and Akademini Kiado, Hungary, 1986.

Nicklas, B.; Ambrosius, W.; Messier, S.; Miller, G.; Penninx, B.; Loeser, R.; Palla, S.; Bleecker, E.; Pahor, M. Diet-induced weight loss, exercise, and chronic inflammation in older, obese adults: A randomized controlled clinical trial. *Am. J. Clin Nutr.* 2004, 79, 544–551.

Nielsen, G.H.; Nielsen, D.; Forge, T. 2017. Chapter 9: Environmental limiting factors for cherry production. In: J. Quero-García, A. Iezzoni, J. Puławska and G. Lang (Eds.) Cherries: Botany, Production and Uses. CABI, 189–222.

Niu, Y.; Zhang, X.; Xiao, Z.; Song, S.; Jia, C.; Yu, H. Characterization of taste-active compounds of various cherry wines and their correlation with sensory attributes. *J. Chromatogr.* B. 2012, 902, 55–60.

Olmstead, J.; Iezzoni, A.; Whiting, M. Genotypic differences in sweet cherry fruit size are primarily a function of cell number. *J. Am. Soc. Hort. Sci.* 2007, 130, 697–703.

Ouzounis, T.; Lang, G.A. Foliar applications of urea affect nitrogen reserves and cold acclimation of sweet cherries (*Prunus avium* L.) on dwarfing rootstocks. *HortScience* 2011, 46, 1015–1021.

Pacifico, S.; Di Maro, A.; Petriccione, M.; Galasso, S.; Piccolella, S.; Di Giuseppe, A.; Scortichini, M.; Monaco, P. Chemical composition, nutritional value and antioxidant properties of autochthonous *Prunus avium* cultivars from Campania Region. *Food Res. Int.* 2014, 64, 188−199.

Padilla-Zakour, O.I.; Tandon, K.S.; Wargo, J.M. Quality of modified atmosphere packaged Hedelfingen and Lapins sweet cherries (*Prunus avium* L.). *HortTechnology.* 2004, 14, 331−337.

Pal, M.D.; Mitre, I.; Asănică, A.C.; Sestras, A.F.; Peticilă, A.F.; Mitre, V. The influence of rootstock on the growth and fructification of cherry cultivars in a high density cultivation system. *Not. Bot. Horti. Agrobot. Cluj Napoca.* 2017, 45, 451−457.

Papadopoulos, N.T. Fruit fly invasion: Historical, biological, economic aspects and management. In: Shelly, T., Vargas, R. and Epsky, N. (eds.) Trapping and Detection, Control, and Regulation of Tephritid Fruit Flies. Springer, Dordrecht, the Netherlands, 2014, pp. 219−252.

Papadopoulos, N.T.; Köppler, K.; Beliën, T. Invertebrate and vertebrate pests: biology and management. In: Quero-García, J., Iezzoni, A., Pulawska, J. and Lang, G. (eds.) Cherries: Botany, Production and Uses. CABI International, Oxfordshire OX10 8DE, UK. 2017, pp. 305–364.

Patten, K.D.; Patterson, M.E.; Kupferman, E. Reduction of surface pitting in sweet cherries. Postharvest Pomol. News 1983, 1, 15−19.

Perry, R.L. Cherry rootstocks. In: Rom, R.C. and Carlson, R.F. (eds.), Rootstocks for Fruit Crops. Wiley, New York, 1987, pp. 217−264.

Petracek, P.D.; Joles, D.W.; Shirazi, A.; Cameron, A.C. Modified atmosphere packaging of sweet cherry (*Prunus avium* L., cv. Sams) fruit: Metabolic responses to oxygen, carbon dioxide, and temperature. *Postharvest Biol. Tec.* 2002, 24, 259−270.

Petriccione, M.; De Sanctis, F.; Pasquariello, M.S.; Mastrobuoni, F.; Rega, P.; Scortichini, P.; Mencarelli, F. The effect of chitosan coating on the quality and nutraceutical traits of sweet cherry during

postharvest life. *Food Bioprocess Technol.* 2014, 8, 394−408.

Poiana, M.A.; Moigradean, D.; Dogaru, D.; Mateescu, C.; Raba, D.; Gergen, I. Processing and storage impact on the antioxidant properties and colour quality of some low sugar fruit jams. *Rom. Biotechnol. Lett.* 2011, 16, 6504−6512.

Poll, L.; Petersen M.; Nielsen, G. Influence of harvest year and harvest time on soluble solids, titrateable acid, anthocyanin content and aroma components in sour cherry (*Prunus cerasus* L. cv. "Stevnsbær"). *Eur. Food Res. Technol.* 2003, 216, 212−216.

Proebsting, E. The interaction between fruit size and yield in sweet cherry. *Fruit Var. J.* 1990, 44, 169−172.

Prvulović, D.; Popović, M.; Malenčić, D.; Ljubojević, M.; Barać, G.; Ognjanov, V. Phenolic Content and Antioxidant Capacity of Sweet and Sour Cherries. Studia UBB Chemia, 2012, LVII 175−181.

Puławska, J.; Gétaz, M.; Kałuzna, M.; Kuzmanović, N.; Obradović, A.; Pothier, J.F.; Ruinelli, M.; Boscia, D.; Saponari, M.; Végh, A.; Palkovics, L. Bacterial diseases. In: Quero-García, J., Iezzoni, A., Pulawska, J. and Lang, G. (eds) Cherries: Botany, Production and Uses. CABI International, Oxfordshire OX10 8DE, UK. 2017, pp. 365–385.

Quero-García, J.; Iezzoni A.; Puławska, J.; Lang, G. Cherries: Botany, Production and Uses. In: Quero-Garcían, J.; Iezzoni, A.; Pulawska, J.; Lang, G. (eds) Boston, MA: © CABI, Oxfordshire OX10 8DE, UK, 2017, II-IX pp.

Radičević, S.; Cerović, R.; Marić, S.; Đorđević, M. Flowering time and incompatibility groups: Cultivar combination in commercial sweet cherry (*Prunus avium* L.) orchards. *Genetika.* 2011, 43, 397−406.

Radičević, S.; Marić, S.; Cerović, R. S-allele constitution and flowering time synchronization-Preconditions for effective fertilization in sweet cherry

(*Prunus avium* L.) orchards. *Rom. Biotechnol. Lett.* 2015, 20, 10997–11006.

Radunic, M.; Jazbec, A.; Pecina, M.; Čosić, T.; Pavičic, N. Growth and yield of the sweet cherry (*Prunus avium* L.) as affected by training system. *Afr. J. Biotech.* 2011, 10, 4901–4906.

Raja, W.H.; Un nabi, S.; Kumawat, L.; Sharma, O.C. Preharvest fruit drop: A severe problem in apple. *Indian Farmer.* 2017, 4, 609–614.

Remón, S.; Ferrer, A.; Marquina, P.; Burgos, J.; Oria, R. Use of modified atmospheres to prolong the postharvest life of Burlat cherries at two different degrees of ripeness. *J. Sci. Food Agric.* 2000, 80, 1545–1552.

Remón, S.; Venturini, M.E.; Lopez-Buesa, P.; Oria, R. Burlat cherry quality after long range transport: Optimisation of packaging conditions. *Innov. Food Sci. Emerg.* 2003, 4, 425–434.

Ren, J.; Leng, P. Role of abscisic acid and ethylene in fruit maturation of sweet cherry. *Acta Hortic. Sinica.* 2010, 37, 199–206.

Renick, L.J.; Cogal, A.G.; Sundin, G.W. Phenotypic and genetic analysis of epiphytic Pseudomonas syringae populations from sweet cherry in Michigan. *Plant Dis.* 2008, 92, 372–378.

Richardson, D.G. Rain-cracking of Royal Ann sweet cherries: Fruit physiological relationships, water temperature, orchard treatments, and cracking index. *Acta Hortic.* 1998, 468, 677–682.

Ridker, P. High-sensitivity C-reactive protein: potential adjunct for global risk assessment in primary prevention of cardiovascular disease. *Circulation.* 2001, 103, 1813–1818.

Robinson, T.L.; Domínguez, L.I. Comparison of the modified Spanish Bush and the Tall Spindle cherry production systems. *Acta Hortic.* 2014, 1058, 45–53.

Rodrigues, L.; Morales, M.; Fernandes, A.; Ortiz, J. Morphological characterization

of sweet and sour cherry cultivars in a germplasm bank at Portugal. *Genet. Resour. Crop Evol.* 2008, 55, 593–601.

Rojas-Argudo, C.; Pérez-Gago, M.B.; del Río, MA. Postharvest quality of coated cherries cv. Burlat as affected by coating composition and solids content. *Food Sci. Technol. Int.* 2005, 11, 417–424.

Romano, G.S.; Cittadini, E.D.; Pugh, B.; Schouten, R. Sweet cherry quality in the horticultural production chain. *Stewart Postharvest Rev.* 2006, 6, 1–9.

Roper, T.; Loescher, W. Relationships between leaf area per fruit and fruit quality in Bing sweet cherry. *HortScience.* 1987, 22, 1273–1276.

Roper, T.R.; Frank, G.G. Planning and Establishing Commercial Apple Orchards in Wisconsin. University of Wisconsin Cooperative Extension, Balsam Lake, Wisconsin, USA, 2004, p. 24.

Rosa, U.A.; Cheetancheri, K.G.; Gliever, C.J.; Lee, S.H.; Thompson, J.; Slaughter, D.C. An electro-mechanical limb shaker for fruit thinning. *Comput. Electron. Agric.* 2008, 61, 213–221.

Rosyara, U.R.; Bink, M.C.A.M.; Weg, E.; Zhang, G.; Wang, D.; Sebolt, A. Fruit size QTL identification and the prediction of parental QTL genotypes and breeding values in multiple pedigreed populations of sweet cherry. *Mol. Breed.* 2013, 32, 875–887.

Salazar, J.A.; Ruiz, D.; Campoy, J.A.; Sánchez-Pérez, R.; Crisosto, C.H.; Martínez-García, P. Quantitative trait loci (QTL) and mendelian trait loci (MTL) analysis in *Prunus*: A breeding perspective and beyond. *Plant. Mol. Biol. Rep.* 2014, 32, 1–18.

Salazar-Gutiérrez, M.R.; Chaves, B.; Anothai, J.; Whiting, M.; Hoogenboom, G. Variation in cold hardiness of sweet cherry flower buds through different phenological stages. *Sci. Hortic.* 2014, 172, 161–167.

Sánchez, R.; Sánchez, M.; Corts, R. Agromorphological characterization of traditional Spanish sweet cherry (*Prunus avium* L.), sour cherry (*Prunus cerasus* L.) and duke cherry (*Prunus* x *gondouinii* Rehd.) cultivars. *Span. J. Agric. Res.* 2008, 6, 42–55.

Sansavini, S.; Lugli, S. New sweet cherry cultivars developed at the University of Bologna. *Acta Hortic.* 2005, 667, 45–52.

Sansavini, S.; Lugli, S. Sweet cherry breeding programmes in Europe and Asia. *Acta Hortic.* 2008, 795, 41–58.

Schmitz-Eiberger, M.; Blanke, M. Bioactive components in forced sweet cherry fruit (*Prunus avium* L.), antioxidative capacity and allergenic potential as dependent on cultivation under cover. *LWT-Food Sci. Technol.* 2012, 46, 388–392.

Schuster, M. Incompatible *S-genotypes* of sweet cherry cultivars (*Prunus avium* L.). *Sci. Hort.* 2012, 148, 59–73.

Schuster, M.; Tobutt, K. Screening of cherries for resistence to leaf spot, *Brumeriella jaapii. Acta Hortic.* 2004, 663, 239–244.

Scorza, R.; Hammerschlag, F.; Stone Fruits. In Biotechnology of Perennial Fruit Crops. CAB International, Wallingford, UK, 1992, pp. 277–301.

Seeram, N.; Momin, R.; Nair, M.; Bourquin, L. Cyclooxygenase inhibitory and antioxidant cyanidin glycosides in cherries and berries. *Phytomedicine.* 2001, 8, 362–369.

Seeram, N.; Zhang, Y.; Nair, M. Inhibition of proliferation of human cancer cells and cyclooxygenase enzymes by anthocyanidins and catechins. *Nutr. Cancer.* 2003, 46, 101–106.

Sekse, L. Fruit cracking mechanisms in sweet cherries (*Prunus avium* L.)—a review. *Acta Hortic.* 1998, 468, 637–648.

Serradilla, M.; Lozano, M.; Bernalte, M.; Ayuso, M.; López-Corrales, M.; González-Gómez, D. Physicochemical and bioactive properties evolution during ripening of "Ambrunés" sweet cherry cultivar. *LWT-Food Sci. Technol.* 2011, 44, 199–205.

Serradilla, M.; Hernández, A.; López-Corrales, M.; Ruiz-Moyano, S.; Córdoba, M.; Martín, A. Composition of the cherry (*Prunus avium* L. and *Prunus cerasus* L.; Rosaceae). In: Simmonds, M.S.J. and Preedy, V.R. (eds) Nutritional Composition of Fruit Cultivars. Academic Press, London, 2016, pp. 127–147.

Serradilla, M.J.; Lozano, M.; Bernalte, M.J.; Ayuso, M.C.; Margarita López-Corrales, M.L.; González-Gómez, D. Physicochemical and bioactive properties evolution during ripening of Ambrune sweet cherry cultivar. *LWT-Food Sci. Technol.* 2010, 44, 199–205.

Serrano, M.; Díaz-Mula, H.M.; Zapata, P.J.; Castillo, S.; Guillén, F.; Martínez-Romero, D. Maturity stage at harvest determines the fruit quality and antioxidant potential after storage of sweet cherry cultivars. *J. Agric. Food Chem.* 2009, 57, 3240–3246.

Serrano, M.; Guillen, F.; Martínez-Romero, D.; Castillo, S.; Valero, D. Chemical constituents and antioxidant activity of sweet cherry at different ripening stages. *J. Agric. Food Chem.* 2005, 53, 2741–2745.

Seymour, E.; Singer, A.; Bennink, M.; Bolling, S. Cherry-enriched diets reduce metabolic syndrome and oxidative stress in lean Dahl-SS rats. Experimental Biology Meeting, 2007.

Sharma, K.; Cachi, A.M.; Sedlák, P.; Skřivanová, A.; Wünsch A. S-genotyping of 25 sweet cherry (*Prunus avium* L.) cultivars from the Czech Republic. *J. Hortic. Sci. Biotechnol.* 2016, 91, 117–121.

Sharma, M.; Jacob, J.K.; Subramanian, J.; Paliyath, G. Hexanal and 1-MCP treatments for enhancing the shelf life and quality of sweet cherry (*Prunus avium* L.). *Sci. Hort.* 2010, 125, 239–247.

Shen, X.; Zhao, K.; Liu, L.; Zhang, K.; Yuan, H.; Liao, X.; Wang, Q.; Guo, X.; Li, F.; Li, T. A role for PacMYBA in ABA-regulated anthocyanin biosynthesis in red-coloured sweet cherry cv. Hong Deng (*Prunus avium* L.). *Plant Cell Physiol.* 2014, 55, 862–880.

Shirasawa, K.; Isuzugawa, K.; Ikenaga, M.; Saito, Y.; Yamamoto, T.; Hirakawa, H. The genome sequence of sweet cherry (*Prunus avium*) for use in genomics-assisted breeding. *DNA Res.* 2017, 24, 499–508.

Simon, G. Review on rain induced fruit cracking of sweet cherries (*Prunus avium* L.), its causes and the possibilities of prevention. *Int. J. Hort. Sci.* 2006, 12, 27–35.

Sitarek, M. Incompatibility in sweet cherry trees on dwarfing rootstocks. *Agronomijas Vestis* 2006, 9, 140–145.

Sitarek, M.; Grzyb, Z.S. Growth, productivity and fruit quality of "Kordia" sweet cherry trees on eight clonal rootstocks. *J. Fruit. Ornam. Plant. Res.* 2010, 18, 169–176.

Sooriyapathirana, S.S.; Khan, A.; Sebolt, A.M.; Wang, D.; Bushakra, J.M.; Allan, A.C. QTL analysis and candidate gene mapping for skin and flesh colour in sweet cherry fruit (*Prunus avium* L.). *Tree Genet. Genomes* 2010, 6, 821–832.

Sotiropoulos, T.; Petridis, A.; Koukourikou-Petridou, M.; Koundouras, S.; Therios, I.; Koutinas, N.; Kazantzis, K.; Pappa, M. Efficacy of using rain protective plastic films against cracking of four sweet cherry (*Prunus avium* L.) cultivars in Greece. *Int. J. Agric. Innov. Res.* 2014, 2, 2319–1473.

Sotirov, D.K. Performance of the sweet cherry cultivars "Van" and "Kozerska" on clonal rotstocks. *Acta Hortic.* 2015, 1099, 727–733.

Southwick, S.M.; Uyemoto, J. Cherry Crinkle-Leaf and Deep Suture Disorders. Publication 8007. Communications Service, University of California, Oakland, California, 1999.

Souza, V.; Pereira, P.; da Silva, T.; de Oliveira Lima, L.; Pio, R.; Queiroz, F. Determination of the bioactive compounds, antioxidant activity and chemical composition of Brazilian blackberry, red raspberry, strawberry, blueberry and sweet cherry fruits. *Food Chem.* 2014, 156, 362–368.

Stojanović, M.; Milatović, D.; Kulina, M.; Alić-Džanović, Z. Susceptibility of sweet cherry cultivars to rain induced fruit cracking in region of Sarajevo. *Agroznanje.* 2013, 14, 179–184.

Sun, S.; Jiang, W.; Zhao, Y. Comparison of aromatic and phenolic compounds in cherry wines with different cherry cultivars by HS-SPME-GC-MS and HPLC. *Int. J. Food Sci. Technol.* 2012, 47, 100–106.

Sun-Waterhouse, D. The development of fruit-based functional foods targeting the health and wellness market: A review. *Int. J. Food Sci. Technol.* 2011, 46, 899–920.

Szabó, Z.; Nyéki, J.; Soltész, M. Frost injury to flower buds and flowers of cherry varieties. *Acta Hortic.* 1996, 410, 315–321.

Szpadzik, E.; Matulka, M.; Jadczuk-Tobjasz, E. The growth, yielding and resistance to spring frost of nine sour cherry cultivars in central Poland. *J. Fruit Ornam. Plant Res.* 2009, 17, 139–148.

Tezotto-Uliana, J.V.; Fargoni, G.P.; Geerdink, G.M.; Kluge, R.A. Chitosan applications pre- or postharvest prolong raspberry shelf-life quality. *Postharvest Biol. Technol.* 2014, 91, 72–77.

Thomas V.; Saraswathy Amma, C.K.; Sethuraj, M.R..Early fruit drop in tree crops with special reference to rubber (*Hevea brasiliensis*)—A review. *Feddes Rep.* 1993, 104, 395–402.

Thomidis, T.; Exadaktylou, E. Effect of a plastic rain shield on fruit cracking and cherry diseases in Greek orchards. *Crop Prot.* 2013, 52, 125–129.

Tian, S.P.; Jiang, A.L.; Xu, Y.; Wang, Y.S. Responses of physiology and quality of sweet cherry fruit to different atmosphere in storage. *Food Chem.* 2004, 87, 43–49.

Tomás-Barberán, F.; Espín, J. Phenolic compounds and related enzymes as determinants of quality in fruits and vegetables. *J. Sci. Food Agric.* 2001, 81, 853–876.

Tomás-Barberán, F.; Ruiz, D.; Valero, D.; Rivera, D.; Obón, C.; Sánchez-Roca, C.; Gil, M. Health benefits from pomegranates and stone fruit, including plums, peaches, apricots and cherries. In: Skinner, M. and Hunter, D. (eds) Bioactives in Fruit: Health Benefits and Functional Foods. Wiley, Hoboken, New Jersey, 2013, pp. 125–167.

Tsuda, T.; Horio, F.; Uchida, K.; Aoki, H.; Osawa, T. Dietary cyanidin 3-O-β-D-glucoside-rich purple corn colour prevents obesity and ameliorates hyperglycemia in mice. *J. Nutr.* 2003, 133, 2125–2130.

TUIK. Turkish Statistical Institute. 2018. http://www.tuik.gov.tr

Tukey, H.B.; Young, J.O. Histological study of the developing fruit of the sour cherry. *Bot. Gaz.* 1939, 100, 723–749.

United States Department of Agriculture. 2017. Fresh Peaches and Cherries: World Markets and Trade. Available: https://apps.fas.usda.gov/psdonline/circulars/StoneFruit.pdf. (accessed April 2, 2018).

USDA-ARS (US Department of Agriculture, Agricultural Research Service) (2018). USDA nutrient database for standard reference, Release 28. http://www.nal.usda.gov/fnic/foodcomp (accessed April 2, 2018).

Usenik, V.; Fabčič, J.; Štampar, F. Sugars, organic acids, phenolic composition and antioxidant activity of sweet cherry (*Prunus avium* L.). *Food Chem.* 2008, 107, 185–192.

Usenik, V.; Fajt, N.; Mikulic-Petkovsek, M.; Slatnar, A.; Stampar, F.; Veberic, R. Sweet cherry pomological and biochemical characteristics influenced by rootstock. *J. Agric. Food Chem.* 2010, 58, 4928–4933.

Usenik, V.; Kastelec, D.; Stampar, F. Physicochemical changes of sweet cherry fruits related to application of gibberellic acid. *Food Chem.* 2005, 90, 663–671.

Usenik, V.; Stampar, F. The effect of environmental temperature on sweet cherry phenology. *Eur. J. Hort. Sci.* 2011, 76, 15.

Usenik, V.; Stampar, F.; Petkovsek, M.; Kastelec, D. The effect of fruit size and fruit colour on chemical composition in "Kordia" sweet cherry (*Prunus avium* L.). *J. Food Compost. Anal.* 2015, 38, 121–130.

Usenik, V.; Zadravec, P.; Stampar, F. Influence of rain protective tree covering on sweet cherry fruit quality. *Eur. J. Hortic. Sci.* 2009, 74, 49–53.

Uyemoto, J.K.; Kirkpatrick, B.B. X-disease phytoplasma. In: Hadidi, A., Barba, M., Candresse, T. and Jelkmann, W. (eds) Virus and Virus-like Diseases of Pome and Stone Fruits. APS Press, St Paul, Minnesota, 2011, pp. 243–245.

Velickova, E.; Winkelhausen, E.; Kuzmanova, S.; Alves, V.D.; Moldao-Martins, M. Impact of chitosan-beeswax edible coatings on the quality of fresh straw-berries (*Fragaria ananassa* cv. Camarosa) under commercial storage conditions. *LWT-Food Sci. Technol.* 2013, 52, 80–92.

Verma, M.K. Cherry Production Technology, 2014. https://www.researchgate.net/publication/282365690_Cherry_production_Technology (accessed March 22, 2018).

Vidrih, R.; Hribar, J.; Sekse, L. Cherry seeds as a source of nutritionally important fatty acids. *Acta Hortic.* 2014, 1020, 165–172.

Walton, V.; Lee, J.: Bruck, D.: Dreves, A. Recognizing fruit damaged by spotted wing drosophila (SWD), *Drosophila suzukii*. EM 9021. Oregon, USA: Oregon State University Extension Service, 2010.

Wang, H.; Zhang, K.; Su, H.; Naihaoye. Identification of the *S-genotypes* of several sweet cherry (*Prunus avium* L.) cultivars by AS-PCR and pollination. *Afr. J. Agric. Res.* 2010, 5, 250–256.

Wang, L.S.; Stoner, G.D. Anthocyanins and their role in cancer prevention. *Cancer Lett.* 2008, 269, 281–290.

Wang, Y.; Long, L.E. Respiration and quality responses of sweet cherry to different atmospheres during cold storage and shipping. *Postharvest Biol. Technol.* 2014, 92, 62–69.

Wang, Y.; Xie, X.; Long, L.E. The effect of postharvest calcium application in hydro-cooling water on tissue calcium content, biochemical changes, and quality attributes of sweet cherry fruit. *Food Chem.* 2014, 160, 22–30.

Wani, A.A.; Singh, P.; Gul, K.; Wani, M.H.; Langowski, H.C. Sweet cherry (*Prunus avium*): Critical factors affecting the composition and shelf life. *Food Pack Shelf Life* 2014, 1, 86–99.

Webster, A. Taxonomic classification of sweet and sour cherries and a brief history of their cultivation. In: Webster, A. and Looney, N. E. (eds) Cherries: Crop Physiology, Production and Uses. CAB International, Wallingford, UK, 1996, pp. 3–24.

Webster, A.D. Dwarfing rootstocks for plums and cherries. *Acta Hortic.* 1980, 114, 201–207.

Webster, A.D.; Schmidt, H. Rootstocks for sweet and sour cherries. In: Webster, A.D. and Looney, N.E. (eds) Cherries: Crop Physiology, Production and Uses. CAB International, Wallingford, UK, 1996, pp. 127–163.

Wenden, B.; Campoy, J.A.; Lecourt, J.; López Ortega, G.; Blanke, M.; Radičević, S.; Schüller, E.; Spornberger, A.; Christen, D.; Magein, H.; Giovannini, D.; Campillo, C.; Malchev, S.; Peris, J.M.; Meland, M.; Stehr, R.; Charlot, G.; Quero-Garcia, J. A collection of European sweet cherry phenology data for assessing climate change. *Sci. Data.* 2016, 3, 160108.

Westwood, M.N. Mahaleb × Mazzard hybrid cherry stocks. *Fruit Varieties J.* 1978, 32, 39–39.

Whiting, M.; Lang, G.; Ophardt, D. Rootstock and training system affect sweet cherry growth, yield, and fruit quality. *HortScience.* 2005, 40, 582–586.

Whiting, M.D.; Lang, G.A. Bing sweet cherry on the dwarfing rootstock Gisela 5: Thinning affects fruit quality and vegetative growth but not CO_2 exchange. *J. Am. Soc. Hort. Sci.* 2004, 129, 407–415.

Whiting, M.D.; Ophardt, D. Comparing novel sweet cherry crop load management strategies. *HortScience.* 2005, 40, 1271–1275.

Whiting, M.D.; Ophardt D.; McFerson J.R. Chemical blossom thinners vary in their effect in sweet cherry fruit set, yield, fruit quality and crop value. *HortTechnology.* 2006, 16, 66–70.

Wilcox, W.F.; Mircetich, S.M. Effects of flooding duration on the development of *Phytophthora* root and crown rots of cherry. *Phytopathology.* 1985, 75, 1451–1455.

Wills, R.; Scriven, F.; Greenfield, H. Nutrient composition of stone fruit (*Prunus* spp.) cultivars: Apricot, cherry, nectarine, peach and plum. *J. Sci. Food Agric.* 1983, 34, 1383–1389.

Wójcik, P.; Akgül, H.; Demirtaş, I.; Sarısu, C.; Aksu, M.; Gubbuk, H. Effect of preharvest spays of calcium chloride and sucrose on cracking and quality of Burlat sweet cherry fruit. *J. Plant Nutr.* 2013, 36, 1453–1465.

Wójcik, P.; Morgaś, H. Impact of postharvest sprays of nitrogen, boron and zinc on nutrition, reproductive response and fruit quality of "Schattenmorelle" tart cherries. *J. Plant Nutr.* 2015, 38, 1456–1468.

Wójcik, P.; Wójcik, M. Effect of boron fertilization on sweet cherry tree yield and fruit quality. *J. Plant Nutr.* 2006, 29, 1755–1766.

Xu, J.; Ikeda, K.; Yamori, Y. Upregulation of endothelia nitric oxide synthase by cyanidin-3-glucoside, a typical anthocyanin pigment. *Hypertension.* 2004, 44, 217–222.

Yıgıt, D.; Baydas, E.; Güleryüz, M. Elemental analysis of various cherry fruits by wavelength dispersive X-ray fluorescence spectrometry. *Asian J. Chem.* 2009, 21, 2935–2942.

Yin, X.; Seavert, C.F.; Turner, J.; Núnez-Elisea, R.; Cahn, H. Effects of polypropylene groundcover on nutrient availability, sweet cherry nutrition, and cash costs and returns. *HortScience.* 2007, 42, 147–151.

Yoshimura, Y.; Nakazawa, H.; Yamaguchi, F. Evaluation of the NO scavenging activity of procyanidin in grape seed by use of the TMAPTIO/NOC 7ESR system. *J. Agric. Food Chem.* 2003, 51, 6409–6412.

Zhang, C.; Whiting, M. Plant growth regulators improve sweet cherry fruit quality without reducing endocarp growth. *Sci Hortic.* 2013, 150, 73–79.

Zheng, X.; Yue, C.; Gallardo, K.; McCracken, V.; Luby, J.; McFerson, J. What attributes are consumers 592 looking for in sweet cherries? Evidence from choice experiments. *Agric. Resour. Econ. Rev.* 2016, 593, 45, 24–42.

Zohary, D.; Hopf, M. Fruit trees and nuts. In: Zohary, D. and Hopf, M. (eds.) Domestication of Plants in the Old World. The Origin and Spread of Cultivated Plants in West Asia, Europe and the Mediterranean Basin. Oxford University Press, 2000, pp. 142–191.

CHAPTER 7

KIWIFRUIT

VISHAL S. RANA* and GITESH KUMAR

Department of Fruit Science, Dr Y.S. Parmar University of Horticulture and Forestry-Nauni, Solan, Himachal Pradesh 173230, India

Corresponding author. E-mail: drvishal_uhf@rediffmail.com

ABSTRACT

The kiwifruit or Chinese gooseberry is a dioecious and deciduous vine native to the China. It has gained enormous popularity in the recent past due to wide adaptability, unique blend of taste and high medicinal value. The kiwifruit has been assessed as a nodal fruit for diversification of mid-hill horticulture which has a wider scope in the domestic as well as international market. It is commercially cultivated in China, Italy, New Zealand, Iran, Greece, and Chile. Although the kiwifruit was introduced in the 1960s in India, its commercial importance was realized during the last three decades. Himachal Pradesh is the pioneer state to demonstrate its commercial cultivation in India, but Arunachal Pradesh, Nagaland, Mizoram, and Sikkim are now the leading kiwifruit-producing states of the country. Most of the cultivars are the descendants of the seedling which fruited in 1910. The nurserymen raised a large number of seedlings subsequently and selected superior types for further multiplication. The commercially grown important pistillate cultivars are Allison, Bruno, Hayward, Hort16A, and Monty. The knowledge about the current status of the scientific cultivation of kiwifruit has been discussed in this chapter in the light of the available literature.

7.1 GENERAL INTRODUCTION

The kiwifruit (*Actinidia deliciosa* Chev.] is one of the very few temperate fruit crops to have been domesticated in the 20th century (Liang, 1983; Ferguson, 1990). It is native to the Yangtze Valley of south and central China. The kiwifruit is also known as Chinese gooseberry,

China's miracle fruit and horticulture wonder of New Zealand. However, this fruit is known as kiwifruit after commercialization in New Zealand owing to its resemblance with the bird kiwi (*Apteryx* spp.) which is the emblem of New Zealand. In China, the kiwifruit is also known by different names like *mihoutao* and *yangtao*, but it is globally traded as kiwifruit (Ferguson, 1984).

FIGURE 7.1 An overview of kiwifruit orchard at Dr YS Parmar University, Solan, Himachal Pradesh, India (personal communication, 2019).

Over the past four decades, the kiwifruit has gained much popularity in many countries of the world owing to its refreshing delicate flavor, pleasing aroma, high nutritive value, abundant antioxidants, health benefits, and economic viability (Figure 7.1). In fact, no other fruit has gained so much popularity in such a short period in the history of commercial horticulture (Rana and Rana, 2003). The full economic potential of this fruit was exploited in New Zealand. Now its commercial cultivation has been extended to several other countries namely, Italy, Greece, Chile, Japan, France, Germany, USA, and Australia.

7.2 AREA AND PRODUCTION

In the world, the kiwifruit is cultivated in an area of 2,47,793 ha with the production of 40,38,872 tonnes with an average productivity of 16.30 ton/ha. The area and production under kiwifruit cultivation have increased at a steady pace since 2000 up to 2017 (Table 7.1 and Figure 7.2). Approximately, 93.7% of the world kiwifruit production is contributed by China, Italy, New Zealand, Iran, Greece, and Chile. The area and production under top 10 kiwifruit-producing countries are presented in Table 7.2 and Figure 7.3. The highest productivity of 35.18 ton/ha is in New Zealand (FAOSTAT, 2018).

Kiwifruit

TABLE 7.1 Year-wise Production of Kiwifruit in the World

Year	Area (ha)	Yield (ton/ha)	Production (tons)
2000–2001	127,032	14.86	1,888,158
2001–2002	129,836	15.27	1,981,997
2002–2003	136,132	14.97	2,037,572
2003–2004	139,125	14.51	2,019,319
2004–2005	144,331	15.95	2,301,972
2005–2006	146,543	15.97	2,339,791
2006–2007	151,248	16.60	2,511,397
2007–2008	156,740	16.26	2,549,311
2008–2009	162,037	16.94	2,745,006
2009–2010	168,666	16.56	2,792,539
2010–2011	172,359	16.46	2,837,310
2012–2013	186,336	16.38	3,052,348
2013–2014	239,680	14.44	3,460,218
2014–2015	224,113	16.30	3,652,526
2015–2016	262,977	15.60	4,101,274
2016–2017	279,104	15.49	4,323,338
2017–2018	247,793	16.30	4,038,872

Source: FAOSTAT (2018) (www.fao.org).

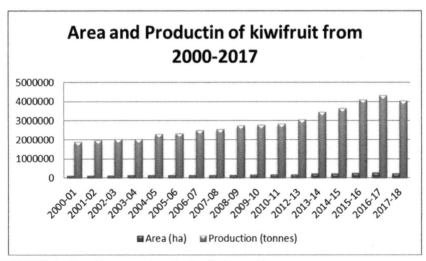

FIGURE 7.2 Trend in area and production in the world.
Source: FAOSTAT (2018) (http://www.fao.org).

TABLE 7.2 Top 10 Leading Kiwifruit Producing Countries

Country	Area (ha)	Production (tons)
China	165,728	2,024,603
Italy	26,403	541,150
New Zealand	11,705	411,783
Iran	10,771	311,307
Greece	9,200	274,600
Chile	8,720	224,916
France	3,798	65,632
Turkey	2,744	56,164
Portugal	2,650	35,411
Japan	1,826	24,456

Source: FAOSTAT (2018) (www.fao.org).

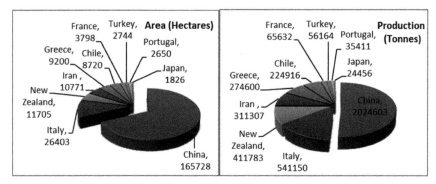

FIGURE 7.3 Area and production of top ten leading kiwifruit producing countries.
Source: FAOSTAT (2018) (http://www.fao.org).

In India, kiwifruit is grown in an area of 4000 ha with the production of 12,000 metric tonnes. Over the past five years, the area under kiwifruit production in India has not increased much, but the production has shown a steady increasing trend (Table 7.3 and Figure 7.4). Arunachal Pradesh, Nagaland, Mizoram, Sikkim, Himachal Pradesh, and Jammu & Kashmir are the major kiwifruit-producing states of the country (Table 7.4).

TABLE 7.3 Year-wise Production of Kiwifruit in India

Year	Area (000 ha)	Production (000 MT)
2011–2012	3	6
2012–2013	4	7
2013–2014	5	8
2014–2015	5	9
2015–2016	4	11
2016–2017	4	11
2017–2018	4	12

Source: NHB (2017) (www.nhb.gov.in)

Kiwifruit

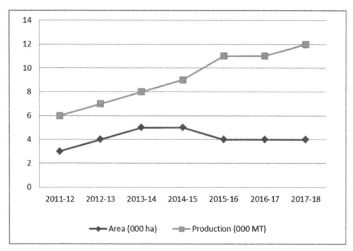

FIGURE 7.4 Area and production of kiwifruit in India.
Source: NHB (2017) (www.nhb.gov.in).

TABLE 7.4 State-wise Production of Kiwifruit in India (2015–2016)

State	Area (000 ha)	Production (000 MT)
Arunachal Pradesh	3.38	6.05
Nagaland	0.23	2.44
Mizoram	0.33	1.02
Sikkim	0.17	0.79
Himachal Pradesh	0.12	0.34
Jammu & Kashmir	0.01	0.01
Total	4.25	10.65

Source: NHB (2016) (www.nhb.gov.in).

7.3 MARKETING AND TRADE

The kiwifruit has a wider scope in the domestic as well as international market. As the apple has revolutionized the economy of farmers in high hills, kiwifruit is a boon to the farmers in the mid and low hills owing to very high economic returns. The kiwifruit cultivation has an important advantage that it comes in the market from October to December when practically no other fresh fruit is available to compete with it. Another major benefit of its cultivation is that the fruit is harvested hard and can be transported to long distances without using any sophisticated packaging material. Although there is a great potential for its cultivation in India, its cultivation yet has not gained momentum because of poor productivity and lack of transfer of cultivation technology to the farmers. Consequently, the kiwifruit produced in our country is not able to compete with the fruits imported from New Zealand, Australia, Italy, and other countries (Figure 7.5).

TABLE 7.5 Export and Import of Kiwifruit in India

	Export		Import	
Year	Quantity (Tons)	Value (1000 US$)	Quantity (Tons)	Value (1000 US$)
2010	0	0	2112	3734
2011	20	14	3886	6880
2012	4	9	5460	9149
2013	6	3	5492	9395
2014	1	1	6575	12,973
2015	4	1	12,389	22,154
2016	15	25	24,481	32,161

Source: FAOSTAT (2018) (www.fao.org).

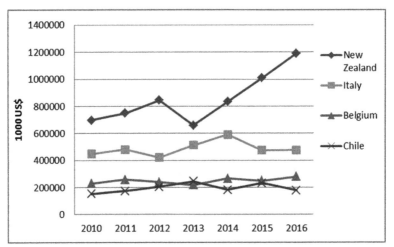

FIGURE 7.5 Export value of major kiwifruit exporting countries of the world.
Source: FAOSTAT (2018) (www.fao.org)

7.4 COMPOSITION AND USES

The kiwifruit is unique in many ways; while most other fruits are attractive in appearance, it is dull brown in color similar to sapota (*Achras zapota*). But the flesh in cross-section is surprisingly very beautiful. It is light green in color with a decorative pattern of lighter colored rays radiating from the center and embedded in between are many small, soft, and dark seeds. It has delicate, distinctive

Kiwifruit

melon-like flavor which to many is suggestive of strawberry, rhubarb, and gooseberry (Smith, 1961). Fruit has refreshing and delicate flavor, pleasing aroma, and high nutritive and medicinal values. Fruit composition varies greatly with climate, cultivar, and stage of maturity at which it is analyzed. Beever and Hopkirk (1990) have analyzed the composition of kiwifruit at maturity and reported that it is a rich source of carbohydrates (17.5%), vitamin C (80–120 mg/100 g), and mineral nutrients like calcium (16–51 mg/100 g), magnesium (10–32 mg/100 g), nitrogen (93–163 mg/100 g), phosphorus (22–67 mg/100 g), potassium (185–576 mg/100 g), and sulfur (16 mg/100 g).

The kiwifruit is also known to contain a protein-dissolving enzyme *actinidin*, belonging to the family, proteases to which papain extracted from papaya belongs. It can be used for tenderization of meat. However, the *actinidin* reacts with milk protein and results in undesirable products. Hence, the eating of kiwifruit is not advisable with milk or milk products. Rana et al. (2011a) have isolated the protein from kiwifruit which exhibited antifungal activity against *Fusarium oxysporum* and *Rhizoctonia solanii* causing several diseases in plants. Numerous edible seeds of kiwifruit containing vitamin E and Ω-3 fatty acids have a potential of a natural blood thinner.

A study performed at the University of Oslo in Norway reported that consuming 2–3 kiwifruit daily for 28 days significantly reduced platelet aggregation and blood triglyceride levels, potentially reducing the risk of blood clots.

7.5 ORIGIN AND DISTRIBUTION

The kiwifruit is native to China and the center of origin of the genus *Actinidia* is in the mountain region of southwestern China. Kiwifruits have a very recent and interesting history. It has been reported to occur naturally as deciduous fruiting vine in Sichuan, Yunnan, Shaanxi, Henan, and Hunan provinces of China (Ferguson, 1984). Geographically, it is distributed from Siberia to Southeast Asia through China and Japan. However, the kiwifruit was introduced into Japan, Russia, Europe, the USA, and New Zealand at the beginning of the 20th century and elsewhere rather recently. The commercial cultivation of the kiwifruit gained momentum after 1960 and is now being cultivated at a large scale in China, Italy, New Zealand, Iran, Greece, Chile, France, Turkey, Portugal, and Japan. The kiwifruit is known by different names like *Yang Tao* (sunny peach) or *Mihou tao* (macaque peach) in South China.

The genus *Actinidia* is remarkably wide spread from latitude 50°N to the equator from cold temperate or arctic forest to the tropics. However, only a few taxa occur in North, and none is found in dry, Western, or most provinces of China, and it seems best suited to warm, humid environments (Liang, 1983). The geographical center of the genus lies in the mountains and hills of South China between the Chang Jiang (Yangtze River) Basin and the Xi Jiang Basin—a narrow band between latitude 25°N and 30°N. A small number of species are found in almost all the countries bordering China that includes Siberia, Korea, Japan, Indochina, Thailand, Indian Himalayas, Malaya, and Indonesia. In New Zealand, the cultivar Hayward was cultivated first around 1924 by Hayward Wright in Avondale. This new species led to the beginning of commercial production around 1940. In 1991, a new variety of the golden, yellow kiwifruit was developed. It was originally named "Hort16A" and was bred in an orchard owned by Horticultural Research in Te Puke.

In India, the kiwifruit was introduced in 1960, and it was first planted at Lalbagh garden in Bangalore as an ornamental as well as fruit plant where it did not fruit. Later on, in the year 1963 under the leadership of late Dr Harbhajan Singh, it was introduced to India from the USA by the plant Introduction Division of Indian Agricultural Research Institute.

This was introduced in its regional station located at Phagli, Shimla of Himachal Pradesh. In 1970, an appreciable number of fruits was harvested, and Dadlani et al. (1971) highlighted its importance. Later on, few cultivars of the kiwifruit, namely, Hayward, Bruno, Monty, Abbot, and Tomuri were introduced from New Zealand, and plants were raised through cuttings that were subsequently planted. Realizing its commercial potential and suitability in the year 1985, sufficient plant materials of all the seven cultivars, namely, Allison, Bruno, Monty, Abbott, Hayward (female), Tomuri (male), and Allison (male) were planted in Solan and Kullu districts of Himachal Pradesh. Now, these plantations are in peak bearing and, on average, about 60–70 kg of fruit/ vine is harvested. Subsequently, the kiwifruit cultivation was extended in the mid hills of Jammu & Kashmir, Uttarakhand, Arunachal Pradesh, Nagaland, Sikkim, Mizoram, Meghalaya, and Manipur.

7.6 BOTANY AND TAXONOMY

According to Lindley (1836), the kiwifruit belongs to the genus *Actinidia* which has been derived from the Greek word *aktis* meaning "ray" due to numerous radiating divisions of the style and placed in the family Actinidiaceae. Liang (1983) further revised the genus and recognized over 50 species and 100 distinct taxa. Recently, Li et al. (2007) reported that the genus has 54 species and 21 varieties. The *Actinidia deliciosa* (*Actinidia chevalier*) var. *deliciosa* and *A. chinensis* are the only two species bearing edible fruits. Among these, *Actinidia deliciosa* (*A. deliciosa* Chev.) C.F. Liang et A.R. Ferguson is cultivated all over the world.

The kiwifruit has been known by a variety of botanical names. The *Actinidia chinensis* was widely accepted for a longer period of time but subsequently *Actinidia chinensis* var. *hispida* was also used. Now, the *Actinidia deliciosa* is the popularly accepted name. These changes in botanical nomenclature have occurred due to a clear understanding of the relationship between the kiwifruit and its relatives in China (Ferguson, 1990). The early writers and plant explorers have recognized two variants of *Actinidia chinensis*. These were given as the Chinese name *yangtao* for a smooth, less hairy fruit and *mao yangtao* for a hairy fruit. The first detailed distinction between botanical characteristics of two variants of *Actinidia chinensis* was made by A. Chevalier. The soft-haired variant, which had smooth skinned fruit, retained the name of *Actinidia chinensis*; the other, which had long, stiff hairs generally persistent on the fruit, was

named as *A. deliciosa* (A. Chev.) C.F. Liang et A.R. Ferguson (Ferguson, 1985). However, Li (1952) described another variant from Taiwan that did not resemble with the foresaid variants and later on was recognized as a separate species, *Actinidia setosa* (Li) C.F. Liang et A.R. Ferguson.

The buds of *Actinidia chinensis* have a small base and are almost spherical, exposed, and covered by only bud scales. There are very short yellow brown downy hairs. The young growing tips of the stem are yellowish green. The older stems are generally reddish brown and relatively straight. The leaves are obovate with small and thick, almost leathery. The leaf margin is flat, the lower surfaces are densely covered by short grayish white, stellate hairs. Flowering shoots are generally 4–5 cm long and thinly covered with soft white hairs that are readily shed. The fruits are spherical, soft, grayish white with a smooth surface and scattered yellowish brown spots. The fruit pulp is yellow, greenish yellow or green. The *Actinidia deliciosa* buds have a larger base and protruding, which is almost completely submerged in the bark with only small aperture. The hairs are long and grayish white. The young growing tips of the stem are covered with crimson hairs. The branches are dark brown, thick, and prone to twisting. The leaves are obovate like *Actinidia chinensis* but are larger and thinner. The leaf margins are undulating with long serration, and the lower surface of the leaves is densely covered with long grayish brown stellate hairs. The petiole is relatively long, purplish red in color, and densely covered with brown hairs. The flowering shoots are 15–20 cm long bearing numerous flowers. The flowers are larger than *Actinidia chinensis*; pistillate flowers are usually larger than the staminate flowers. The fruits are elongated to ovoid or cylindrical. The long, hard yellow brown hairs are not readily shed, but if lost, numerous brown spots are seen on the fruit surface. The fruit pulp is dark green or jade green in color (Ferguson, 1990).

The chromosome number of *Actinidia chinensis* has been reported to be $2n = 2x = 58$, $4x = 116$ and that of *Actinidia deliciosa* $2n = 4x = 116$, $6x = 174$, $8x = 232$, occasionally $12x = 358$ (Huang, 2016).

7.7 VARIETIES AND CULTIVARS

The kiwifruit is a dioecious plant and bears staminate and pistillate flowers on the separate plants. The kiwifruit cultivars are the descendants of a progeny of some seedlings, which fruited in 1910.The nurserymen raised a large number of seedlings subsequently and selected superior types, which were vegetatively propagated and distributed to the growers. The commercially grown important pistillate cultivars are, Abbott, Allison Bruno, Elmwood, Greensil, Hayward, Monty, Qinmei, and Wilkins Super; and the staminate cultivars are Allison, Matua, and Tomuri. The cultivars cannot be distinguished from each other on the basis of morphological characters of stem, shoot, and leaves. Flowering and fruit maturity periods are also not sharply distinct as these often overlap. The cultivars available in India belong to *Actinidia deliciosa*. The brief description of some commercial cultivars is given below.

7.7.1 PISTILLATE CULTIVARS

7.7.1.1 ABBOTT

This is an early flowering and maturing cultivar. Its medium size and oblong fruits are covered with dense hairs. Fruits are very sweet in taste with lower ascorbic acid content than other cultivar and have medium titratable acidity. It requires less chilling hours and can be successfully grown in mid and foot hill regions.

7.7.1.2 ALLISON

It resembles very much to the cultivar, Abbott, except that its fruits are slightly broader in length. The fruits are medium in size and slightly tapering at both ends. The petals of Allison flowers are overlapping and crimped along the margins. It is a prolific bearer and matures earlier than the cultivar Hayward. This is the most commercial cultivar in India.

7.7.1.3 BRUNO

Bruno fruits are easily distinguished from the fruits of other cultivars due to their more fruit length and higher ascorbic acid content. The fruits are suitable for processing purpose and raising seedling rootstocks for nursery production. They have a poor shelf life as compared to Allison.

7.7.1.4 HAYWARD

This cultivar was selected as a chance seedling by Hayward Wright. It is the most popular cultivar throughout the kiwifruit growing regions of the world because it has large fruit size, attractive oval shape, and maximum storage quality. The central core (columella) is larger than other cultivars. The fruits are born singly at the nodes. This is comparatively a shy bearing in India and late maturing. It requires a late flowering pollinizer and close planting for producing commercial crop. It can also be grown from mid hills to high hills because of high chilling requirement.

7.7.1.5 MONTY

It is also a late flowering cultivar, but the fruit maturity is not late due to the short growing period. This fruit is oblong in shape and larger in size. It is a prolific bearer, but its shelf life is poorer than all other commercial cultivars.

7.7.2 STAMINATE CULTIVARS

7.7.2.1 ALLISON (MALE)

It is the most popular and synchronizes with almost all pistillate cultivars of the kiwifruit.

7.7.2.2 MATUA

Hairs of peduncle are short, and flowers are borne in groups of usually 3 (1–5), early in the season. This cultivar is a good pollinizer for early and mid-season pistillate cultivars.

7.7.2.3 TOMURI

Hairs of peduncle are long, and flowers appear usually in groups of 5 (1–7), late in the season. It is a good pollinizer for the cultivar Hayward (Figure 7.6).

FIGURE 7.6 Important kiwifruit cultivars (personal communication, 2019).

7.7.3 NEW VARIETIES

The major focus in developing new varieties of the kiwifruit in China has been on the selection and breeding of natural *Actinidia* germplasm resources through several surveys. The initial efforts were on the selection and breeding, particularly on *A. chinensis* var. *chinensis* for developing superior quality fruits than the global cultivar Hayward (*A. chinensis* var. *deliciosa*). A number of new cultivars were released mainly from *A. chinensis* var. *chinensis* and few from *A. chinensis* var. *deliciosa*. The Chinese cultivars were mainly those selected from the wild accessions. The important cultivar Hayward was introduced from New Zealand. The cultivars of *A. chinensis* var. *chinensis* account for more than 255 of the Chinese kiwifruit plantings (Huang and Ferguson, 2003). Huang (2016) have reported some new Chinese yellow-fleshed kiwifruit cultivars (*A. Chinensis* var. *Chinensis*) such as Wuzhi No. 3, Ganmi No. 1 (Syn. Zaoxian), Ganmi No. 2 (Syn. Kuimi), Ganmi No. 3 (Syn. Jinfeng), Lushanxiang, and Yixiang that were selected from wild populations.

Kiwifruit 429

The domestication of *A. chinensis* var. *deliciosa* started in New Zealand over more than 50 years. Almost, all kiwifruit cultivars in commerce including Hayward belong to *Actinidia deliciosa*. The fruits of these cultivars are distinguishable from each other. The first yellow-fleshed kiwifruit to enter in international trade was launched as HORT-16A or the golden kiwifruit in 2000. The fruit of this cultivar has smooth brown skin and a beak shape at the stem end with golden yellow flesh. The fruit is less tart than the green kiwifruit. This cultivar was developed by Zesperi Ltd, New Zealand, and is marketed as Zesperi® Gold. The fruit of this cultivar has been known to fetch a higher market price due to less hairiness. In 2010, another variety called ENZA Red was developed, which is originally a cultivar of the Chinese variety Hong Yang. However, it has a short shelf life, which may restrict its commercialization.

7.8 BREEDING AND CROP IMPROVEMENT

The kiwifruit has become a mainstream fruit crop and is grown for high nutritional content and unique attractive internal color and appearance. The fruit flavor is mild, variable and is considered too sour by some consumers. The very positive response by consumers to the flavor of the new cultivar Hort16A suggests that there may be significant market opportunities for different flavored kiwifruit, including more intense, aromatic, and sweeter types. Compared to many mainstream fruits such as apples and bananas, the kiwifruits are not convenient to eat. It is not easy to judge the optimum stage of maturity, which can suggest as the proper eating stage. The period of optimum ripeness is quite short. Most of the consumers do not eat the skin, and the fruit has to be peeled or cut open. New cultivars with edible or peelable skin could be much more attractive to future consumers. All the factors that can affect breeding objectives should be known to the breeders for taking into consideration the target markets (Seal, 2003).

There is lack of evidence of useful variation in *Actinidia* for resistances to the biotic and abiotic stresses. To find such resistances, there is the requirement of screening of large populations. As "Hayward" is a good cultivar, it may be more rewarding initially to improve management practices and developing rootstocks that are more tolerant to adverse soil conditions. The future breeding work can be undertaken in order to identify the donor having some desirable traits such as size, shape, lack of hairs, color, flavor, high sugars and ascorbic acid contents, early maturity, low chill types, and drought tolerance. For achieving these objectives, the exploration of evolution center of the kiwifruit is required for the collection of the entire gene pool, and the material should be evaluated in situ so that it can provide some genetic resources for direct use without wasting the time. The detailed evaluation of entire genetic resources should be undertaken in order to identify the desirable traits.

The prediction of the best breeding strategies for the kiwifruit is difficult due to little previous work and very less knowledge regarding the pattern of inheritance. The effects of dioecism are that the good fruiting types cannot be crossed directly and the contribution of a particular male parent attributes such as fruit quality is usually overlooked. Polyploidy and problems of maintaining sufficient populations of large and vigorous vines that require support structures, training, and pruning are some other difficulties. Chance seedlings of *A. deliciosa* sometimes produce good quality fruits like "Hayward" and other cultivars. The controlled cross between selected *A. deliciosa* parents and F_1 seedling populations is another approach to develop cultivars whose fruit mature before the global cultivar Hayward. Researchers have identified and crossed some early flowering staminate and pistillate parents. Mass selection also appears to be a promising breeding method.

7.9 SOIL AND CLIMATE

The kiwifruit can be grown on a wide range of soils with adequate soil moisture. Well-drained, sandy–loam soil with a good amount of organic matter is ideal for the cultivation of the kiwifruit. Heavy clay soil with poor drainage is not suitable. A soil pH of 5.5–6.5 is considered ideal for vine growth and fruit production, but higher pH up to 7.3 affects adversely due to Mn deficiency.

The kiwifruit plant is very hardy and can withstand a wide range of climatic conditions. It is a deciduous vine, which can tolerate freezing temperature during dormancy. The cultivars Matua and Hayward were more affected than Bruno and Monty, whereas Abbott and *Actindia arguta* remained undamaged by freezing at –15°C (Costa et al., 1985). The plants do not die at –23 °C and recover growth after severe cold injury (Testolin and Messina, 1987). The frost during spring and autumn is, however, very injurious that kills immature shoots and fruit buds and causes bark splitting at the ground level (Gremminger et al., 1982), whereas at a higher elevation, frost or early snowfall in the season causes leaf fall, resulting into improper fruit ripening. The exact chilling requirement of the kiwifruit is not known. Brio et al. (1985) reported the chilling requirement of 1200 hours at 0 °C –10 °C for bud break, but they could not observe the strong relationship between chilling duration and flower bud break. Lawes (1984) found that 600 hours of winter chilling is sufficient for bud break, but 850 hours chilling was required for the more flower bud numbers.

In India, the kiwifruit can be grown successfully at 900–1800 m above mean sea level. Long days and high temperature are very much conducive to shoot growth and dry matter accumulation. A rainfall of about 120–150 cm well distributed throughout the growing period is sufficient for proper growth and development. As the kiwifruit plant has vigorous growth and viny habit, it does not resist strong winds during growing period (April–May). Therefore, a wind break plantation is essential well before the planting of vines. This can be easily and cheaply accomplished by planting *Populousnigra*, *Sesbania agyptica*, and *Salix babilonica* and later on controlled their growth by topping and trimming to make a good shelter.

7.10 PROPAGATION AND ROOTSTOCK

Kiwifruit plants are raised through cuttings, grafting, and budding. The tissue culture method of propagation can also be used for a large-scale multiplication of plants. Seedlings of some cultivated cultivars, namely,

Bruno and Abbott are commonly used as rootstocks for the kiwifruit. Seeds of these two varieties are preferred because of good germination and strong seedling vigor. Kiwifruit vines are mainly propagated by grafting and budding on the seedling rootstock. Although these methods take about two years to develop a plant, they are easiest, economical, and can be used for large-scale multiplication. Before sowing, seeds are stratified for 30–35 days at 0 °C–5 °C to break dormancy. Stratified seeds are sown in sand beds during February. Germinating seedlings are very sensitive to direct sunlight so they must be protected by creating a shade. Seedlings are transplanted in the nursery beds during July–August, which attain a graftable size within a year. Tongue grafting and chip budding done in the last week of January give better success. Nursery-management practices such as weeding, irrigation, staking of grafted plants are performed at regular intervals.

Individual vines in an orchard often vary in their growth and cropping. In Italy, "DI" rootstock, produced from a staminate plant by micropropagation, has been widely used for this reason. This rootstock has been reported to be tolerant to calcareous soils. Another rootstock "TR2" induces precocious fruiting in "Hayward" without any deleterious effects on the vine growth or fruit storage life. However, Bruno seedlings are the most widely used rootstock in New Zealand and even in India because this cultivar gives high and uniform germination, and strong vigor. The growers are sometimes reluctant to use cutting-grown plants either as the fruiting plants or as rootstocks because they grow more rapidly and have been found to be more productive.

Commercial planting on seedling rootstock can be screened to determine whether the superior cropping of some plants is due to superior rootstocks. Populations of seedling plants from breeding programs are also being screened for reduced vigor or compact growth habit. Rootstocks need not be of the same species as the scion, and different *Actinidia* species are being tested to assess their suitability as rootstocks for "Hayward." The use of a cold-resistant rootstocks could overcome the winter damage to trunks that occurs in colder districts. However, the affinity of "Hayward" scions and various rootstocks is not yet certain as it is possible that graft incompatibility between *Actinidia* species will be expressed only after a number of years.

Another suitable and rapid method of propagation is cutting. Hardwood, semihardwood, and softwood cuttings are the different types of stem cuttings used to raise own rooted plants. Raising of plants by cutting is quick, less expensive,

and require less space and skill. The ideal cutting should have short internodes and should be 0.5–1.0 cm thick, about 10–15 cm in length with at least 3–4 buds. Cuttings should be taken from the middle portion of the current season's growth shoot during July in the case of semihardwood cuttings and one-year-old shoot during January–February in the case of hardwood cuttings. After preparation, the cuttings are dipped in the 5000 ppm IBA solution for 10–15 s. The rooting medium containing forest soil + sand + soil (1:2:1) has been found best as it resulted in higher rooting percentage and other root characteristics (Rana and Babita, 2016)

7.11 LAYOUT AND PLANTING

An ideal land for kiwifruit cultivation should have a gentle slope. Steep slopes should be converted into terraces for plantation. For the successful establishment of vineyard, it is essential that the soil is thoroughly prepared. Pits preparation and filling of pits by mixing well-rotten farm yard manure should be completed at least 15 days before the planting. The planting is done during the dormant season. Plant spacing depends on the system of training, variety, irrigation facilities, and soil fertility. In general, planting is done at a spacing of 6 m from plant to plant and 4 m from row to row on the T-bar

Kiwifruit

Trellis system. In the Pergola system of training, a spacing of 6 m × 6 m is recommended for getting better fruit production. The soil should be firmly placed around the roots.

The kiwifruit is a dioecious plant; therefore, interplanting of a male plant is essential for proper fruit production. Adequate pollination is essential for getting good-sized fruits. In India, two male cultivars, namely, Tomouri and Allison (♂) are generally interplanted. The male plants should be evenly spread throughout the orchard, so that every female plant is in the direct sight of a male plant. Planting male and female plants in the 1:9 ratio is advocated.

7.12 IRRIGATION

The kiwi plants require much water due to vigorous vegetative growth and larger leaf size, vine habit, and high humidity in the natural habitat. In general, fully grown vine requires 80–100 L of water for total daily transpiration from the 16 to 17 m² canopy area during summer. Due to scanty and irregular distribution of rainfall in most of kiwifruit-growing areas, the irrigation is important means for increasing the productivity of fruits. Young trees should be irrigated at a 2–3 days interval, whereas bearing trees are to be irrigated at 20% depletion of soil moisture from field capacity at a 5–6 days interval during summer to get better-sized fruits. Clean cultivation with mulching of the tree basin with 15 cm thick hay grass is recommended for kiwifruit orchards.

As the kiwifruit requires heavy water for its cultivation, there is need to screen out various cultivars that can be easily adaptable to those areas where water availability is very less. Pratima et al. (2017) have reported that the deficit irrigation results in an increase in the chlorophyll stability index and decrease in the relative water content of kiwifruit cultivars, namely, Allison, Hayward, Abbott, Monty, and Bruno. Among different cultivars, Bruno was found to exhibit more resistance to deficit irrigation, whereas the cultivar Hayward exhibited the least resistance to deficit irrigation.

7.13 NUTRIENT MANAGEMENT

The kiwifruit makes much vegetative growth and yields heavily, requiring an adequate nutrient application. The symptoms of nutrient deficiencies are not often seen in kiwifruit vines. However, to keep the vine healthy and productive, they must be adequately and regularly supplemented with macro- and micronutrients. The concentration of nutrients in kiwifruit leaves varies significantly during the growing season. The most appropriate stage of growth for detecting nutrient disorder is prior to the fruit set. Leaf sampling should be done at the same physiological growth stage each year for determining the nutrient status of vines. Two to three youngest fully expanded leaves on the current season canes from at least 20 vines throughout the orchard should be taken as a sample. The leaf sample collection after the fruit set has not been found as the appropriate stage for diagnosing nutrient disorders (Clark et al., 1986). The optimum ranges for macronutrients, namely, N (3.5%–3.9%), P (0.6%–0.7%), K (2.65%–2.75%), Ca (1.35%–1.45%), Mg (0.30%–0.35%), and S (0.50%–0.55%) and for micronutrients, namely, Mn (85–95 ppm), Fe (115–150 ppm), Zn (55–70 ppm), Cu (20–30 ppm), and B (18–30 ppm) have been reported by Clark et al. (1986) in kiwifruit leaves, sampled four weeks after leaf emergence.

The young vines have a very high capacity for the absorption of N and K; very high P concentration in leaves does not affect the growth of the vines. Kumar et al. (1995) also advocated that the 850–950 g N, 500–600 g P, and 800–900 g K coupled with 20–30 kg well-rotten farm yard manure should be regularly mixed after the five years of the planting under Indian agroclimatic conditions. Nitrogen fertilizer is applied in two equal dressings: first, slightly before bud burst (March–April); and the second dose can be given on the commencement of monsoon where irrigation facilities are not available. The fertilizer should be applied only when the sufficient soil moisture is there; otherwise, the root damage can occur. Ideally, there should be some rain after each application to mix the fertilizer in the soil; otherwise, the suitable arrangement of irrigation is essential. Suggestions on local soil type requirements should always be sought, and whenever necessary, this schedule may be modified according to status of the soil.

7.14 TRAINING AND PRUNING

The kiwifruit plant is a vine-like grape, meaning thereby that it requires a similar training structure for heavy fruiting. In fact, the vine is more vigorous than grape and, therefore, the training structure should be stronger than that of grapes. Training of the kiwifruit vine is, therefore, an important aspect and requires constant attention while the plants are developing. The main aim of training is to establish and maintain an ideal framework of main branches and the bearing shoots. Training also facilitates soil management, pollination, and harvesting operations. The supporting structures (fences) are erected even before planting the vines or thereafter as early as possible. A single-wire fence is commonly adopted, though another wire is sometimes provided by some growers, and then structure takes the form of Kniffen system. In this system, a tensile wire of 2.5 mm thickness is strung on the top of the pillars which are 1.8–2 m high above the ground. The pillars are made up of wood, concrete, or iron. A cross-arm of 1.5 m length for outrigger wires is fixed on the pole. This training is known as T-bar trellis or the telephone system. The two laterals coming from the main branch are trained. In another system called pergola or bower system, a flat topped network of criss-cross wires is erected to train kiwifruit vines. However, this system is costly and difficult to manage but gives a higher fruit yield.

The pruning is one of the most important and significant aspect of kiwifruit vine management in obtaining consistently good production with quality fruits, which should be done with great precision and care. Successful pruning is to prevent the vines becoming dense, tangled, and allow access for bees during the flowering period for effective pollination, penetration of light, insecticide, and fungicide sprays. Air movement around and through the vines minimizes the fungal disease such as *Botrytis*. The kiwifruit vine bears fruits on the current season growth, which arises from one-year-old shoot. The ideal fruiting shoots have short internodes and well-developed buds.

The pruning in kiwifruit should be done in such a way that the fruiting areas are available in current season growth. To achieve this, a pruning cycle by replacing laterals for 3–4 years is followed. In the beginning, a lateral arising from the main trunk is cut back in winter to provide a sufficient space for 4–5 fruiting shoots at 4–5 bud intervals between the shoots. The strong upright shoots arising at undesirable points are pruned during spring. This is true with the cultivar Hayward where the shoots of only medium vigor bears fruits. The summer pruning is performed to head back the fruiting shoots and thinning out of overlapping shoots. In fact, summer pruning is the selection and encouragement of correct laterals to bear fruits in the subsequent years and expose the vines to the sun. Summer pruning by pinching the terminal portion of bearing shoots at petal fall has resulted in higher yields of better quality fruits (Rana et al., 2011b).

7.15 INTERCROPPING AND INTERCULTURE

In the kiwifruit, the intercropping is feasible during the initial 3–4 years. The benefit of growing intercrops is to occupy the vacant space and supplement additional income to the farmers. It also ensures water and soil conservation, weed control, and improvement of the orchard environment and soil fertility. The intercrops such as soya beans, cowpeas, and turmeric, and vegetable crops such as tomato and zinger can be grown in the kiwifruit orchard. However, the basins of the kiwifruit vines should be mulched with 15 cm thick hay grass, especially under the rainfed situation. The growing of green manure crops is also effective for the organic nutrition of the kiwifruit orchards, and preferably leguminous plants should be selected as they are nitrogen fixers and improve the soil nutrient status.

7.16 FLOWERING AND FRUIT SET

The flowering begins after 2–3 years of planting, but good sizeable crops are borne after 4–5 years. The cultivar Hayward takes more time to produce worthwhile crop. The low production of crop in this cultivar is due to its lower percentage of flowering shoots, fewer flowering axils per shoot, and lesser flower per shoot (Brundell, 1975). The flowers are borne in the axil of 4–5 basal buds of the current season's growth. This may be single or in inflorescence. The initiation and development of floral primordia are late in the kiwifruit as compared to other deciduous crops. The flowers are developed in late summer and the differentiation of floral parts takes place only during the current spring

similar to that of grapes. Polito and Grant (1984) reported that in the Hayward kiwifruit, the dormant buds contained undifferentiated primordial in the leaf axils. The primordial remained undifferentiated until bud break and approximately two months before the full bloom when they became trilobed as bracts leading to the initiation of lateral floral primordial. However, the lateral flowers often aborted in this cultivar and the acropetal development of terminal flowers proceeded rapidly. Five to seven sepals were initiated, followed by a whorl of 5–7 petals. The stamen initiation was centripetal, and the anthers developed to produce a nonviable pollen. The centripetal portion of the floral apex was converted into a large compound pistil with numerous hollow styles, terminating by an open stigma. The amount and duration of flowering is influenced by a large number of internal and external factors such as the crop load in the preceding fruiting season, severity of pruning, leaf area, nutrition, availability of water, chilling, and prevailing temperature.

The anthesis in the kiwifruit takes place from the last week of April and is over by the third week of May (Rathore, 1984) under Shimla conditions of Himachal Pradesh. The flowers open first in the cultivar Abbott and last in

Hayward, whereas other cultivars have an intermediate trend. The application of 4% Dormex at 40 days prior to the anticipated date of bud break has been reported to advance the bud break by one week and fruit set by five days (Babita and Rana, 2013). Pollination is the most important factor determining the economic return to the kiwifruit orchardist. The kiwifruit is functionally dioecious. The female flowers, though have the stamens, do not produce a viable pollen (Ford, 1971; Eynard and Gay, 1978). Hence, for proper pollination and fertilization to trigger the development of the fruit, the planting of staminate cultivars in the proper ratio is essential. Generally, most of the pistillate flowers develop normally into fruits in all the cultivars as there is practically no fruit drop. The kiwifruit of acceptable commercial size should contain more than 1000 seeds (Hopping, 1976; Davison, 1977).

The role of pollinating insects is vital in kiwifruit, being a dioecious plant. The honeybees act as a pollinator in the kiwifruit, but sometimes the optimum crop load is not achieved because the flower of the kiwi plant is not a good source of nectar. The lack of pollination leads to economic losses to the farmers. Under such conditions, the hand pollination has been found to be beneficial. The adequate pollination demands the synchronization of the pistillate and staminate flowers. Hopping and Jerram (1979) have reported that under day/night temperature of 24 °C/8 °C, the pollen germinated and most tube penetrated into style within 7 hours of pollination. The pollen from fresh male flower is considered better for fertilization (Ford, 1971).

7.17 FRUIT GROWTH, DEVELOPMENT, AND RIPENING

The growth period of the kiwifruit is very long as compared to other

Male Flower

Hand Pollination

deciduous fruits. Rathore (1981) reported a cyclic pattern of fruit growth in four cultivars of the kiwifruit. He further reported that the kiwifruit grows rapidly in the first phase of fruit growth for about seven weeks after fertilization and becomes slow in the second phase for the next five weeks but again accelerates from the 13th week onward, continuing at a steady rate up to the 31st week. He could not, however, observe marked differences among the third, fourth and fifth phases as was observed by Pratt and Reid (1974). A triple sigmoid curve of volume growth in Abbott, Bruno, and Hayward cultivars has been reported by Pratt and Reid (1974) and Scienza et al. (1983). Several researchers have reported the major increase in fruit weight and fruit volume in the first 10 weeks after anthesis (Pratt and Reid, 1974; Hopping, 1976; Okuse and Ryugo, 1981). Three different phases of growth, namely, rapid growth during the first 8–9 weeks (Stage I), slow growth for about 3 weeks (Stage II), and then a second period of growth (Stage III) for 5–10 weeks have been recorded by several scientists (Pratt and Reid 1974; Hopping, 1976; Rana and Rana, 2003). However, there are conflicting reports in the literature on the overall pattern of fruit development which probably varies between the cultivars.

The time course changes in the physicochemical composition of the kiwifruit during growth and development have been analyzed (Okuse and Ryugo, 1981; Ben-Arie et al., 1982; Fuke and Matsuoka, 1982; Lees, 1982; Reid et al., 1982; Matsumoto et al., 1983). Among the chemical constituents, the changes occur mainly in carbohydrates, organic acids, tannins, polyphenolics, cytokinin-like substances, and pigments. Total sugars are low, and starch is the main carbohydrate in early stages of growth which gets hydrolyzed as the fruits reach maturity (Okuse and Ryugo, 1981; Fuke and Matsuoka, 1982). A marked decrease in starch concentration and a consequent increase in sugars were found between 17 and 20 weeks of fruit growth (Okuse and Ryugo, 1981; Reid et al., 1982). Tannins and polyphenolic substances are low in mature fruits and types of cytokinin change with the development of fruit (Okuse and Ryugo, 1981).

The kiwifruit does not follow a typical respiration pattern of either climacteric or nonclimacteric fruits. Wright and Heatherbell (1967) considered it a nonclimacteric fruit, but when the fruit ripened in a natural condition, it exhibited the climacteric peak of respiration and, therefore, is a climacteric type (Pratt and Reid, 1974; Zhang et al., 1985). Endogenous ethylene production is very low in the kiwifruit, but as small as 0.1 ppm ethylene can trigger ripening in a cold storage

Kiwifruit 439

where the effect decreases with low temperature (Harris, 1981; Reid and Harris, 1977; McDonald and Harman, 1982).

7.18 HARVESTING AND YIELD

The stage of maturity at which the fruit is harvested is of paramount importance in the kiwifruit which is harvested hard and there is no perceptible change in color. The determination of fruit maturity is also very important for obtaining good price from the produce. The early harvesting of the kiwifruit results in the development of poor flavor and aroma. On the other hand, if the fruits are left on the vine, they will mature, ripen, and eventually abscise from the plant at a slightly over-ripe stage. Individual fruits on the vine are observed to ripen at quite different times. The irregularity of natural ripening pattern makes it impractical to harvest and cause damages during harvesting, grading, and packing operations. Keeping all the factors in mind, the physiologically mature fruits can be harvested and allowed to ripen off the vine.

The kiwifruits having 6.2% TSS are ideal for harvesting. The delay in harvesting deteriorates their storability. They are easily harvested by snapping off the fruit at the abscission layer at the base of the stalk. Under Indian conditions, the fruits are plucked from mid-October to the last week of December, depending upon the climatic conditions and cultivars. Harvesting coincides with the commencement of the dormant season, which means that allowing fruit late on the vines will make them prone to frost and birds damage. In a study on "Allison" kiwifruit, flesh firmness, total soluble solids, total sugars were considered as dependable indices of maturity, which could be used in conjunction with days from full bloom for assessing optimum maturity (Rana et al., 2003).

A large number of factors such as cultivars, training system, pollination, nutrition, age, irrigation, and climatic conditions are responsible to determine the yield. Under Indian agro-climatic conditions, the approximately 60–120 kg/plant yield can be achieved at the age of seven years of planting. The plant grown on trellis produce about 25 ton/ha on an average yield after seven years of planting in New Zealand, and comparable yields were also obtained in the USA (Smith, 1961).

7.19 POSTHARVEST HANDLING AND STORAGE

After harvesting, the kiwifruit should be individually inspected for quality and graded into different sizes on the basis of fruit size or weight before packaging. These operations take place after harvesting, which is very much essential for assuring high-quality commodity to export market. This can be done manually by the visual inspection of fruit. During the grading, the caution must be taken to remove the damaged, blemished or bruised, diseased, or misshapen fruit; otherwise, they contaminate the other fruits during packing and transportation. In India, there are no packaging and grading standards for the kiwifruit. However, the fruits after harvest are categorized into three grades, namely, A Grade (>80 g), B Grade (50–80 g), and C Grade (<50 g). The fruits fetch price according to the grade, but definitely, international market demands a fruit weighing more than 100 g.

The importance of the kiwifruit among the growers of the world is partly attributed to its excellent keeping quality, which is very much essential for export to get maximum economic return for a longer period of time. Kiwifruits are having the essential qualities and can be stored for a quite long time even at room temperature with slightly judicious management of temperature, humidity, and ethylene concentration. In a cold climate, storage at room temperature after wrapping the fruits in film provided with ethylene absorbent has been found very economical. In controlled atmospheric storage, it can be stored for more than six months. The kiwifruit when shrink-wrapped in heat shrinkable films like polyolefin (13μ), low-density polyethylene (25μ), and especially Cryovac (9μ) reduces the respiration as well as ethylene evolution rate, which increases the shelf life of the kiwifruit (Sharma et al., 2012). Postharvest treatment with polyamines such as spermine and spermidine has been found to extend the shelf life of the kiwifruit at 22 ± 4 °C temperature and $65\% \pm 5\%$ relative humidity (Jhalegar et al., 2012). Park et al. (2015) reported that the kiwifruit preconditioned at 20 °C for two days with RH 85% in a growth chamber coupled with 1-methylcyclopropene treatment extended the shelf life.

7.20 PROCESSING AND VALUE ADDITION

A number of processed products from the kiwifruit have been prepared in different kiwifruit growing countries of the world. Like most fruits, the kiwifruit undergoes changes in chemical, physical, nutritional, and flavor characteristics during processing; particularly the color, flavor, and texture of the processed kiwifruit are different from those of the fresh kiwifruit. The kiwifruit has a hairy skin that must be peeled before processing. Several peeling methods have been investigated by food scientists. Beutel et al. (1976) reported that a gas-flame technique for peeling the kiwifruit was impractical, whereas lye peeling worked satisfactorily. The partially soft fruit was peeled in a 15% lye solution for 90 s at boiling temperature, and then washed in cold water. The lye-peeled fruit retained a higher concentration of ascorbic acid than the fruit peeled by hand.

The fruits required for commercial canning in the sliced form should have a soluble solid concentration in the range of 10°–13 °Brix. Fruits are normally canned in the 35 °Brix sugar solution in acid-resistant cans and seamed under vacuum. The flavor of the canned kiwifruit is quite different from that of the fresh product. The freezing of the kiwifruit as a whole peeled fruit, slices or dices, is also practiced, which differs a little from the freezing of other fruits. Beutel et al. (1976) suggested a method by dipping fruit slices in a solution of sucrose (12%), ascorbic acid (1%), and malic acid (0.25%) for 3 min prior to freezing in order to inhibit enzymatic and nonenzymatic changes during processing and storage.

Benítez et al. (2013) have reported that Aloe vera coating of the sliced kiwifruit reduced the respiration rates and microbial spoilage. The texture of the uncoated samples deteriorated more rapidly than the treated slices during storage. Aloe vera coating also improved the quality of stored kiwifruit slices. The best results were obtained with 5% coating, indicating that this may be a healthy alternative coating for a fresh-cut kiwifruit. Besides this, several other products such as, juice, nectar, wine, jam, leather, and candy can also be prepared from the kiwifruit.

7.21 DISEASE, PESTS, AND PHYSIOLOGICAL DISORDERS

Fortunately, diseases in the kiwifruit do not cause much economic losses. Among the known diseases of the kiwifruit in the world, these are caused by the bacteria and the fungi. The brief description of various diseases is described as under.

7.21.1 FUNGAL DISEASES

7.21.1.1 ROOT ROT

This disease is caused by soil borne fungi, *Phytophthora* spp. The infection of these fungi first appears in early summer and results in delayed bud break, wilting of leaves, and reduction in leaf size. The infection of fungus first being by invading and rotting outer roots, and the rots spread toward the crown.

7.21.1.2 ARMILLARIA ROOT ROT

Yellowing of leaves and decline in vine vigor are the initial symptoms of infection by these wood-rooting fungi, and slowly the vine usually dies within one or two years. The symptoms on vine with above ground have rotten, pulp wood, with white mycelium under the bark and one or more of the main roots along the crown. Slowly as the rot progresses, white fungal mat advances the bark up to the lower trunk. The main characteristics of the disease are black fungal strands, which grow away and can infect roots of adjacent plant. To prevent the spread of the diseases, the fossils of infected trees must be burned and fumigation of infested soil or by putting the barrier between infected and healthy plants can avoid the further destruction of orchard.

7.21.1.3   LEAF SPOTS

This is caused by several fungi named *Alternaria alternate, Botryosphaeria spp., Cladosporium spp., Colletotrichum acuitatum, Epicoccumn purpurascens, Fusarium spp., Glomerella cinglata, Penicillium spp., Phomaexiyna and Phomopsis spp.* White lesions appear on the leaves which later on become extensive brown and irregular in shape (Hawthorne et al., 1982). It seems that these fungi can colonize only injured leaf tissue and are unable to attack healthy tissue directly. Safe and judicious management practices that minimize leaf damage in the kiwifruit orchard should also reduce the infection of fungal leaf spotting.

7.21.1.4   LEAF INFECTION

Botrytis cinerea and *Sclerotinia sclerotiorum* can invade leaves to cause large spreading lesions (Pennycook, 1985). During the wet summer weather which is very much congenial for its growth in wounded leaf tissue spread into healthy leaf from infected tissue.

7.21.1.5   STORAGE ROT

The disease caused by fungus *Botrytis cinerea* during storage. The rots begin at the stem and advance toward the distal end, while damaged or bruised fruits become infected at any point of injury. Flesh of the diseased fruit is glassy and water soaked. The affected part is darker than the healthy part, and at an advanced stage, white fluffy mycelium may spread to the adjacent healthy fruits causing secondary infection. Spray of fungicide at late blossom to check the infection on flower parts and again before harvest to protect the fruit in storage (Pennycook, 1985).

7.21.2   BACTERIAL DISEASES

7.21.2.1   BACTERIAL LEAF SPOT AND BACTERIAL BLOSSOM BLIGHT

It is caused by *Pseudomonas virdiflava* and the leaves infected from late spring onward. Initial symptoms of the bacterium are dark lesions (Young, 1984). Later on, the lesions may become extensive and the necrotic tissue eventually disintegrates. The effect of infestation on vine vigor is not significant, but the pathogen can also rot blossom called blossom blight and reduce the yield. Early symptoms of blossom blight are brown, sunken areas as sepals of an unopened flower bud, but these lesions may not develop further. When infection occurs within the bud, the petals are yellow oranges as the flowers open and the inner tissues are seen to be dark brown and rotten. In heavily infested female flowers, all anthers and filaments are rotten and styles are stunted often brown. Blossoms with infection are usually shed, but lightly infected ones may develop into small or distorted fruits.

The pathogens infect both kiwifruit leaves and flowers. It is likely that the bacteria are present on the vine throughout the year. Wet weather or rain during bud break and early flowering favors the infection. No satisfactory control of the disease has yet been found, but the spraying of bactericides before the bud break certainly reduces the infection uptown some extent.

7.21.3 PESTS

Initially, kiwifruit cultivation is considered to be pest free. This is because of the small extent of plantings and partly because the fruit is destined at the local market which would tolerate minor imperfections. Slowly after the development of export market, the growers became aware that many blemishes caused the pest of kiwifruits were introduced. Important insect pests that cause damage to the kiwifruit are leaf roller, greedy scale, red spider mite, and thrips. A leaf-roller caterpillar usually eats leaves and when population builds up in the absence of any control, it damages the fruit by boring into the pulp and the fruit may drop prematurely. The spray of any suitable insecticide after petal fall followed by another spray after one month will be sufficient to control this pest. Greedy scale, mites, and thrips can similarly be controlled.

7.21.4 NEMATODES

Cobb (1929) and Siddiqui et al. (1973) reported that the nematodes (eelworms) that are often associated with kiwifruit roots are mainly root knot nematode (*Meloidogyne incognita*). Besides this, several other nematodes were also found associated like lesion nematode (*Pratylnchus*), stubby root nematode (*Paratrichodorus*), and stunt nematode. Root-knot nematode causes a characteristic distortion or knotting and galling of the kiwifruit root system. Heavily infested roots are much shorter than normal and have fewer branches. Before planting the kiwifruit orchard, it is essential to ensure that the planting material should be free from nematode, which has been grown under hygienic nursery practices. This leads the plants to have the best possible chance of good establishment and rapid early growth.

7.21.5 NONPATHOGENIC DEFECTS

Among abiotic stresses, damages by frost, hail, sun scald, and wind are common in kiwifruit cultivation. At the outset, an unseasonal spring frost can cause the distortion of the young fruitlet with the fruit tapering toward the blossom end, which is markedly concave. Hail besides damaging shoots causes typical pitting, scarring, and distortion of fruit. The severity of symptoms will depend on the size and duration of the hailstorm. Damaged fruits are scarred with unsightly blemishes, although the lesions will often heal. Fruit exposed to the sun are prone to sun scald. This leaves a typical sunken, brown, leathery scar over the affected surface, rendering the fruit useless even for processing. Sometimes, the strong winds during the spring season damage the young shoots, which lead to the loss of fruit bearing shoots. In addition to fruit rejections due to pests or disease infestation, there are a number of rejection categories resulting from nonpathogenic causes that are important in the fruit-grading operation. These categories include the following.

7.21.5.1 FLATS

A flat is a fruit whose width is greater than its length. Flats are difficult to pack in the standard single-layer tray. Flat fruits are more likely to be produced on the two proximal flowering buds of a shoot. Various influences that retard vegetative growth around the time of fruit cause this disorder. These influences include tipping during normal summer pruning and the application of growth retardant materials at this time.

7.21.5.2 DROPPED SHOULDER

The fruit appears lop-sided with the shoulder at one side sloping away abnormally. The cause of this condition, which can occur on both large and small fruit, is not completely known. It may be the result of inadequate pollination. A sloped angle at 15° is considered the limit for export fruit.

KEYWORDS

- **kiwifruit**
- **Chinese gooseberry**
- *Actinidia* spp.
- **kiwi berry**
- *Actinidiaceae*

REFERENCES

Babita; Rana, V.S. Effect of Dormex on bud break, flowering and yield in kiwifruit cv. Hayward in mid hill zone of Himachal Pradesh. *Journal of Horticultural Sciences* 2013, 8, 47–50.

Beever, D.J.; Hopkirk, G. *Fruit development and fruit physiology*. In *Kiwifruit Science and Management*. Warrington, I.J.; Weston, G.C., Eds.; Ray Richards (in association with New Zealand Soc. Hort. Sci.), New Zealand, 1990; pp. 97–121.

Ben-Arie, R.; Gross, J.; Sonego, L. Changes in ripening-parameters and pigments of the Chinese gooseberry (kiwi) during ripening and storage. *Scientia Horticulturae*. 1982, 18, 65–70.

Benitez, S.; Achaerandio, I.; Sepulcre, F.; Pujola, M. Aloe vera based edible coatings improve the quality of minimally processed "Hayward" kiwifruit. *Postharvest Biology and Technology*. 2013, 81, 29–36.

Beutal, J.A.; Winter, F.H.; Manners, S.C.; Miller, M.W. A new crop for California: Kiwifruit. *California Agriculture*. 1976, 30, 5–7.

Brio, Y.; Erez, A.; Bravdo, B.A. The effect of winter climate on bud dormancy of kiwifruit (*Actinidia chinensis* Planch). *Alon Hanotea*. 1985, 39, 1017–1026.

Brundell, D.J. Flower development of the Chinese gooseberry (*Actinidia chinensis* Planch). II. Development of the flowering bud. *New Zealand Journal of Botany*. 1975, 13, 485–496.

Clark, C.J.; Smith, G.S.; Prasad, M.; Cornforth, I.S. Eds. *Fertilizer Recommendations for Horticultural Crops Grown in New Zealand*. Ministry of Agriculture and Fisheries, Wellington, New Zealand, 1986; pp. 23–25.

Cobb, N.A. Root knot nematode on *Actinidia chinensis, A. arguta. A purpurea. Plant Disease Reporter, Supplement*. 1929,73, 377.

Costa, G.; Biassi, R.; Enynard, I.; Bergamini, A.; Testolin, R.; Messina, R. Danni da freddo all Actinidia. *Revista di Frutticoltura de ortofloricoltura*. 1985,47, 13–15.

Dadlani, S.A.; Singh, B.P; Kazim, M. Chinese gooseberry a new fruit. *Indian Horticulture*.1971,16, 13–15.

Davison, R.M. *Flowering and pollination in kiwifruit*. In *Proc. Citrus and Subtrop. Seminar, Waitangi*. New Zealand Ministry of Agriculture and Fisheries, Whangarei,1977; p. 6.

Eynard, I.; Gay, G. *Alcuniaspetti dell allegagionenell "actinidia"*. In *Incontro Frutticolo 1;* Eynard ,I., coord.; Universita di Torino, Torino, 1978; pp. 131–136.

FAOSTAT. www.faostat.org (accessed March 10, 2019), 2018.

Ferguson, A.R. A kiwifruit by any other name. *New Zealand Fruit and Produce*. 1985, 26, 28–30.

Ferguson, A.R. Botanical nomenclature: *Actinidia chinensis, Actinidia deliciosa* and *Actinidia setosa*. In: Warrington, I.J.; Weston, G.C. Eds.; *Kiwifruit: Science and Management*. Ray Richards (in association with the New Zealand Society for Horticultural Science), Auckland, 1990; pp. 36–57+2 plates.

Ferguson, A.R. Kiwifruit: A botanical review. *Horticultural Reviews*. 1984, 6, 1–64.

Ford, I. Chinese gooseberry pollination. *New Zealand Journal of Agricultural Research*. 1971, 122, 34–35.

Fuke, Y.; Matsuoka, H. Changes in sugar, starch, organic acid and free amino acid contents of kiwi fruits during growth and after ripening (in Japanese; seen in abstract only). *The Japanese Society for Food Science and Technology.* 1982, 29, 642–648.

Gremminger, U.; Husistein, A.; Aeppli, A. Reaction of young *A. chinensis* plants o frost. Schweizerische Zeitschrift Fiirobst. *Obst und weinbau.* 1982,118, 110–115.

Harris, S. Ethylene and kiwifruit. *The Orchardist of New Zealand.* 1981, 54, 105.

Hawthrone, B.T.; Rees- George, J.; Samuels. G.J. Fungi associated with leaf spots and post harvest fruit rots of kiwifruit (*Actinidia chinensis*) in New Zealand. *New Zealand Journal of Botany.* 1982, 20, 143–150.

Hopping, M.E.; Jerram, E.M. Pollination of kiwifruit (*Actinidia chinensis* Planch.): Stigma style structure and pollen tube growth. *New Zealand Journal of Botany.* 1979, 17, 233–240.

Hopping, M.E. Structure and development of fruits and seeds in Chinese gooseberry (*Actinidia chinensis* Planch.). *New Zealand Journal of Botany.* 1976, 14, 63–68.

Huang, H.W. *Kiwifruit: The Genus Actinidia.* Science Press, Beijing. 2016; p. 334.

Huang, H.W.; Ferguson, A.R. Kiwifruit (*Actinidia chinensis* and *A. deliciosa*) plantings and production in China. *New Zealand Journal of Crop and Horticultural Science.* 2003, 31, 197–102.

Jhalegar, Md. J.; Sharma, R.S.; Pal, R.K.; Rana, V. Effect of postharvest treatments with polyamines on physiological and biochemical attributes of kiwifruit (*Actinidia deliciosa*) cv. Allison. *Fruits.* 2012, 67, 13–22.

Kumar, J.; Rana, S.S.; Verma, H.S. The exotic kiwi has found a home in the mid hills of Himachal Pradesh. *Indian Horticulture.* 1995, 40, 28–29.

Lees, H.M.N. Kiwifruit (*Actinidia chinensis*) Hayward variety: increase in fruit size from March to June, during three seasons. *The Orchardist of New Zealand.* 1982, 55, 72–75.

Li, H.L. A taxonomic review of the genus *Actinidia. Journal of the Arnold Arboretum.* 1952, 33, 1–61.

Li, J.Q.; Li, X.W.; Soejarto, D.D. Actinidiaceae. In: *Flora of China*, vol. 12. Wu, Z.Y.; Raven, P.H.; Hong, D.Y., Eds.; Science Press/Missouri Botanical Gardens, Beijing/St. Louis, MO, 2007; pp. 334–360.

Liang, C.F. On the distribution of *Actinidias. Guihaia*, 1983, 3, 229–248.

Lindley, J. A natural system of botany; or a systematic view of the organization, natural affinities, and geographical distribution, of the whole vegetable kingdom. 2nd ed. Longman, London, 1836.

Matsumoto, S.; Obara, T.; Luh, B.S. Changes in chemical constituents of kiwifruit during postharvest ripening. *Journal of Food Science.* 1983, 48, 607–611.

McDonald, B.; Harman, J.E. Controlled- atmosphere storage of kiwifruit. I. Effect on firmness and storage life. *Scientia Horticulturae.* 1982, 17, 113–123.

NHB. *Area and Production of Horticulture Crops for 2017–18* (Final). http://nhb.gov. in (accessed March 10, 2019), 2017.

NHB. *Horticulture Crops Estimates for the Year 2015–16.* http://nhb.gov.in (accessed March 10, 2019), 2016.

Okuse, I.; Ryugo, K. Compositional changes in the developing "Hayward" kiwifruit in California. *Journal of the American Society for Horticultural Sciences.* 1981, 106, 73–76.

Park, Y.S.; Myeng, H.I.; Gorinstein, S. Shelf life extension and antioxidant activity of "Hayward" kiwi fruit as a result of prestorage conditioning and 1-methylcyclopropene treatment. *J Food Sci Technol.* 2015, 52, 2711–2720.

Pennycook, S.R. Fungal fruit rots of *Actinidia deliciosa* (Kiwifruit). *New Zealand Journal Experimental Agriculture.* 1985, 13, 289–299.

Polito, V.S.; Grant, J.A. Initiation and development of pistillate flowers in

Actinidia chinensis. Scientia Horticulturae. 1984, 22, 365–371.

Pratima, P.; Sharma, N.; Rana, V.S. Effect of deficit irrigation on chlorophyll stability index, relative water content and anatomical characteristics of different kiwifruit cultivars. *Indian Journal of Ecology.* 2017, 44, 704–711.

Pratt, H.K.; Reid, M.S. Chinese gooseberry: Seasonal patterns in fruit growth and maturation, ripening, respiration and the role of ethylene. *Journal of the Science of Food and Agriculture.* 1974, 25, 747–757.

Rana, N.; Walia, A.; Rana, V. Isolation of an antifungal protein from kiwifruits. *International Journal of Food and Fermentation Technology.* 2011a, 1, 129–131.

Rana, V.; Rana, N. Studies on fruit growth and organic metabolites in developing kiwifruit. *Indian Journal of Plant Physiology* 2003, 8, 138–140.

Rana, V.S.; Babita. Effect of rooting media and IBA on rooting behaviour and vegetative growth of kiwifruit (*Actinidia deliciosa* Chev.). *Indian Journal of Ecology* 2016, 43, 343–345.

Rana, V.S.; Basar, J.; Rehalia, A.S. Effect of severity of summer pruning on the vine characteristics, fruit yield and quality of kiwifruit. *Acta Horticulturae.* 2011b, 913, 393–399.

Rana, V.S.; Chopra, S.K.; Singh, N. Standardization of optimum harvest maturity in relation to storage performance in "Allison" kiwifruit (*Actinidia chinensis*). *Indian Journal of Agricultural Sciences* 2003, 73, 256–260.

Rathore, D.S. Physico-chemical evaluation of the fruits of four cultivars of Chinese gooseberry *Actinida chinensis*. *Indian Journal of Horticulture.* 1981, 32, 62–65.

Rathore, D.S. Propagation of Chinese gooseberry from stem cuttings. *Indian Journal of Horticulture.* 1984, 41, 237–239.

Reid, M.S.; Harris S. Factors affecting the storage life of kiwifruit. *The Orchardist of New Zealand.* 1977, 50, 76–77.

Reid, M.S.; Heatherbell, D.A.; Pratt, H.K. Seasonal patterns in chemical composition of the fruit of *Actinidia chinensis*. *Journal of the American Society for Horticultural Sciences* 1982, 107, 316–319.

Scienza, A.; Visai, C.; Conca, E.; Valenti, L. Relationship between development, fruit ripening and the presence of endogenous hormones in *Actinidia chinensis*. In: *Attidel II IncontroFrutticolo SOI Sull 'Actinidia,* Udine, Udine, Italy, Regional Center for Agricultural Experimentation for Friuli-Venezia Giulia and the Fruit-Growing Section of SOI. 1983; pp. 401–421.

Seal, A.G. The plant breeding challenges to making kiwifruit a worldwide mainstream fresh fruit. *Proceedings of the International Symposium on Kiwifruit, ISHS 2003. Acta Hort.,* 2003, 610.

Sharma, R.R.; Pal, R.K.; Rana, V. Effect of heat shrinkable films on storability of kiwifruits under ambient conditions. *Indian Journal of Horticulture.* 2012, 69, 404–408.

Siddiqui, I.A.; Sher, S.A.; French, A.M. *Distribution of Plant Parasitic Namatodes in California.* Division of Plant Industry, California Department Food and Agriculture, Sacramento. 1973; p. 168.

Smith, R.L. Exotic crops; Chinese gooseberry. *Western Fruit Growers.* 1961, 115, 19–20.

Testolin, R.; Messina, R. Winter cold tolerance of kiwifruit. A survey after winter frost injury in Northern Italy. *New Zealand Journal Experimental Agriculture.* 1987, 15, 501–504.

Wright, H.B.; Heatherbell, D.A. A study of respiratory trends and some associated physio-chemical changes of Chinese gooseberry fruit *Actinidia chinensis* (Yang-Tao) during the later stages of development. *New Zealand Journal of Agricultural* 1967, 10, 405–414.

Young, J.M. Little light at the end of the bud rot tunnel. *Southern Horticulture* (N.Z.) 1984, 13, 12–14.

Zhang, S.; Meng, S.; Li, Y.; Zhang, J.; An, H.; Cai, D.; Fu, K. Studies on respiratory climacteric and ethylene release during storage of Chinese gooseberries (*Actinidia chinensis* Planch.) *Acta Horticulturae Sinica*. 1985, 12, 95–100.

CHAPTER 8

STRAWBERRY

G. QUINTERO-ARIAS[1], J. VARGAS[1], J. F. ACUÑA-CAITA[1], and
J. L. VALENZUELA[2*]

[1]*Departamento de Ingeniería Civil y Agrícola. Universidad Nacional de Colombia. Bogotá, República de Colombia*

[2]*Departament of Biology and Geology,
Campus of International Excellence (ceiA3), CIAIMBITAL,
Universidad de Almería, 04120 Almería, Spain*

Corresponding author. E-mail: jvalenzu@ual.es

ABSTRACT

This document presents a summary of aspects to be taken into account for the production of strawberry crop—starting from a general description of crop with data from areas and places of production, going through the taxonomy and physiology, to advance with details of preparation and growing planning, which involve aspects of nutrition and response to nutrients, irrigation design and fertilization. It also includes some concepts related to pests, diseases, and crop management as pruning to prevent its appearance. Some aspects related to harvest and postharvest are analyzed from the physiological and commercial maturity vision.

8.1 GENERAL INTRODUCTION

Strawberries are a very popular dessert with a great demand due to its flavor, aroma, and attractiveness. Its pleasant flavor is one of the reasons that its cultivation goes back to antiquity, in the Roman Empire, and perhaps also in the classical Greek, although it is not easy to find ancient references about the cultivation of this specie. Strawberry was also cultivated in the Mediaeval Age, where it was considered as a powerful aphrodisiac. But over the following centuries, the growing area expanded and an era of splendor began. When, after the discovery of America, new species of the genus *Fragaria* arrived in Europe, then took place a natural hybridization and the beginning of breeding programs that gave rise to numerous hybrids and cultivars. Theses strawberry plants, which are now cultivated in all over the world, are the ancestors of all modern cultivars.

Strawberries grow better in areas where there are cold winters. Their edaphic requirements are different according to the cultivars. In general, they avoid clay soils since they need a good drainage and a high organic matter content. In order to obtain bountiful harvests, irrigation and fertilization programs are essential, but different factors such as the cultivar, the type of soil, the type of irrigation system used, the climatic conditions, as well as pest control must always be taken into account. Handling all of these factors well can lead to successful cultivation.

Strawberries are the most economically important soft fruit worldwide due mainly to their intense red color, crunchy and juicy texture, and delicious flavor that make the strawberry very attractive and easy to eat. But, apart from this, strawberry is also used in the agrifood industry.

8.2 AREA AND PRODUCTION

Annual worldwide production of strawberries can reach over 4.3 million tons. More than 75 countries have produce strawberry; however, the bulk of this production is largely concentrated in the Northern Hemisphere. The biggest producers of strawberries are China and the USA, followed by Turkey, Spain, Egypt, and Mexico (Figure 8.1). According to the Food and Agriculture Organization of the United Nations, the United States of America with a worldwide production of about

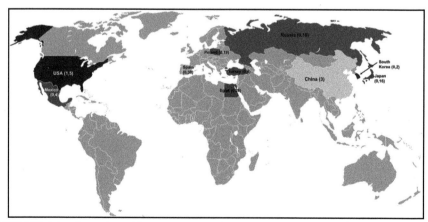

FIGURE 8.1 Countries with a greater strawberry production of 150,000 tons per year. Production in millions of tons is shown in the brackets (FAO, 2017).

30% represents by far the largest producer of strawberry (FAO, 2017). The USA has been very successful in achieving higher productivity and improved competitiveness, mainly due to efficient agricultural techniques. The production estimate of the United States is 56.4 tons/ha in their nearly 25,000 ha land, leading to a total production of more than 1400 million tons. On the contrary, Turkey with the 12,000 ha area reach the production of only 390,000 tons due to lower productivity, that is, 28.5 tons/ha. Spain with an area dedicated to strawberry about 7500 ha has a production of nearly 42 tons/ha, which made up to 380,000 tons (Figure 8.2).

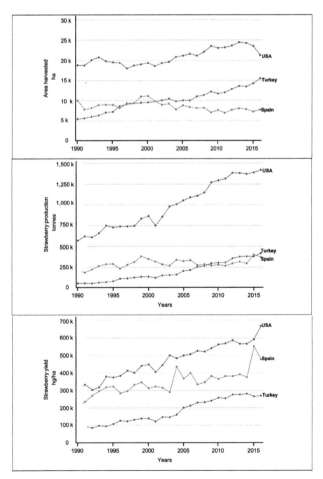

FIGURE 8.2 Area, production and yield for the USA, Turkey, and Spain during the 1990–2015 period (FAO, 2017).

8.3 MARKETING AND TRADE

The trade of strawberries, either fresh or processed, has shown an upward trend in recent years, despite presenting significant price variability, exchange volumes, and overall value of trade. It is estimated that worldwide, the largest volume of strawberry exports occurs between the months of February and May, whereas the minimum export occurs between July and October. For the European market, there are peaks of production between the months of February and May, whereas between the months of July and December, there is a shortage of the product in markets. The main consumers in these markets are Germany, France, the United Kingdom, Italy, Belgium, Portugal, Poland, and Russia that appears in recent years as a rising target. As with all commodities, prices of strawberries fluctuate over time, and the global average price of fresh strawberries was almost double that of frozen strawberries (USD 2.95 and 1.58 per kg, respectively). Strawberry consumption is close to 1.6 kg per capita per year in the European countries; however, in the United States of America, it is higher, reaching almost 3.7 kg per capita per year. In the main exporting countries, more than 70% of the production is intended to the fresh market, while the remainder 30% is destined to processed fruit, mainly frozen. Spain is the world's largest exporter of fresh strawberry with almost 300,000 tons per year dedicated to export. In fact, close to 75% of the total strawberry imported by European countries come from Spain. The United States is the world's second largest strawberry exporter with 140,000 tons per year; however, it is paradoxically the largest strawberry importer in the world with more than 160,000 tons strawberries imported per year. Other countries with a high import ratio are Germany, Canada, France, and the United Kingdom. These countries, along with the USA, represent about 43% of world imports. The market screening clearly shows the existence of a growing demand; consequently, world strawberry production is growing significantly, although the sown area has been reducing at the same time (Strick, 2007). The volume of international trade has been growing almost constantly, and it is expected to exceed USD 2600 and 675 million in terms of fresh and frozen strawberries by 2029.

8.4 COMPOSITION AND USES

The importance of the strawberry is not only due to its commercial interest but also due to its nutritional values. Strawberry is a source of vitamin C, antioxidants, fiber, and phenolic compounds; all of these are necessary for a healthy diet (Tulipani et al., 2008). Indeed, strawberries are richer in vitamin C than citrus, and 150 g of strawberry supplies approximately 150% of daily requirement (Crespo et al., 2010). Moreover, strawberry helps regulate blood pressure due to its high levels of magnesium and potassium. In addition, the high content of niacin in strawberry helps to increase the levels of high-density lipoprotein which has been found to decrease the risk of cardiovascular diseases (Azodanlou et al. 2003; Mahmood et al., 2012). Strawberry is also a fruit with a low sugar content, a hypo-caloric fruit which only contribute a calorie content of 24 kcal per 100 g. In this way, strawberries are recommended to be included into a varied and balanced diet which is a prerequisite for good health (Sesso et al., 2007; Tudor et al., 2015). The chemical composition of a strawberry is as follows: 89.6% of water, 7% of carbohydrates, 0.7% proteins, 0.5% lipids, and 2.2% of fiber. The fruit contains 2.6% glucose, 2.3% fructose, and 1.3% sucrose, and with regard to minerals, it contains mostly potassium (164 mg/100 g), phosphorus (21 mg/100 g), and calcium (21 mg/100 g). Anthocyanins contributes to the bright red color of the fruit and are associated with strong antioxidant activity.

In addition, strawberry is a perfect fruit for the processing industry, and a variety of products derived from it such as preserves, flavoring, concentrates, jellies, purees, processed sliced berries, juices, and extracts are very high. Moreover, strawberries are used as ingredients in other food products and snacks such as yoghurts, ice cream, sweets, gummy candies, cakes, breakfast cereals, and chocolate bars.

TABLE 8.1 Strawberry Nutrition Facts

Kilocalories	35	Fiber	2.2 g
Water	85 ml	Potassium	190 mg
Protein	0.7 g	Magnesium	12 mg
Fat	0.5 g	Provitamin A	5 mg
Carbohydrates	7 g	Vitamin C	60 mg
Vitamin E	0.23 mg	Vitamin B6	0.06 mg
Phosphorus	26 mg	Iron	0.46 mg
Calcium	21.5 mg	Folate	20 mg
Total phenols	58–210 mg	Total anthocyanins	55–145 mg

Source: USDA (2017).

8.5 ORIGIN AND DISTRIBUTION

The scientific name for the genus of strawberry is *Fragaria*, and it belongs to the family Rosaceae. The most ancient record about strawberry was wrote by Pliny in 23–79 AD (Darrow, 1966). In North America, *Fragaria virginiana* (*F. virginiana*) was cultivated by pioneers because of the two reasons: first, it was a native strawberry; and second, it has the ability to withstand drought and severe weather conditions. This was brought to Europe in the early years of 17th century where other species such as *Fragaria vesca* (*F. vesca*) was cultivated in France and Northern Europe, whereas in Russia mainly *Fragaria moschata* (*F. moschata*) was cultivated. Perhaps the most important event in relation to strawberry took place 300 years ago when a French Navy sailor came back to France from Chile and Peru with some specimens of *Fragaria chiloensis* (*F. chiloensis*). Then these plants were spread and put in contact with *F. virginiana* and both were cultivated together. A natural hybrid came up with the result of a hardier plant with big and better fruits. This hybrid was known as *Fragaria × ananassa* which has since been spread around the world. *Fragaria × ananassa* is octoploid, whereas wild strawberries are diploid, and the increased ploidy results in higher production and larger fruits. By contrast, with this crossing, some lower organoleptic qualities were also reached; however, their fruits showed bright red color and juicy texture (Khoshnevian et al., 2013).

Strawberry is one of the most popular crops since there is a great demand of this fruit—both for fresh consumption and for the fruit-processing industry. This has made it a characteristic crop in many areas of the world, as can be deduced from the data presented in Sections 8.2 and 8.3.

8.6 BOTANY AND TAXONOMY

Strawberry (*Fragaria × ananassa*) is a typical hardy, perennial, and roseate plant. It is defined taxonomically as a species with:

Vegetal kingdom
Order: Rosales
Family: Rosaceae
Subfamily: Rosoideae
Genre: *Fragaria*
Species: *Fragaria sp.*

8.6.1 ROOT SYSTEM

Strawberry has fasciculated roots composed of a percentage of greater volume dedicated to support the plant and other parts to rootlets, which absorb nutrients and water for the growth of the plant (Strand, 2008). This set is detached from the stem with threads of the same size, and distributed at depths of up to 30 cm, although roots have been found up to 80 cm (Flórez Faura and Mora Cabeza, 2010; Ruíz and Piedrahita, 2012). Each year the primary root is replaced by new roots at successively higher levels on the crown.

8.6.2 STEM (CROWN)

The stalk of the strawberry is known as a crown, because of its rosette shape covered by basal leaves, which have axillary buds, capable of generating different organs for plant (leaves, stolons, flowers, or new crowns). They form in the crop depending on the nutritional and environmental conditions (Ruíz and Piedrahíta, 2012).

8.6.3 STOLON

It is a prostrate stem formed from the crown, which can form a new crown that becomes a daughter plant around three weeks. A vigorous plant can produce up to 15 stolon systems, but the vigor of each new plant is different from the mother plant (Flórez Faura and Mora Cabeza, 2010; Ruíz and Piedrahíta, 2012)

8.6.4 LEAVES

The leaves of the strawberry plant are composed of three serrated edge leaflets and two stipules. They are arranged in a spiral phyllotaxis, where the fifth leaf is situated above the first one. This arrangement ensures the highest sunlight interception; moreover, the leaves have a high stomata density on their lower surface, which means that the strawberry's plant has a very efficient capability to absorb carbon dioxide from the atmosphere. However, this high stomatal density can be as much of a handicap to face water-shortage conditions

8.6.5 FLOWER AND FRUIT

Strawberry's flower has five green sepals (sometime more than 5) and five white petals, numerous stamens, and a receptacle, usually known as a torus that bears a vast number of pistils. Cultivated varieties are only hermaphrodite flowers (Ruíz and Piedrahíta, 2012). When the flower's pistil develops, it becomes a small fruit called achene that bears only one seed. The achenes are the small and hard structures which dot the fleshy strawberry fruit; therefore, the strawberry "fruit" is not a fruit itself but an aggregate fruit. Then, the fruit is defined as a polyaquenium, since it is the union of hundreds of small achenes that are the product of the fertilized ovum and de edible egg. After fertilization, the receptacle grows and develops, and their shape depends on the variety and the fruit appears to have several shapes: oval, conical, or balloon-shaped fruits. The receptacle color varies from pale red to intense red and bright red. This parameter becomes an index of quality at the time of commercialization (Perkins-Veazie, 1995).

8.7 CULTIVARS

There are about 1000 varieties of strawberry around the world. Differences between cultivars are based in the vigor, resistance to pathogens, and fruit characteristics (size, color, shape, flavor, etc.) (López-Aranda et al., 2011). Cultivars can be classified according to their photoperiod as short-day, long-day, and neutral-day plants (Hummer and Hancock, 2009; Serçe and Hancock, 2005). Strawberry's short-day plants are also called June-bearing plants, and they bloom when days have less than 12 hours of light. The most of cultivars belong to this class such as Allstar, Cavendish, Chandler, Sparkle, and Wendy. Strawberry's long-day plants are also called ever-bearing strawberry plants, although they produce two harvests per year: one during the spring and another in late summer. Whereas most June-bearing strawberry cultivars are of the *Fragaria × ananassa* variety, the ever-bearing strawberry cultivars are of the *F. vesca* variety. The third class includes strawberry plants that bloom regardless of the hour of light; they are the day-neutral strawberries. Plants of this class have a fruit quality and productivity higher than other class, and they should be grown as annual plants. Albion is one of the most known cultivars that belongs to this class.

The number of varieties is enormous, and new varieties appear every year. It would be impossible to give a complete list of existing varieties of strawberries. Fortunately, there is a tool that allows us to know the main characteristics of many varieties. It is an interactive list available at https://strawberryplants. org/strawberry-varieties/.

Among the main varieties we can remark: Alba, Albión, Allstar, Antilla, Atlas, Benton, bolero, Camarosa, Cambridge Favourite, Candonga, Cavendish, Chandler, Charlotte, Earlyglow, Flair, Florentina, Fortuna, Kent, Lucía, Monterey, Pandora, Primoris, Rainier, Royal Sovereing Sabrina, San Adreas, Splendor, Sweet Ann, and Ventana.

8.8 BREEDING AND CROP IMPROVEMENT

Fragaria × ananassa is the most popular strawberry crop, although other species such as *F. chiloensis*, *F. vesca*, *F. viridis*, *F. moschata*, and *F. virginiana* have less importance. *F. vesca* is an alpine forest strawberry, which was very popular in the 16th and 17th centuries until it was replaced by a species from America. Currently, *F. vesca* is cultivated in home gardens due to their small but very delicious fruits. *F. moschata* was also grown during the late 15th and 16th centuries, and their fruits present a musky flavor, whereas *F. viridis*, a whit green fruit, was being used for ornamental purposes. *F. virginiana*, known as a wild strawberry, is spread in North America, Canada, and Alaska. Its fruits are very aromatic but small. Neither of these species has currently important from a commercial point of view.

F. chiloensis was domesticated by the Mapuche, an indigenous tribe who lives in the southern land in what is now Chile. During the Spanish colonization period, *F. chiloensis* was introduced in Europe. In the 18th century, in France, there was an accidental hybridization between *F. chiloensis* and *F. virginiana* that gave rise to *Fragaria × ananassa*.

F. vesca has 14 chromosomes and it is a diploid specie, whereas *Fragaria × ananassa* has 56 chromosomes and it is an octoploid specie. Table 8.2 includes several species of the world indicating their ploidy and main agronomic characteristics.

TABLE 8.2 Strawberry Species of the World with Their Ploidy and Main Agronomic Characteristics

Specie	Ploidy	Main Characteristic	Distribution
F. vesca	2×	Fruits are red, soft, long and aromatic. Self-compatible; tolerant to adverse conditions such as drought, cold, heat; and resistance to several diseases	Worldwide
F. viridis		Self-incompatible with bearing fruits of color greenish pink, firm, and aromatic	Asia and Europe
F. nilgerrensis		Fruits are tasteless but pink; self-compatible; and resistant to pests like aphids and a lot of diseases	Southeastern Asia
F. daltoniana		Fruits red brilliant but tasteless; self-compatible	The Himalayas
F. nipponica		Fruit unpleasant; plant tolerant to cold weather; self-incompatible	Japan
F. orientali	4×	Slight aromatic and soft fruits	East Russia and West China
F. moupinensis		Orange–red fruits, with less flavor	North China
F. ×bringhurstii	5×	Intermediate characteristics if *F. vesca* and *F. chiloensis*	California

TABLE 8.2 *(Continued)*

Specie	Ploidy	Main Characteristic	Distribution
F. moschata	6×	Fruit dark to light red, with a musky flavor; plant tolerant to flood soil, cold, and shade; resistant to mildew	European Siberia
F. chiloensis	8×	Fruit dark red and internal flesh white; much tolerant to a wide range of conditions due to rusticity	Chile and West-North America
F. virginiana		Fruit dark red or scarlet, with internal flesh white, very aromatic; soft to deep red or scarlet fruit; much tolerant to a wide range of adverse conditions	North America, including Canada and Alaska
F. × ananassa		Fruits spherical or oblong, very big, and red	Worldwide

Strawberry is one of the crops with many cultivars; in fact, nearly 100 news cultivars are registered each year (UPO, 2019). However, the genetic basis to obtain this high number of modern cultivars is narrow. Only 10 parental genotypes were used to obtain new cultivars until 1990. Although the number of genotypes is scarce, it has not been a problem, since being an octoploid, this species allows us to preserve a very high genomic variability, as has been demonstrated using randomly amplified polymorphic DNA (Gambardella et al., 2000) and as is also shown by Gil-Ariza et al. (2009). In more than 200 years of breeding, very few number of genetic diversities have been produced. The fact that wild germplasm has been used is another factor that has contributed significantly to maintaining genetic diversity.

Currently, there are a lot of breeding programs. Some programs are financed by public funds, but others are financed by private companies. All these programs have the goal to improve *Fragaria × ananassa.* In Europe, the main effort for the breeding program is located in France, the Netherlands, the United Kingdom, Italy, and Spain; in America, the main breeding programs are carried out in the USA (California, Florida, and Maryland), Canada (Ontario and Quebec); while in Chile, in Talca, there is a breeding program with *F. chiloensis.*

8.9 SOIL AND CLIMATE

8.9.1 SOIL

The strawberry crop requires a careful soil preparation because its optimal development is in loose

Strawberry

soils, sandy or loamy sand, with good drainage (Tagliavini et al., 2005). For planting of mother plants, a process must be carried out where first of all a root and leaf area cut is made to stimulate new vegetative shoots, then it is disinfected in a solution of fungicides and insecticides, and finally, it is transplanted leaving them right in the perforations of the quilted (Roussos et al., 2012). It must be taken into account that the application of nutrients in these soils is carried out in small quantities and with a greater frequency (Horneck et al., 2011). The strawberry crop takes place in moderately acid soils, with pH values between 5.5 and 6.5 (Tagliavini et al., 2005). Acid soils can restrict microbial activity, reduce the availability of essential nutrients, and cause aluminum toxicity in the soil, delaying root growth, and restricting access to water and nutrient intake. In addition, it is necessary to conduct electrical conductivity (EC) tests to evaluate the levels of affectation by salts present in the soil. Since the strawberry crop is considered sensitive to salinity, it requires an EC of less than 1 dS/m. The crop requires high soil fertility, which is why it prefers soils with high cation exchange capacity, defined as the total capacity of a soil to maintain interchangeable cations. This influences the stability of the soil structure, its availability of nutrients, the pH, and the reaction

of soils to fertilizers and amendments. Additionally, it requires an amount of organic matter greater than 1%, ideally 2%–3%, and a carbon–nitrogen ratio (C/N) of close to 10. For most crops, the ideal values are as follows: nitrates 10–50 mg/kg and ammonia 0–5 mg/kg. (Hochmuth and Albregts, 1994). Regarding the content of phosphorus in soil, we find that, for strawberries, it is considered that <10 ppm of P is very low, 10–15 ppm is low, 16–30 ppm is medium, 31– 60 ppm is high, and > 60 ppm is very high. Similarly, if the P is used in the form of P_2O_5 (phosphoric anhydride), then values higher than 50 ppm are recommended for strawberries (Peverill et al., 1999).

Soil microorganisms play the key role in mineralizing organic sulfur; therefore, the biological activity and factors that influence the microbial populations (temperature and humidity) will determine the rate of S available for the plant (Peverill et al., 1999). In general, for most crops, the sulfur level between 37 and 78 ppm is considered adequate, values lower than 23 ppm are considered low, and values higher than 116 ppm are considered high (Hochmuth and Albregts, 1994).

8.9.2 CLIMATE

The vegetative and reproductive growth of strawberry is regulated by a set of environmental and physiological signals. Conditions such as light intensity, light quality, photoperiod, temperature, and availability of water in the soil have a greater or lesser influence on the growth of leaves, crowns, roots, stolons, inflorescences, flowers and fruits, and consequently, on the yield of crop and the quality of fruit. On the other hand, the response of genotypes varies widely, depending on the conditions and the culture environment (Darnell et al., 2003; Darrow, 1966; Hancock, 1999).

There are times in the year that naturally offer unfavorable weather conditions for planting plants for fruit production, either due to high temperatures and heavy rains that occur at the end of summer or due to low temperatures that occur at the end of winter and early spring, which affect the establishment of newly transplanted crops. In other cases, the affectations occur during periods of flowering and fruiting, when they coincide with moments of high temperature and high rainfall, with the frosts of spring and autumn or with periods of strong winds or hail (Demchak 2009; Demchak & Hanson, 2013; Hancock, 1999; Herrington et al., 2011; Hummer and Hancock, 2009; Singh et al.,2012). The optimum temperature range for strawberry growth is between 10 °C and 26 °C (Ledesma et al., 2008)—results that agree with those obtained by Verheul et al. (2007) of 18 °C diurnal and 12 °C nocturnal.

8.10 PROPAGATION

In strawberries, the method most used for propagation is by means of stolons. That means, the propagation is generated naturally favoring the rooting of organs such as stolons, crowns, and meristem culture (Bish et al., 1997). However, for breeding processes, the reproduction of strawberry plants via seed is used through which hybrid plants are obtained (Frankel and Galun, 2012). When there are species or cultivars that cannot reproduce sexually, cloning is the only way. Horizontal aerial stems are used with alternating long and short internodes that generate adventitious roots, known as stolons (Davies et al., 1994; Wilk et al., 2009), allowing the new plant to explore with its own root system, soils beyond the root system of the mother plant (Husaini and Neri, 2016). The production of mother plants in nurseries of countries of temperate zones is done by stolons, due to long days and high temperatures (Bish et al., 2001; Cantliffe et al., 2003). The daughter plants obtained become self-sufficient between two and three weeks after continuing to be united to the mother plant by stoloniferous filaments. During this period, the daughter plants can exchange carbon, water, mineral nutrients, and biochemical components through vascular transport (Holzapfel and Alpert, 2003). After being cut, the new plant obtained from the mother plant is stored in cold rooms at −2 °C until the time of sowing arrives. The cold accumulated plant during this time allows it to generate the first flowering which is generally weak, along with a high production of stolons, which must be cut to increase vigor and production of crowns (Lieten et al., 1995).

8.10.1 PROPAGATION BY CROWN DIVISION

It is a method used mainly in varieties with little production of stolons, but with high production of secondary crowns. For this reason, it is only implemented in mother plants with well-developed crowns. In order to stimulate this development, fertilizers with good nitrogenous sources, high amounts of watering, and constants are made in order to promote the formation of secondary roots in each of the crowns.

8.10.2 USE OF ROOTING TRAYS

Stolons of 1 cm of root are used, which are cut from the mother plant with small pruning shears, and are taken to plastic rooting trays, where substrates such as peat and coconut fiber are used. After 30–40 days, plants can be taken to the field (Durner et al., 2002; Lieten, 1998).

8.10.3 MICROPROPAGATION

Strawberry micropropagation was introduced in 1974 and was widely used by the European seed companies due to the decrease in the incidence of pathogenic diseases of soil, such as the case of *Verticillium*, *Phytophtora cactorum*, or *Phytophthora Fragarie*; and due to the increase in the number of stolons by the mother plant (Boxus, 1974; Boxux and Larvor, 1987; Boxus et al., 1984). It is a method based on the cultivation of plant tissues and on various principles such as cellular totipotency (inherent ability of a single cell to provide the genetic material necessary to develop a completely new plant) through a massive multiplication in vitro. This type of in vitro culture is shown as an alternative for obtaining plants free of pathogens, together with a high multiplication rate.

8.10.3.1 PHASES OF MICROPROPAGATION

Murashige in 1974 initially proposed a procedure that allows us to produce new uniform plants, free of diseases and in relatively quick times, and divided them into three stages: the first stage is known as Stage I that comprises the establishment of an aseptic crop; the second stage (Stage II) comprises the multiplication of propagules, and the third stage (Stage III) comprises the preparation for the restoration of plants in the soil (Murashige, 1974). However, a more developed procedure based on Murashige's model was proposed mainly for a commercial establishment (Debergh and Maene, 1981). This procedure has the following stages: the preparation of mother plants under hygienic conditions (Stage 0), the establishment of aseptic cultures (Stage I), the production of viable propagules (Stage II), the preparation for field growth (Stage IIIa), and transfer of plants to the field (Stage IIIb). In the first two stages, it is recommended that the "mother" plant has the typical characteristics of variety and is free of diseases (George et al., 2008). Additionally, any culture medium that presents contamination before or after sowing must be eliminated, disinfecting or sterilizing the work area, both in cutting and extraction of explants, and in sowing to the culture medium, without causing a phytotoxicity (Sathyanarayana and Varghese, 2007), making cuts to the explants with differentiated organs to extract tissues that may be new explants (George et al., 2008). In the third stage, the rooting and elongation of plant structures

must be guaranteed. It is highly recommended that elongation be done in vitro and rooting ex vitro, generating a reduction in costs, as well as physiologically and structurally better root formation (hardened plants) than those developed in vitro, besides avoiding damage to roots when transplanted to the field (Sathyanarayana & Varghese, 2007). In strawberries, an elongation of the aerial part in vitro is performed first through an auxin–cytokinin ratio of 6-benzylaminopurine, benzyl adenine of 0.1 mg/l and indole-3-butyric acid of 0.2 mg/l for three weeks (Moradi et al., 2011). Once the plants have completed this stage, they are in ideal conditions to be transferred to the field.

8.11 LAYOUT AND PLANTING

8.11.1 SELECTION AND PREPARATION OF LAND

Strawberry cultivation is demanding in terms of the characteristics of soils it requires for its development. For this reason, it is found in soils preferably where they have not been planted with Solanaceae since there are several pathogens that affect both Solanaceae and strawberry. A flat terrain is recommended. Although they can also be found on gentle to moderate slopes, they should be loose and homogeneous, so a preparation is recommended, whether with disk plough, subsoiler, or chisel plough, depending on the type of soil. Simultaneously, with the preparation, acidity corrections, weed control, and other preparations can be made (Flórez Faura and Mora Cabeza, 2010).

8.11.2 CONSTRUCTION OF BEDS

The production of strawberries in soil requires the construction of beds, which allows adequate drainage, proper development of roots more than 30 cm deep, and facilitates culture- and crop-management tasks. These beds have a height of 40 cm, built in the shape of a trapezoid, where the main base is 70 cm and the lower base is 50 cm. Normally, a distance between 40 and 50 cm between beds is left. In each bed, there are two rows of plants with a 30–40 cm gap between them. The same distance is maintained for the irrigation lines, thus placing two lines in each bed.

8.11.3 MULCH OR PLASTIC MULCH

To reduce the levels of evaporation in soil by generating an impermeable barrier, increase productivity by maintaining soil temperature, prevent growth of weeds and direct contact between the fruit and the ground,

and cover the beds with plastic films forming the mulch (Medellín et al., 2013). There are different types of padding that vary both in size and color, generating differences in duration, thermal conductivity, and light reflection, using 1.25 caliber plastic with a duration between 18 and 24 months (Flórez Faura and Mora Cabeza, 2010). These authors indicate that the use of black plastic in the quilts generates a thermal gain in the first 5 cm of soil, whereas green plastics and thermal coffee accumulate more heat that can increase up to 3 °C to 30 cm depth of the soil under minimum environmental temperatures of 17 °C and maximum of 20 °C–24 °C. Casierra et al. (2011, 2014) evaluated the effect of plastic pads of red, yellow, blue, black (control), silver, and green colors on the growth and quality of the strawberry fruit sown at 2500 m above sea level in Colombia, without finding significant differences between the red padding and black witness in most factors. However, the red mulch showed higher values in the leaf area, fresh fruit weight, and fruit length, the silver mulch recorded the lowest values in pH, fruit dry weight, total soluble solids ratio/total titratable acidity, whereas the black mulch showed the greatest value of this relationship. A greater absorption of red light far away by phytochrome may be the cause of the positive effect of the red mulch, which influences the enzymes that affect flavor and sweetness. Based on these results, it is worth highlighting the benefits of using the red mulch for strawberry cultivation despite scarce use in the crop (Casierra et al., 2011).

8.11.4 SOWING

The number of plants per hectare varies depending on the plantation frame between 40,000 and 69,000 plants. These values also vary depending on the type of terrain, climatic conditions of the area, variety, and size of the plant. To sow mother plants, known as "*Frigo plants*" because they are frozen, it is necessary to give a time between 24 and 48 hours for thawing, before subjecting them to disinfection, which is carried out submerging the plants in fungicidal solutions for 5 min. For plants that are propagated by stolon, this procedure is carried out only in the case of leaving the root naked when passing from the tray to the ground. Normally, the plants are sown in the traditional pattern, a quincunx, giving a distance between plants of between 25 and 35 cm, depending on the variety, texture and fertility of soil, accumulation of cold hours by the plant, and use of different structures such as greenhouses, and micro- and macrotunnels (Flórez Faura and Mora Cabeza, 2010).

8.12 IRRIGATION

Plants obtain water from the soil through their roots, ascending through their conductive vessels of the xylem, and then distribute to different organs, especially, and in great proportion, to leaves. There water passes from the liquid to gaseous state, released by the stomata, what is known as the transpiration process. Irrigation plays an important role in the development of cultivation. The roots of strawberries develop 90% in the first 30–40 cm of depth. That is why they must be permanently supplied with water, thereby decreasing the effect of possible percolation and drainage. The amount of water to apply depends on variety, agroclimatic conditions, foliar area, stage in which the crop is, among others. It also depends fundamentally on the texture of soil, the quantity and quality of water to be used, control of soil salinity and penetration of water into soil, and irrigation time.

Plans sown in 1 ha can consume between 4000 and 6000 m^3 of water per year (Krüger et al., 1999). The irrigation system that best adapts to the sowing conditions is the drip, since it presents a greater efficiency in the application of water resources, managing low flow rates, and avoiding percolation. When it is a hydroponic crop, or when it does not use soil but substrates, the water requirement of the plant is better controlled. This type of irrigation can also be used for fertigation (application of fertilizers).

The success and development of strawberry production depend mostly on fertilization and irrigation. That is why with an irrigation according to the needs of the plant, in addition to gaining an increase in the size of fruit, the yield of crop as well as the quality of the product is increased (Yuan et al., 2004). Although these systems optimize the use of water by saving this resource in great quantity, there may be problems of outcrop of salts, since these are not washed by the low flow of the system.

The water–soil–plant relationship is fundamental for the development and production of the different crops. According to Morillo (2015), in this relationship, there is a fourth fundamental and decisive element in the decision-making of irrigation: the environment. That is why, to determine the water needs of crop, a model based on the water–soil–plant–environment relationship is necessary: the water retention capacity of soil (water–soil), the climatic demand given by the potential evapotranspiration (environment–soil), the crop coefficient (water–plant), and finally the consumptive use of water for cultivation (the water–soilplant system).

Strawberry crops need a regular supply of water to obtain an abundant and quality harvest. Irrigation needs become more critical when the crop is established in arid and semi-arid zones. The crops should be watered either through rainfall or direct supplements. Traditionally, furrow irrigation and the sprinkler system have been the more usual techniques to water the crop. However, the widespread use of mulching, as well as raised bed, these methods are not convenient because of the supply of an excess of water to the plants and moreover the efficiencies of these techniques are very low and it requires a large volume of water to produce 1 kg strawberries. This efficiency is contrary to sustainable agriculture; therefore, other irrigation techniques such as drip irrigation are replacing the old ones. Drip irrigation is one of the most used techniques of irrigation due to its high efficiency in the use of water, although there are certain mistaken perceptions that crops with drip irrigation offer fruits of poorer quality. Currently, there are a lot of different strategies adopted for drip irrigation management, with all of them oriented to increase the efficiency of watering. In this sense, Èger et al. (1999) recommend scheduling the irrigation according to tensiometer measurement as well as a climatic water balance model. If the crop takes place under greenhouse, then Yuan et al. (2004) pointed out that for the strawberry crop grown under plastic greenhouse a pan factor of 1–1 is the optimal guideline for irrigation since in case you use a smaller factor, production will be reduced (Bleiholder et al., 2001).

8.13 NUTRIENT MANAGEMENT

8.13.1 REQUIREMENTS AND EXTRACTION OF NUTRITION IN STRAWBERRY CROP

Each crop presents different nutritional needs throughout its development, depending on the phenological stage of the plant. For this reason, the objective of the nutritional management is to provide conditions and adequate amounts of fertilizers for a good development. The nutritional requirements of crop can be divided into several stages, taking into account the scale of Biologische Bundesanstalt, Bundessortenamt und Chemische Industrie development in strawberry (Meier, 2001) that defines the following stages of development: rooting, vegetative growth, flowering, and fruit filling. For each of the stages mentioned, the plant demands the same nutrients, although in different proportions and quantities. Currently, fertilization recommendations for strawberry crop vary depending on the varieties, latitudes, and tests performed by the related authors. Table 8.3 shows the extractions of nutrients according to several authors.

TABLE 8.3 Extractions of Macronutrients (Expressed in kg/ha) in the Strawberry Crop

Authors	N	P_2O_5	K_2O	Ca	Mg
Bottoms et al. (2013)	229	53,8	162	–	–
Molina et al. (1993)	312	43	285	224	50
Verdier (1987)	150	125	400	–	–
Lieten and Misotten (1993)	125	40	190	78	23
Haifa (2017)	150	150	240	120	60

Source: Results by different authors in kg/ha of nutrient extracted for eight months.

In order to comply with the requirements of crop, some authors recommend supplying the plants with the equivalent of all nutrients captured by them, but supply in a gradual manner and depending on the growth and development of the crop. For example, Haifa (2017) recommends supply 233 kg/ha of ammonium nitrate (34% N), 344 kg/ha of superphosphate (25% P_2O_5), 376 kg of potassium sulfate (50% K_2O), 192 of dolomite (26% CaO), and 119 kg/ha of magnesium sulfate

Strawberry

(16% MgO), as a basis for the fertilization of the crop, which is projected for a duration of 7–8 months. Table 8.4 shows the absorption of the main elements during three phases of crop development in detail (Stage I: from 1 to 9 weeks after sown; Stage II: from 10 to 18 weeks after sown; Stage III: from 19 to 30 weeks after sown). This study was performed in a crop with a planting density of 50,000 plants/ha.

TABLE 8.4 Absorption of Nutrients by Strawberry in Three Different Stages of Crop Development[a]

Macronutrient	N			P			K			Ca			Mg		
Stage	I	II	III	I	II	III	I	II	III	I	II	III	I	II	III
kg/h	12	106	195	2	13	28	13	97	174	6	54	164	2	19	29
Total		312			43			285			224			50	
Micronutrient	Fe			Cu			Zn			Mn					
Stage	I	II	III	I	II	III	I	II	III	I	II	III			
g/h	500	1925	10550	23	76	223	45	221	1138	130	845	2535			
Total		12975			321			1404			3510				

[a] Macronutrients are expressed in kg/ha and micronutrients are expressed in g/ha.

The nutrient extractions reported by Molina et al. (1993) (Table 8.4) over 30 weeks of culture show a particular behavior for each element; the demand for N was always increasing. Although the authors reported the peak of vegetative growth in Stage II with the maximum development of the leaf area and stolons, the high demand of nitrogen for the production phase of Stage III is due to the foliar renewal that presented crop after the first productive peaks (El-Arabi et al., 2003).

8.13.2 METHODS OF FERTILIZATION IN STRAWBERRY

One of the key factors that determine the realization of agronomic and economic objectives of fertilization is the manner in which fertilizers are applied, thereby achieving high yields with reasonable costs in agricultural production systems (Pauletti et al., 2010). There are three different methods: foliar fertilization, edaphic fertilization, and fertigation.

8.13.3 EDAPHIC FERTILIZATION AND FERTIGATION

Edaphic fertilization is the direct application of fertilizer superficially or incorporated into the soil. After its application, the fertilizers are released and become available to be taken up by the roots of plants.

Fertigation is the process by which mineral fertilizers are added to irrigation water and distributed in crop through the irrigation system (Kafkafi and Tarchitzky, 2011). Generally, fertigation is carried out by means of drip systems; however, in some crops, this can be applied using other types of irrigation, such as sprinkling or microsprinkling. This system must present minimum conditions of equipment for the incorporation of fertilizers into the irrigation system, a filtering system, mixing tanks and devices for injecting the fertilizer into the irrigation system.

8.13.4 FOLIAR FERTILIZATION

The uptake of nutrients by leaves and other aerial parts of the plants is regulated by the epidermal cells of the outer walls of the aerial tissues. These walls are covered by the layers of wax, pectin, hemicellulose, and cellulose that protect the leaf from the excessive loss of organic and inorganic solutes by rain (Salas, 2002; Tagliavini and Toselli, 2005). Foliar fertilization has become a common and important practice for producers, because it corrects the nutritional deficiencies of plants, favors the good development of crops, and improves the yield and product quality (Fageria et al., 2009; Trinidad-Santos and Aguilar-Manjarrez, 1999). It is generally considered as a technique to supply nutrients quickly to plant in order to correct nutritional deficiencies. This characteristic is especially important for crops, which perform better when their demand for nutrients is fully satisfied throughout the entire growth cycle (Tagliavini and Toselli, 2005).

8.14 FLOWERING AND FRUIT SET

The main characteristic of the strawberry fruit is that it is an infructescence. Botanically it is considered an etaerio, that is to say a floral receptacle thickened and fleshy, turned into fruit, on which a high quantity of achenes coming from an apocarpous flower is inserted. In this flower, there are from 5 to 10 sepals, on average, 5 petals, 20–35 stamens (microsporophiles), and 500–800 pistils, arranged in a spiral around the receptacle. The male organs develop first and then the female organs. Stamens form pollen grains (microspores) that mature before flowering or during the opening of the anthers. After the opening of these, the pollen is able to pollinate. In the receptacle appear many pistils, arranged in spiral. Each pistil is made up of carpels, stigma, and ovaries; the stigma is rough and sticky which allows a better reception of pollen grains, better pollination, and fertilization (Selamovska, 2014).

FIGURE 8.3 Pollination of a strawberry flower (left). Receptacle in development, one can still see the remains of the styles of flowers. (Photo courtesy: J. L. Valenzuela)

In natural conditions, the flower is pollinated entomophilically, usually by bees or bumblebees (Figure 8.3). The success of pollination is based on insects making several visits to the flower due to the pistils mature sequentially, and therefore several visits are necessary to achieve a well-formed fruit (Chagnon et al., 1989; Zapata et al., 2014). In greenhouse crops also, bumblebees are used for pollination. In this case, it must be kept in mind that if the flowering is not abundant, the bumblebees will make too many visits to the flower to get more pollens. A shortage of flowers may lead to an "overvisiting" the available flowers with the result of damaging the receptacle which, when ripe, manifests the damage in the form of blotched spots on the septum of the fruit (Kopper, 2017). The fertilization takes place about 24 or 48 hours after the pollination, and immediately after, the petals begin to dry and fall, and the pistils begin to dry. After fertilization, continues the formation of the zygotes, the embryogenesis, and the formation of the seeds, as well as the formation of physiologically mature fruits. The shape and size of the fruit depend on the number of fertilized pistils; so if the fertilization is carried out partially, deformed, small, and poor-quality fruits are obtained (Selamovska, 2014). It is, therefore, necessary to pay attention to pollination and consider it as a high yield factor in strawberry (Figure 8.4).

FIGURE 8.4 Abnormal fruits due to a poor pollination.
(Photo courtesy: J. L. Valenzuela)

Since almost 70 years ago, it has been demonstrated that in strawberry the presence of auxins is necessary for the perfect development of the receptacle. Nitsch in the 1950s demonstrated that by removing the achenes from a strawberry receptacle, it is developed abnormally, and the application of auxins acts as a substitute for the achenes, stimulating the growth of the receptacle (Nitsch, 1950, 1955). He demonstrated that auxins are essential for the growth of the receptacle and that achenes are the source of auxins.

In strawberry fruit development, seven phases are distinguished, but theses phases are denominated in one way or another according to different authors, but always the different phases are divided on the bases of coloration and size of the receptacle. These phases begin with flowering and end with strawberry red ripening (Fai et al., 2008; Symons et al., 2012). Figures 8.5 shows these phases.

FIGURE 8.5 Stages of development of the receptacle.

Auxin levels in each phase vary, thus reaching an auxin peak before the white phase, to decrease as the strawberry matures. It is believed that there is a coordinated action between the achenes and the receptacle to ensure the maturation of the achenes before the ripening of the fruit. The decrease in the auxin levels triggers maturation, since the *de novo* synthesis of messenger RNAs associated with maturation is induced (Benítez-Burraco et al., 2003; Castillejo et al., 2004; Fai et al., 2008). Thus, when the auxin levels are high, the growth of the receptacle is stimulated, and the maturity and development of the coloration are inhibited. After the decline of the auxin levels, growth stops, and ripening processes are stimulated (Symons et al., 2012). The necessary participation of gibberellins in the growth, development, and ripening of the receptacle began to be studied in 1969, when Thompson in 1969 achieved the development of nonpollinated strawberries by applying gibberellins, alone or in conjunction with auxins (Thompson, 1969). Subsequent studies in tomato and *Arabidopsis* indicate that the auxins have a role in such a way that they stimulate the production of gibberellins during fruit setting (Fuentes et al., 2012). The participation of auxins and gibberellins in the development of the receptacle has been exhaustively studied by Kang and collaborators in 2013 (Kang et al., 2013). These authors have studied the transcripts that are produced in the early stages of development of the receptacle of flowers pollinated manually; transcripts from nonpollinated flowers; transcripts from emasculated flowers to which gibberellic acid 3 (GA3) was added; transcripts from flowers that were treated with Naphthalatamic acid (an inhibitor of polar transport of auxins); transcripts from flowers treated with Naphtalene acetic acid (NAA, a synthetic auxin); and transcripts from flowers treated with NAA and GA3. The results show how the length of the receptacle of the flowers treated with NAA and GA3 were similar to that of the flowers pollinated manually (Figure 8.6). On the contrary, there is a noticeable decrease in the length of the receptacle both in the nonpollinated flowers and in those that were applied with the auxin transport inhibitor (Kang et al., 2013).

Moreover, these authors showed that the receptacle of the flowers treated with auxins or with gibberellins only does not reach the size of pollinated flowers or that of those treated with auxins and gibberellins together. It is shown that achenes participate decisively in the development of the receptacle, and that all achenes cooperate together (Kang et al., 2013).

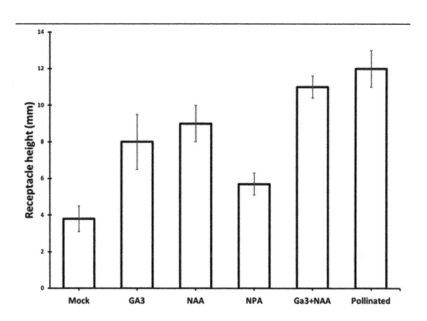

FIGURE 8.6 Parthenocarpic fruit development induced by auxin and GA.

8.15 FRUIT CHARACTERISTICS AND QUALITY CRITERIA

A strawberry of high quality must have a series of characteristics as it is to present a uniform red color according to its variety. Moreover, the fruit must be firm, flavored, and does not present damages, defects of shape, or diseases. It is recommended that they have a minimum soluble solids content of 7 °Brix and a titratable acidity of 0.8%. Given that ripeness is, in principle, associated with the coloring provided by the anthocyanins, strawberry marketing standards in different countries are usually based on this parameter to define the quality of the product. Thus, in the United States of America, the state department of agriculture classifies strawberries as Category 1 when they present a minimum of three quarters of their colored surface, whereas Category 2 will be those that present not less than half of its surface (USDA, 2006). In the same sense is the commercialization directive of the European Union, although in this case, it does not make reference to a percentage of color when it classifies the strawberries as an extra category. It does indicate the percentage of whitish stain that strawberries of Categories I and II can present, which is one-tenth and one-fifth, respectively. By not indicating otherwise, it is understood that strawberries of the extra category cannot present any whitish stain (Comision de las Comunidades Europeas, 2002). In Colombia, the Colombian Technical Norm NTC 4103 appears. It indicates that the coloration must be homogeneous and in agreement with the state of maturity. Although this standard is for Chandler strawberries, it can be extensive for other varieties (INCOTEC, 1997) and clearly indicates the homogeneity in terms of coloration. There are also several works where color charts are established to indicate the stages of maturity and all of them pointing

out that strawberry of the highest quality should have uniform red coloration (Cámara de Comercio de Bogotá, 2015; Flórez Faura and Mora Cabeza, 2010; Patiño Sierra et al., 2014).

The selection of the strawberry by the color will depend on the cultivar and the preference of the market, although in general it cannot be collected in the stages in that it does not allow them to reach an adequate degree of maturity or in a state of overmaturity when the coloring is very intense and the fruit is soft. However, the intensity of coloration varies from one cultivar to another, but it must be kept in mind that an intense red color is not always associated with a lower fruit firmness. Jeong et al. (2016) studied four cultivars with different color intensities, from soft red to intense red, and found that the firmness of the fruit was similar in all the cultivars and that the one that showed the least firmness was a variety whose fruits were red intense in ripening. At harvest, when picking up the fruits, we must pay close attention to the management of the fruit, since strawberry is a fruit that is easily damaged. In the same way, the management must be careful after harvest, during later handling. For this reason, it tends to collect in a state that often rubs the immature state. With large areas of the whitish fruit, a state that allows it to withstand subsequent handling and suffering less damage is not the right state of maturity and the strawberry lacks in flavor and sugars, since it does not increase its sugar content after harvesting and also the color evolution after harvesting is very slow, so it is difficult to achieve an adequate coloration if it is collected in an immature state (Figure 8.7).

FIGURE 8.7 Strawberries harvested at different ripening stages.

The importance of harvest at the appropriate state of maturity is reflected in the following study. Strawberries from the cultivars Chandler, Big Bear, and Sweet Charlie were harvested in four stages of development, from the state in which they begin to color, to completely red and stored at 1 °C for eight days. The physical and chemical changes that determine strawberry quality were studied, and these changes were compared with those that occur with strawberries that were kept in the plant until their complete development and maturation (Nunes et al., 2006). It was found that the strawberries harvested in the state of three-fourth of their complete coloration (corresponding to Stages 6 and 7 of the color charts, as shown in Figure 8.7) reached the same pH values, titratable acidity, vitamin C content, and total soluble content, during storage than those strawberries that were harvested at the appropriate maturity stage and completely red, with the advantage that the fruits harvested in the de-coloring state were firmer and red after storage. In this sense, Nunes et al. (2006) indicated that the strawberries harvested in three-fourth of coloration can be stored for a period of time greater than those harvested completely in red. But it must be kept in mind that any stage of development prior to three-fourth of red coloration means that the strawberry has not ripen, has not reached adequate coloration during the storage period, and will be of poor quality after harvest. In addition, in the study of Nunes et al. (2006), strawberries were kept for eight days in a cold storage and after these eight days, it was found that similar values were reached to those presented by strawberries ripened in the plant and in a completely red state.

8.16 HARVESTING AND YIELD

The harvest of strawberries is a critical moment since they must be picked at the optimum time, because it is a nonclimacteric fruit. Strawberry fruits should be harvested when 75% of their surface has turned red and the fruit is still firm. A common mistake among farmers is to harvest the fruits in an immature state, with the belief that if the fruit is very red, it is more susceptible to mechanical damage during postharvest handling, but in reality, what they do is collect the fruits with less flavor.

Strawberries are much perishable and after harvesting, within two or three days, they become rot under natural environmental conditions, so the temperature of storage is another critical point, and the postharvest life depends directly on the cold storage. As the temperature rises, the fruits become soften very quickly and the postharvest life becomes very short. A fruit harvested in full maturity and maintained at room temperature deteriorates by 80% in only 8 hours. A good, postharvest handling implies that fruits must be picked when the sun rises, in the early morning, transported to the processing site as quickly as possible, and kept in the shade till they reach the packing house and then kept in a cool place until final processing. But the harvest is not everything, at the moment of harvest, other processes of great importance begin: selecting the fruit, packing it, transporting it, and storing it properly, to present a good product in the market.

The best storage conditions have to control the relative humidity; otherwise, a low relative humidity implies wilting, weight loss, and dehydration of the fruits, while a high relative humidity is a great advantage for the development of microorganisms and rot. In relation with freshness, this variable is important because it allows the fruit look better or promotes spoil by molding growth or other undesirable characteristics. Therefore, it is very important to have the relative humidity under control. The most recommended relative humidity fluctuates between 85% and 95%. Moreover, a precooling could be essential to keep the quality of berries, and only if the strawberry will be consumed within 24 hours after picking, avoid precooling. Berries are much perishable and require a longer period in a cold storage, at the recommended temperature of around 0 °C.

A strawberry fruit of high quality is linked not only to their appearance (color and shape) but also to their firmness, flavor, odor, and nutritional values. The harvest index is related to color, and it is not recommendable to harvest the fruit with less than three-fourth of the fruit skin of berry being pink or red. It is recommended to harvest the fruit with in a soluble solid content of at least 7 °Brix and a titratable acidity around 0.8%. In Section 8.15, we can find some quality criteria.

Due to nonclimacteric character, the ethylene production is very low (<0.1 μl C_2H_4/kg/h), and berries do not respond to ethylene treatment. For this reason, they must be picked in a maturity stage—very close to ripe. Despite their insensibility to ethylene, it is convenient avoid ethylene sources near of fruits during handling and storage to minimize the incidence of pathologies.

8.17 PROTECTED/HIGHTECH CULTIVATION

The growth yield and quality of strawberry are regulated by a complex interaction of environmental and physiological signals. The intensity and quality of light, photoperiod, temperature, and water availability influence the physiological process in numerous ways, such as photosynthesis and gas exchange, water relationship, or photomorphological processes (Blanke and Cooke, 2004; Casierra-Posada and Vargas, 2007; Casierra-Posada et al., 2012). It is possible to influence and modify these variables through the use of crop-protection systems such as plastic mulch, macrotunnels, and greenhouses (Wang and Camp, 2000).

8.17.1 MULCHING

Mulching is very common in strawberry crop. There are different types of mulching films that vary in sizes and colors, with differences in duration, thermal conductivity, and light reflection (Flórez Faura and Mora Cabeza, 2010). These authors indicated that the use of black plastic generates a thermal gain in the first 5 cm of soil, while green and brown plastics accumulate more heat (up to 3 °C at 30 cm depth under minimum environmental temperatures of 17 °C and maximum from 20 °C to 24 °C) (Figure 8.8).

FIGURE 8.8 Mulching.
(Photo courtesy: J. L. Valenzuela)

Other mulches used are the white–black film, where the white color faces up reflects toward the leaves up to 65% of the photosynthetically active radiation (PAR), thus increasing the photosynthesis and development processes. The silver–black film where the silver color faces up reflects up to 27% of visible radiation and approximately 30% of the UV radiation, enhancing the red color of fruits. In both films, the black color faces down, thereby blocking the entry of light and suppresses the germination of weeds.

8.17.2 TUNNELING

The tunnels are simple structures of arches in galvanized steel, which serve as support for a plastic film that generates the greenhouse effect, allowing reasonable levels of ventilation, thanks to the ease of handling the covers. Their advantage is that they are easy to remove and relocate to land, they are modular, and can be built with the crop adapting to topography. In addition, its cost is less than that of a greenhouse. Its objective is to maintain a favorable environment for cultivation through protection against rain, hail, and short frosts, avoiding the presence of free water in the leaves and therefore reducing possible levels of fungi. This is why it is necessary to know the temperature and humidity ranges in the area where they are installed to program the handling of covers, and radiation levels to select the appropriate plastic. Due to greenhouse, the temperature pattern is similar to that of the external conditions, changing the speed of heating and cooling more rapidly, obtaining interior temperatures greater by 3 °C–5 °C

and relative humidity 10% higher. This means that in times of drought or summer, the temperature can rise up to 25 °C–27 °C, which would cause the problems of blanching or softening the fruit. For this reason, when this situation is expected in the months of high solar radiation, low cloud cover, and little rain (summer), the plastic should be raised to center to facilitate air circulation and balance temperature with the outside during the day. On the other hand, in times of rain with low temperature, the plastic film and the structure maintain a higher temperature than outside, which is favorable for the crop production (Rubio et al., 2014; Singh et al., 2006).

The wind is a variable which is also directly affected in macrotunnels. It is, therefore, important that at the time of installation, your location allows a permanent ventilation. They should be oriented in the predominant direction of the winds, in such a way that there is a necessary air flow inside. It is also important to remember that there is a lateral ventilation because the plastic is located 80 cm from the ground.

With respect to PAR also, there is a noticeable influence on the environment in the tunnel. Inside the tunnel, under tropical conditions, the maximum PAR reaches about 700 μmoles/m^2/s while outside it reaches 1,200 μmoles/m^2/s. For this reason, all climatic variables must be balanced in favor of crop, making permanent measurements.

8.17.3 GREENHOUSES

With the increasing demand of strawberry in the market, it has been opted to carry out the sowing of the strawberry under greenhouses. These types of practices generate appropriate environmental conditions for an optimal development of the strawberry plant, guaranteeing a quality product. In addition, greenhouse production is three times as much as the production from open-pit cultivation systems (Demchak, 2009). The control of temperature, humidity, and PAR varies according to the design of each greenhouse, the plastic cover film, and the equipment available for this purpose (thermal screens, fans, and humidification systems). Not all greenhouses operate in the same way, and their designs are based on the needs of the plantation and the areas where they are located.

8.18 POSTHARVEST HANDLING AND STORAGE

Strawberry is a fruit very sensitive to mechanical damage; hence, it is obvious that you must avoid hitting them and dump them from large baskets to smaller ones. Therefore, it is advisable to collect it directly in the packing baskets weighing between 250 and 500 g. In 2015, a study to quantify the damages suffered by strawberries in the harvest and marketing chain was carried out. For the realization of this study, the strawberries were hand-picked and were brought in boxes to the laboratory where they were packed in plastic containers of 10 cm wide, 17 cm long and 10 cm high, each one containing 60 strawberries distributed in three layers (Aliasgarian and Ghassemzadeh, 2015). A transport from a distance of 55 km to the market was simulated. The results of the study show that the moment of harvest is key to the reduction of damage, since it is where the percentage of damage is inferred to the product is higher. Reaching at this stage 51% of the damages were reported, whereas at the stages of packaging and commercialization, the damages reached 17% and 32%, respectively (Aliasgarian and Ghassemzadeh, 2015). These results clearly show that given the high susceptibility to damage during harvesting, it is essential to use the appropriate equipment and trained personnel, as well as careful packing and refrigerated transport to avoid vibrations, as many of the damages that are found during transport are due to bruises and abrasions that are caused by the friction between the fruits. Also, in the case of strawberries, much of the damage that produces bruises is due to compression. So, it is not a good strategy to pile many fruits in the same box (Holt and Schoorl, 1982).

Because strawberries are much perishable, they require careful handling and strictly follow good postharvest handling practices. In order to maintain postharvest life at its optimum level, it is necessary that after harvesting field heat is removed as quickly as possible and the cold chain is maintained during transportation and commercialization. Strawberries are not sensitive to cold, so they can be kept at very low temperatures, close to 0 °C, but try to avoid freezing. The low temperatures in this case maintain the quality, prolong the postharvest life, and reduce the fungal infection. It must be kept in mind that the increase in coloration that occurs after the harvest is significantly delayed when the strawberries are kept cold. In addition, after harvesting, strawberries undergo a radical cut of the hydric supply—the reason why the transpiration will be one of the first problems to face. With the transpiration, not only water is lost and therefore weight, but also the fruit deteriorates in its appearance, it loses shine and smoothness, and the quality diminishes quickly. The loss of weight can reach almost 9% in three days if the fruits are not kept under refrigeration. If they are kept at temperatures close to 0 °C at least in the first 48 hours, it can be negligible. However, transpiration increases drastically when it reaches room temperature after the period of refrigeration (Shin et al., 2007). The similar circumstances should also be created when the commercialization processes begin and when strawberries are put on sale. In this case, the use of refrigerated countertops and cabinets is essential in order to extend postharvest life and maintain quality by reducing water loss and avoiding the appearance of withered fruit (Mitcham, 2016). In addition to the transpiration, another physiological process contributes to the deterioration of the strawberry after harvesting is the respiration, which is obviously slowed down by cold storage. The respiratory rate varies widely from one cultivar to another, but in general terms, the production of CO_2 kg/h between 6 and 10 ml can be estimated if the strawberries are conserved at 0 °C. Figure 8.8 can be quintupled if the temperature reaches 10 °C and becomes 10 times higher when the strawberries are kept at 20 °C. As the increase in deterioration is linked to the respiratory rate, the loss of quality in the strawberry is very high when the temperature rises a few degrees above the optimum of conservation (Mitcham, 2016). Strawberries do not produce a high amount of ethylene due to their nonclimacteric character, with the production of <0.1–1 C_2H_4/kg/h. In addition, they do not respond to the ethylene treatments, so they must be collected in a state very close to their optimum maturity, although it is convenient to eliminate ethylene from the environment during conservation to minimize the incidence and development of diseases (El-Kazzaz et al., 1983).

8.19 DISEASES AND PESTS

Increases in strawberry production are limited by several factors, among which are the losses generated by pests and diseases (Abrol and Shankar 2012). This generated an increase in the use of pesticides with results still to be improved, since in developing countries losses are found in the order of 40%–50% compared with 25%–30% losses of developed countries (Parsa et al., 2014). For this reason, over the past decades, the paradigm in the protection of crops against pests and diseases has been the "integrated pest management" (IPM), summarized in the following eight steps (Barzman et al., 2015):

1. Prevention and suppression
2. Monitoring
3. Decision-making
4. Prioritization of nonchemical methods
5. Selection of pesticides
6. Reducing the use of pesticides
7. Antiresistance strategies and
8. Evaluation.

This is to achieve the objective of pest control through the use of integrated strategies that have a viable relationship between the cost-effectiveness and the components, and that in turn are friendly to the environment (Parsa et al., 2014).

Strawberry plants are susceptible to threat from various pests and diseases. Adopt *precautionary* and *protective* measures to avoid jeopardize production. Below is a brief overview of the main pests and diseases of strawberries.

18.9.1 ARTHROPODS PEST

18.9.1.1 CYCLAMEN MITE (PHYTONEMUS PALLIDUS)

From the Tarsonemidae family, due to its small size (<0.25 mm), it is difficult to see it with the naked eyes. Its eggs and larvae are translucent and once they mature the color turns translucent and bright orange (Zalom et al., 2014). The development of the pest is favored in temperatures between 15 °C and 21 °C and with percentages of relative humidity between 60 and 80 (Cloyd, 2010). It prefers young tissues, attacking leaves and flowers not yet deployed (Cloyd, 2010). The symptoms of damage are atrophied leaves and flowers, small and twisted, causing dwarfism in the plant when the problem is severe (Figure 8.9).

FIGURE 8.9 Plant affected by Cyclamen mite.
(Photo courtesy: J. F. Acuña)

18.9.1.1.1 Management

Taking into account that it is very difficult to eliminate the pest, it is possible to carry out preventive treatments before planting by submerging the plants in water at a temperature of 44 °C for 30 min (Hoy, 2016). As with some chemicals, most predators that feed on this pest are not economically effective for implementation; however, control has been reported with the predatory mite *Amblyseius californicus* (*Neoseiulus californicus*) when the cyclamen mite population is low (Cloyd, 2010; Dara, 2016; Easterbrook et al., 2001).

18.9.1.2 SPIDER (TETRANYCHUS URTICAE)

It is one of the main pests in the world due to its resistance, high fecundity, and short development time (Fraulo and Liburd, 2007). It is distinguished because on its back it has two dark spots on its side and its body (Zalom et al., 2014). It generates a yellowing of the infested leaves, and when the damage is severe, necrotic spots develop on the leaves (Fasulo and Denmark, 2000) (Figure 8.10).

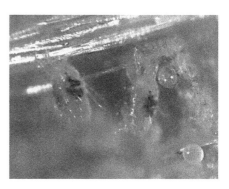

FIGURE 8.10 Spyder (*Tetranychus urticae*)
(Photo courtesy: J. F. Acuña)

18.9.1.2.1 Management

They are considered economic thresholds to carry out a control. For this, predators such as *Phytoseiulus persimilis* and *Amblyseius californicus* (*Neoseiulus californicus*) are used, which are the most commonly used commercially. The cultural control of this plague consists of maintaining a vigorous plant that does not present problems of hydric stress. For chemical control it is possible to use the following active ingredients: bifenazate, acequinocyl, spiromesifen, etoxazole, fenpyroximate, hexythiazox, abamectin, etc. (Dara, 2016).

18.9.2 DISEASES

18.9.2.1 GRAY MOLD OR BOTRYTIS (BOTRYTIS CINEREA)

The most known disease worldwide of strawberry is gray mold, where the applications for its control are made weekly (Leroch et al., 2013). In general, the damage begins in the fruit with small brown lesions under the calyx, and quickly the lesion increases and becomes gray as mycelium and sporula grow. The easy dispersal of spores through the air, workers, tools, animals, etc. makes it difficult to control. The spores can remain in a state of dormancy until the appropriate conditions for their development (humid and cold environments) (Figure 8.11).

FIGURE 8.11 Gray mold (*Botrytis cinerea*)
(Photo courtesy: J. F. Acuña)

18.9.2.1.1 Management

At the cultural level, the inoculum of this pathogen can be reduced by the defoliation of dead material and infected fruits. Additionally, the establishment of greenhouses or macrotunnels effectively reduces this problem. Active ingredients such as pyraclostrobin/boscalid, fenhexamid, cyprodinil/fludioxonil, thiophanate-methyl, iprodione, captan and thiram are used for chemical control (Dara, 2016). The biological control of Botrytis has also been reported with the use of some antagonists such as *Aureobasidium pullulans* and *Candida oleophila*, and simultaneous applications with compatible fungicides can increase its effectiveness (Ippolito and Nigro, 2000). In other studies, control effectiveness has also been reported by the bacteria *Bacillus licheniformis* and the fungus *Trichoderma* (Kim et al., 2007).

18.9.2.2 Anthracnose

Anthracnose is a disease caused mainly by *Colletotrichum acutatum* or species of the same genus. It is recognized by small brown spots on stolons and petioles that lead to the plant death. It can affect all the structures of the plant. The flowers once infected, dry quickly. In fruits, the lesions turn dark brown and then orange when sporulation occurs. When the fungus infects the crown, it causes wilting and death of the whole plant (Baroncelli et al., 2015; Turechek and Heidenreich, 2012).

The optimum temperature for its development is from 27 °C to 32 °C, so in humid environments with high temperatures, the disease increases rapidly.

18.9.2.2.1 Management

Solarization or fumigation of the soil is suggested to destroy the *Colletotrichum inoculum*, since it can survive up to nine months without a host plant. For chemical control, rotate between the active ingredients cyprodinil/fludioxonil, captan and azoxystrobin (Bolda et al., 2016). For biological control, different antagonistic fungal and bacterial characteristics are reported (Ji et al., 2013; Yamamoto et al., 2015).

KEYWORDS

- **crop conditions**
- **diseases and pest**
- **flowering**
- **postharvest**

REFERENCES

Abrol, D.P. & Shankar, U. (2012). *Integrated Pest Management: Principles and Practice* (D. P. Abrol & U. Shankar, Eds.). Jammu: CABI.

Aliasgarian, S., & Ghassemzadeh, H. (2015). Mechanical damage of strawberry during harvest and postharvest operations. *Acta Technologica, 18*(1), 1–5.

Azodanlou, R., Darbellay, C., Luisier, J.-L., Villettaz, J.-C., & Amadò, R. (2003). Quality assessment of strawberries (*Fragaria* species). *Journal of Agricultura and Food Chemistry, 51*(3), 715–721.

Baroncelli, R., Zapparata, A., Sarrocco, S., Sukno, S. A., Lane, C. R., Thon, M. R., Vanncci, G., Holub, E. & Sreenivasaprasad, S. (2015). Molecular diversity of anthracnose pathogen populations associated with UK strawberry production suggests multiple introductions of three different *Colletotrichum* species. *PLoS One, 10*(6), e0129140.

Barzman, M., Bàrberi, P., Birch, A. N. E. et al. (2015). Agron. *Sustainable Development, 35*, 1199.

Bleiholder, H., Weber, F. E., Feller, C. et al. (2001). Growth stages of mono- and dicotyledonous plants. *Growth stages of mono- and dicotyledonous plants* (pp. 37–40) (U. Meier, Ed.). Braunschweig, Germany: Federal Biological Research Centre for Agriculture and Forestry.

Benítez-Burraco, A., Blanco-Portales, R., Redondo-Nevado, J., Bellido, M. L., Moyano, E., Caballero, J., & Muñoz-Blanco, J. (2003). Cloning and characterization of two ripening- related strawberry (*Fragaria × ananassa* cv. Chandler) pectate lyase genes. *Journal of Experimental Botany, 54*(383), 633–645.

Bish, E. B., Cantliffe, D. J., & Chandler, C. K. (2001). A system for producing large quantities of greenhouse-grown strawberry plantlets for plug production. *HortTechnology, 11*(4), 636–638.

Bish, E. B., Cantliffe, D. J., Hochmuth, G. J., & Chandler, C. K. (1997). Development of containerized strawberry transplants for Florida's winter production system. *Acta Horticulturae, 439*, 461–468.

Blanke, M. M., & Cooke, D. T. (2004). Effects of flooding and drought on stomatal activity, transpiration, photosynthesis, water potential and water channel activity in strawberry stolons and leaves. *Plant Growth Regulation, 42*(2), 153–160.

Bolda, M. P., Zalom, F. G., Koike, S. T., Westerdahl, B. B., Fennimore, S. A., Larson, K. D., … Strand, L. L. (2016). *Pest Management Guidelines: Strawberry* (Vol. 87). California: University of California.

Bottoms, T., Bolda, M., Gaskell, M., & Hartz, T. (2013). Determination of strawberry nutrient optimum ranges through diagnosis and recommendation integrated system analysis. *HortTechnology, 23*(3), 312–318.

Boxus, P., Damiano, C., & Brasseur, E. (1984). Strawberry. In D. A. Ammirato, P. V. Evans, W. R. Sharp, & Y. Yamada (Eds.), *Handbook of plant cell culture* (Vol. 3, pp. 453–486). New York, NY: Macmillan.

Boxus, P. H. (1974). The production of strawberry plants by in vitro micro-propagation. *Journal of Horticultural Science, 49*(3), 209–210.

Boxus, P., & Larvor, P. (1987). *In vitro culture of strawberry plants* (L. P. Boxus, Ed.). Luxembourg: Office for Official Publications of the European Communities.

Cámara de Comercio de Bogotá. (2015). Strawberry Manual (Original title in Spanish). Cámara de Comercio de Bogotá, Núcleo Ambiental SAS, Bogotá , Colombia.

Cantliffe, D. J., Shaw, N., Jovicich, E., Rodriguez, J. C., Secker, I., & Karchi, Z. (2003). Plantlet size affects growth and development of strawberry plug transplants. *Proceedings of the Florida State Horticultural Society, 116*, 105–107.

Casierra-Posada, F., Peña-Olmos, J. E., & Ulrichs, C. (2012). Basic growth analysis in strawberry plants (*Fragaria* sp.) exposed to different radiation environments. *Agronomía Colombiana, 30*(1), 25–33.

Casierra-Posada, F., & Vargas, Y. F. (2007). Growth and fruit production in strawberry cultivars (Fragaria sp.) affected by waterlogging. (Original title in Spanish). *Revista Colombiana de Ciencias Hortícolas, 1*(1), 21–32.

Casierra-Posada, F., & Vargas, Y. F. (2007). Crecimiento y producción de fruta en cultivares de fresa (Fragaria sp.) afectados por encharcamiento. *Revista Colombiana de Ciencias Hortícolas, 1*(1), 21–32.

Casierra, F., Fonseca, E., & Vaughan, G. (2011). Fruit quality in strawberry (*Fragaria* sp.) grown on colored plastic mulch. *Agronomía Colombiana, 29*(3), 407–413.

Castillejo, C., de la Fuente, J. I., Iannetta, P., Botella, M. Á., & Valpuesta, V. (2004). Pectin esterase gene family in strawberry fruit: Study of FaPE1, a ripening-specific isoform. *Journal of Experimental Botany, 55*(398), 909–918.

Chagnon, M., Gingras, J., & De Oliveira, D. (1989). Effect of honey bee (*Hymenoptera*: Apidae) visits on the pollination rate of strawberries. *Journal of Economic Entomology, 82*(5), 1350–1353.

Cloyd, R. A. (2010). *Broad mite and cyclamen mite: Management in greenhouses and nurseries* (Agricultural Experiment Station and Cooperative Extension Service). Kansas State University.

Comision de las Comunidades Europeas. (2002). Regulation (EEC) No 843/202 laying down the marketing standard for strawberries and amending regulation (ECC) No. 899/87.

Crespo, P., Bordonaba, J. G., Terry, L. A., & Carlen, C. (2010). Characterisation of major taste and health-related compounds of four strawberry genotypes grown at different Swiss production sites. *Food Chemistry, 122*(1), 16–24.

Dara, S. K. (2016). Managing strawberry pests with chemical pesticides and non-chemical alternatives. *International Journal of Fruit Science, 16*(Suppl. 1), 129–141.

Darnell, R. L., Cantliffe, D. J., Kirschbaum, D. S., & Chandler, C. K. (2003). The physiology of flowering in strawberry. *Horticultural Reviews*, *28*, 325–349.

Darrow, G. M. (1966). *The strawberry. History, breeding and physiology*. New York, NY: Holt, Rinehart & Winston.

Davies, F. T., Davis, T. D., & Kester, D. E. (1994). Commercial importance of adventitious rooting to horticulture. In *Biology of adventitious root formation* (pp. 53–59). Boston, MA: Springer.

Debergh, P. C., & Maene, L. J. (1981). A scheme for commercial propagation of ornamental plants by tissue culture. *Scientia Horticulturae*, *14*(4), 335–345.

Demchak, K. (2009). Small fruit production in high tunnels. *HortTechnology*, *19*(1), 44–49.

Demchak, K., & Hanson, E. J. (2013). Small fruit production in high tunnels in the US mall fruit production in high tunnels in the US. *Acta Horticulturae*, *987*, 41–44.

Durner, E. F., Poling, E. B., & Maas, J. L. (2002). Recent advances in strawberry plug transplant technology. *HortTechnology*, *12*(4), 545–550.

Easterbrook, M. A., Fitzgerald, J. D., & Solomon, M. G. (2001). Biological control of strawberry tarsonemid mite *Phytonemus pallidus* and two-spotted spider mite *Tetranychus urticae* on strawberry in the UK using species of *Neoseiulus* (*Amblyseius*) (*Acari*: Phytoseiidae). *Experimental and Applied Acarology*, *25*(1), 25–36.

Èger, E. K., Schmidt, G., & Èckner, U. B. (1999). Scheduling strawberry irrigation based upon tensiometer measurement and a climatic water balance model. *Scientia Horticulturae*, *81*, 409–424.

El-Arabi, S., Ghoneim, I. M., Shehata, A. I., & Mohamed, R. A. (2003). Effects of nitrogen, organic manure and biofertilizer applications on strawberry plants. I-vegetative growth, flowering and chemical constituents of leaves. *Journal of Agriculture and Environmental Sciences*, *2*(2), 36–67.

El-Kazzaz, M. K., Sommer, N. F., & Fortlage, R. J. (1983). Effect of different atmospheres on postharvest decay and quality of fresh strawberries. *Phytopathology*, *73*(2), 282–285.

Fageria, N. K., Filho, M. B., Moreira, A., & Guimarães, C. M. (2009). Foliar fertilization of crop plants. *Journal of Plant Nutrition*, *32*(6), 1044–1064.

Fai, A., Hanhineva, K., Beleggia, R., Dai, N., Rogachev, I., Nikiforova, V. J., … Aharoni, A. (2008). Reconfiguration of the achene and receptacle metabolic networks during strawberry fruit development. *Plant Physiology*, *148*(2), 730–750.

FAO. (2017). http://faostat. fao.org/.

Fasulo, T. R., & Denmark, H. A. (2000). Two-spotted spider mite, *Tetranychus urticae* Koch. *UF/IFAS Featured Creatures* (EENY-150).

Flórez Faura, R., & Mora Cabeza, R. A. (2010). Strawberry (Fragaria x ananassa Duch.): Production and postharvest handling (Original title in Spanish). Bogotá: Editorial Universidad Nacional de Colombia.

Frankel, R., & Galun, E. (2012). *Pollination mechanisms, reproduction and plant breeding* (Vol. 2). Basingstoke, UK: Springer Science & Business Media.

Fraulo, A. B., & Liburd, O. E. (2007). Biological control of two spotted spider mite, *Tetranychus urticae*, with predatory mite, *Neoseiulus californicus*, in strawberries. *Experimental and Applied Acarology*, *43*(2), 109–119.

Fuentes, S., Ljung, K., Sorefan, K., Alvey, E., Harberd, N. P., & Østergaard, L. (2012). Fruit growth in *Arabidopsis* occurs via DELLA-dependent and DELLA-independent gibberellin responses. *The Plant Cell*, *24*(10), 3982–3996.

Gambardella, M., Cadavid-Labrada, A., & Diaz, V. P. (2000). Isozyme and RAPD characterization of wild and cultivated native Fragaria in Southern Chile. *Acta Horticulturae*, *567*, 81–84.

George, E. F., Hall, M. A., & de Klerk, G. J. K. (2008). *Plant propagation by tissue culture* (3rd ed.). Basingstoke, UK: Springer Science & Business Media.

Gil-Ariza, D.J., Amaya, I., López-Aranda, J.M., Sánchez-Sevilla, J.F., Botella, M.A., & Valpuesta, V. (2009). Impact of plant breeding on the genetic diversity of cultivated strawberry as revealed by expressed sequence tag-derived simple sequence repeat markers. *Journal of the American Society for Horticultural Science, 134,* 337–347.

Haifa. (2017). Strawberry crop guide: Haifa nutrition recommendations. Retrieved July 7, 2020, from http://www.haifa-group.com.

Hancock, J. F. (1999). Strawberries. *Crop production science in horticulture* (Vol. 11). Wallingford, UK.: CAB International.

Herrington, M. E., Hardner, C., Wegener, M., Woolcock, L. L., & Dieters, M. J. (2011). Rain damage to strawberries grown in southeast Queensland: Evaluation and genetic control. *HortScience, 46*(6), 832–837.

Hochmuth, G. J., & Albregts, E. (1994). *Fertilization of strawberries in Florida.* Service Institute of Food and Agriculture Sciences, EDIS: University of Florida Cooperative Extension.

Holt, J. E., & Schoorl, D. (1982). Strawberry bruising and energy dissipation. *Journal of Texture Studies, 13*(3), 349–357.

Holzapfel, C., & Alpert, P. (2003). Root cooperation in a clonal plant: Connected strawberries segregate roots. *Oecologia, 134*(1), 72–77.

Horneck, D. A., Sullivan, D. M., Owen, J. S., & Hart, J. M. (2011). *Soil test interpretation guide.* (Oregon State University Extension Service, EC 1478).

Hoy, M. A. (2016). *Agricultural acarology: Introduction to integrated mite management.* Boca Raton, FL: CRC Press.

Hummer, K. E., & Hancock, J. (2009). Strawberry genomics: Botanical history, cultivation, traditional breeding, and new technologies. In *Genetics and genomics of Rosaceae* (pp. 413–435). New York, NY: Springer.

Husaini, A. M., & Neri, D. (2016). *Strawberry: Growth, development and diseases.* Jammu: CABI.

INCOTEC. (1997). Colombian Tecnhical Regulation. NTC 4103.: Fresh fruits: Strawberry, cultivar Chandler (Original title in Spanish). Bogotá: Instituto Colombiano de Normas Técnicas y Certificación.

Ippolito, A., & Nigro, F. (2000). Impact of preharvest application of biological control agents on postharvest diseases of fresh fruits and vegetables. *Crop Protection, 19*(8–10), 715–723.

Jeong, H. J., Choi, H. G., Moon, B. Y., Cheong, J. W., & Kang, N. J. (2016). Comparative analysis of the fruit characteristics of four strawberry cultivars commonly grown in South Korea. *Korean Society for Horticultural Science, 34,* 396–404.

Ji, M., Yang, J., Wu, X., Xiao, T., Yao, K., & Zhuang, Y. (2013). Biocontrol of strawberry anthracnose caused by *Colletotrichum fragariae. Agricultural Science & Technology, 14*(11), 1569.

Kafkafi, U. & Tarchitzky, J. (2011). *Fertigation: A tool for efficient fertilizer and water management.* Paris, France: International Fertilizer Industry Association (IFA) & International Potash Institute.

Kang, C., Darwish, O., Geretz, A., Shahan, R., Alkharouf, N. & Liu, Z. (2013). Genome- scale transcriptomic insights into early-stage fruit development in woodland strawberry *Fragaria vesca. The Plant Cell, 25*(6), 1960–1978.

Khoshnevisan, B., Rafiee, S., & Mousazadeh, H. (2013). Environmental impact assessment of open field and greenhouse strawberry production. *European Journal of Agronomy, 50,* 29–37.

Kim, J. H., Lee, S. H., Kim, C. S., Lim, E. K., Choi, K. H., Kong, H. G., ..., Moon, B. J. (2007). Biological control of strawberry gray mold caused by *Botrytis cinerea* using *Bacillus licheniformis* N1 formulation. *Journal of Microbiology and Biotechnology*, *17*(3), 438–444.

Kopper. (2017). Fresa. Retrieved 7 July, 2020, from at https://www.koppert.es.

Krüger, E., Schmidt, G., & Brückner, U. (1999). Scheduling strawberry irrigation based upon tensiometer measurement and a climatic water balance model. *Scientia Horticulturae*, *81*(4), 409–424.

Ledesma, N. A., Nakata, M., & Sugiyama, N. (2008). Effect of high temperature stress on the reproductive growth of strawberry cvs. "Nyoho" and "Toyonoka." *Scientia Horticulturae*, *116*(2), 186–193.

Leroch, M., Plesken, C., Weber, R. W. S., Kauff, F., Scalliet, G., & Hahn, M. (2013). Gray mold populations in German strawberry fields are resistant to multiple fungicides and dominated by a novel clade closely related to *Botrytis cinerea*. *Applied and Environmental Microbiology*, *79*(1), 159–167.

Lieten, F. (1998). Recent advances in strawberry plug transplant technology. *Acta Horticulturae*, *513*, 383–388.

Lieten, F., Kinet, J.-M., & Bernier, G. (1995). Effect of prolonged cold storage on the production capacity of strawberry plants. *Scientia Horticulturae*, *60*(3–4), 213–219.

Lieten, F., & Misotten, C. (1993). Nutrient uptake of strawberry plants (cv. *Elsanta*) grown on substrate. *Acta Horticulturae*, *348*, 299–306.

López-Aranda, J. M., Soria, C., Santos, B. M., Miranda, L., Domínguez, P., & Medina- Mínguez, J. J. (2011). Strawberry production in mild climates of the world: A review of current cultivar use. *International Journal of Fruit Science*, *11*(3), 232–244.

Mahmood, T., Anwar, F., Iqbal, T., Bhatti, I. A., & Ashraf, M. (2012). Mineral composition of strawberry, mulberry and cherry fruits at different ripening stages as analyzed by inductively coupled plasma-optical emission spectroscopy. *Journal of Plant Nutrition*, *35*(1), 111–122.

Medellín, L. A. C., Rivera, D. C. A., Caicedo, D. R., Rativa, C. M. G., & Trujillo, M. M. P. (2013) Assessment of Materials for Strawberry Mulching Grown under Greenhouse. (Original title in Spanish). *Revista Facultad de Ciencias Básicas*, *9*(1), 8–19.

Meier, U. (Ed.). (2001). *Growth stages of mono- and dicotyledonous plants*. Braunschweig, Germany: Federal Biological Research Centre for Agriculture and Forestry.

Mitcham, E. J. (2016). Strawberry. In M. Gross, K.C. Wang, & C.Y. Salveit (Eds.), *The commercial storage of fruits, vegetables, and florist and nursery stocks* (pp. 559– 561). Washington, DC: Agricultural Research Service, USDA.

Molina, E., Salas, R., & Castro, A. (1993). Growth curve and nutrient absorption in strawberries (Fragaria x ananassa Duch. Cv. Chandler) grown in Alajuela. (Original title in Spanish). Agronomía Costarricense., *17*(1), 63–67.

Moradi, K., Otroshy, M., & Azimi, M. R. (2011). Micropropagation of strawberry by multiple shoots regeneration tissue cultures. *Journal of Agricultural Technology*, *7*(6), 1755–1763.

Morillo, J. G. (2015). Towards precision irrigation in strawberry cultivation in the Doñana area. (Original title in Spanish). University of Córdoba, Spain, Córdoba, Spain.

Murashige, T. (1974). Plant propagation through tissue cultures. *Annual Review of Plant Physiology*, *25*(1), 135–166.

Nitsch, J. P. (1950). Growth and morphogenesis of the strawberry as related to auxin. *American Journal of Botany*, *37*, 211–215.

Nitsch, J. P. (1955). Free auxins and free tryptophane in the strawberry. *Plant Physiology, 30,* 33–39.

Nunes, M. C. N., Brecht, J. K., Morais, A. M., & Sargent, S. A. (2006). Physicochemical changes during strawberry development in the field compared with those that occur in harvested fruit during storage. *Journal of the Science of Food and Agriculture, 86*(2), 180–190.

Parsa, S., Morse, S., Bonifacio, A., Chancellor, T. C., Condori, B., Crespo-Pérez, V., Shaun L. Hobbs, A., Kroschel, J., Ba, M. N., Rebaudo, F., Sherwood, S. G., Vanek, S. J., Faye, E., Herrera, M. A., & Dangles, O. (2014). Obstacles to integrated pest management adoption in developing countries. *Proceedings of the National Academy of Sciences, 111*(10), 3889–3894.

Patiño Sierra, D. I., García Valencia, E. I., Abello, E., Quejada Rovira, O., Rodríguez Mariaca, H. D., & Arroyave Tobón, I. C. (2014). Technical Manual of Strawberry Crops under Good Agricultural Practices (Original title in Spanish). Medellín, Colombia: Francisco Vélez Litografía.

Pauletti, V., Serrat, B., Motta, A., Favaretto, N., & Anjos, A. (2010). Yield response to fertilization strategies in no-tillage soybean, corn and common bean crops. *Brazilian Archives of Biology and Technology, 53*(3), 563–574.

Perkins-Veazie, P. (1995). Growth and ripening of strawberry fruit. *Horticultural Reviews, 17,* 267–297.

Peverill, K. I., Sparrow, L. A., & Reuter, D. J. (1999). *Soil analysis: An interpretation manual.* Clayton, Australia: CSIRO Publishing.

Roussos, P. A., Triantafillidis, A., & Kepolas, E. (2012). Strawberry fruit production and quality under conventional, integrated and organic management. *Acta Horticulturae, 926,* 541–546.

Rubio, S. A., Alfonso, A. M., Grijalba, C., & Pérez, M. M. (2014). Determination of the production costs of strawberry cultivated in an open field and with a high tunnel. *Revista Colombiana de Ciencias Hortícolas, 8*(1), 67–69.

Ruíz, R., & Piedrahíta, W. (2012). Manual for the growing of fruit trees in the tropics (Original title in Spanish). Produmedios, Bogotá. 2012. pp 474–495. (G. Fisher, Ed.). Bogotá (Colombia): Produmedios.

Salas, R. E. (2002). Diagnostic tools to establish recommendations for leaf fertilization (Original title in Spanish). Fertilización foliar, principios y aplicaciones. (G. Meléndez & E. Molina, Eds.). Laboratorio de Suelos y Foliares CIA/UCR.

Sathyanarayana, B. N., & Varghese, D. B. (2007). *Plant tissue culture: Practices and new experimental protocols.* New Delhi, India: I. K. International.

Selamovska, A. (2014). Strawberry: Factor of high yield. In N. Malone (Ed.), *Strawberries: Cultivation, antioxidant properties and health benefits* (1st ed., pp. 121–144). Hauppauge: Nova Publishers.

Serçe, S., & Hancock, J. F. (2005). The temperature and photoperiod regulation of flowering and runnering in the strawberries, *Fragaria chiloensis, F. virginiana,* and *F. × ananassa. Scientia Horticulturae, 103*(2), 167–177.

Sesso, H. D., Gaziano, J. M., Jenkins, D. J. A., & Buring, J. E. (2007). Strawberry intake, lipids, C-reactive protein, and the risk of cardiovascular disease in women. *Journal of the American College of Nutrition, 26*(4), 303–310.

Shin, Y., Liu, R. H., Nock, J. F., Holliday, D., & Watkins, C. B. (2007). Temperature and relative humidity effects on quality, total ascorbic acid, phenolics and flavonoid concentrations, and antioxidant activity of strawberry. *Postharvest Biology and Technology, 45*(3), 349–357.

Singh, A., Syndor, A., Deka, B. C., Singh, R. K., & Patel, R. K. (2012). The effect of microclimate inside low tunnels on off-season production of strawberry (*Fragaria × ananassa* Duch.). *Scientia Horticulturae, 144*, 36–41.

Singh, R., Asrey, R., & Kumar, S. (2006). Effect of plastic tunnel and mulching on growth and yield of strawberry. *Indian Journal of Horticulture, 63*(1), 18–20.

Strand, L. L. (2008). *Integrated pest management for strawberries*. Davis, CA: UCARN Publications.

Strick, B. C. (2007). Berry crops: Worldwide area and production systems. In Y. Zhao (Ed.), *Berry fruit value added products for health promotion* (1st ed., pp. 3–49). Boca Raton, FL: CRC Press.

Symons, G. M., Chua, Y.-J., Ross, J. J., Quittenden, L. J., Davies, N. W., Reid, J. B. (2012). Hormonal changes during non-climacteric ripening in strawberry. *Journal of Experimental Botany, 63*(13), 4741–4750.

Tagliavini, M., Baldi, E., Lucchi, P., Antonelli, M., Sorrenti, G., Baruzzi, G., & Faedi, W. (2005). Dynamics of nutrients uptake by strawberry plants (*Fragaria × ananassa* Dutch.) grown in soil and soilless culture. *European Journal of Agronomy, 23*(1), 15–25.

Tagliavini, M., & Toselli, M. (2005). Foliar application of nutrients. In D. Hillel, J. L. Hateld, D. S. Powlson, C. Rosenzweig, K. M. Scow, M. J. Singer, & D. L. Sparks (Eds.), *Encyclopedia of soils in the environment* (pp. 53–59). Cambridge, MA: Elsevier Academic Press.

Thompson, P. A. (1969). The effect of applied growth substances on development of the strawberry fruit: II. Interactions of auxins and gibberellins. *Journal of Experimental Botany, 20*(3), 629–647.

Trinidad-Santos, A., & Aguilar-Manjarrez, D. (1999). Foliar fertilization, an important support for crop yield. (Original title in Spanish). Terra Latinoamericana 17: 247–255. Terra Latinoamericana, 17, 247–255.

Tudor, V., Manole, C. G., Teodorescu, R., Asanica, A., & Barbulescu, I. D. (2015). Analysis of some phenolic compounds and free radical scavenging activity of strawberry fruits during storage period. *Agriculture and Agricultural Science Procedia, 6*, 157–164.

Tulipani, S., Mezzetti, B., Capocasa, F., Bompadre, S., Beekwilder, J., Ric de Vos, C. H., Capanoglu, E., Bovy, & Battino, M. (2008). Antioxidants, phenolic compounds, and nutritional quality of different strawberry genotypes. *Journal of Agricultural and Food Chemistry, 56*(3), 696–704.

Turechek, B., & Heidenreich, C. (2012). Strawberry anthracnose. Retrieved July 7, 2020, from https://www.cabdirect.org/cabdirect/abstract/20137803883.

UPO. (2019). *International union for the protection of new varieties of plants.* Retrieved July 7, 2020, from https://www.upov.int/about/es/.

USDA. (2006). *United States Standards for grades of strawberries*. Retrieved July 7, 2020, from https://www.ams.usda.gov.

USDA. (2017). *USDA food composition databases*. Retrieved July 7, 2020, from https://www.ams.usda.gov.

Verdier, M. (1987). Strawberry crop in temperate climates. Huelva, Spain. (Original title in Spanish). Ediciones Agrarias S.A.

Verheul, M. J., Sønsteby, A., & Grimstad, S. O. (2007). Influences of day and night temperatures on flowering of *Fragaria × ananassa* Duch., cvs. Korona and Elsanta, at different photoperiods. *Scientia Horticulturae, 112*(2), 200–206.

Wang, S. Y., & Camp, M. J. (2000). Temperatures after bloom affect plant growth and fruit quality of strawberry. *Scientia Horticulturae, 85*(3), 183–199.

Wilk, J. A., Kramer, A. T., & Ashley, M. V. (2009). High variation in clonal vs. sexual reproduction in populations of the wild strawberry, *Fragaria virginiana*

(Rosaceae). *Annals of Botany, 104*(7), 1413–1419.

Yamamoto, S., Shiraishi, S., & Suzuki, S. (2015). Are cyclic lipopeptides produced by Bacillus amyloliquefaciens S13-3 responsible for the plant defence response in strawberry against Colletotrichum gloeosporioides? *Letters in Applied Microbiology, 60*(4), 379–386.

Yuan, B., Sun, J., & Nishiyama, S. (2004). Effect of drip irrigation on strawberry growth and yield inside a plastic greenhouse. *Biosystems Engineering, 87*(2), 237–245.

Zalom, F. G., Bolda, M. P., Dara, S. K., & Joseph, S. (2014). *UC IPM pest management guidelines: Strawberry.* University of California statewide integrated pest management program. Oakland: UC ANR Publication (No. 3468).

Zapata, I. I., Villalobos, C. M. B., Araiza, M. D. S., Solís, E. S., Jaime, O. A. M., & Jones, R. W. (2014). Effect of strawberry pollination by *Apis mellifera* L. and *Chrysoperla carnea* S. on fruit quality (Original title in Spanish). Nova Scientia, 7(13), 85-100., 7(13), 85–100.

CHAPTER 9

MULBERRY

JER-CHIA CHANG* and YI-HSUAN HSU

Department of Horticulture, National Chung Hsing University 145, Taichung 40227, Taiwan, Republic of China

Corresponding author. E-mail: jerchiachang@dragon.nchu.edu.tw

ABSTRACT

Mulberry (*Morus* spp.) has been regarded as the sole crop of its economic importance in sericulture. The domestication of mulberry started several thousand years ago, and it has been extensively grown globally, especially in China, India, Japan, and Turkey. Mulberry is wind pollinated and highly heterozygous and evolved a complex population as a result. Over 68 species have been identified, while the main species cultivated are white (*M. alba*), black (*M. nigra*), red (*M. rubra*) mulberry. In addition to be cultivated for its foliage to feed the silkworm (*Bombyx mori*), mulberry has the increased potential in horticultural, food industry, and human health utilization worldwide, for example, the multiple uses in berry production due to its tasty and high phytochemicals for health benefit, and landscaping utilization owing to their special and attractive canopy shapes in a few varieties. This chapter covers origin and history, distribution and production area, botany and taxonomy, reproductive characteristics, breeding and cultivars, planting and cultivation, propagation, and pest and disease control. Moreover, uses and nutrition value, harvest and postharvest, packaging and transport, and trade and marketing are involved.

9.1 GENERAL INTRODUCTION

The mulberry plants are deciduous (most species) or evergreen (few species) woody plants that belong to the phylum *Angiospermae*, class Dicotyledonae, order Rosales, family Moraceae, and genus *Morus* (Linné, 1753; Westwood, 1993). Mulberry trees were initially cultivated for sericulture, and the primary species for this purpose, the white mulberry (*Morus alba*), was introduced to many areas via human activities (Figure 9.1) (Minamizawa, 1976; Sharma et al., 2000). Since 419 AD, it has spread along the Silk Route to India, South Korea, Japan, Europe, and Africa, and reached as far as the Americas. Today, white mulberry and other 67 *Morus* plants are widely distributed around the globe and primarily concentrated between 10° South and 50° North in latitude (Minamizawa, 1976; Sharma et al., 2000; Vijayan et al., 2012; Zhao et al., 2007a).

Although mulberry trees are well known for their versatile uses in sericulture, in providing forage and medicine or as fruit trees, landscaping, and as gardening plants, sericulture remains the main purpose of mulberry cultivation (Sànchez, 2000). The breeding of mulberry has, therefore, long focused on foliage quality and generated thousands of varieties and accessions. However, the development of varieties for horticulture to be used as fruit trees and landscaping/gardening plants has rarely been attempted (Minamizawa, 1976; Sànchez, 2000; Vijayan et al., 2009; Vijayan et al., 2011a, b; Vijayan et al., 2012; Doss et al., 2012).

Mulberry fruits have high antioxidant activity (Chen, 2005) and can be eaten fresh or processed as mulberry jelly, fruit vinegar, fruit whine, or fruit juice. Mulberry leaves can be used for making tea or be ground as seasoning. Mulberry trees can also be harnessed as crops for making functional health supplements. The white epidermis of mulberry root and the leaves, fruits, and stems of the mulberry tree all have health-promoting effects according to ancient manuscripts such as *Bencaogangmu* in China (Li, 1596). In recent years, scientists have discovered that

FIGURE 9.1 *Morus alba* route map spread along from China to the world (adapted from Minamizawa, 1976; Sharma et al., 2000; Google Map, 2014).

mulberry trees contain substances capable of inhibiting intestinal α-glycosidases, such as 1-dexoynojiri-mycin (1-DNJ) and fagomine (Asano et al., 1994). Mulberry leaves contain γ-aminobutyric acid (γ-GABA), which can help control hypertension and display antifatigue activity (Chen et al., 2016). In addition, some mulberry varieties have beautiful and elegant appearance, such as *Morus latifolia* var. "Unryuu" and *M. alba* var. "Shidareguwa" and are thus suitable for garden landscaping (Aroonpong and Chang, 2015; Chang et al., 2014b, 2018; Yamanouchi et al., 2009).

Currently, there are more than 40 countries worldwide that engage in economic mulberry cultivation, with a total cultivation area of 1.01 million ha. China, India, and Turkey are the major producers (Liu et al., 2017; Central Silk Board, Ministry of Textiles, Government of India, 2018; Turkish Statistical Institute, 2017). Presently, the main purpose of mulberry cultivation remains sericulture, followed by berry production. Other less popular purposes include using mulberry as livestock forage, medicine crop, and ornament plants (Sànchez, 2000; Ercisli and Orhan, 2007).

9.2 AREA AND PRODUCTION

The current world total cultivation area for mulberry is over 1.01 million ha, and this mainly includes China (789 million ha) (Nation Cocoon and Silk Coordination Office, China, 2017), India (217 million ha) (CENTRAL SILK BOARD, Ministry of Textiles, Government of India, 2018), Japan, South Korea, the USA, and Taiwan (Vijayan et al., 2011a; Dandin, 2014). China produced 635.7 million tons of silkworm cocoons in 2017. Guangxi Province is the most important area of production with a mulberry cultivation area of 212 million ha and a cocoon production of 318 million tons (50.0%), followed by Sichuan Province with a cultivation area of 130 million ha and a cocoon production of 81 million tons (12.8%), and Yunnan Province with a cultivation area of 99 million ha and a cocoon production of 48 million tons (7.5%) (Table 9.1) (Nation Cocoon and Silk Coordination Office, China, 2017).

In India, the main area of production is Karnataka, with a total cultivation area of 80,873 ha and a cocoon production of 61.4 million tons (Satish, 2014).

Turkey is the primary producer of mulberry fruits. In 2017, Turkey had a total cultivation area of 2,269 ha that yielded a total of 74,383 tons. Diyarbakir is the main area of production that accounts for 15.9% of the total yield of Turkey, followed by Malatya (10.9%), Elazığ (7%), and Erzurum (7%) (Table 9.2) (Turkish Statistical Institute, 2017).

494 Temperate Fruits: Production, Processing, and Marketing

TABLE 9.1 Mulberry Production Area, Silkworm Cocoons Production and Proportion of Production in Major Provinces of China in 2017

Ranking	Major Provinces of China	Production Area (1000 ha)	Silkworm Cocoons Production (tons)	Proportion of Production (%)
1	Guangxi	212.3	318,000	50.0
2	Sichuan	130.0	81,170	12.8
3	Yunnan	98.6	47,481	7.5
4	Guangdong	24.9	39,609	6.2
5	Jiangsu	32.4	38,871	6.0
6	Zhejiang	40.0	20,470	3.2
7	The rest of provinces	250.5	90,084	14.3
	Total	788.7	635,685	100

Source: Nation Cocoon and Silk Coordination Office, China (2017) (http://scyxs.mofcom.gov.cn/).

TABLE 9.2 Mulberry Berries Production Area, Production, Yields in Major Counties of Turkey in 2017

County	Area of Compact Fruits (ha)	Berries Production (tons)	Yield (kg/tree)
Diyarbakır	1055.8	11,844	24.0
Malatya	52.5	8097	56.7
Elazığ	61.0	5235	44.0
Erzurum	229.7	5208	93.4
Ankara	51.1	4267	55.7
Erzincan	39.7	4066	36.8
The rest of the counties	779.7	35,666	–
Total	2269.5	74,383	31.4

Source: Turkish Statistical Institute (2017) (https://biruni.tuik.gov.tr/medas/?kn=92&locale=en).

9.3 MARKETING AND TRADE

In 2016, total silkworm cocoon export quantity was 6,217 tons in the world, and India, China, and Vienna were the primary exporters. Total silkworm cocoon export value reached 69,556 million USD in the world. By export value, China was the largest importer that accounted for 50% of export value. The world silkworm cocoon total import quantity was 6,488 tons and China was the primary importer. Total silkworm cocoon import quantity reached 66,957 million USD in the world. By import value, Italy was the largest importer that accounted for 36.3% of the world total import value (Table 9.3) (FAOSTAT, 2016).

In 2016, total silk export quantity was 8962 tons in the world and China was the primary exporter. Total silk export value reached 378,957 million USD in the world, by export value, China was the largest exporter that accounted for 83.3% of the world total export value. Total silk import quantity was 9167 tons in the world, which is mainly attributed to India and Romania. Total silk import value reached 411,045 million USD in the world. By import value, India and Romania were the largest importer that accounted for 38.4% and 20.6% of the world's total import value, respectively (Table 9.4) (FAOSTAT, 2016).

Turkey is the major importer and exporter of mulberry fruits, with an export quantity of 3195 tons and an import quantity of 2,177 tons (Turkish Statistical Institute, 2017).

TABLE 9.3 Export and Import Amount of Silkworm Cocoons in Major Countries in 2016

	Export Quantity		Export Value		Import Quantity		Import Value	
Ranking	Country	Tons	Country	1000 USD	Country	Tons	Country	1000 USD
1	India	1968	China	34,783	China	3,392	Italy	24,327
2	China	1456	India	14,939	Italy	827	Germany	9584
3	Vietnam	855	Germany	8097	Germany	471	China	8633
4	Uzbekistan	733	Italy	4473	Vietnam	289	India	5245
5	Germany	296	Vietnam	2170	Korea	288	Japan	3712
6	The rest	909	The rest	5094	The rest	1221	The rest	15,474
	World	6217	World	69,556	World	6488	World	66,975

Source: FAOSTAT (2016) (http://www.fao.org/faostat/en/#data).

TABLE 9.4 Export and Import Amount of Silk Raw in Major Countries in 2016

	Export Quantity		Export Value		Import Quantity		Import Value	
Ranking	Country	Tons	Country	1000 USD	Country	Tons	Country	1000 USD
1	China	6,927	China	315,678	India	3,757	India	157,686
2	Vietnam	427	Italy	18,347	Romania	1,610	Romania	84,485
3	Italy	341	Vietnam	16,858	Vietnam	741	Italy	39,234
4	Malaysia	274	Malaysia	11,978	Italy	710	Vietnam	37,739
5	USA	264	Romania	7,079	Japan	395	Japan	20,766
6	The rest	729	The rest	9,017	The rest	1,954	The rest	71,135
	World	8,962	World	378,957	World	9,167	World	411,045

Source: FAOSTAT (2016) (http://www.fao.org/faostat/en/#data).

9.4 COMPOSITION AND USES

9.4.1 COMPOSITION

9.4.1.1 FRUIT COMPOSITION

Mulberry fruits contain anthocyanins; polyphenols; flavonoids; malic acid; citric acid; several amino acids; carotenes; vitamins A, C, and E; potassium; phosphate; and calcium (Ercisli and Orhan, 2007; Lee and Hwang et al., 2017; Yuan and Zhao, 2017).

Ercisli and Orhan (2007) compared the fruit compositions of *Morus nigra* (*M. nigra*), *Morus rubra* (*M. rubra*), and *M. alba*, and found that *M. nigra* and *M. rubra* have a higher total acid content (TAC) than *M. alba*, whereas *M. alba* contains more total soluble solids (TSS) than *M. nigra* and *M. rubra*. *M. alba* has the highest TSS/TAC ratio, while there is no significant difference between these two contents in *M. nigra* and *M. rubra*. Therefore, *M. alba* fruits are recommended as processed food, and *M. nigra* fruits may be eaten fresh.

Malic acid is the primary organic acid found in mulberry fruits, whose content is between 1.132 and 4.467 g/100 g fresh weight and is the highest in *M. rubra* and the lowest in *M. nigra*. Other less abundant organic acids include citric acid, tartaric acid, and succinic acid (Gundogdu et al., 2011). However, Özgen et al. (2009) suggested that *M. nigra* and *M. rubra* contain more citric acid than malic acid.

Soluble solid content (SSC) primarily consists of fructose and glucose; for fructose, *M. alba* > *M. nigra* > *M. rubra*, and the content is between 5.4 and 6.3 g/100 g fresh weight, while for glucose, *M. nigra* > *M. alba* > *M. rubra*, and the content is between 6.0 and 7.7 g/100 g fresh weight (Gundogdu et al., 2011). In addition, Özgen et al. (2009) noted that *M. nigra* and *M. rubra* contain very less quantities of sucrose, that is, only 0.04 g/100 g fresh weight.

The content of vitamin C is between 11.3 and 24.4 mg/100 g fresh weight with *M. alba* containing the highest value. All three mulberries are low in total fat; with fat content between 0.85% and 1.1%, *M. alba* has the highest value, followed by *M. nigra* (0.95%) and *M. rubra* (0.85%). Linoleic acid (*cis*-C18:2 ω6) is the primary fatty acid with a content range of 43.4%–61.9%, while palmitic acid (C16:0) is the second abundant fatty acid with a content range of 12.1%–24.8% (Ercisli and Orhan, 2007).

Ascorbic acid content is between 19.4 and 22.4 mg/100 ml with *M. rubra* at the lower limit. Both *M. nigra* and *M. rubra* have very high levels of the total polyphenols and total flavonoids [1035–1422 mg gallic acid equivalent/100 g fresh weight and 219–276 mg QE (quercetin equivalents)/100 g fresh weight] (Ercisli and Orhanm, 2007).

As for minerals, K is the most abundant with a content range from 834 mg/100 g (*M. rubra*) to 1,668 mg/100 g (*M. alba*), followed by N, P, Ca, Mg, Na, Fe, Mn, Zn, and Cu (Ercisli and Orhan, 2007).

9.4.1.2 LEAF

The contents of soluble nonprotein nitrogen in mulberry leaves (*M. alba*) account for 26.1% of its total nitrogen and the total true protein represents 14.4% of total dry matter. The major protein constituents are prolamins, albumins, globulins, and glutelins, with each accounting for 44.1%, 11.1%, 9.7%, and 8.5% of total dry matter, respectively; 26.6% of the total true protein is insoluble proteins (Kandylis et al., 2009).

9.4.1.3 PHYTOCHEMICALS IN ALL PARTS OF PLANT

Many parts of the mulberry tree such as the bark, stems, leaves, roots, and fruits contain special compounds. For instance, 1-deoxynojirimycin (DNJ) can lower blood glucose levels and thus prevent diabetes (Kimura et al., 2007). Resveratrol is one of the derivatives of stilbenes, and as a plant secondary metabolite, it has antiviral (Campagna and Rivas, 2010), antioxidant (Gülçin, 2010), neuroprotective (Bastianetto et al., 2015), vascular dilatory (Novakovic et al., 2006), antitumor,

and anti-inflammatory properties (Tili and Michaille, 2011), and it can also prevent cardiovascular diseases (Zordoky et al., 2015); GABA is a nonprotein amino acid that not only functions as a major inhibitory neurotransmitter in the central nervous system (Kimura et al., 2002), but also performs anti-hypertensive (Yang et al., 2012) and antifatigue (Chen et al., 2016) activities. All these compounds may be extracted as medicine and/or health supplements.

9.4.2 USES

9.4.2.1 SERICULTURE

The most important use of mulberry trees around the world is to rear the mulberry silkworm (*Bombyx mori*) for sericulture, of which China and India are the major producers. Mulberry leaves are the only food for silkworm caterpillars, which are raised to spin cocoons as a raw material for silk. The silk is then woven into products such as cloth, clothes, and scarves (Das, 2014).

9.4.2.2 WOOD

Mulberry wood can be used in handcrafts, furniture, and as a wooden material for sport equipment such as tennis bats; the slender stems can be used for weaving baskets (Sànchez, 2000).

9.4.2.3 LANDSCAPING

Mulberry trees are drought tolerant and can withstand pruning. They can be planted as street trees, green fence, bonsai, or garden landscaping trees. They are highly valued ornamental plants in Europe, the USA, Japan, South Korea, and China. The "Unryuu" mulberry (*M. latifolia* var.) with Z-bending shoots/canopy and "Shidagreguwa" mulberry (*M. alba* var.) with weeping shoots/canopy are both common landscaping mulberry varieties (Aroonpong and Chang, 2015; Chang et al., 2014b; Zhao et al., 2007; Vijayan et al., 2011a; Sànchez, 2000; Yamanouchi et al, 2009).

9.4.2.4 FUNCTIONAL FOOD

All the parts of a mulberry tree (white epidermis of the root, leaves, tree bark, etc.) may be used as medicine. It can inhibit intestinal α-glycosidase activity (Kimura et al., 2007). It contains γ-GABA to aid against fatigue (Chen et al., 2016). The tyrosinase-inhibiting contents of mulberry can inhibit melanin biosynthesis and thus exert a skin-lightening effect (Lee et al., 2002).

9.4.2.5 FORAGE

Mulberry leaves can be used to make tea or be ground as seasoning. Moreover, mulberry leaves are highly digestible, contain high protein and low fiber content, and are rich in nutrients, and thus can be used as silkworm food and supplementary the protein source for the forage of ruminant livestock (Yao et al., 2000).

9.4.2.6 BERRY (FRUIT)

Mulberry fruit may be eaten fresh or processed into fruit jelly, fruit vinegar, wine, and fruit juice. It is rich in polyphenols that can eliminate 2,2-diphenyl-1-picrylhydrazyl free radicals (antioxidant), effectively inhibit α-glucosidase activity, and lower blood glucose level to alleviate diabetes (Wang et al., 2013). The anthocyanins, polyphenols, and flavonoids in mulberry fruits have high antioxidant capacity as well as anti-inflammatory and antitumor properties (Chen et al., 2006; Zafra-Stone et al., 2007). Mulberry fruits also contain resveratrol, which is an antioxidant and is also effective in preventing cardiovascular diseases, further extending the importance of mulberry as a health food (Tseng, 2010).

9.5 ORIGIN AND DISTRIBUTION

Members of the Moraceae family first appeared in the mid-Cretaceous (Zerega et al., 2005), originated from tropical Asia and the islands of the Indian and Pacific Oceans (Berg, 2001; Rohwer, 1993), and later invaded the low-altitude areas of Southwest China and Himalaya mountains (Sànchez, 2000; Vijayan et al., 2004b). However, mulberry species in Japan and China exhibit the highest diversity (Sharma et al., 2000). Plants of genus *Morus* are anemophilous and natural hybridization is very common. In addition, they are highly adaptive and can survive and reproduce even under extreme geographical and climatic conditions. Consequently, within this genus, a numerous species and ecotypes with complex morphological and genetic diversity have evolved (Zhao et al., 2007b).

China is one of origin for mulberry trees, and they have been important economic crops for China since ancient times. Sericulture was once a common scene in the countryside of South China for several dynasties. The mulberry trees later spread along the Silk Route to Central Asia and were eventually introduced to Europe and the Americas. Today, mulberry trees are widely distributed and cultivated between 50° North and 10° South in latitude around the globe, with China and India being the top countries for cultivation. The primary species for cultivation are the white mulberry (*M. alba*), black mulberry

(*M. nigra*), and red mulberry (*M. rubra*) (Sànchez, 2000; Vijayan et al., 2004b). However, along with the identification and exploitation of the germplasm, other species are attracting attention as well; the examples include the Guangdong mulberry (*Morus atropurpurea*), Lu mulberry (*M. latifolia*), Long mulberry (*Morus laveigata*), and Mountain mulberry (*Morus bombycis*) (Minamizawa, 1976; Vijayan et al., 2011a, 2011b). China and Japan are the countries with the richest germplasm in species and varieties of mulberry (Vijayan et al., 2011a; 2011b) (Figures 9.2 and 9.3).

9.6 BOTANY AND TAXONOMY

9.6.1 PLANT BOTANICAL

The taproot is thick with well-developed lateral roots and root hairs. The epidermis of young roots is bright yellow, while that of mature roots is brown yellow with conspicuous lenticels. The newly grown stems are pale green, whereas the mature ones may be brown, taupe, or brown yellow (Minamizawa, 1976). The buds are enclosed by brown, dark brown, dark red, or black bud scales (Vijayan et al, 2011a) and can be divided into primary and secondary buds or leaf and flower buds. The primary bud develops from a node

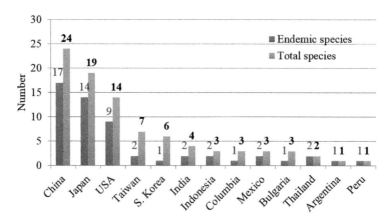

FIGURE 9.2 Distribution of the mulberry species in different countries (adapted from Chang et al., 2014a; Chang et al., 2014b; Vijayan et al., 2011a).

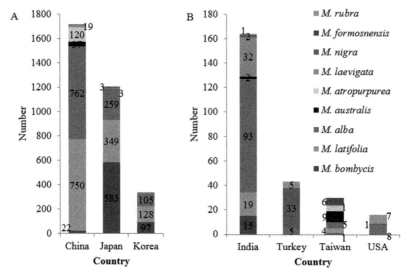

FIGURE 9.3 Number (A) and species distribution of mulberry germplasm in major countries (adapted from Chang, 2006; Chang et al., 2014a, 2014b; Kafkas et al., 2008; Vijayan et al., 2011a, 2011b).

at the leaf axil, while secondary buds laterally form beside the primary bud.

The new leaves are pale green, whereas the mature leaves are light or dark green. The leaf surface may be smooth or rough (Chang, 2006; Vijayan et al., 2011a). The leaf margin may be entire or cleft; and the leaf tip can be long caudate,

caudate, acute, or obtuse. The leaf base can be truncate, rounded, or cordate and the phyllotaxy follows a spiral with an angle of 1/2° or 2/5°. The leaf margin has sharp or rounded teeth (Minamizawa, 1976). The stem may be erect, semi-erect, weeping, or Z-bending (Chang, 2006; Chang et al., 2014c; Zhao et al., 2007a; Vijayan et al., 2011a).

FIGURE 9.4 Fruiting habits of mulberry.

Inflorescences develop at the leaf axils on newly formed or current season's shoots in spring emerged from the lateral buds of the previous year or two-year-old woody branches (Figure 9.4). The staminate flowers form a catkin (Figure 9.5A), whereas the pistillate flowers form a spike (Figure 9.5B) (Liu et al., 1994). Most mulberry species are dioecious and diclinous, while a few are monoecious and monoclinous (Figure 9.5C) (Vijayan et al., 2011a). Each pistillate flower (drupelet) has four perianth segments (which form the edible part of the fruit), and the ovary is spherical with one ovule inside (Figure 9.5D). Each staminate flower has four perianth segments surrounding four stamens, whose filament curves inward before blooming, with two anthers attached to the tip of the filament. At maturation, the curved filaments rapidly straighten outward the anthers break and release the yellow pollen grains, which are carried away by wind.

Notably, sex alternation within a plant may occur and sex expression may be affected by the environmental factors; such as high temperature and long daylight favor staminate flowers, whereas low temperature and short daylight favor pistillate flowers (Tikader et al., 1995; Vijayan et al., 2011a). However, further research might be needed to document how sex is determined by genetic and cytology.

The berry is a multiple fruit comprised of many drupelets (Figure 9.5E). Berries may be oblong or elongated (Figure 9.5F and G). Each berry may contain 50–70 drupelets, but sometimes can have as many as 180 drupelets. Seeds are present or absent inside the drupelets. Ripe berries are white, red, or dark purple. The red and dark purple berries are rich in anthocyanins. The berry contains approximately 90% of water and is succulent and delicious (Chang et al., 2004; Chang et al., 2014a; Vijayan et al., 2011a).

FIGURE 9.5 Morphology of flowers and fruits in some mulberry accessions in Taiwan. (A) Staminate flower, called catkin (Taisang No. 3). (B) Pistillate flower, called spike (Shiaying). (C) Monoclinous flower (68H22033). (D) Pistillate flower has four perianth segments and one ovule for each drupelet (Shiaying). (E) Multiple fruits (Shiaying). (F) Fruits oblong (Miaoli No. 1). (G) Fruits elongated (Elongated Fruit No. 1).

9.6.2 TAXONOMICAL

The conventional taxonomy of mulberry was established by Linné (1753), who recognized seven species (white mulberry, black mulberry, red mulberry, etc.) in the genus *Morus* based on the fruit and leaf traits. Koizumi (1917) used pistillate flowers as the main trait for classification and divided mulberries into the following two main groups according to the style length in Japan: the dolichostylae (style is longer than 0.5 cm) and the macromorus (style is shorter than 0.5 cm). These main groups were further divided into subgroups of pubescents and papillosae on the basis of their stigma morphology. After considering the natural distribution, the genus *Morus* was revised to contain 30 species and 10 varieties. This system is currently adopted by many countries as the reference for mulberry taxonomy (Minamizawa, 1976). However, the style length is only applicable for the monoecious pistillate species (varieties).

Chang et al. (2014b) surveyed the 27 mulberry varieties (accessions) of Taiwan and conducted numerical taxonomy analysis and clustering analysis using principal component analysis (PCA). The PCA analyzed nine representative traits, including four quantity traits (leaf blade width, leaf petiole width, length/width ratio of leaf blade, and chilling requirement for dormancy), and five quality traits (bud morphology during dormant period and growth period, color of mature leaf, morphology of leaf margin and leaf tip), and grouped the Taiwan varieties (accessions) into three clusters. The first cluster contains *M. laevigata* that clearly diverges from all other mulberry species (accessions); the second cluster contains *M. australia*, *M. formosensis*, *M. atropurpurea*,

Mulberry

and *M. bombycids* that clearly show closer relation to each other than to the rest of the group; and the third cluster contains *M. alba* and *M. latifolia* that are similar in appearance (Table 9.5) (Chang et al., 2014b). These achievements might be crucial for helping breeders and taxonomists to effectively identify species in the field without relying on sexual traits. Additionally, their results demonstrate that vegetative characteristics and chilling can be used for plant classification. Besides, Chang et al. (2018) further indicated that the internode length of shoot could be used as an effective/quick index for identifying the polyploidy of mulberry plant in the field/wild

Sharma et al. (2000) were the first to publish a diversity analysis on the genus *Morus* using amplified fragment length polymorphism of DNA-based markers. Later, other investigators also published analysis results on the germplasm pool of different countries using various molecular marker techniques (Awasthi et al., 2004; Botton et al., 2005; Kafkas et al., 2008; Zhao et al., 2007a). All results indicate high genetic diversity among mulberry species that can be clearly grouped into clusters of cultivar and wild species. There are discrepancies within each cluster, depending on which the analysis method was used (Awasthi et al., 2004; Vijayan

TABLE 9.5 Classification of *Morus* in Taiwan Using Vegetative Characters and Chilling Requirements as Identification Traits Without Limitation of Sexual Types

Species		Traits
M. laevigata	1.	Ovate acute-like bud during the dormant period; arrow-like bud during vegetative growth
	2.	Wider petiole; leaf length/width ratio >1.4; a densely acute margin
	3.	Dark green leaf
	4.	Higher chilling requirement
M. atropurpurea	1.	Longideltoid bud during the dormant period; arrow-like bud during vegetative growth
	2.	Leaf length/width ratio >1.4; longer leaf blade
	3.	Light green leaf with a serrate margin; acuminate apex
	4.	Medium chilling requirement
M. bombycis	1.	Longideltoid bud during the dormant period; arrow-like bud during vegetative growth
	2.	Leaf length/width ratio >1.4; longer leaf blade
	3.	Dark green leaf with a serrate margin; caudate to long caudate apex
	4.	Higher chilling requirement

504 Temperate Fruits: Production, Processing, and Marketing

TABLE 9.5 *(Continued)*

Species	Traits
M. australis	1. Longideltoid bud during the dormant period; arrow-like bud during vegetative growth 2. Leaf length/width ratio >1.4; longer leaf blade 3. Dark green leaf with a serrate margin; caudate to long caudate apex. 4. No chilling requirement, evergreen 5. Green growing bud
M. formosensis	1. Longideltoid bud during the dormant period; arrow-like bud during vegetative growth 2. Leaf length/width ratio >1.4; longer leaf blade 3. Dark green leaf with a serrate margin; caudate to long caudate apex 4. No chilling requirement; evergreen 5. Red to purple growing bud
M. alba	1. Orthodeltoid bud during the dormant period; rosette bud during vegetative growth 2. Leaf length/width <1.4; round leaf blade; acute or obtuse apex 3. Higher chilling requirement 4. Thin leaf blade; obtuse margin; shallowly or deeply cordate base
M. latifolia	1. Orthodeltoid bud during the dormant period; rosette bud during vegetative growth. 2. Leaf length/width ratio <1.4; round leaf blade; acute or obtuse apex 3. Higher chilling requirement 4. Thick leaf blade; crenate margin; deeply cordate base

Source: Chang et al. (2014b).

et al., 2006), but they are in general accordance with the conventional taxonomical structure.

9.7 VARIETIES AND CULTIVARS

Development of most varieties has specifically focused on sericulture for several centuries; however, releases of varieties for berry production have slowly increased recently in China (Table 9.6), Japan (Table 9.7), India (Table 9.8), Turkey (Table 9.9), Korea (Table 9.10), the United States (Table 9.11), and Taiwan (Table 9.12).

The important varieties of *M. alba, M. nigra* and *M. rubra* in the United States (Reich, 2008), and of *M. atropurpurea* and *M. laevigata* in Taiwan (Chang, 2006; Chang and Chang, 2010; Chang et al., 2014a, 2018), and their properties with fruit

Mulberry 505

TABLE 9.6 Traits of Major Cultivated Mulberry Varieties for Berry Production in China

Variety	Traits
"Da 10" (*M. atrpurpurea*)	A natural triploid ($2n = 3x = 42$), fruit large, high yield, purplish-black fruits, flavor rich, early mature variety and foliage (30,000 kg/ha leaf yield), and fruit utilization (fruit yield 4.78 kg/tree)
"Baiyuwang"	Fruit 3.5–4 cm long, fruit yield 4.56 kg/tree, white fruits, juicy rich, good tasting, SSC 20%, and high drought tolerant and chilling
"Hongguo" series	Fruit long shape, purple–black fruits, good tasting, and high disease resistant and chilling
"Jialing No. 30"	Tetraploid ($2n = 4x = 56$), a selection from "Zhongsang 5801" by colchicine, leaf yield 3 kg/tree, fruit weight 1 kg/tree, purple fruits, SSC 9%, and foliage and fruit utilization
"Jinqiang No. 63"	Selection from progeny of "5801" × "CL," fruit weight 1.9 kg/tree, purple fruits, SSC 14%–15%
"Four seasons" (*M. atrpurpurea*)	Ever-bearing, high yield, and high resistance
"Langsang No. 1" (*M. latifolia*)	Branches are naturally curve, sericulture, ornamental and berry use, and making functional health supplements food

Source: Huo (2000), Wang et al. (2011), Liu and Wang (2012), Zhang et al. (2013), Dandin (2014), and Teng et al. (2014).

TABLE 9.7 Traits of Major Cultivated Mulberry Varieties in Japan

Species	Variety	Sex	Ploidy, Traits, and Origin
M. alba	Shinichinose	♀♂	Diploid, every bud of nodal can flower, selection from progeny of "Ichinose" × "Kokuso 21," high chilling requirement, branches upright and foliage utilization
	Ichinose	♀	Diploid, drought tolerant, and high yield
	Kairyonezumigaeshi	♂	Diploid, high chilling requirement, drought tolerant, and high yield
M. bombycis	Kenmochi	♀	Diploid, high chilling requirement. and cold resistant
M. latifolia	Mizumihaiteku	♂	Diploid, high chilling requirement, branches upright and high yield.
M. microphylla	Himeguwa	♀	Small tree with rosette canopy and tiny fruit.

Source: Minamizawa (1976), Dandin (2014), and Chang et al. (2018).

TABLE 9.8 Traits of Major Cultivated Mulberry Varieties in India

Species	Variety	Sex	Leaf Yields (MT ha^{-1} yr^{-1})		Traits, Ploidy, Uses, and Origins
			Under Irrigated Condition	Under Rainfed Condition	
M. indica					
	Kanva-2	♀	30–35	10–12	Diploid, widely cultivated in Southern India, selection from natural and variability, leaf un- lobed, and foliage and fruit utilization.
	S-13	♂	–	–	Diploid, selection from open pollinated hybrids progeny of "Kanva-2," and foliage utilization.
	S-34	♂	–	15	Diploid, selection from progeny of "S30"×"Berc776," and foliage utilization.
	S-36	–	38–45	–	Selection from Berhampore local accessions by chemical mutagenesis, leaf un-lobed, produce profuse branching, leaf high yield and foliage utilization.
	S-54	–	38–45	–	Selection from Berhampore local accessions by EMS chemical mutagenesis, and foliage utilization.
	V-1	♂	60–70	–	Selection from progeny of "S30" "Berc776," leaf un- lobed, branches fewer, leaf high yield and foliage utilization.
M. sinensis					
	MR-2	–	30–35	–	Diploid, selection from open pollinated hybrid population, and foliage utilization.
Unidentified					
	DD	–	35–40	–	Selection from Dehra Dun variety and foliage utilization.

Source: Datta (2000) and Vijayan et al. (2004a, 2011a).

Mulberry 507

TABLE 9.9 Traits of Major/Promising Cultivated Mulberry Varieties for Berry Production in Turkey

Species	Variety	Traits of Berry
M. nigra	Gumushaci karadut 8	Black fruits, fruit weight 1.8 g, high soluble solids, high juice yield, and medium seed formation
	Hikmet 1	Black fruits, fruit weight 1.6 g, high soluble solids, high acidity, high juice yield, and medium seed formation
	Erzincan karadut 16	Black fruits, fruit weight 1.6 g, high soluble solids, high acidity, high juice yield, and medium seed formation
	MN1	Black fruits, high soluble solids, and high acidity
	MN4	Black fruits, high soluble solids, and high acidity
M. rubra	Dortyol mor dut	Red fruits, fruit weight 1.9 g, high soluble solids, low acidity, and medium seed formation
	Mersin mor dut	Red fruits, fruit weight 1.5 g, high soluble solids, low acidity, and low seed formation
	MR1	Purple fruits, high soluble solids, and medium acidity
	MR4	Purple fruits, high soluble solids, and medium acidity
	R6	Red fruits, low soluble solids, and low acidity
M. alba	Gumushaci beyaz	Fruit weight 3.1 g, high soluble solids, low acidity, low juice yield, and low seed formation
	Tosya beyaz	Fruit weight 3.0 g, high soluble solids, low acidity, and low seed formation
	Angut 3	Fruit weight 2.6 g, high soluble solids, low acidity, and medium seed formation
	Yediveren	Fruit weight 2.6 g, high soluble solids, medium acidity, and low seed formation
	Angut 4	Fruit weight 1.6 g, high soluble solids, low acidity, and seedless formation
	Lokum dut	Fruit weight 1.2 g, high soluble solids low acidity, low juice yield, and seedless formation

Source: Özgen et al. (2009), Ercisli et al. (2010), and Yilmaz et al. (2012).

508 Temperate Fruits: Production, Processing, and Marketing

TABLE 9.10 Yield and Traits of Berry in Cultivated Mulberry Varieties in Korea

Variety	Yield (kg/tree)	Traits of Berry with/without Origin and Ploidy
"Cheongnosang"	13.0	Fruit weight 2.0–2.5 g; 13–17.7 °Brix; high yield and high soluble solids
"Chungil" (*M. alba*)	–	Diploid; fruit weight 2.2 g; 14.6 °Brix; yield 338 kg/10 ha
"Jeolgokchosaeng" (Chungbuk)	9.5	Fruit weight 4.5 g; 15.8 °Brix; high yield
"Jukcheonchosaeng"	8.7	Fruit weight 3.0 g; 17.0 °Brix; high soluble solids
"Palcheongsipyung"	8.9	Fruit weight 1.8–3.4 g; 14.6–18.5 °Brix; high soluble solids
"Sangmaru" (*M. alba*)	–	Tetraploidy; selection from "Chungil" by colchicine; fruit weight 3.8 g; 14.2 °Brix; yield 358 kg/10 ha
"Susungppong"	14.2	Fruit weight 3.1–3.3 g; 10.4–14.6 °Brix; high yield
"Suwonnosang"	10.8	Fruit weight 3.6 g; 16.1 °Brix; high yield

Source: Kim et al. (2005), Kim et al. (2006), and Sung et al. (2015).

TABLE 9.11 Traits of Berry in Cultivated Mulberry Varieties in the United States

Species	Variety	Traits of Berry
M. nigra	"Black Persian"	Fruit large (over 2.5 cm long and almost 2.5 cm wide), juicy rich, subacidic flavor, and poor drought tolerant
	"Downing"	Purple–black fruits, good tasting, fruit maturation, fruit ripening period from June to September, and poor chilling injury
	"Illinois Everbearing"	Black fruits, good-tasting, fruit ripening period from June to August, and chilling injury tolerant (0 F–30 °F)
M. alba	"Pakistan"	Originated in Pakistan, cordate leaf, fruit over 8 cm long and poor chilling injury
	"Weeping"	Special tree shape and black fruits
M. rubra	"Johnson"	Black fruits; fruit large, juicy, and with "rich vinous" flavor

Source: Reich (2008).

Mulberry

TABLE 9.12 Origin, Yield and Traits of Cultivated Mulberry Varieties in Taiwan

Species and Variety	Origin (Year/originator)	Use*	Availability/ Year of release	Sex	Yield / tree	Traits
M. atropurpurea						
"Taisang No. 19"	1957/Accession selected in Taipei	B	Commercially available/1950s	♀	18.9 Kg	Fruit with purple black, weight: 5 g; soluble solids 5 °Brix, acidity: 0.9%
"Shiaying"	2004/Accession selected in Shiaying, Tainan by Jer-Chia Chang	B	Commercially available /2008	♀	19.5 Kg	Fruit with purple black, weight: 5.8 g; 14 °Brix, acidity: 0.7%
"Miaoli No. 1"	1983/Accession selected in Dahu, Miaoli by Ming-Wen Chang/ bred by Jer-Chia Chang	B	Commerically available/ 2006	♀	21.4 Kg	Fruit with purple black, weight: 6.1 g; 7.7 °Brix, acidity: 0.8%
"Miaoli No. 2"	1985/ bred by Jinn-Tsair Lin Taisang No. 19×Taisang No. 3	B	Commercially available /2012	♀	High yield	Fruit with purple black, weight: 5.1 g; 10.9 °Brix, acidity: 0.6%
M. laevigata						
"Elongated Fruit No. 1"	1981/indirectly introduced from South Africa by Deng-Jen Wu	B	Commercially available/1990s	♀	5.6 kg/ triploid	Fruit with purple red, weight: 6.5 g; 20.3 °Brix, acidity: 0.7%
M. alba (possible)						
"Black Diamond"	2000/Selected by Jeng-Yuan Huang in Shuilin, Yunlin	B	Commercially available/2010	♀	High yield	Fruit with purple black, weight: 6.7 g; 8 °Brix, acidity: 0.6%
M. australis						
"Taisang No. 1" (Sakuma)	1919/Accession selected	F	Commercially available / Renamed as "Taisang No. 1" since1945	♂	815 g (leaf)	Evergreen

510 Temperate Fruits: Production, Processing, and Marketing

TABLE 9.12 *(Continued)*

Species and Variety	Origin (Year/originator)	Use*	Availability/ Year of release	Sex	Yield / tree	Traits
Unidentified						
"Taisang No. 2"	1967/bred by Chung-Chih Hsieh 44C005 × 46C022	F	Commercially available/ 1978	♀	1667 g	Evergreen
"Taisang No. 3"	1967/bred by Chung-Chih Hsieh 44C005 × 46C020	F	Commercially available/ 1978	♂	1492 g (leaf)	Evergreen
M. alba						
"Shidareguwa"	1980/ from Japan	L	Commercially available/1980	♀	–	Small fruit (2–3 g); soft and weeping canopy
M. latifolia						
"Unryuu"	1985/ from Japan	L	Commercially available/1980	♂	–	Z-bending canopy

* F, Foliage; B, berry; L, Landscaping and gardening.

Source: Chang (2006) and Chang et al. (2014c, 2018).

characteristics are indicated in the following.

The fruit of *M. alba* is high in sugar but lacks sourness. The flesh is firm. Fruit size varies; for example, the fruits of the "Pakistan" variety may be longer than 8 cm. Many varieties were derived from the Russian mulberry (*M. alba* "Tatarica"), including "Ramesy's White," "White Russian," "Bames," and "Victoria." However, the fruits of these varieties are generally low in quality.

The fruit of *M. nigra* is large and succulent with a proper combination of sweet and sour tastes, making it the best fruit that can be eaten fresh. The fruit quality is consistent across varieties. "Black Persian" and "Noir of Spain" are common varieties; their fruits are longer than 2.5 cm, approximately 2.5 cm wide, big, and succulent with a slightly sour taste.

The fruit of *M. rubra* is dark red or black with a nice flavor similar to that of the black mulberry. It is mainly cultivated in the USA, such as the "Johnson" variety, which produces big black fruits with nice taste.

The fruit of *M. atropurpurea* is large due to a high number of druplets, and medium sweet, and slightly sour. This species has a high yield of dark purple fruits that are suitable for eating fresh or making processed food and is the primary variety for fruit cultivated in the Southeast Asia, such as the "Miaoli No. 1," "Shiaying," and "Taisang No. 19" of Taiwan.

The fruit of *M. laevigata* is long with a large number of drupelets, and its color ranges from deep red to dark purple. Fruit sweetness is high, while sourness is low, and it is suitable for eating fresh. The "Elongated Fruit No. 1" of Taiwan is an example. But being triploid, the fruit yield of this variety is low due to a massive preharvest drop.

9.8 BREEDING AND CROP IMPROVEMENT

With regard to breeding, the mulberry has the following characteristics:

1. Most are monoecious and thus do not need emasculation for crossing.
2. Some mulberry species have interspecific crossing compatibility.
3. The juvenile period is short or can be shortened.
4. The flowering season is adjustable or the plants are capable of off-season flowering.
5. Mulberry species have polyploid traits, $2n = 1x–22x = 14–308$; most are diploids such as *M. alba* and *M. indica*, but triploids, tetraploids, hexaploids, octaploids, and dexapolids have also been reported (Minamizawa, 1976; Ram Rao et al., 2013; Reich, 2008). Moreover, there is a naturally occurring haploid (*M. notabilis*) (Yu et al., 1996; He et al., 2013).

All these could increase the potential of breeding by induced mutation (Chang, 2006; Chang et al., 2014c).

The objectives of breeding include strong foliage growth, high fruit yield, high fruit quality, storage and transportation tolerance, diverse fruiting season, resistance or tolerance to popcorn disease (*Ciboria shirana*), resistance or tolerance to the mulberry beetle (*Apriona rugicollis*), and attractive appearance (for ornamental purposes) (Table 9.13).

There are three breeding methods available that are discussed in the following sections.

9.8.1 TRADITIONAL BREEDING

The procedure of traditional breeding is complex, labor intense, and time-consuming (Vijayan et al., 2012). At present, China is actively conducting cross-breeding among foliage mulberries (Yang et al., 1992; He et al., 2013). Although

512   Temperate Fruits: Production, Processing, and Marketing

TABLE 9.13 Expected Traits for Sericulture, Berry and Ornamental Uses of Mulberry Breeding in Taiwan

Desirable Traits	Foliage	Berry	Ornamental
1. Sex limitation		♀	
2. Vigorous	✓	✓	✓
3. High quality of berry		✓	
(1) High sweetness (over 15 °Brix)		✓	
(2) Good taste		✓	
(3) High antioxidant		✓	
(4) Low acidity (<0.8%)		✓	
(5) Large fruit (over 3 g per fruit)		✓	
(6) Medium or low seed formation		✓	
4. High functional composition of leaves, roots and stems	✓	✓	
5. Abundant yield (over 15 kg for berry and 30 kg for foliage per adult plant, respectively)	✓		
6. Diversified season/evergreen	✓		
7. Storage and transport tolerant	✓	✓	
8. Mulberry beetle resistant or tolerant	✓	✓	✓
9. Swollen (Popcorn) disease and rust resistant or tolerant	✓	✓	✓
10. Attractive canopy appearance (Weeping, Z-bending or others)			✓

Source: Chang (2006).

some fruit varieties have been obtained, such as "Hongguo No. 2" and "Jialing No. 20," they are only of medium quality. Some other newly bred varieties, such as "Da 10," "Baiyuwang," "Jialing No. 30," and "Jinqiang No. 63" (Table 9.6), are regarded to be superior but have yet to be popularized.

In Taiwan, potential genotypes were selected from chance seedling, local genotype, and hybrid progeny that passed the Distinctness, Uniformity, and Stability tests (Figure 9.6) before being released several fruit varieties, such as "Miaoli No. 1" (Chang, 2008), "Elongated Fruit No. 1" (Chang and Chang, 2010), and "Shiaying" (Chang et al., 2014a); and foliage varieties, such as "Taisang No. 1," "Taisang No. 2," and "Taisang No. 3" (Table 9.12) (Shiesh, 1969; Lin and Lu, 2001; Chang, 2006).

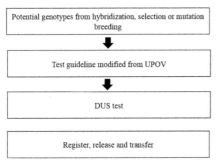

FIGURE 9.6 Breeding process of mulberry in Taiwan.
UPOV, the International Union for the Protection of New Varieties of Plants; DUS, Distinguish, Uniform, and Stable

Japan has bred 17 hybrid varieties, such as the "Astubaroku"; screened 11 natural bud mutants, such as the "Ten Characters"; and selected 120 natural triploid varieties, such as "Red Wood" (Minamizawa, 1976). In India and Turkey, the primary varieties currently cultivated were mostly selected from locally collected cultivars and wild varieties; only a few were obtained by cross-breeding. Among the former, there is essentially no horticultural variety (Vijayan et al., 2012), while among the latter, there are some fruit varieties/promising genotypes such as "N4" and "R6," "MN4" and "MR6," and "Gamushaci beyaz" of mulberry (Table 9.9) (Kafkas et al., 2008; Ercisli et al., 2010; Yilmaz et al., 2012).

In summary, traditional breeding has been quite successful in developing varieties that have high foliage yield, high adaptive ability, and good leaf quality, and are suitable for special cultivation technologies. Unfortunately, the existed commercial varieties for berry production were rare and principally selected from local genotype or spontaneous mutation; therefore, the major impediment for the development of the horticultural industry is the availability of improved cultivars. Up to date, almost all of the breeding achievements in mulberry are still directed toward foliage rather than horticultural usage. A large-scale and intensive breeding program specifically for horticultural utilization is expected.

9.8.2 MUTATION BREEDING

The current cultivated varieties of mulberry can easily be artificially induced to become polyploids. This may be done by radiating buds before sprouting with γ rays or by directly treating dormant buds with colchicine. For instance, the South Korean tetraploid "Sangmaru" (*M. alba* L.) was bred from the diploid "Chungil" (*M. alba* L.) through induced mutagenesis (Sung et al., 2015). Keeping tissue culture-derived plantlets in a medium containing 0.1% colchicine for 24 hours or treating seedlings with 0.2% colchicine for 48 hours can successfully induce tetraploids (Chakraborti et al., 1998; Guo et al., 1989). Mulberry is quite resistant to radiation. Diploid buds are relatively sensitive to radiation, whereas seeds are less sensitive (Minamizawa, 1976). When developing mulberry buds are subjected to radiation-induced mutagenesis, there is a good chance that tetraploids can be created and further radiation can effectively increase the mutation rate (Minamizawa, 1976). The mutants obtained by radiation or colchicine treatment are actually chimeras containing both normal and mutated tissues. Because the proliferation rate of mutant cells is less than that of normal cells, regression may occur after a few years. That can be prevented by repeatedly cutting away branches, which may promote the sprouting of completely mutated stems (Minamizawa, 1976).

9.8.3 BIOTECH-ASSISTED BREEDING

In recent years, biotech has provided several powerful tools and techniques that can potentially help to overcome the problems encountered in conventional breeding (Vijayan et al., 2012). For instance, we can now develop transgenic systems using the micropropagation technique for breeding mulberry varieties (accessions) with high environmental tolerance. Thus, biotech has become another important tool for mulberry-variety improvement (Vijayan, 2011a; Vijayan et al., 2012).

However, biotech-assisted breeding is still in its infancy, and currently research mainly focuses on breeding foliage mulberry varieties, while breeding for horticulture variety has not been attempted. In the future, a molecular marker database may be established before breeding for new varieties; such a database can facilitate the selection of superior parent plants for hybridization and allow early selection on hybrid progeny (Vijayan, 2009; Vijayan et al, 2018).

Vijayan et al. (2006) had established marker-assisted selection of genes linked to the known phenotypic traits, such as internode length, foliage yield, and foliage protein content by inter simple sequence repeats. The mapping of quantitative trait loci and even functional genomic should be soon to be undertaken to boost the crop improvement of mulberry in the future (Chang et al, 2014a).

9.9 SOIL AND CLIMATE

Mulberry trees are highly adaptive. For example, white mulberry can grow in arid, polluted, and–infertile soil. However, *M. nigra* is less tolerant toward unfavorable soil conditions. Mulberry trees are primarily distributed between 10° South and 50° North in latitude, and their favorable temperature for reproduction ranges between 23 °C and 27 °C (Minamizawa, 1976). They should not be planted in areas that may experience late spring frost to facilitate sprouting and fruiting. Favorable cultivation conditions include the annual precipitation of 600–2500 cm, plenty of sunlight with day length 9–13 hours, temperate or subtropical areas at an altitude lower than 2000 m. The soil should be well-drained loam or sandy loam with a thick and rich surface layer, the soil pH should be 6.2–6.8, and the slope should be within 10° (Chang and Liu, 2006; Reich, 2008).

Mulberry cultivation for berry should be avoided in areas where the wet season occurs between fruiting and harvest for preventing the incidence of popcorn disease (Chang, 2008; Chang and Chang, 2010; Chang et al, 2014a).

9.10 PROPAGATION AND ROOTSTOCK

Besides hybridization breeding, mulberry trees are asexually propagated for maintaining the traits of each variety. Both air layer and cutting (with dormant woody twigs or new shoots) can generate numerous good plantlets in a short period of time. Grafting is seldom used because of inferior survival (Chang and Liu, 2006). For varieties or accessions that are hard to propagate by the conventional cuttage and air layering methods, such as "Shidagreguwa" mulberry, tissue culture micro-propagation technique may be used (Aroonpong and Chang, 2015).

9.10.1 AIR LAYER

Make a 0.8–1 cm girdling at the distal half of a mature shoot approximately 1 cm away from a bud; after the latex flow stops, tightly wrap the cut with wet sphagnum moss (soaked the previous night) and transparent tape; after 35–45 days, when the fibrous roots cover half of the sphagnum moss, the air layer is ready to be cut off and planted into the field or pot (Chang and Liu, 2006).

9.10.2 CUTTING WITH DORMANT AND WOODY TWIGS

At the end of winter and before mulberry sprouts (around December in Taiwan), select mature shoots as cuttings; cut the shoot into 15 cm segments, each with 3–5 buds; cut the top flat; and make a slant cut at the base below a node. Place the cuttings in moist soil, and sprouting and rooting will start in approximately 2–3 weeks; after 4 months the plantlet will be ready for planting (Chang and Liu, 2006).

9.10.3 CUTTING WITH NEW AND GREEN SHOOTS

New shoots approximately 15–20 cm long may be taken as cuttings, keeping only two leaves at the top. Half of a leaf should be trimmed away if it is too big, and the trimmed cutting should be immediately placed in water for preventing leaves from withering; the cutting may be soaked in 10 ppm IBA for 30 min for promoting rooting. Maintain humidity after placing the cutting; after 4 months the plantlet will be ready for potting or field planting (Xu, 2001).

9.10.4 TISSUE CULTURE

Some of the mulberry varieties that are hard to develop new roots from, such as "Unryuu," "Shidagreguwa," and "Elongated Fruit No. 1," may be propagated through tissue culture (Lu, 2002, 2003; Aroonpong and Chang, 2015). For example, efficient plantlet production method for micropropagation of "Shidareguwa" mulberry using shoot tips has been successfully developed for the first time. In addition, higher root induction was achieved *in vitro* from regenerated shoots cultured on MS medium without plant growth regulators. (Aroonpong and Chang, 2015).

9.11 LAYOUT AND PLANTING

Field planting should be carried out after dormancy or in spring when the buds break because during this period the mulberry is dormant with sufficient stored nutrient and a low metabolism rate; thus, the survival rate after planting is high. February to March in spring and September in autumn are proper times for planting (Minamizawa, 1976).

Mulberry cultivation for foliage purpose may be planted with row spacing as 150–180 cm, and spacing in the row as 60–90 cm. On average, 6000–11,000 seedling can be planted in the 1 ha area; such compact planting can achieve a high yield and is easy for mechanical management and harvesting.

Mulberry cultivation for berry purpose may be planted with the row spacing versus spacing in the row as 4×4 m² or 4×5^2 m, and management and harvest facilitation depend on the canopy growth of current. In addition, to ensure well fruiting and fruit characteristics, 5% of the plants as pollinizer with cross-compatible varieties (e.g., "Taisang No. 1" and "Taisang No. 3" have only the staminate flower in Taiwan for "Miaoli No. 1") to be inter planted in the orchard (Chang and Liu, 2006; Chang, 2008). However, some species of plants that set fruit with good fruit quality resulting from the absence of pollinizers/pollination have been primarily identified in Taiwan (Chang et al., 2016).

9.12 IRRIGATION

Mulberry plants require plenty of water during bud emerge, shoot growth, flowering and setting, and fruit development in the spring. Mulberry trees vigorously grow during the summer and require plenty of water. Insufficient soil water at this period causes growth decline, leaf hardening, and leaf turning yellow. Therefore, proper irrigation is important; common irrigation methods include flood irrigation, borderstrip irrigation, and sprinkler irrigation.

In late autumn, plant growth slows and water requirement decreases as well. During the wet and typhoon seasons, drain ditches should be dug to help drainage. In low-lying areas, mulberry trees may be planted on raised stands to prevent flooding. Inappropriate drainage could cause root hypoxia and damage, and thus affect the plant's ability to absorb water and nutrients (Lee et al., 1994).

Siddalingaswamy et al. (2007) reported that plants are spaced at 60 × 60 cm^2 and drip irrigated at the 0.6 CPE/Epa level could save 33%–66% of water and increase 25% of shoot yield in mulberry.

9.13 NUTRIENT MANAGEMENT

Mulberry orchards primarily require organic fertilizer, and the amount and timing of fertilization depend on tree vigor and growth season. Generally, once the orchard has been established, long-acting organic fertilizers may be applied in December (after leaf falling and before sprouting) and May (after pruning); 5000 kg/ha may be applied to the soil by intertillage. In addition, short-acting fertilizers may be applied 20–30 days before the mulberry trees sprout.

Mulberry trees require large quantities of nitrogen fertilizer to support shoot and leaf growth, while phosphate and potassium fertilizers can improve foliage quality and promote tree vigor. Moreover, potassium fertilizer may be provided during the fruiting period to promote fruit growth and increase fruit sugar content (Chang and Liu, 2006; Xue and Chang, 2014).

Recommended doses of NPK fertilizer in mulberry varies with rainfed and irrigated/semi irrigated conditions. N: P: K (kg/ha/year) = 100: 50: 50 is applied in rainfed region, whereas N: P: K (kg/ha/year) = 300: 120: 120 is applied in irrigated/semi irrigated region (TNAU AGRITECHPORTAL Sericulture, 2014). Besides, N: P: K (kg/ha/year) application for rainfed region is 150: 50: 50 in acid soil (pH 4.3–5.0) for increasing quality and yield of leaves in mulberry (Singh et al., 2016).

Application of biofertilizers, for example, bacterial (*Azotobacter* and *Azospirillum*) and vesicular-arbuscular mycorrhiza fungi in semi-acid conditions has been reported to effectively reduce 50% cost/dose of chemical fertilizer such as nitrogen and phosphorus (Ram Rao et al., 2007).

9.14 TRAINING AND PRUNING

Mulberries are perennial trees; the branch may grow 1.5–2 m/year. Economic cultivation requires cultivation to obtain an appropriate tree shape and size. After field planting, the training requires to have 3–4 main trunks in the same year. During the second year, 6–9 main branches may be maintained to grow approximately 60–90 cm above the ground. In the third year, the tree should have 12–18 branches 90–120 cm above the ground to complete the tree shape (Chang, 2006; Chang and Liu, 2006; Chang et al., 2014c).

Pruning should be scheduled in early May, after having harvested the fruits in March and April. At the canopy, approximately 120 cm above the ground, all branches from previous years that form the fruiting shoots would be cut back to control the height of tree and allow new branches to grow (Figure 9.7) (Chang and Liu, 2006).

FIGURE 9.7 Plants of mulberry after postharvest training and pruning in late spring in Taiwan.

9.15 INTERCROPPING AND INTERCULTURE

The primary intercropping plants for mulberry trees are winter green manure crops that grow after mulberry trees have shed their leaves. Those crops are usually sowed in September and include alfalfa, garden pea, and horse bean. The mulberry trees thrive in summer; during this period, people may take the advantage of the abundant sunlight available after summer pruning to intercrop fast-growing crops that can be harvested in a short time. Corkwood tree and mung bean are common summer green manure crops. However, green manure intercropping may compete with mulberry trees for water and nutrients, and thus demands careful water and nutrient management (Xue and Chang, 2014).

9.16 FLOWERING AND FRUIT SET

With regard to the flowering season, for varieties such as "Miaoli No. 1," "Taisang No. 19," and "Elongated Fruit No. 1" the period of dormancy lasts approximately 2 months and blooming lasts for half a month. The flowering season and harvest season vary between mulberry varieties. For instance, "Miaoli No. 1" and "Taisang No. 19" bloom 5–7 days earlier than "Elongated Fruit No.1," and their harvest time is also 5–7 days earlier than that of the latter (Chang and Chang, 2010).

In dry air, the mulberry pollen looks like a short tube, approximately 25–35 μm long. The pollen swells into a sphere with 2–3 germ pores. Although there are several germ pores, when a pollen grain germinates on the stigma of pistil, the pollen tube will only grow out of one germ pore. The optimal temperature for pollen germination and pollen tube growth is 25 °C–30 °C, the germination rate varies between 40.1% and 70.5% (Minamizawa, 1976), which is somewhat higher than those of "Taisang No. 1" and "Taisang No. 3" from a preliminary report proposed by Chang and Chung (2005).

Mulberry pollen remains viable for 10–13 days when stored in dark and cool places. For long-term storage, the pollen may be sealed in a container with desiccant for 60 days. Alternatively, the staminate branch may be cut in February and stored under 0 °C–5 °C. The stored branch can then be brought under a higher temperature to encourage sprouting and flowering before pollination; this also allows the flowering-phase adjustment of staminate to facilitate hybrid pollination (Minamizawa, 1976). However, pollen viability and storage may dramatically vary with species, accessions, and environment, and little has been know about these, and they, including viability in vivo, need further examination.

Most mulberry species undergo double fertilization and only a few are capable of parthenocarpy, therefore hybrid progeny for breeding purpose needs to be examined (Minamizawa, 1976; Reich, 2008; He et al., 2013). Chang (2006) speculated that mulberry may exhibit apomixis in addition to double fertilization and parthenocarpy because the trees of "Taisang No.19" that underwent off-season production in late autumn and early winter did not have a pollen source for pollination and fertilization but were still able to grow fruits that contained viable seeds. However, this speculation needs further confirmation using artificial pollination and bagging trials. Griggs and Iwakiri (1973) discovered parthenocarpy in a few white mulberry trees, and similar phenomena has also been observed in other varieties such as "Shidagreguwa," "Gaoqiao," and "Chunri" (Minamizawa, 1976) and "Shiaying" (Chang et al., 2016). Moreover, the varieties "Illinois," "Everbearing," and "Hicks" also do not require pollination to bear fruits (Reich, 2008).

With regard to the fruiting rate, "Tufts" mulberry and "Hicks" mulberry with open or artificial pollination can achieve fruiting rates of 87.5%–89% and 62.5%–66.5%, respectively. If the flowers are bagged before blooming to prevent heterogeneous pollen, then the fruits will be 100% seedless, whereas most pollinated drupelets contained seeds (Figure 9.8A), and the fruit size is not significantly different from the fruits from open or artificial pollination.

The normal development of nonpollinated fruits suggests parthenocarpy in these varieties (Figure 9.8B). (Griggs and Iwakiri, 1973).

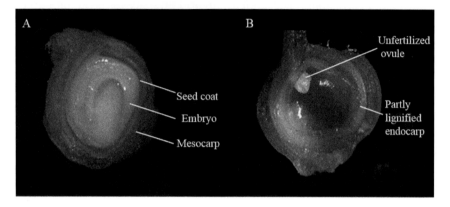

FIGURE 9.8 Ultrastructure of ovary in mulberry with (A) pollination and (B) without pollination.

9.17 FRUIT GROWTH, DEVELOPMENT, AND RIPENING

Mulberry fruits are multiple fruits formed from many florets on a rachis. The perianth expands and becomes red and fleshy, each floret has 4 sepals surrounding 1 ovary, and the stigma is bifid and protrudes out of the sepals.

Fruit development can be divided into the following three continuous periods:

1. Cell division;
2. seeds grow slowly and harden; and
3. fruit flesh rapidly grows until ripening.

Griggs and Iwakiri (1973) studied "Tufts" mulberries (*M. rubra*) and found that fruit development followed a double-sigmoid curve with two well-defined periods of rapid growth. However, seed development showed a single-sigmoid curve with continuous growth during the plateau stage of ovary growth curve. Chang et al. (2004) preliminarily pointed out that the fruit development of "Taisang No. 19" and "Miaoli No. 1" mulberries exhibited a single-sigmoid curve, and fruit size growth, wet weight, sugar content, acidity decline, and the absolute growth rate all increased at 5–7 weeks after full bloom. By the seventh week, the fruits almost reached the final weight and were ready for harvest. Fruit color started with green, began to turn red after the fourth and fifth weeks, and became dark purple between the seventh and eighth weeks. Fruit harvesting lasted for approximately half a month.

Maturation periods vary with species and cultivars due to their chilling requirement. For example, "Miaoli No. 1" blooms in middle-to-late January, the fruits ripe from mid-March to mid-April; "Taisang No. 19" and "Elongated Fruit No. 1" bloom in late January and their harvest seasons are from mid-March to mid-April; and "Shiaying" blooms in late January and is harvested between late March to mid-April. The fruit ripening dates for above four varieties are all from March to April (Chang et al., 2010; 2014a).

Liu et al. (2015) measured the changes in respiration rate and ethylene production during the fruit development of "Jialing No. 40" mulberry and detected a peak in both measurements that appeared 26 days after full bloom, suggesting those are climacteric fruits. However, in follow-up experiments, the "Jialing No. 40" fruits took 34 days after full loom were treated with hetero-geneous ethylene, and the endog-enous ethylene production slightly decreased; the postharvest activity of aminocyclopropanecarboxylate oxidase was temporarily inhibited as well.

Tseng (2010) studied the in-storage respiration rate and ethylene production of the fruits of "Miaoli No. 1" (*M. atropurpurea*) and "Taisang No. 19" that had underwent total color conversion (dark purple) and showed that both parameters gradually declined to a plateau during the period of storage, without the respiration rate and ethylene production peaks that are characteristic for nonclimacteric fruits. The author thus suggested that these fruits are not climacteric and may belong to the garden product type with high respiration rate (20–40 mg CO_2/kg/h, at 5 °C) and medium ethylene production (1–10 µL C_2H_4/kg/h, at 20 °C).

Lee et al. (2017) measured the biochemical changes during fruit ripening of white mulberry. Along with fruit ripening, lightness (L value) decreases, redness (a value) first increases and then decreases, yellowness (b value) decreases, hardness decreases, SSC increases, and the acidity decreases. Soluble sugars consist of glucose, fructose, and sucrose, among which the glucose and fructose show similar changes and their levels are higher than sucrose. K, Ca, and Mg are the main minerals found in fruits and their levels decrease during ripening. Fruit hardness is primarily correlated with pectin and calcium ions; the insoluble protopectin tightly binds to cellulose and hemicellulose and is hydrolyzed during fruit ripening into soluble pectic acid; calcium ion promotes the cross-linkage of pectin complex. Among the phenol contents, some organic acids decrease along with fruit ripening. However, due to the active biosynthesis and accumulation of anthocyanins (mainly cyanidin-3-glucosid/rutinoside), total phenols increase during fruit ripening.

9.18 FRUIT RETENTION AND FRUIT DROP

Chang et al. (2004) preliminarily investigated eight-year-old "Taisang No. 19," "Miaoli No.1," and "Elongated Fruit No. 1" mulberry trees for the number of fruits per shoot under an open pollinated condition during the fruit development period and the drupelets and seed numbers per fruit at the time of ripening. In all three varieties, the number of fruits per shoot decreased along with fruit development. During harvest, the numbers (setting ratio) were 6.5 (92.7%), 4.5 (73.1%), and 0.6 (12.8%), respectively, but the rates of inviable (hollow) seeds were 1.6%, 2.2%, and 90.2%, respectively, suggesting that the triploid "Elongated Fruit No. 1" mulberry is prone to fruit dropping with poor fruit retention due to nonpollination and fertilization.

9.19 HARVESTING AND YIELD

9.19.1 FRUIT RIPENING INDEX

Harvested fruits should be near ripened, with 80% of the skin having turned dark purple. Although the quality of such fruits, as evaluated by weight, total phenols content, and anthocyanins, is inferior to the completely ripened fruits, but TSS and acidity do not significantly differ from the ripened fruits, and the fruits are slightly harder, which increases storability (Chang and Shiesh, 2012).

9.19.2 FRUIT YIELD

Mulberry fruits are thin skinned, vary in ripening time, and require human judgment to determine the degree of ripening, and are therefore generally harvested by hand. The yield varies across varieties (accessions); *Morus atropurpurea* Roxb varieties such as "Shiaying," "Miaoli No. 1," and "Taisang No. 19" may yield 18.9–21.4 kg/tree, wheras "Elongated Fruit No.1" yields only 5.6 kg/tree (Chang et al., 2014a).

9.20 PROTECTED/HIGHTECH CULTIVATION

9.20.1 FORCING

Hydrogen cyanamide and the garlic extract solution have preliminarily

9.20.2 NET CULTIVATION

For preventing bird and insect attack during fruit maturation period owning to high sweetness berry in "Elongated Fruit No. 1," field net house has been generally applied in Taiwan (Chang and Chang, 2010).

9.21 PACKAGING AND TRANSPORT

Due to high water content, mulberries are perishable, making transportation and storage extremely difficult. Therefore, except for fruit-picking tourism, few fresh berries are sold on general markets (Chang and Liu, 2006).

Chang (2012) stored fully ripe "Miaoli No. 1" berries at 5 °C in polyvinyl chloride (PVC) box covered with plastic food wrap, polystyrene (PS) plastic box, polyethylene (PE) Ziploc bag, and lipless PVC box (control), respectively. Weight loss increases with storage time. The fruits in lidless PVC box lost water quickly due to the lack of cover; the weight loss was high and the fruit hardness decreased. PS packaging is porous, resulting in a high weight loss of 15%. The sealed PE and PVC packaging effectively prevented water loss, thus resulting in minimal weight loss and maintained relatively high fruit hardness. Due to respiration, the acidity increased initially, whereas the sugar content decreased. As the temperature decreased, so did the acidity, while SSC increased; the SSC and acidity were lowest in PVC boxes. However, the impermeable PE bag and PVC box packaging led to the increased rotting rate, which reached 20% and 30% on the sixth day of storage, respectively. The permeability of packaging material affects internal humidity, which in turn affects the rotting of stored berries; fruit rotting increases in less permeable packaging. Therefore, for mulberry storage, packaging materials with good permeability should be used, and internal humidity should be kept low while maintaining appropriate ventilation for decreasing fruit rotting rate.

9.22 POSTHARVEST HANDLING AND STORAGE

Precooling is one of the important postharvest treatments for fruits. Its purpose is to rapidly remove field heat, lower physiological metabolism in garden crops, and prolong their shelf life. The fully ripened berries of "Miaoli No. 1" and the 80%–90% ripened berries of "Da 10" were precooled by vacuum cooling to lower the center temperature of fruits to 5 °C and stored under 0 °C at a relative humidity of 80% (Tseng, 2010; Han et al., 2017). The precooled "Miaoli No. 1" berries showed no rotting after three days, whereas the control group berries that had been delayed for 3–4 hours before room cooling had a rotting rate of 19%. By the ninth day of storage, the rotting rate of the precooled group was 37%, whereas the rotting rate of the control group reached 59% (Tseng, 2010). As for the 80%–90% ripened berries of "Da 10," the respiration rate and rotting rate of precooled berries were significantly lower than that of berries without precooling. The precooled "Da 10" berries maintained better color, titratable acidity, and SSC. Moreover, polyphenol oxidase activity was lower than that of the control group as well (Han et al., 2017).

Chemical preservatives such as sodium benzoate, sodium propionate, and sodium sorbate may prolong the shelf life of mulberries but may raise health concerns. Precooled "Da 10" berries fumigated with 2 ppm ozone for 30 min under 5 °C and 80% relative humidity have a significantly lower respiration rate than that of the control group that slows down nutrient consumption and metabolism and help maintain fruit hardness (Han et al., 2017). The general fruit quality of the ozone-treated group is significantly better than that of the control group. Ozone significantly inhibits the degradation of fruit epidermal tissue, maintains fruit shape, inhibits bacterial infection, and prevents evaporation, and thus lowers the rotting rate while improving fruit quality (Han et al., 2017). Chitosan-*g*-Caffeic Acid can decrease the rotting rate and weight loss of 90% ripened berries of "Seedless Dashi" (*Morus multicaulis* L.), maintain fruit quality, improve antioxidant capacity, and maintain cell integrity (Yang et al., 2016).

9.23 PROCESSING AND VALUE ADDITION

Mulberry fruits are highly nutritional, but due to the difficulty of storing fresh fruits, they are generally used as processed food. Processed mulberry products have great diversity, including mulberry wine, mulberry honey, dry mulberry fruit, mulberry vinegar, mulberry juice, mulberry ice cream, mulberry jelly, preserved mulberry, and mulberry jam.

The following precautions should be taken while making mulberry juice. First, pay attention to fruit intactness during washing and remove any rotten parts. The berries should be screened after cleaning to get rid of unripe berries. The newly extracted juice should be heated immediately after filtering for 5–10 min to prevent browning. The

Mulberry

original juice should be diluted with water at a 1:1 ratio to acquire favorable sourness, aroma, and proper color. The optimal final sweetness should be 10–13 °Brix, which may be slightly raised to 14–16 °Brix if honey is used to adjust taste (Wu, 2011).

When making mulberry jam, as the fruits are concentrated by heating to approximately 80 °C, if the acidity is not high enough, 0.5% citric acid may be added to adjust pH to 3. Stir the mixture till it is homogeneous before adding 3–4 loads of sugar. The jam must be pasteurized before bottling. The criterion for bottling is the thickness of the jam. One may drip 1–2 drops of cold water into the jam being cooked: if the cold water does not spread, it means the jam is ready (Wu, 2011).

The procedure for brewing mulberry wine is a little more complicated. The process is as follows: washing, selecting, crushing, filling in tank, fermentation, decanting, aging, decanting, fining, filtration, aging, blending, storing, cold filtration, fine filtration, pasteurization, and bottling. Fermentation temperature should be maintained below 26 °C, the sugar and acid contents need to be adjusted so that sugar should be 150 g/l, and citric acid 1 g/l. Aging may last 3–6 months till the wine appears crystal clear (Zhang et al., 2018).

9.24 DISEASE, PEST, AND PHYSIOLOGICAL DISORDERS

Mulberry trees may suffer fruit and leaf diseases, as well as leaf and branch pests, as described below (Liu, 1994).

9.24.1 FRUIT DISEASE

Popcorn (swollen fruit) disease [(*Mitrula shiraiana* (P. Henn.) ItoetImai, *Ciboria shirana* (P. Hen)]

The pathogen is an ascomycete fungus, which overwinters in soil as sclerotium. In January and February, during the mulberry blooming season, the sclerotium in soil grows into apothecium. The asci on apothecium release ascospores, which are transmitted by wind to infect mulberries (Figure 9.9A). The pathogen may be of *Mitrula shiraiana* (P. Henn.) ItoetImai or *Ciboria shirana* (P. Hen). The apothecium of *Mitrula shiraiana* (P. Henn.) ItoetImai has thinner heads, the affected berries turn gray upon ripening, and the fruitlets dwindle with brown spots. The apothecium of *Ciboria shirana* (P. Hen) is bowl shaped, and the affected berries are enlarged and turn milky white or gray upon ripening. The diseased fruits contain black sclerotia, which fall to the ground with the fruit. In the next spring, the released ascospores may infect the entire orchard by wind.

Control:

1. Clear weeds before mulberry blooming.
2. Collect and burn all diseased fruits.
3. Apply lime (1200 kg/acre) once a year between January and April.
4. Use a tiller to shallow till the field to eliminate the sclerotia in soil to prevent primary infection. Alternatively, calcium cyanamide may be applied (1000 kg/acre) to adjust soil acidity and inhibit the pathogen.
5. Cover the orchard with a silver gray or blue plastic membrane to prevent the sclerotia from falling into ground, germinating into apothecia, and releasing ascospores to be transmitted by wind.
6. Adjust the reproduction season so that mulberries do not ripe in wet seasons may also decrease the chance of infection.

FIGURE 9.9 Main disease and pest of mulberry. (A) Popcorn disease and (B) mulberry beetle.

9.24.2 LEAF DISEASE

9.24.2.1 MULBERRY RUST

Pathogen: *Aecidium mori* Barclay, which is a basidiomycete fungus.

Symptoms: This disease can occur all year round; affected parts include leaves, new shoots, buds, and fruits. The primary sources of infection are pale orange basidiospores, which is transmitted by wind to infect mulberry leaves and spread the disease. The infected tissues swell and may all become enlarged and deformed. In severe cases, new leaves and shoots wither and die and affected parts turn dark brown and are easy to break.

Control:

1. Cut away infected one-year-old branches and promptly remove and burn infected buds, leaves, and shoots.
2. Improve the pruning of mulberry trees, harvest the leaves in time to feed silkworms, and decrease the chance of infection.
3. Select disease-resistant mulberry varieties for cultivation.
4. The proper application of calcium compounds such as calcium cyanamide or calcium dihydrogenphosphate during tillage can not only improve soil acidity/basicity and increase yield, but also effectively control mulberry rust.

9.24.3 INSECT PEST

9.24.3.1 MULBERRY PSYLLID (PAUROCEPHALA PSYLLOPTERA)

Usually erupts in autumn when the weather is hot and dry, the nymphs often aggregate on the underside of mulberry leaves and new shoots to suck sap, which can cause buds and leaves to turn yellow, wither, and fall. Moreover, their white powdery excretions contain honeydew, which likely induces mold growth, and thus affects leaf photosynthesis and tree vigor.

Control:
1. Plant the trees in newly established fruit mulberry orchards along the north–south direction and avoid being close to high and dense objects such as dikes, walls, high-raised field ridges, houses, and bamboo bushes.
2. Pay attention to reserve space at the center and around the orchard as farm trails and ventilation routes. This will help to avoid creating a favorable environment for mulberry psyllid.
3. Place emphasis on the weed management of mulberry orchards for ensuring the penetration of sunlight and good ventilation, apply surface irrigation or sprinkle irrigation as required to interfere with the mating and reproduction of mulberry psyllid to control its population density.
4. Protect and utilize *Menochilus sexmaculatus*, *Lemnia saucia*, 3-banded (*Harmonia octomaculata*), and 13-spotted (*Hippodamia tredecimpunctata*) (Linnaeus) ladybirds, as well as the Anpin lacewings and mantises, for promoting predation on mulberry psyllids.

9.24.3.2 MULBERRY LONGHORN BEETLE (APRIONA GERMARI)

Mulberry longhorn beetles are holometabolous insects. Its life cycle has the following four stages: egg, larva, pupa, and adult. It is a common insect pest and has one generation in a year. The adults usually emerge in May and June and lay eggs in mulberry twigs that are 1 cm in diameter or older than one year. The adults are most active in June and die out in August (Figure 9.9B). The adult beetle uses its mandibles to make a U-shaped cut into bark and wood and lays eggs into the cut at night. Each pair can lay about 100 eggs. The smaller end of the egg points upward, while the bigger head end points downward. The egg stage lasts for 10–16 days. After hatching, the larva eats its way downward into the xylem, forming a tunnel with holes open to outside for ventilation and excretion. The larva stays in the xylem until April or May next year before moving into the root. The mature larva turns its head from downward to upward as preparation for emerging from the tree as adult. The larva then plugs both ends of its pupa chamber with wood dust before pupating. The pupa stage lasts for 14–20 days before the adult emerges. The affected branch loses a circle of cortex, which may either be broken by strong wind at the site of damage or wither because water and assimilation products fail to reach the new shoot. The tree that suffers severe damage in the xylem will first grow weaker, then become senescent, and eventually die.

Control:

1. Kill the adults: during summer and autumn, survey the orchard in the morning or evening to catch and kill adult beetles.
2. Remove eggs: search for eggs to remove if U-shaped cuts in branches are found.
3. Kill the larvae: if feces are spotted near branches, use a thin iron or copper wire to probe into excretion holes and kill the larvae.
4. Pruning: Prune mulberry trees between March and August every year to maintain a low profile for preventing egg laying and larva development.

5. Renewal of old trees: for more than 10-year-old trees with declined vigor, perform collar pruning at 5–10 cm above ground to force the growth of new shoots.

9.24.3.3 MULBERRY WHITEFLY (TRIALEURODES MORI)

Mulberry whiteflies have yellow bodies; the wings are covered with opaque white powder. Its eggs are small and black and can be found on new leaves. The eggs take 10–15 days to hatch. The nymphs and adults often live on the back of leaves to suck sap. In severe cases, the leaves will decolorize and curl, or even wither and fall. Infestation often occurs in areas that are hot and dry, or hot and wet with poor ventilation.

Control:

1. Manage mulberry orchards by natural farming to maintain ecological equilibrium between insect species. Use sprinkle irrigation equipment to irrigate the orchard in the evening to interfere and wash away mulberry whiteflies. These measures can keep the whitefly population below economic damage levels.
2. Avoid creating a warm and hot environment, reserve space at the center and around the orchard as farm trails and ventilation routes.

3. Protect and utilize spider mite destroyers (*Stethorus punctum*), lacewings, and other predatory ladybird beetles as natural enemies against the whitefly.
4. When planting mulberry trees, one should manage a north–south-orientated arrangement in addition to terrain considerations.
5. Maintain the soil humidity of mulberry orchard between 70%–80%, irrigate when it is hot and dry, or when it has not rained for 7 days in a row in summer, otherwise irrigate once every 10 days.

KEY WORDS

- **Mulberry (*Morus* spp.)**
- **taxonomy**
- **fruit**
- **flower**
- **berry**
- **fruiting**
- **nutrition**
- **breeding**
- **varieties**
- **cultivation**
- **orchard**
- **pest**
- **disease**
- **harvest**

ACKNOWLEDGEMENTS

We thank Lan-Yen Chang, Yen-Hua Chen and Wan-Ling Wu, for their technical assistance with this work.

REFERENCES

Álvarez, I. and J. F.Wendel. 2003. Ribosomal ITS sequences and plant phylogenetic inference. Mol. Phylogenet Evol. 29(3): 417–434.

Asano, N. K., Oseki, E. Tomioka, H. Kizu, and K. Matsui. 1994. N-containing sugars from *Morus alba* and their glycosidase inhibitory activities. Carbohydr. Res. 259(2): 243–255.

Aroonpong, P. and J. C. Chang. 2015. Micropropagation of a difficult-to-root weeping mulberry (*Morus alba* var. Shidareguwa): a popular variety for ornamental purposes. Sci. Hortic. 194: 320–326.

Awasthi, A. K., G. M. Nagaraja1, G. V. Naik, S. Kanginakudru, K. Thangavelu, and J. Nagaraju. 2004. Genetic diversity and relationships in mulberry (genus *Morus*) as revealed by RAPD and ISSR marker assays. BMC Genet. 5: 1–9.

Bastianetto, S., C. Ménard, and R. Quirion. 2015. Neuroprotective action of resveratrol. Biochim. Biophys. Acta 1852: 1195–1201.

Berg, C. C. 2001. Moreae, Artocarpeae, and Dorstenia (Moraceae). New York Botanical Garden Press, New York.

Botton, A., G. Barcaccia, S. Cappellozza, R. Da Tos, C. Bonghi, and A. Ramina. 2005. DNA fingerprinting sheds light on the origin of introduced mulberry (Morus spp.) accessions in Italy. Genet. Resour. Crop Evol. 52: 181–192.

Campagna, M. and C. Rivas. 2010. Antiviral activity of resveratrol. Biochem. Soc. Trans. 38: 50–53.

Chakraborti, S. P., K. Vijayan, B. N. Roy, and S. M. H. Qadri. 1998. *In vitro* induction of tetraploidy in mulberry (*Morus alba* L.). Plant Cell Rep. 17: 799–803.

Chang, J. C. and J. P. Chung. 2005. Morphology of anther and pollen grains, and *in vitro* pollen viability of mulberry. J. Chinese Soc. Hort. Sci. 50(4): 588–589 (abstract in Chinese).

Chang, J. C., R. Z. Wang, S. Y. Li, and Y. Q. Luo. 2004. Setting and development of fruit in mulberry. J. Chinese Soc. Hort. Sci. 50(4): 546 (abstract in Chinese).

Chang, J. C. 2006. Taxonomy and cultivar improvement of mulberry in Taiwan. J. Taiwan Soc. Hort. Sci. 52(4): 377–392 (in Chinese with English summary).

Chang, J. C. and Y. T. Liu. 2006. Growth, physiology and cultivation of mulberry in Taiwan. Agr. World 275: 46–54 (in Chinese).

Chang, J. C. 2008. "Miaoli No. 1" mulberry: a new cultivar for berry production. HortScience 43: 1594–1595.

Chang, J. C. and M. W. Chang. 2010. "Elongated Fruit No. 1"mulberry: an elite cultivar for fresh consumption. J. Amer. Pomol. Soc. 64: 101–105.

Chang, J. C., M. W. Chang, and L. Y. Chang. 2014a. "Shiaying" mulberry: a promising cultivar for fresh consumption and processing. J. Amer. Pomol. Soc. 68(1): 33–39.

Chang, L. Y., K. T. Li, W. J. Yang, J. C. Chang, and M. W. Chang. 2014b. Phenotypic classification of mulberry (*Morus*) species in Taiwan using numerical taxonomic analysis through the characterization of vegetative traits and chilling requirements. Sci. Hortic. 176: 208–217.

Chang, L. Y., W. L. Wu, P. Aroonpong, C. Y. Chung, and J. C. Chang. 2014c. Mulberry production and research in Taiwan: from sericulture to horticulture and human health. Hortic. NCHU 39(4): 1–11.

Chang, J.C., Y. H. Hsu, and Y. C. Chu. 2016. Existence of various fruiting mechanisms in mulberry (Morus spp.). Abstract. p. 65. II ISHS International Workshop on Floral Biology and S-Incompatibility in Fruit Species, 23–26 May, 2016, Murcia, Spain.

Chang, L.Y., K. T. Li, W. J. Yang, M. C. Chung, J. C. Chang, and M. W. Chang. 2018. Ploidy levels and their relationship with vegetative traits of mulberry (*Morus* spp.) species in Taiwan. Sci. Hortic. 235: 78–85.

Chang, Y. L. 2012. Effects of cultivation on practices and postharvest treatment on the fruit quality of mulberry (*Morus* spp.). Department of Horticulture National Chung Hsing University Master Thesis (in Chinese with English summary).

Chang, Y. L. and C. C. Shiesh. 2012. The characteristics of fruit of "Miaoli No. 1" mulberry depending on maturity stage. Hortic. NCHU 37(4): 11–19 (in Chinese with English summary).

Chen, P. N., S. C. Chu, H. L. Chiou, W. H. Kuo, C. L. Chiang, and Y. S. Hsieh. 2006. Mulberry anthocyanins, cyanidin 3-rutinoside and cyanidin 3-glucoside, exhibited an inhibitory effect on the migration and invasion of a human lung cancer cell line. Cancer Lett. 235(2): 248–259.

Chen, H., X. He, Y. Liu, J. Li, Q. Y. He, C. Y. Zhang, B. J. Wei, Y. Zhang, and J. Wang. 2016. Extraction, purification and anti-fatigue activity of γ-aminobutyric acid from mulberry (*Morus alba* L.) leaves. Sci. Rep. 6: 18933.

Chen, T. Y. 2005. Antioxidant activity of fruits produced in Taiwan as determined by FRAP assay. Department of Horticulture College of Bioresources and Agriculture National Taiwan University Master Thesis (in Chinese with English summary).

Chung, C. Y. and J. C. Chang. 2016. A preliminary study on off-season production in "Elongated Fruit No. 1" Mulberry. Horti. NCHU 41(4): 27–40 (in Chinese with English summary).

Dandin, S. B. 2014. Mulberry (*Morus* spp.) - An unique gift of nature to the mankind. The 23rd International Congress on Sericure & Silk Industry. Bangalore, India.

Das, Kakoli. 2014. Silk in fashion-Trends. Indian Silk 53: 77–79.

Datta R. K. 2000. Mulberry cultivation and utilization in India. FAO Electronic Conference on Mulberry for Animal Production. http://www.fao.org/docrep/005/X9895E/x9895e04.htm #TopOfPage

Dos, S. G., S. P. Chakraborti, S. Roychowdhuri , N. K. Das, K. Vijayan, and P. D. Ghosh. 2012. Development of mulberry varieties for sustainable growth and leaf yield in temperate and subtropical regions of India. Euphytica 185: 215–225.

Ercisli, S. and E. Orhan. 2007. Chemical composition of white (*Morus alba*), red (*Morus rubra*) and black (*Morus nigra*) mulberry fruits. Food Chem. 103: 1380–1384.

Ercisli, S., M. Tosun, B. Duralija, S. Voća, M. Sengul, and M. Turan. 2010. Phytochemical content of some black (*Morus nigra* L.) and purple (*Morus rubra* L.) mulberry genotypes. Food Technol. Biotechnol. 48 (1): 102–106.

FAOSTAT Data. 2016. http://www.fao.org/faostat/en/#data

Functioning of central silk board & note on sericulture. 2018. CENTRAL SILK BOARD, Ministry of Textiles, Govt. of India.

Google Map. 2014. https://www.google.com/maps

Griggs, W. H. and B. T. Iwakiri. 1973. Development of seeded and parthenocarpic fruits in mulberry (*Morus rubra* L.). J. Hortic. Sci. 48(1): 83–97.

Guo, Z. X., Wang S. H. D. D. Su, J. Y. Wu, and P. H. Chen. 1989. Studies on induction of hybrid mulberry seedlings to tetraploid plants by using colchicine. Acta Sericol. Sinica 15(1): 13–17 (in Chinese).

Gundogdu, M., F. Muradoglu, R. I. G. Sensoy, and H. Yilmaz. 2011. Determination of fruit chemical properties of *Morus nigra* L., *Morus alba* L. and *Morus rubra* L. by HPLC. Sci. Hortic. 132: 37–41.

Gülçin, İ. 2010. Antioxidant properties of resveratrol: a structure-activity insight. Innov. Food Sci. Emerg. Technol. 11: 210–218.

Han, Q., H. Y. Gao, H. J. Chen, X. J. Fang, and W. J. Wu. 2017. Precooling and ozone treatments affects postharvest quality of black mulberry (*Morus nigra*) fruits. Food Chem. 221: 1947–1953.

He, N. G., C. Zhang, X. W. Qi, S. C. Zhao, Y. Tao, G. J. Yang, T. H. Lee, X. Y. Wang, Q. G. Cai, D. Li, M. Z. Lu, S. T. Liao, G. Q. Luo, R. J. He, X. Tan, Y. M. Xu, T. Li , A. C. Zhao, L. Jia, Q. Fu, Q. W. Zeng, C. Gao, B. Ma, J. B. Liang, X. L. Wang, J. Z. Shang, P. H. Song, H. Y. Wu, L. Fan, Q. Wang, Q. Shuai, J. J. Zhu, C. J. Wei, K. Zhu-Salzman, D. C. Jin, J. P. Wang, Tao Liu, M. D. Yu, C. M. Tang, Z. J. Wang, F. W. Dai, J. F. Chen, Y. Liu, S. T. Zhao, T. B. Lin, S. G. Zhang, J. Y. Wang, J. Wang, H. M. Yang, G. W. Yang, J. Wang, A. H. Paterson, Q. Y. Xia, D. F. Ji, and Z. H. Xiang. 2013. Draft genome sequence of the mulberry tree *Morus notabilis*. Nature Commun. 4: 1–9.

Huo, Y. K. 2000. Mulberry cultivation and utilization in China. FAO Electronic Conference on Mulberry for Animal Production. http://www.fao.org/docrep/005/X9895E/x9895e04.htm #TopOfPage

Kafkas, S., M. Özgen, Y. Doğan, B. Özcan, S. Ercişli, and S. Serçe. 2008. Molecular characterization of mulberry accessions in Turkey by AFLP markers. J. Amer. Soc. Hort. Sci. 133: 593–597.

Kandylis, K., I. Hadjigeorgiou, and P. Harizanis. 2009. The nutritive value of mulberry leaves (*Morus alba*) as a feed supplement for sheep. Trop. Anim. Health Prod. 41: 17–24.

Kim, H. B., G. B. Sung, and S. W. Kang. 2005. Evaluation of fruit characteristics according to mulberry breeding lines for fruit production. Korean J. Crop Sci. 50: 224–227 (in Korean).

Kim, H. R., Y. H. Kwon, H. B. Kim, and B. H. Ahn. 2006. Characteristics of mulberry fruit and wine with varieties. J. Korean Soc. Appl. Biol. Chem. 49(3): 209–214. (in Korean)

Kimura, M., K. Hayakawa, and H. Sansawa. 2002. Involvement of γ-aminobutyric acid (GABA) B receptors in the hypotensive effect of systemically administered GABA in spontaneously hypertensive rats. Jap. J. Pharmacol. 89: 388–394.

Kimura, T., K. Nakagawa, H. Kubota, Y. Kojima, Y. Goto, K. Yamagishi, S. Oita, S. Oikawa, and T. Miyazaw. 2007. Food-grade mulberry powder enriched with 1-deoxynojirimycin suppresses the elevation of postprandial blood glucose in humans. J. Agric. Food Chem. 55(14): 5869–5874.

Koidzume, G. 1917. Koidzume taxonomical discussion on *Morus* plants. Bull. Imp. Seric. Exp. Stat. 3: 1–62.

Lee, S. H., S. Y. Choi, H. Kim, J. S. Hwang, B. G. Lee, J. J. Gao, and S. Y. Kim. 2002. Mulberroside F isolated from the leaves of *Morus alba* inhibits melanin biosynthesis. Biol. Pharm. Bull. 25(8): 1045–1048.

Lee, Y. C. and K. T. Hwang. 2017. Changes in physicochemical properties of mulberry fruits (*Morus alba* L.) during ripening. Sci. Hortic. 217: 189–196.

Lee, Y. S., Z. H. Chen, and W. Y. Huan. 1994. Sericulture. China Agricultural Press. Beijing (in Chinese).

Li. S. Z. 1596. Bencaogangmu. Jinling Press (in Chinese).

Lin J. T and Lu H. S. 2001. Study on mulberry production in Taiwan. Sci. Agri. 49: 281–287 (in Chinese with English summary).

Linné, C. 1753. Morus, p.968. In: Linné, C. (ed.). Species plantarum. Vol. 2. Stockholm, Sweden.

Liu, C., W. Xiang, Y. Yu, Z. Q. Shi, X. Z. Huang, and L. Xu. 2015. Comparative analysis of 1-deoxynojirimycin contribution degree to α-glucosidase inhibitory activity and physiological distribution in *Morus alba* L. Ind. Crops Prod. 70: 309–315.

Liu, Y. C., F. Y. Lu, and C. H. Ou. 1994. Trees of Taiwan. Monographic Publication No.7.College of Agriculture, National Chung-Hsing University Taichung, Taiwan, Republic of China.

Liu, J. and X. Wang. 2012. Research on the progress of mulberry variety of "Longsang No. 1." For. Eng. 28(1): 17–23 (in Chinese).

Liu, W. Q., Y. L. Yu, R. J. Liu. 2017. Analysis on operation of Chinese cocoon silk industry in 2016 and prospect in 2017. J. Silk 54(6): 1–7 (in Chinese).

Liu, Z. C. 1994. Control of insect pest in mulberry. Council of Agriculture, Taipei, Taiwan (in Chinese).

Lu, M. C. 2002. Micropropagation of *Morus latifolia* Poilet using axillary buds from mature trees. Sci. Hortic. 96: 329–341.

Lu, M. C. 2003. Tissue culture of *Morus laevigata* Wall using axillary buds. J. Chinese Soc. Hort. Sci. 49(3): 251–258 (in Chinese with English summary).

Minamizawa, K. 1976. Moriculture: basic and application. Houmeishia Publishing Tokyo. Japan.

Ministry of Commerce of the People's Republic of China Department of Market Operation and Consumption Promotion (Nation Cocoon and Silk Coordination Office). 2017. http://scyxs.mofcom.gov. cn/

Novakovic, A., L. G. Bukarica, V. Kanjuh, and H. Heinle. 2006. Potassium channels-mediated vasorelaxation of rat aorta induced by resveratrol. Basic Clin. Pharmacol. Toxicol. 99: 360–364.

Orhan, E. and S. Ercisli. 2010. Genetic relationships between selected Turkish mulberry genotypes (*Morus* spp) based on RAPD markers. Genet. Mol. Res. 9(4): 2176–2183.

Özgen, M., S. Serç, and C. Kaya. 2009. Phytochemical and antioxidant properties of anthocyanin-rich *Morus nigra* and *Morus rubra* fruits. Sci. Hortic. 119: 275–279.

Ram Rao, D. M., J. Kodandaramaiah, M. P. Reddy, R. S. Katiyar, and V. K. Rahmathulla. 2007. Effect of VAM fungi and bacterial biofertilizers on mulberry leaf quality and silkworm cocoon characters under semiarid conditions. Caspian J. Env. Sci. 5(2): 111–117.

Ram Rao, D. M., K. Jhansilakshmi, P. Saraswathi, A. A. Rao, S. R. Ramesh, M. M. Borpuzari, and A. Manjula. 2013. Scope of pre-breeding in mulberry crop improvement—A review. Weekly Sci. Int. Res. J. 1(6): 1–18.

Reich, L. 2008. *Morus* spp. mulberry, p. 504–507. In: Janick, J. and R. E. Paull (eds.). The encyclopedia of fruit and nuts. Cambridge University Press, Cambridge.

Rohwer, J. G. 1993. Moraceae, p. 438–453. In: Kubitzki, K., J. G. Rohwer, and V. Bittrich (eds.). The Families and Genera of Vascular Plants II. Flowering Plants · Dicotyledons: Magnoliid. Hamamelid and Caryophyllid Families, Springer, Germany.

Sànchez, M. D. 2000. World distribution and utilization of mulberry, potential for animal feeding. FAO Electronic conference on mulberry for animal production. http://www.fao.org/docrep/005/X9895E/x9895e02.htm

Satish, G. 2014. Silk production scenario in Karnataka: recent trends. Indian Silk 53: 60–62.

Sharma, A., R. Sharma, and H. Machii. 2000. Assessment of genetic diversity in a Morus germplasm collection using fluorescence-based AFLP markers. Theor. Appl. Genet. 101: 1049–1055.

Shiesh, Z. Z. 1969. Hybridization breeding of mulberry. Sericulture Experimental Station, Miaoli ,Taiwan (in Chinese)

Singh, G. S., R. L. Ram, M. Alam, and S. N. Kumar. 2016. Soil test based fertilizers recommendation of NPK for mulberry (*Morus alba* L.) Farming in .Acid Soils of Lohardaga, Jharkhand, India. Int .J. Curr. Microbiol. App. Sci. 5(6): 392–398.

Sung, G. B., Y. S. Kim, K. Y. Kim, S. D. Ji, and H. B Kim. 2015. Characteristics of mulberry cultivar "Sangmaru" (Morus alba L.) for mulberry fruit production. J. Seric. Entomol. Sci. 53(2): 110–117 (in Korean).

Siddalingaswamy, N., U. D. Bongale, Basavaiah, S. B. Dandin, S. N. Narayanagowda, and R. M. Shivaprakash. 2007. A study on the efficiency of micro-irrigation systems on growth, yield and quality of mulberry. Indian J. Seric. 46(1): 76–79.

Teng, C. T., J. S. Teng, and S. K. Teng. 2014. Release of "Jinqiang No. 63" mulberry for berry purpose—a technical report. Sichuan Canye 42(2): 12–15, 46 (in Chinese).

Tikader, A., K. Vijayan, M. K. Raghunath, S. P. Chakroborti, B. N. Roy, and T. Pavankumar. 1995. Studies on sexual variation in mulberry (*Morus* spp.). Euphytica 84(2): 115–120.

Tili, E. and J. J. Michaille. 2011. Resveratrol, MicroRNAs, inflammation, and cancer. J. Nucleic Acids. 2011: 102431.

TNAU AGRITECHPORTAL Sericulture. Sericulture Technology:: Mulberry Cultivation. http://agritech.tnau.ac.in/sericulture/seri_mulberry%20cultivation.html

Tseng, W. B. 2010. Studies on phosharvest physiology and handling techniques of mulberry fruits (*Morus atropurpurea* Roxb.). Department of Horticulture College of Bioresources and Agriculture

National Taiwan University Master Thesis (in Chinese with English summary).

Turkish statistical institute. 2017. http://www.turkstat.gov.tr/Start.do

Vijayan, K., A. K. Awasthi, P. P. Srivastava and B. Saratchandra. 2004a. Genetic analysis of Indian mulberry varieties through molecular markers. Hereditas 141: 8–14.

Vijayan, K., P. P. Srivastava, and A. K. Awathi. 2004b. Analysis of phylogenetic relationship among five mulberry (*Morus*) species using molecular markers. Genome 47: 439–448.

Vijayan, K., P. P. Srivatsava, C. V. Nair, A. K. Awasthi, A. Tikader, B. Sreenivasa, and S. R. Urs. 2006. Molecular characterization and identification of markers associated with yield traits in mulberry using ISSR markers. Plant Breed. 125: 298–301.

Vijayan, K. 2009. Approaches for enhancing salt tolerance in mulberry (*Morus* L) -a review. Plant Omics J. 2(1): 41–59.

Vijayan, K., A. Tikader, W. G. Zhao, C. V. Nair, S. Ercisli, and C. H. Tsou. 2011a. *Morus*, p. 75–95. In: Kole, C. (ed.). Wild Crop Relatives: Genomic and Breeding Resources: Tropical and Subtropical Fruits. Springer Heidelberg, Dordrecht.

Vijayan, K., B. Saratchandra, and J. A. T. de Silva. 2011b. Germplasm conservation in mulberry (*Morus* spp.). Sci. Hortic. 128: 371–379.

Vijayan, K., P. P. Srivastava, P. J. Raju, and B. Saratchandra. 2012. Breeding for higher productivity in mulberry. Czech J. Genet. Plant Breed. 48(4): 147–156.

Vijayan, K., G. Ravikumar, and A. Tikader. 2018. Mulberry (*Morus* spp.) breeding for higher fruit production, p 89–130. In: Al-Khayri, J. M., S. M. Jain, and D. V. Johnson (eds.). Advances in Plant Breeding Strategies: Fruits. Vol. III. Springer, Switzerland.

Wang, X. L., M. D. Yu, C. Lu, C. R. Wu, and C. J. Jing. 2011. Study on breeding and photosynthetic characteristics of new polyploidy variety for leaf and fruit-producing mulberry (*Morus* L). Sci. Agric. Sin. 44(3): 562–569 (in Chinese).

Wang, Y. H., L. M. Xiang, C. H. Wang, C. Tang, and X. G. He. 2013. Antidiabetic and Antioxidant Effects and phytochemicals of mulberry fruit (*Morus alba* L.) polyphenol enhanced extract. Plos One 8(7): e71144.

Westwood, M. N. 1993. Temperature-Zone Pomology: Physiology and Culture (Third Edition). Timber Press, Portland Oregon.

Xue, C. M. and Z. X. Chang. 2014. Key technologies for cultivation of mulberry. Jin Dun Publishing House. Beijing (in Chinese).

Xu, Y. P. 2001. Propagation of mulberry for ornamental uses. Miaoli District Agricultural Research and Extension Station, Council of Agriculture, Executive Yuan Press, Taiwan, ROC.

Yamanouchi, H., A. Koyama, H. Machii, T. Takyu, and N. Muramatsu. 2009. Inheritance of a weeping character and the low frequency of rooting from cuttings of the mulberry variety "Shidareguwa." Plant Breed. 128: 321–323.

Yang, C. F., B. B. Han, Y. Zheng, L. L. Liu, C. L. Li, S. Sheng, J. Zhang, J. Wang, and F. Wu. 2016. The quality changes of postharvest mulberry fruit treated by chitosan-g-caffeic acid during cold storage. J. Food Sci. 81(4): 881–888.

Yang, J. H., X. H. Yang, and C. G. L. 1992. Polyploidy breeding and its advances of mulberry. Sci. Seric. 18(3): 195–201 (in Chinese)

Yang, N. C., K. Y. Jhou, and C. Y. Tseng. 2012. Antihypertensive effect of mulberry leaf aqueous extract containing γ-aminobutyric acid in spontaneously hypertensive rats. Food Chem. 132(4): 1796–1801.

Yao, J., B. Yan, X. Q. Wang, and J. X. Liu. 2000. Nutritional evaluation of mulberry leaves as feeds for ruminants. Livestock Research for Rural Development. 12(2): 9–16. http://www.cipav.org.co/lrrd/lrrd12/2/yao122.htm

Yilmaz, K. U., Y. Zengin, S. Ercisli, M. N. Demirtas, T. Kan, and A. R. Nazli. 2012. Morphological diversity on fruit characteristics among some selected mulberry genotypes from TURKEY. J. Anim. Plant Sci. 22(1): 211–214.

Yu, M. D., Z. H. Xiang, L. C. Feng, Y. F. Ke, X. Y. Zhang, and C. G. Jing. 1996. The discovery and study on a natural haploid *Morus notabilis* Schneid. Acta Sericol. Sinica 22(2): 67–71.

Yuan, Q. X. and L. Y. Zhao. 2017. The mulberry (*Morus alba* L.) Fruit-A review of characteristic components and health benefits. J. Agric. Food Chem. 65: 10383–10394.

Zafra-Stone, S., T. Yasmin, M. Bagchi, A. Chatterjee, J. A. Vinson, and D. Bagch. 2007. Berry anthocyanins as novel antioxidants in human health and disease prevention. Mol. Nutr. Food Res. 51: 675–683.

Zerega, N. J., W. L. Clement, S. L. Datwyler, and G. D. Weiblen. 2005. Biogeography and divergence times in the mulberry family (Moraceae). Mol. Phylogenet. Evol. 37(2): 402–416.

Zhang, J. Y. Z, G. J. Xie, D. D. Huang, S. G. Sun, X. Zhang, S. R. Shi, and S. J. 2018. Dynamic changes of main physical and chemical indexes during main fermentation process in mulberry wine. Sci. Technol. Food Ind. 14: 18–28 (in Chinese).

Zhang, X. X. J. Z. Guo, and D. M. Guo. 2013. Study on the characteristics of growth and fruiting of different fruit mulberry varieties. North. Horti. 1: 21–23 (in Chinese).

Zhao, W., Y. Wang, T. Chen, G. Jia, X. Wang, J. Qi, Y. Pang, S. Wang, Z. Li, Y. Huang, Y. Pan, and Y. H. Yang. 2007a. Genetic structure of mulberry from different ecotypes revealed by ISSRs in China: an implication for conservation of local mulberry varieties. Sci. Hortic. 115: 47–55.

Zhao, W., Z. Zhou, X. Miao, Y. Zhang, S. Wang, J. Huang, H. Xiang ,Y. Pan, and Y. Huang. 2007b. A comparison of genetic variation among wild and cultivated *Morus* species (Moraceae: *Morus*) as revealed by ISSR and SSR markers. Biodivers. Conserv. 16: 275–290.

Zordoky, B. N. M., I. M. Robertson, and J. R. B. Dyck. 2015. Preclinical and clinical evidence for the role of resveratrol in the treatment of cardiovascular diseases. Biochim. Biophys. Acta 1852: 1155–1177.

CHAPTER 10

CHESTNUT

GABRIELE L. BECCARO[1,2*], DARIO DONNO[1,2],
MICHELE WARMUND[3], FENG ZOU[4], CHIARA FERRACINI[1],
PAOLO GONTHIER[1], and MARIA GABRIELLA MELLANO[1,2]

[1]*Department of Agricultural, Forest and Food Sciences,
University of Turin, Grugliasco, Turin, Italy*

[2]*Chestnut R&D Centre, Cuneo, Italy*

[3]*Division of Plant Sciences, University of Missouri, Columbia, SC, USA*

[4]*Key Laboratory of Cultivation and Protection for Non-wood Forest trees,
Central South University of Forestry and Technology, Changsha,
People's Republic of China*

Corresponding author. E-mail: gabriele.beccaro@unito.it

ABSTRACT

In the *Castanea* genus, some traits are strongly variable (e.g., morphological and ecological traits, vegetative and reproductive habits, resistance to biotic, and abiotic stresses). Over the centuries, human migrations strongly influenced the present range of distribution of the species, and many regions in Asia and Europe have a centuries-old growing tradition. In particular, Romans greatly contributed to the spread of this species in the European countries. Moreover, in some Western areas, local domestication and spontaneous spread of the tree after the last glacial period were also observed (Beccaro et al., 2012, *Silvae Genetica* **2012**, *61*, 292). This chapter presents the state of knowledge in chestnut production from a nursery to a market. This nut crop provides a calorie-rich carbohydrate food sources and is profitable for producers. Additionally, the cultivation of chestnut improves the soil and helps trees sequester carbon. From Mediterranean Europe to Chile, from China to Australia, chestnut production is greatly increasing, especially in the last several decades, and the establishment of high-density orchards, increasing worldwide, is a relatively recent innovation.

10.1 ORIGIN AND DISTRIBUTION, BOTANY, AND TAXONOMY

The Fagaceae family is often divided into 5 or 6 subfamilies that are generally accepted to include 8–10 genera. According to the Angiosperm Phylogeny Group III (APG III) classification, this family includes 7 genera (*Castanea*, *Castanopsis*, *Fagus*, *Lithocarpus*, *Quercus*, *Trigonobalanis*, and *Chrysolepis*) and almost 700 species. The genus *Nothofagus* is included in the list, according to Royal Botanic Gardens of Kew, or attributed to a specific family Nothofagaceae, according to the APG III classification.

The genus *Castanea* is widespread in the Boreal Hemisphere and includes as many as 13 species according to the classification given by Bounous and Marinoni (2005). The natural distribution of the European chestnut (*Castanea sativa* [*C. sativa*]) includes the European and Mediterranean countries and is mostly used for the nut and timber. In Asia (China, Korea, Japan, and Vietnam), *Castanea crenata* (*C. crenata*) (Japanese chestnut), and *Castanea mollissima*, *Castanea seguinii* (*C. seguinii*), *Castanea*

davidii (*C. davidii*), and *Castanea henryi* (*C. henryi*) (Chinese hestnuts) (Willow leaf or pearl chestnut) occur. *C. henryi* is cultivated in Zhengjiang, Fujian, and Hunan Provinces in South China as a fresh nut, whereas *C. crenata* and *C. mollissima* are cultivated for nuts, and *C. seguinii* and *C. davidii* are cultivated for firewood. In North America, the native range of *Castanea dentata* (*C. dentata*) is from Maine to Florida, while *Castanea pumila ashei* or *C. ashei* is found in southeastern states (Mellano et al. 2012). *Castanea floridana* (Florida chinkapin), *C. ashei* (Ashe chinkapin), *Castanea alnifolia* (Creeping chinkapin), and *C. paucispina* are the native plants of South USA.

The most important species, cultivated for nut production, are *C. mollissima* (Chinese chestnut), *C. sativa* (European chestnut, Spanish chestnut, Sweet chestnut), *C. crenata* (Japanese chestnut), producer of large size nuts, and their hybrids. The European chestnut is also a source of timber and other forest products, and it is prized as an ornamental in the landscape (Figure 10.1). Other researchers have addressed the origin and history of the European chestnut (Pereira-Lorenzo et al., 2012). An initial diversification within Asia (early Eocene) followed by an intercontinental dispersion and divergence between the Chinese species and the European one (middle Eocene) was hypothesized by several studies based on cpDNA sequencing analysis (Lang et al., 2007).

FIGURE 10.1 *C. mollissima* plantation in northern China (A), European traditional plantation (B), high-density plantation of *C. sativa* hybrids (C), traditional plantation in northern Italy (D) and traditional plantaion in Switzerland (E)

10.2 AREA, PRODUCTION, MARKETING, AND TRADE

World production of chestnut has been growing since 1990 (Figure 10.2). In 2016, worldwide production of chestnuts was estimated in 2.2 million tons (Food and Agriculture Organization of the United Nations, 2018), primarily from Asia and Europe. In Asia the largest producers are China (1.9 million tons, mainly *C. mollissima*), Republic of Korea (56,000 tons, mainly *C. crenata*), and Japan (26,200 tons, mainly *C. crenata*). Increased production in China is due to expanded acreage of plantations and improved nut yields. Other leading producers are Turkey (65,000 tons), the main producer in Europe, Italy (51,000 tons), Greece (32,000 tons), Portugal (27,000 tons), Spain (16,000 tons), and France (8600 tons) (Food and Agriculture Organization of the United Nations, 2018). In comparison to FAO data, the Livre Blanc de la Chataigne Europeenne (2017) reports lower quantities for some countries (e.g., ~25,000 tons/year for Italy, 40,000 tons/year for Portugal, 32,000 tons/year for Spain), and higher values for others (e.g., 12,000 tons/year for Greece). New plantations in Australia, Chile, and the USA have also increased global production. The US chestnut production (~340 tons in 2012) is concentrated in Michigan, Pennsylvania, Oregon, Florida, California, Missouri, and Ohio (Fulbright et al., 2009; Schnelle, 2016; Warmund, 2014). High-density production (150–200 tree/ha) in Chile was about 3000 tons in 2016 (Food and Agriculture Organization of the United Nations, 2018).

From 1993 to 2015, the main countries exporting chestnuts were

China, Italy, Republic of Korea, Portugal, and Spain; in the same 20 years, Japan was the main importer, followed by France, China, and Italy. In 2015, China exported more than 39,000 tons of chestnuts and imported 16,000 tons. Japan was the second largest Asian country for importation, with about 10,400 tons imported in 2015.

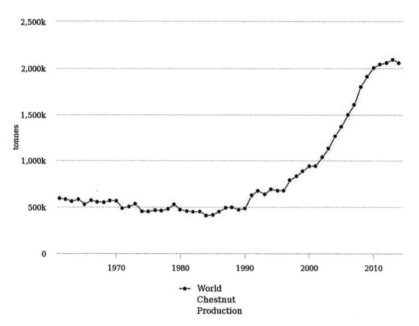

FIGURE 10.2 World chestnut production trend from 1961 to 2016 (Food and Agriculture Organization of the United Nations, 2018) (Data source: FAOSTAT).

10.3 CULTIVARS

Today chestnut diversity is the result of the co-evolution between the *Castanea* trees and the human cultivation of its species. Over the centuries, large quantities of heterogeneous chestnuts have been produced for fresh consumption or to be preserved as a staple during winter in rural areas (Conedera et al., 2014). Currently, superior cultivars are selected for plantations to supply high-quality nuts required by consumers (e.g., regular in shape, good flavor and texture, sweet, easy to peel, without pellicle intrusion or hollow kernels, and no shell splitting) (Bounous, 2004).

Nut descriptors can be used to understand the traditional uses of a variety, but phenomorphological observations are ineffective for genotype identification, because the nut is strongly influenced by environmental factors and cultural practices. Therefore, the cultivars are usually characterized phenotypically and genotypically using microsatellite markers (random-amplified polymorphic DNA, simple sequence repeat, inter-simple sequence repeat, amplified fragment length polymorphism, Sequence-related amplified polymorphism) (Gobbin et al., 2007; Torello Marinoni et al., 2013; Viéitez and Merkle, 2005).

Many excellent *C. sativa* cultivars are grown in France, Greece, Italy, Portugal, and Spain that present hundreds of named cultivars selected for different uses (e.g., candying, roasting, drying, flour, or fresh consumption).

In Italy, highly prized cultivars are the Marrone type ("Chiusa Pesio," "Val Susa," "Castel del Rio," and "Marradi"). Marrones are characterized by their large size and are sold for fresh market and candying (*marrons glacés*). Additionally, "Gabbiana," "Pastinese," and "Carpinese" cultivars, selected in Italy, have small, but very sweet and easy to peel nuts suitable for drying and for flour (Bounous, 2014).

In France, recommended cultivars for new plantations are *C. sativa* cultivars, such as "Marron de Chevanceaux," "Marron de Goujounac," "Sardonne," and "Comballe," which produce marron nuts between 12 and 18 g.

Interspecific hybrids (*C. sativa* × *C. crenata*), resistant to ink diseases and suitable for fresh consumption, such as "Bouche de Betizac," "Maridonne," "Marigoule," and "Precoce Migoule," have been selected by French researchers (Bourgeois, 2004; Pereira-Lorenzo and Ramos-Cabrer, 2004).

In Portugal, "Longal," one of the most ancient cultivars, found growing throughout the chestnut regions (Trás-os-Montes, northern–eastern Portugal), is considered the best cultivar for the agri-food industry. "Judía" and "Martaínha," due to their larger nut size, are usually promoted for fresh consumption (Ribeiro et al., 2007). Important cultivars in Spain are "Parede," "Longal," "Amarelante," "Negral," "Injerta," and "Verata" propagated profusely during the last 300 years (Pereira-Lorenzo et al., 2001, 2019).

In Japan, most commercial cultivars are Japanese chestnuts (*C. crenata*). Only few cultivars like "Riheiguri" are Japanese × Chinese hybrids. More recent plantings of grafted trees (*C. crenata* on *C. crenata* rootstock) are precocious (bearing 4th–5th year after planting). *C. crenata* trees also produce large size nuts with a marked hilum and are sweeter than the European ones. "Ishizuki," "Tsukuba," "Tanzawa," and "Ginyose" cultivars are also resistant to canker blight and ink disease (Tanaka et al., 2005).

Reciprocal crosses of *C. sativa* × *C. crenata* have been obtained in France by *Institut national de la recherche agronomique* (National Institute of Agricultural Research) to evaluate ink disease and canker blight resistance. Some Euro-Japanese hybrids are used on their own roots or as a rootstock. Trees of these hybrids have early ripening nuts and are more vigorous than *C. crenata trees*. Their nuts are suitable for fresh consumption and industrial processing, and they usually mature earlier than European ones. Euro-Japanese hybrids prefer low altitudes, deep soils, and frost-free sites (Osterc et al., 2008).

High-density orchards of *C. mollissima* have a relatively small tree size and bear a large crop of nuts at a young age. This species is the most blight resistant, although it may be affected by *Cryphonectria*. Fruits are usually easy to peel with a nut weight between 10 and 30 g. More than 300 *C. mollissima* cultivars or ecotypes are described in China with subtropical types usually with larger nuts than northern ones. Large nuts are processed and exported to southeastern Asia. Superior cultivars include "Qingzha," "Duanzha," "Jiaoza," "Jiujiazhong," "Jiandingyouli," "Tanqiaobanli," "Qiancidabanli," "Yanshanzaofeng," "Zhenandabanli," and "Daiyuezaofeng" grafted onto *C. mollissima* (Fernandez-Lopez, 2011) seedlings.

These cultivars were selected for commercial use in China.

10.4 BREEDING AND CROP IMPROVEMENT

First hybridizations were made in 1884 in the USA, in 1926 in Spain, and in 1929 in Japan. Hybrid clones have been released in Japan, the USA, and Europe (France, Spain, and Portugal); and some of them are commercialized for nut, timber, and rootstocks (Pereira-Lorenzo et al., 2012). Genetic resources of chestnut have been collected by different institutions throughout the world. Cultivars, seedlings, and interspecific hybrids are preserved in different institutions in Austria, China, France, Hungary, Korea, Italy, Portugal, Slovak Republic, Slovenia, Spain, Switzerland, Turkey, and the United Kingdom (Bounous and Beccaro, 2002).

Chestnut breeding in Europe began with the production of hybrids resistant to ink disease (*Phytophthora* spp.) to substitute the indigenous species. The chestnut ideotype is a function of the final use (nuts or timber), and production and processing technology (harvesting systems, fresh, or processed uses). For nut production, the most important breeding objectives include the following: good horticultural traits, product quality, suitability to storage and processing, and easy peeling.

For timber, important characters include wood quality, rapid growth, and nonchecking of wood (ring shake). Ease of propagation and resistance to major diseases and pests are common for nut and timber types (Pereira-Lorenzo et al., 2012).

10.5 COMPOSITION AND USES

Chestnut fruits were considered very important food sources in European and Asian rural areas for many centuries. Nuts were used in different ways: roasted or boiled in water or milk and consumed instead of bread or served hot with wine or milk as a soup (Adua, 1998; Bounous et al., 2000). However, a progressive decline in their production in the 1990s was caused by the emergence of severe tree diseases and the rural depopulation. More recently, there has been a resurgence in demand for traditional foods as value-added products (Barreira et al., 2012; Avanzato, 2009). Consumers also have a renewed interest in chestnuts due to their nutritional quality and potential health benefits.

Chestnuts are good sources of starch (up to 70%), minerals (especially potassium and, to a lesser degree, calcium, sodium, and phosphorus), lipophilic vitamins (vitamin E), and appreciable levels of fiber (7%–8%), but contain low amounts of protein (2%–4%) and, unlike other nuts, fat (2%–5%) (Zhu, 2017; Neri, 2010). They also contain antioxidant compounds, such as polyphenols and hydrosoluble vitamins (in particular, C, B1, and B2) (Bounous et al., 2000). Additionally, chestnuts are a source of essential fatty acids, mainly linoleic acid; these compounds present an important activity against cardiovascular diseases in adults and promote the development of the brain and retina of infants (Borges et al., 2007).

10.6 SOIL AND CLIMATE

Site selection is an important factor in the establishment of new high-density and traditional plantations. Generally, low and mid-mountain hilly areas are preferred for many *Castanea* species because they are well sun-exposed and protected by spring frosts. The best soils are acidic (pH 5–6.5), light textured, fertile, and well drained. Heavy clay, high pH, and poorly drained soils negatively affect tree growth and also contribute to the spread of *Phytophthora*-borne diseases. A sufficient water supply for the European cultivars is ensured by a rainfall of over 800–900 mm/ year, while *C. mollissima*, *C. henryi*, and Euro-Japanese hybrids need 1200–1300 mm/year (Bounous and Beccaro, 2002; Guohua and Jingyun, 2001).

10.7 PROPAGATION AND ROOTSTOCKS

Chestnut trees are clonally propagated by grafting, cutting, and tissue culture (Figure 10.3) to guarantee trueness to the type of superior cultivars (Pereira-Lorenzo and Fernandez-Lopez, 1997). Tree uniformity in high-density plantations is essential for establishing trees at close spacings and managing the nut crop.

When grafting trees, the use of clean scion wood is essential for the propagation of pest-free trees. Scion wood collected from plantings that may be infested with Asian chestnut gall wasp (*Dryocosmus kuriphilus* [*D. kuriphilus*]) can be immersed in hot water (52 °C) for 10 minutes to kill overwintering larvae present in chestnut buds (Warmund, 2012). However, the collection of scion wood from areas free of *D. kuriphilus* ensures the propagation of noninfested trees.

Rootstocks influence tree growth and longevity, disease resistance, stress tolerance, and nut size. Seedling rootstocks of a compatible chestnut cultivar are often used to produce grafted trees. Seed size, origin, viability, and germinability are important factors in the production of seedling rootstocks. Rootstocks show important effects on the scion performance and, consequently, they also influence the tree growth and nut size. Seeds are not able to immediately germinate at harvest, and a postharvest treatment (e.g., stratification) is required in order to overcome dormancy. In nature, seeds are often covered by leaf litter in the fall and are naturally stratified. In Europe, propagators place nuts in a sand, peatmoss, or sawdust layer (5–8 cm), in a stratification box in an outdoor sheltered site or in a cold room for three months before germination to avoid extreme temperature fluctuations and fungal decay (Mellano and Donno, 2015).

In early spring, the germinants are then sown at 30 cm apart in outdoor raised beds (25–30 cm in depth) to avoid ink disease (*Phytophthora* spp.). Outdoor beds may also be covered by a black plastic film, leaving a 40 cm access space between the beds (Osterc et al., 2005). For seedling rootstock production in the USA, nuts are stratified at 5 °C in a cold room and then planted in containers for air-root pruning in greenhouses before grafting (Warmund, 2013).

Grafting and budding are mainly used to propagate clones and cultivars performing different techniques as chip budding, whip and tongue, cleft, and bark grafts (Pereira-Lorenzo and Ramos-Cabrer, 2004). Dormant scion wood is collected from mother plants pruned during the previous year. Another propagation method, stooling, is used to multiply some Euro-Japanese hybrids: it produces rootstocks rooted on trees of the parent variety. Chestnut was previously considered difficult to propagate by cuttings, but this method is currently used to produce hybrid trees in France, Italy, Spain, and Portugal (Beccaro et al., 2019). Tree propagation by tissue culture is still difficult (Pereira-Lorenzo and Fernandez-Lopez, 1997).

FIGURE 10.3 (A) *C. sativa* bud grafting, (B) chestnut crown grafting, (C) micropropagation of *C. sativa* hybrid, and (D) plants growing in pots at nursery.

10.8 LAYOUT AND PLANTING

New orchards tend to utilize compact trees grafted onto rootstocks that bear large-size nuts in the fourth year after planting.

The planting distance depends on the cultivar vigor and rootstock. The growth habit of trees is also variable depending on species and cultivars. *C. sativa* trees are generally upright and tall, while those of *C. mollissima* are relatively smaller.

Various pruning techniques are used to maximize light interception for the production of high-quality nuts. Tree spacings for European chestnut cultivars vary from 8 m × 12 m to 12 m × 12 m, while those of less vigorous Euro-Japanese hybrids range from 7 m × 7 m to 10 m × 10 m. Japanese and Chinese cultivars may be planted closer, depending on local site conditions (Bounous and Beccaro, 2002).

Postplanting weed control is important during the establishment phase (first 3–4 years). Organic mulches, such as pine bark, saw dust, grass, straw, and wood chips, can not only conserve moisture for trees and provide the organic matter, but can also increase vole or rodent damage to tree trunks during winter. A weed-free zone (1–2 m under the trees) can also be maintained, using registered herbicides or tilling as an alternative to mulch (Bounous and Beccaro, 2002).

10.9 IRRIGATION

The need for irrigation depends on local pedoclimatic conditions, particularly the annual rainfall. Young chestnut trees are drought intolerant and are difficult to grow without irrigation. The highest tree demand for moisture is from late spring through harvest. Tensiometers are useful to determine when irrigation is necessary. Drip irrigation systems are recommended to optimize water availability (Bounous and Marinoni, 2005).

10.10 NUTRIENT MANAGEMENT

Manure and mineral fertilizers are often incorporated into the soil. In Italy 40–50 ton/ha of cow manure is soil incorporated at a depth of 20–30 cm, resulting in an organic matter content of at least 2%. Mineral fertilizers with phosphorus (10–20 ppm) and potassium (100–150 ppm) are applied as mineral superphosphate and potassium sulfate, respectively. Nitrogen, easily leached away, is applied during the spring (250 g/tree of ammonium sulfate or ammonium nitrate) to stimulate vegetative growth (Table 10.1).

During the adult phase, the most limiting nutrient is nitrogen, which is applied in early spring and in late spring. The nitrogen inputs depend on the plant age (Serdar and Macit, 2009). Table 1 shows an application protocol for the first five years of a *C. sativa* orchard. In the USA, fertilizer application rates for bearing orchards are determined by foliar analysis (Warmund, 2018). On fertile soils, a significant increase in the nut yield occurred when ammonium nitrate was applied at 70 kg/ha on April 1 and June 15, resulting in a foliar nitrogen content of 2.4% (Beccaro et al., 2019).

Table 10.1. Nitrogen Fertilization in the First Five Years of a *C. Sativa* Orchard

Year	Ammonium Nitrate (g)	Application from the Base of the Tree (m)
First	250	1.0
Second	300	1.5
Third	450	2.0
Fourth	600	2.5
Fifth	750	3.0

From the sixth year onward, a broadcast application of 60–80 kg/ha/year of nitrogen can be applied (0.3–0.4 ton/ha of ammonium nitrate), depending on productivity. Phosphorus and potassium can be applied at longer time intervals, based on the plant needs. Potassium deficiencies are common in light texture soils and, if detected, K applications are suggested in a range from 25 kg/ha (6-year-old plants) to 120 kg/ha (over 10-year-old trees) (Bounous, 2014).

10.11 TRAINING AND PRUNING

Pruning is necessary to increase the light interception in the canopy, resulting in a regular nut crop (without biennial bearing), increased productivity, and larger nut size. The traditional open vase pruning system is particularly suitable for European chestnut trees. For this system, the 3–4 branches, equidistantly

Chestnut

spaced around the trunk with wide branch angles, are selected. At the end of the fourth year, trees are usually well shaped, and subsequent pruning is limited to thinning out crowded branches, and eliminating dead, broken, or damaged wood. For high-density plantations (more than 150–200 tree/ha), a pyramidal training system is used in which the central leader is dominant (Bounous and Marinoni, 2005).

10.12 FLOWERING AND FRUIT SET

Chestnut is a monoecious species (pistillate and staminate flowers on the same plant), producing flowers on current season's growth. There are two kinds of inflorescences: the male catkins, located in the middle part of the shoot; and bisexual catkins, located at the middle and terminal portion of the shoot. Female flowers develop singly or grouped in clusters (two or three) on the bisexual catkins. Most of chestnut cultivars are self-sterile and require cross-pollination from another cultivar with a similar period of bloom. Abundant pollen quantities are produced by genotypes with long stamens, whereas others, including *Marrone*, produce sterile male flowers. Generally, chestnut is wind pollinated, but insects may help plant pollination (Avanzato, 2009).

10.13 HARVESTING

Chestnuts are traditionally manually harvested (Figure 10.4), resulting in a high labor cost. Before harvesting litter removal underneath the trees accelerates chestnut collection and nets may also be used. Low-cost harvest equipment, using a paddock vacuum system, has been developed in the USA for small-scale orchards (Warmund et al., 2012). More expensive harvesting machines are commonly used in high-density orchards, such as suction or sweeper harvesters (Formato et al., 2012). Daily harvesting is preferred to reduce the incidence of disease infection.

10.14 POSTHARVEST HANDLING AND STORAGE

Chestnuts are highly perishable and are placed in cold storage immediately after the harvest. Traditional and modern methods are used to maintain nut quality, reduce metabolic activity, delay mold growth, and limit worm-infested nuts (Liu and Bai, 2008).

For curing, nuts are submerged in water (18 °C–20 °C) for 4–7 days and then thoroughly dried. Cured nuts can be stored for 5–6 months without loss of quality. For sterilization, chestnuts are put in water at hot temperature (50 °C) for 45 min, then air dried on sanitized cement floors

FIGURE 10.4 Chestnut harvesting by (A) hand and (B) mechanical way.

or by fans for several days, and finally stored for 3–4 months.

For dried chestnut-based products, the nut moisture content is decreased to 10% of the original fresh weight. In some areas (e.g., in Italy and France), specific cultivars are produced for this use. Dried nuts are peeled and selected before marketing; after selection, they can also be grinded into flour.

For the controlled atmosphere storage, nuts are placed in rooms maintained at 20% CO_2, 2% O_2, 1 °C–2 °C, and 95% relative humidity. This storage method reduces fungal infection and results in a high-quality final product. Whole or peeled chestnuts can be frozen at −40 °C for 12 hours and then stored at −20 °C and 80%–90% relative humidity for more than one year. Generally, this technique is used for the highest quality and valuable chestnuts, such as Marrone types.

10.15 PROCESSING AND VALUE ADDITION

Most of the chestnuts produced are sold to the fresh market, with a high demand at the beginning of the harvest season in the Northern Hemisphere. The "Fresh Fruit and Vegetables (FFV) Standards" of the United Nations Economic Commission for Europe are used by governments, producers, traders, importers, exporters, and other international organizations to define nut quality and sizing, related tolerances and traits, and to regulate marketing. The FFV Standards apply to chestnuts (*C. sativa*, *C. crenata*, and their hybrids) for fresh consumption, whereas sweet chestnuts for industrial processing are excluded from these regulations. While chestnuts are graded into different commercial sizes, "Marrone" types are often distinguished from other chestnuts in

European countries. Italian quality standards designate Marrone as a cultivar or a group of cultivars with specific morphological and sensory traits as an oblong shape, large size, small hilum, light color and lightly raised stripes, easily peeled, rarely double kernel, good flavor, and consistent texture. A simpler definition is used by France. French quality standards classify any chestnut with less than 12% of double embryo as Marrone (Beccaro et al., 2019).

FIGURE 10.5 Roasted chestnut in Bangkok, Thailand market.

Semiprocessed chestnuts are peeled or peeled and frozen or processed in purée, becoming a basic ingredient for many products (Bellini et al., 2004). Whole peeled chestnuts may be water packed in tins, dry packed in glass jars, or boiled and vacuum packed. The best chestnuts and Marrone types (peeled and frozen) are candied for *marrons glacés* or are used in syrup and in alcohol. Chestnut purée is often used for creams in desserts. Dried and roasted chestnuts (Figure 10.5), and flour are packed and sold throughout the year (Adua, 1998).

10.16 PESTS AND DISEASES

10.16.1 PESTS

Carpophagous tortricid moths may represent a serious threat in several chestnut-growing areas in Europe. In particular, three species are considered as major pests of chestnut fruit: *Pammene fasciana* L., *Cydia fagiglandana* (Zeller), and *C. splendana* (Hübner) (Lepidoptera: Tortricidae). These moths are commonly known as the early chestnut moth, intermediate chestnut moth, and late chestnut moth, respectively. In Southern Europe, major damage is detected on sweet chestnut (*C. sativa*), but other common hosts are oaks (*Quercus* spp.) and beech (*Fagus sylvatica*) L. The flight period may occur from June to July for *P. fasciana* adults, and later in the season for *C. fagiglandana* (July–September) and *C. splendana* (August–September). All three species are oligophagous and univoltine, with larval instars developing in the fruit. Larvae penetrate and dig tunnels inside the nut cause an earlier fruit drop to the ground and significant yield loss (Pedrazzoli et al., 2012; Ferracini, 2019). The ethology of these insects is quite similar. Mature larvae overwinter mainly in the soil or under the bark, while the emergence of the adults generally is recorded in the following late spring or summer, with diapause reported for some species (Torrini et al., 2017).

Chemical control is difficult because of the larval endophytic development; for this reason, low impact and biological strategies have been investigated. The seasonal flight period and population density have been monitored by pheromone-baited traps (Bengtsson et al., 2014). In particular, the potential of sexual pheromones in specific control programs, as mating disruption or sexual confusion, and plant volatiles for practical application have been studied (Otter et al., 1996).

Another important key pest in Europe is the chestnut weevil *Curculio elephas* Gyllenhal (Coleoptera: Curculionidae). This univoltine insect develops both on chestnuts and acorns (*Quercus* spp.). Females lay their eggs by piercing the husk with their rostrum and inserting an egg into the hole. The newly emerged larvae penetrate the fruit and feed on the kernel tissues (Figure 10.6).

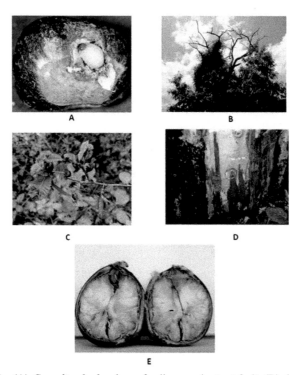

FIGURE 10.6 (A) *Curculio elephas* larva feeding on chestnut fruit, (B) death of branches caused by chestnut blight, (C) galls by *Dryocosmus kuriphilus*, (D) necrosis of the cambial layer associated with ink disease, and (E) nut rot caused by *Gnomoniopsis castaneae*.

Later, they exit the fruit to burrow into the ground to overwinter. Adults mate and lay eggs just after emergence from the ground in late August and September (Desouhant, 1998; Speranza, 1998). The damage may be considerable, depending on the environment and the climatic conditions, with infestation rates up to 90%. As already pointed out for the chestnut moths, control of the larval stage is difficult because larvae occur within the chestnut fruit. The overwintering population could be reduced by killing the individuals occurring in the soil in order to prevent or reduce damage. Proper timing in applying control measures is critical for the containment of both tortrix moths and the weevil when they are not inside the fruit. Entomopathogenic fungi (e.g., *Metarhizium anisopliae* (Metch.) Sorok and *Beauveria bassiana* (Bals.) Vuill.) and nematodes (families Steinernematidae and Heterorhabditidae) have been tested in laboratory trials, especially against *C. splendana* and *C. elephas* last instar larvae, but results in field applications were often unsatisfactory (Torrini et al., 2017; Karagoz et al., 2009). In North America, the lesser chestnut weevil (*Curculio sayi* Gyllenhal) and the larger chestnut weevil (*Curculio caryatrypes* Boheman) are major pests infesting chestnuts, resulting in significant crop loss in mature plantations (Keesey and Barrett, 2008).

In the last decade, the Asian chestnut gall wasp, *D. kuriphilus* Yasumatsu (Hymenoptera: Cynipidae) significantly impacted chestnut production, representing one of the most recent examples of exotic pest accidentally introduced in the European forestry environment. This gall wasp, native to China, established as a pest in Japan, Korea, the USA, Nepal, Canada, and since 2002 in many European countries, affecting chestnut orchards and coppices (*Castanea* spp.). It is a univoltine and thelytokous species, and females lay eggs in buds during summer (Moriya et al., 1989). In the following early spring, greenish-red galls develop on new shoots, suppressing elongation and causing twig dieback (Otake, 1980). Severe reduction of fruiting and yield losses have been estimated to reach up to 85% in Northern Italy (Battisti et al., 2014). The community of native parasitoids recorded on *D. kuriphilus* is mainly composed of chalcid species (Hymenoptera: Chalcidoidea), commonly known to be parasitoids of oak cynipid gall wasps. Since many of these wasps are generalist, and their emergence time does not often coincide with gall development, they proved ineffective biocontrol agents for the control of the gall wasp. However, the exotic parasitoid, *Torymus sinensis* Kamijo (Hymenoptera: Torymidae), native to China, was released into Japan (1975), the USA (in the late 1970s),

Italy (2005), and more recently in several other European countries (Croatia, France, Hungary, Portugal, Slovenia, Spain, and Turkey). It is univoltine, predominantly reproducing amphigonically (a specialized form of sexual reproduction) and it may exhibit a prolonged diapause mainly as late instar larva (Ferracini et al., 2015). After emergence in early spring and mating, *T. sinensis* females lay eggs inside the larval chamber of newly formed galls, usually one egg per host larva. After hatching, the larva feeds ectoparasitically on a *D. kuriphilus* larva and the parasitoid pupates in the host larval chamber during winter.

A density-dependent mortality on *D. kuriphilus* was caused by *T. sinensis* release, and in a few years, it considerably reduced the population density of its superabundant host, both in Japan and Italy. Its high pressure (parasitism rate above 90% in some Northwestern Italian sites) has also affected the native parasitoid community, involving an expansion of its host range with an occasional feeding also on nontarget oak galls, characterized by a high systematic and ecological affinity with *D. kuriphilus* galls (Ferracini et al., 2017, 2018).

Currently, galls are mainly located on young suckers, and their presence is perfectly in line in a view of biological balance, since they represent a useful reservoir for the maintenance of the parasitoid as well. This low presence of both the pest and its parasitoid, reported in Northern Italy 15 years after the first release of *T. sinensis*, demonstrates the successful use of this cost-effective biocontrol agent to manage an exotic invasive pest (Ferracini et al., 2018, 2019).

10.16.2 *DISEASES*

Chestnut is affected by several diseases caused either by fungi or by fungal-like organisms, some of which, historically, have had tremendous effects on the cultivation of *Castanea* species. Few diseases may cause tree decline or mortality, while others are associated with the spoilage of nuts. The former group includes several foliar diseases and more importantly, the ink disease and the chestnut blight.

Since the mid-1800s, serious epidemics of ink disease have been reported on both *C. dentata* and *C. sativa*. This root rot is caused by *P. cambivora* and *P. cinnamomi* (Hayden et al., 2013). Generally, these pathogens cause an initial underground infection of fine roots that may spread to the lower trunk area. Symptoms such as reduced shoot growth and premature leaf fall may occur, sometimes preceding tree death (Hayden et al., 2013). Often, a dark and viscous substance bleeds from the bark at the base of the trunk in association with above-ground colonization by the pathogen. Also, necrosis of the cambial layer and phloem is apparent when bark is removed from the lower trunk and main roots are examined. The spread of the disease is mainly due to the movement of soil harboring pathogen inoculum, including sexual resting spores (i.e., oospores), and to the dissemination of asexual flagellated spores (i.e., zoospores) that can travel actively short distances or passively long distances in flowing water (Hayden et al., 2013). Thus, infections are associated with sites characterized by accumulation of water like flood areas, poorly drained draws, underground water tables, and near unpaved roads or dirt tracks (Hayden et al., 2013). Ink disease is hard to control because pathogens can survive as resting oospores for several years in infected host tissues or in the soil (Vettraino et al., 2001). When establishing new orchards and plantations, a loamy and well-drained soil with high organic contents helps in the prevention of ink disease (Hayden et al., 2013). Also, the establishment and maintenance of a canalization network may help avoiding water flow on the soil and hence the spread of the disease. Before planting, dead and diseased trees that harbor the fungus are removed, including their stumps and root systems. Mulching combined with organic amendments is also effective in controlling the disease (Turchetti et al., 2000), as well as chemical treatments (Gentile et al., 2009; Skoudridakis and Bourbos, 1990). The control of ink disease can be also achieved by using resistant rootstocks. *C. crenata* and the hybrids *C. sativa* × *C. crenata* showed high levels of resistance (EPPO, 2018).

Chestnut blight is caused by *Cryphonectria parasitica*, a fungus native to eastern Asia and introduced in the early 1900s in North America and then into Europe (Heiniger and Rigling, 1994). The disease devastated the native *C. dentata* forests in the USA, as well as *C. sativa* trees in Europe. *C. parasitica* infects the trunk and branches of chestnut (Heiniger and Rigling, 1994). On young trees, infections generally occur in the area of the graft. In coppices, old orchard trees and high forests, infections can be observed at the base of the trunk, along the trunk, or on branches in the crown (Prospero and Rigling, 2013). Chestnut blight in its typical virulent form leads to extensive necrosis and bark cankers. Cankers may rapidly enlarge and girdle the stem/branch, causing the death of plant tissue distal to the infection (Prospero and Rigling, 2013). Orange or reddish pustules (stromata) harboring the sexual (perithecia) and asexual (pycnidia) fruiting bodies of the fungus may develop on the infected bark. Infections occur by means of sexual or asexual spores through wounds or dead host tissues (Prospero and Rigling, 2013).

In many European stands, chestnut trees gradually recovered from the disease (Robin and Heiniger, 2001). This recovery occurred thanks to a phenomenon called hypovirulence (Heiniger and Rigling, 1994). Hypovirulence is caused by a virus (hypovirus, CHV-1), which affects both sporulation abililty and virulence of the fungus (Milgroom and Cortesi, 2004). The hypovirus can spread from infected to noninfected *C. parasitica* individuals via hyphal anastomosis between fungal strains belonging to the same vegetative compatibility (VC) type and via asexual spores (conidia). Therefore, chances for the spread of hypovirulence are higher in fungal populations characterized by low VC-type diversity (Milgroom and Cortesi, 2004). The biological control of chestnut blight is based on the inoculation of virulent cankers with hypovirus-infected *C. parasitica* strains belonging to the same VC type of those that have infected the tree, which may result in the canker healing. Details on the requirements and methods of treatment have been previously described (Prospero and Rigling, 2013). Pruning to remove virulent cankers is recommended. New infections are less likely to occur when tools for grafting and pruning are disinfected during use.

Nut rots of chestnut are associated with pre- or postharvest infection by a number of fungi, including *Ciboria batschiana*, causing a black rot, and *Phomopsis* spp., associated with a mummification of nuts (Washington et al., 1997). Although nuts rots may be locally and occasionally detrimental for nut growers, they have never threatened the cultivation of chestnut. However, since the mid-2000s, an increased incidence of nut rots has been reported in Europe and Australasia which was associated with a previously unknown disease caused by the newly described fungus, *Gnomoniopsis castaneae* (Visentin et al., 2012). Depending on the location and year, substantial yield losses due to *G. castanea* can occur. For instance, disease incidence as high as 93% was reported in northwestern Italy (Lione et al., 2015). Infected nuts show a brown discoloration and a peculiar texture degradation with the kernel appearing sometimes chalky and dehydrated (Visentin et al., 2012). Interestingly, the fungus is also present with high frequency and asymptomatically in buds, bark and green tissues (Visentin et al., 2012; Lione et al., 2016). This fungus has also been reported in association with cankers similar to those caused by *C. parasitica* (Pasche et al., 2016). Sexual spores released from perithecia on burs are responsible for floral infection (Visentin et al.,

2012). Infectious conidia can be released from acervuli that develop on galls of the Asian gall wasp *Dryocosmus kuriphilus* (Maresi et al., 2013). However, infection by means of sexual spores is prevalent in the biology of *G. castaneae* (Sillo et al., 2017). Hence, an effective control strategy could rely on the removal of fallen burs (Sillo et al., 2017).

KEYWORDS

- ***Castanea* spp.**
- **taxonomy**
- **propagation**
- **breeding**
- **genetic resources**
- **chemical composition**
- **pest and disease control**

REFERENCES

Adua, M. In *II International Symposium on Chestnut 494* **1998**, p. 49.

Avanzato, D. *Following chestnut footprints (Castanea spp.): cultivation and culture, folkrore and history, traditions and uses. International Society for Horticultural Science* (ISHS), **2009**.

Barreira, J. C.; Casal, S.; Ferreira, I. C.; Peres, A. M.; Pereira, J. A.; Oliveira, M. B. *Food and Chemical Toxicology* **2012**, *50*, 2311.

Battisti, A.; Benvegnù, I.; Colombari, F.; Haack, R. A. *Agricultural and Forest Entomology* **2014**, *16*, 75.

Beccaro, G. L.; Torello-Marinoni, D.; Binelli, G.; Donno, D.; Boccacci, P.; Botta, R.; Cerutti, A. K.; Conedera, M. *Silvae Genetica* **2012**, *61*, 292.

Beccaro, G.; Alma, A.; Bounous, G.; Gomes-Laranjo, J. *The Chestnut Handbook. Crop & Forest Management* **2019** Taylor & Francis, CRC Press, p. 378. ISBN9780429445606

Bellini, E.; Giordani, E.; Marinelli, C.; Perucca, B. In *III International Chestnut Congress 693* **2004**, p. 97.

Bengtsson, M.; Boutitie, A.; Jósvai, J.; Toth, M.; Andreadis, S.; Rauscher, S.; Unelius, C. R.; Witzgall, P. *Frontiers in Ecology and Evolution* **2014**, *2*, 46.

Borges, O. P.; Carvalho, J. S.; Correia, P. R.; Silva, A. P. *Journal of Food Composition and Analysis* **2007**, *20*, 80.

Bounous, G. In *III International Chestnut Congress 693* **2004**, p. 33.

Bounous, G., Ed. *Il Castagno: Risorsa Multifunzionale in Italia e Nel Mondo*; II ed.; Edagricole: Bologna, Italy, **2014**.

Bounous, G.; Beccaro, G. *FAO-CIHEAM, Nucis Newsletter* **2002**, *11*, 30.

Bounous, G.; Botta, R.; Beccaro, G. *Nucis Newsletter* **2000**, *9*, 44.

Bounous, G.; Marinoni, D. T. *Horticultural Reviews* **2005**, *31*, 291.

Bourgeois, C. *Le châtaignier: un arbre, un bois*; *Forêt privée française* **2004**.

Conedera, M.; Krebs, P.; Tinner, W.; Pradella, M.; Torriani, D. *Vegetation History and Archaeobotany* **2004**, *13*, 161.

Desouhant, E. *Entomologia experimentalis et applicata* **1998**, *86*, 71.

EPPO. *Invasive Species Compendium*. CAB Publishing: Boston, MA, **2018**.

Fernandez-Lopez, J. *Forest Systems* **2011**, *20*, 65.

Ferracini C., 2020. Pests. In: The Chestnut Handbook: Crop & Forest Management (edited by Beccaro, Alma, Bounous, Laranjo), Boca Raton, CRC Press, pp. 317–342.

Ferracini, C.; Bertolino, S.; Bernardo, U.; Bonsignore, C. P.; Faccoli, M.; Ferrari, E.;

Lupi, D.; Maini, S.; Mazzon, L.; Nugnes, F. *Biological Control* **2018**, *121*, 36.

Ferracini, C.; Ferrari, E.; Pontini, M.; Nova Hernández, L. K.; Saladini, M. A.; Alma, A. *BioControl* **2017**, *62*, 445.

Ferracini, C.; Ferrari, E.; Pontini, M.; Saladini, M. A.; Alma, A. *Journal of Pest Science* **2019**, 92, 353.

Ferracini, C.; Gonella, E.; Ferrari, E.; Saladini, M. A.; Picciau, L.; Tota, F.; Pontini, M.; Alma, A. *BioControl* **2015**, *60*, 169.

Food and Agriculture Organization of the United Nations; Rome; **2018**.

Formato, A.; Guida, D.; Lenza, A.; Palcone, C.; Scaglione, S. In *CIGR-AgEng2012 International Conference of Agricultural Engineering. Valencia, Spain,* July **2012**, p. 8.

Fulbright, D.; Mandujano, M.; Stadt, S. In *I European Congress on Chestnut-Castanea 2009 866* **2009**, p. 531.

Gentile, S.; Valentino, D.; Tamietti, G. *Journal of Plant Pathology* **2009**, 565.

Gobbin, D.; Hohl, L.; Conza, L.; Jermini, M.; Gessler, C.; Conedera, M. *Genome* **2007**, *50*, 1089.

Guohua, L.; Jingyun, F. *Acta Ecologica Sinica* **2001**, *21*, 164.

Hayden, K.; Hardy, G. S. J.; Garbelotto, M. In *Infectious Forest Diseases*; Gonthier, P., Nicolotti, G., Eds.; CAB Publishing: Boston, MA, **2013**, p. 519.

Heiniger, U.; Rigling, D. *Annual Review of Phytopathology* **1994**, *32*, 581.

Karagoz, M.; Gulcu, B.; Hazir, S.; Kaya, H. K. *Biocontrol Science and Technology* **2009**, *19*, 755.

Keesey, I. W.; Barrett, B. A. *Journal of the Kansas Entomological Society* **2008**, *81*, 345.

Lang, P.; Dane, F.; Kubisiak, T. L.; Huang, H. *Molecular Phylogenetics and Evolution* **2007**, *43*, 49.

Lione, G.; Giordano, L.; Ferracini, C.; Alma, A.; Gonthier, P. *Acta Oecologica* **2016**, *77*, 10.

Lione, G.; Giordano, L.; Sillo, F.; Gonthier, P. *Plant Pathology* **2015**, *64*, 852.

Liu, Y. M.; Bai, X. A. *Journal of Hebei Agricultural Sciences* **2008**, *1*, 046.

Maresi, G.; Longa, O.; Turchetti, T. *iForest* **2013**, *6*, 294.

Mellano, M. G.; Beccaro, G. L.; Donno, D.; Torello; Boccacci, P.; Canterino, S.; Cerutti, A. K.; Bounous, G. *Genetic Resources and Crop Evolution* **2012**, *59*, 1727.

Mellano, M. G.; Donno, D. *CASTANEA* **2015**, *3*, 8

Milgroom, M. G.; Cortesi, P. *Annual Review of Phytopathology* **2004**, *42*, 311.

Moriya, S.; Inoue, K.; Otake, A.; Shiga, M.; Mabuchi, M. *Applied Entomology and Zoology* **1989**, *24*, 231.

Neri, L.; Dimitri, G.; Sacchetti, G. *Journal of Food Composition and Analysis* **2010**, *23*, 23.

Osterc, G.; Fras, M. Z.; Vonedik, T.; Luthar, Z. *Acta Agriculturae Slovenica* **2005**, *85*, 2.

Osterc, G.; Stefancic, M.; Solar, A.; Stampar, F. *Acta Agriculturae Scandinavica. Section B. Soil and Plant Science* **2008**, *58*, 162.

Otake, A. *Applied Entomology and Zoology* **1980**, *15*, 96.

Otter, C. d.; Cristofaro, A.; Voskamp, K.; Rotundo, G. *Journal of Applied Entomology* **1996**, *120*, 413.

Pasche, S.; Calmin, G.; Auderset, G.; Crovadore, J.; Pelleteret, P.; Mauch-Mani, B.; Barja, F.; Paul, B.; Jermini, M.; Lefort, F. *Fungal Genetics and Biology* **2016**, *87*, 9.

Pedrazzoli, F.; Salvadori, C.; De Cristofaro, A.; Di Santo, P.; Endrizzi, E.; Sabbatini Peverieri, G.; Roversi, P. F.; Ziccardi, A.; Angeli, G. *IOBC/WPRS Bulletin* **2012**, *74*, 117.

Pereira-Lorenzo, S.; Ballester, A.; Corredoira, E.; Vieitez, A. M.; Agnanostakis, S.; Costa, R.; Bounous, G.; Botta, R.; Beccaro, G. L.; Kubisiak, T. L. In *Fruit Breeding*; Springer, **2012**, p. 729.

Pereira-Lorenzo, S.; Diaz-Hernandez, M. B.; Ramos-Cabrer, A. M. *Journal of the American Society for Horticultural Science* **2006**, *131*, 770.

Pereira-Lorenzo, S.; Fernandez-Lopez, J. *Journal of Horticultural Science* **1997**, *72*, 731.

Pereira-Lorenzo, S.; Ramos-Cabrer, A. In *Production Practices and Quality Assessment of Food Crops Volume 1*; New York, NY: Springer, **2004**, p 105.

Pereira-Lorenzo, S.; Ramos-Cabrer, A. M.; Diaz-Hernandez, B.; Ascasibar-Errasti, J.; Sau, F.; Ciordia-Ara, M. *HortScience* **2001**, *36*, 344.

Prospero, S.; Rigling, D. In *Infectious Forest Diseases*; Gonthier, P., Nicolotti, G., Eds.; Boston, MA: CAB Publishing, **2013**, p. 318.

Ribeiro, B.; Rangel, J.; Valentao, P.; Andrade, P. B.; Pereira, J. A.; Bolke, H.; Seabra, R. M. *Food Chemistry* **2007**, *100*, 504.

Robin, C.; Heiniger, U. *Forest Snow and Landscape Research* **2001**, *76*, 361.

Schnelle, M. A. *HortScience* **2016**, *51*, 1339.

Serdar, U.; Macit, I. In *I European Congress on Chestnut-Castanea* **2009**, *866*, 303.

Sillo, F.; Giordano, L.; Zampieri, E.; Lione, G.; De Cesare, S.; Gonthier, P. *Plant Pathology* **2017**, *66*, 293.

Skoudridakis, M.; Bourbos, V. *Comparative effectiveness of metalaxyl and phosethyl A1 to Phytophthora cambivora (Petri) Buism.* **1990**, *55*, 963.

Speranza, S. In *II International Symposium on Chestnut 494* **1998**, p. 417.

Tanaka, T.; Yamamoto, T.; Suzuki, M. *Breeding Science* **2005**, *55*, 271.

Torello Marinoni, D.; Akkak, A.; Beltramo, C.; Guaraldo, P.; Boccacci, P.; Bounous, G.; Ferrara, A.; Ebone, A.; Viotto, E.; Botta, R. *Tree Genetics & Genomes* **2013**, *9*, 1017.

Torrini, G.; Benvenuti, C.; Binazzi, F.; Marianelli, L.; Paoli, F.; Sabbatini Peverieri, G.; Roversi, P. F. *International Journal of Pest Management* **2017**, 1.

Turchetti, T.; Maresi, G.; Nitti, D.; Guidotti, A.; Miccinesi, G.; Rotundaro, G. *Monti e Boschi* **2000**, *51*, 26.

Vettraino, A.; Natili, G.; Anselmi, N.; Vannini, A. *Plant Pathology* **2001**, *50*, 90.

Viéitez, F. J.; Merkle, S. A. *General Editor: Gabrielle J. Persley, The Doyle Foundation, Glasgow, Scotland.* **2005**, 265.

Visentin, I.; Gentile, S.; Valentino, D.; Gonthier, P.; Tamietti, G.; Cardinale, F. *Journal of Plant Pathology* **2012**, *94*, 411.

Warmund, M. In *V International Chestnut Symposium 1019* 2012, p 243.

Warmund, M. R. *Journal of the American Pomological Society* **2013**, *67*, 29.

Warmund, M. R. *Journal of the American Pomological Society* **2014**, *68*, 190.

Warmund, M. R. *Journal of the American Pomological Society* **2018**, *72*, 12.

Warmund, M. R.; Biggs, A. K.; Godsey, L. D. *HortTechnology* **2012**, *22*, 376.

Washington, W.; Allen, A.; Dooley, L. *Australasian Plant Pathology* **1997**, *26*, 37.

Zhu, F. *Food and Bioprocess Technology* **2017**, *10*, 1173.

INDEX

A

Abbé Fetel, 123
Abbott, 426
Actinidia deliciosa. See Kiwifruit
Aecidium mori, 526
Agrobacterium tumefaciens, 290
Allison, 426–427
Allison (male), 427
Alternaria alternata, 87
Alternaria mali, 232
Ambred, 24
American plums, 304–305
Amri, 24
Amrich apple, 24
Anarsia lineatella, 327
Anthracnose, 481–482
Aphis pomi, 234
Apple. *See also* Propagation and
 rootstock
 area and production, 3–5
 biotechnology and integration of
  molecular tools, 31–32
 botanical description, 16–17
 breeding objectives, 27–30
 climate change, impact of, 33–34
 composition of, 7–9
 distribution of production, 4
 EU countries in production, 13
 exports value, 6
 genetic breeding, methodology of,
  30–31
 harvested area and production of, 3
 health and antioxidant research, 10–11
 imports value, 6
 inflorescence, 16
 leading producing countries, 5
 marketing and trade, 5–7
 market uses, 9

 nutrient content values, 7–8
 nutritional/ neutraceutical uses, 9–10
 origin and distribution, 11–13
 overview, 2–3
 packaging and transport, 69–71
 postharvest handling and storage, 71–75
 processed product, 10
 processing and value addition, 75–77
 purpose of pruning, 48–49
 recommended spacing of planting,
  39–40
 soil, 32–33
 species and hybrid species of, 14–15
 taxonomy, 13–15
 training type, 47–48
 varieties and cultivars, 17–18
 world average values, 4
Apple Mosaic Virus, 83
Apple Orchards
 characteristics of different vegetal
  species, 52
Apple packing process, 73
Apple scab, 78–79
Apple starch, 59
Armillariella mellea, 311
Aureobasidium pullulans, 481

B

Bacillus licheniformis, 481
Baldwin apple, 22
Bardacik, 205
Barlett, 121
Beauveria bassiana, 551
Beh Torsh, 194
Bereczcki, 195
Beurré Bosc, 122
Beurré d´Anjou, 121
Boron deficiency in peach, 275

Botryosphaeria dothidea, 165
Botryosphaeria obtusa, 79
Botryosphaeria theobromae, 165
Botrytis cinerea, 481
Botrytis cinerea, 86, 232
Brachycaudus helichrysi, 291
Braeburn, 20
Bramley's Seedling, 22
Bruno fruits, 427

C

Cacanska lepotica, 308
Cacopsylla pyricola, 170
Cameo, 21
Candida oleophila, 481
Candidatus Phytoplasma mali, 85
Canopy management, 49–50
Carpophagous tortricid moths, 549
Castanea mollissima, 539
Castanea sativa, 538
Chaenomeles speciosa, 195
Chaubattia Anupam, 24
Chaubattia Princess, 24
Cherry training systems, 372
Chestnut
  botany and taxonomy, 538–539
  breeding and crop improvement,
    542–543
  composition, 543
  cultivars, 540–542
  diseases, 552–555
  flowering and fruit set, 547
  harvesting, 547
  layout and planting, 545
  marketing and trade, 539–540
  origin and distribution, 538–539
  pests, 549–552
  postharvest handling and storage,
    547–548
  processing and value addition,
    548–549
  production area, 539
  propagation and rootstocks, 544–545

soil and climate, 543
training and pruning, 546–547
uses, 543
world chestnut production trend, 540
Chinese gooseberry. *See* Kiwifruit
Chojuro, 124
Ciboria batschiana, 554
Cladosporium herbarum, 87
Cleaning drop, 60
Cognassier. *See* Quince
Colletotrichum acutatum, 481–482
Colletotrichum fructicola, 87
Conference pear, 123
Conotrachelus crataegi, 234
Coptodisca splendoriferella, 234
Cortland, 21
Corythucha cydoniae, 234
Cotogno. *See* Quince
Cotoneum. *See* Quince
Crataegus atrosanguinea
  red fruits of, 203
Cripps Pink, 20–21
Cryphonectria parasitica, 553
Curculio caryatrypes, 551
Curculio elephas larva, 550
Cyclamen mite, plant affected by, 480
Cydia fagiglandana, 549
Cydia molesta, 89, 170
Cydia pomonella, 87–88, 169–170, 234
Cydia pomonella granulovirus (CpGV),
  170
Cydonia oblonga, 129. *See* Quince

D

Delbarestivale, 23–24
Diplodia seriata, 79, 165
Diseases in plum
  bacterial canker, 325
  brown rot, 325
  crown gall, 325
  crown rot, 326
  oak root fungus, 326
  sharka or plum pox, 326

Index 561

Doyenne du Comice, 122
Dried plums, 324
Dropped shoulder kiwifruit, 444
Dryocosmus kuriphilus, 555

E

Edaphic fertilization and fertigation,
  467–468
Ekmek, 194
Elstar, 20
Empire apple, 21
Epiphyas postvittana, 88–89
Erwinia amylovera, 129
Erwinia amylovora, 82–83, 164, 230, 231
European canker, 80
European plum, 303
European plum cultivars
  characteristics of, 306
Euzophera bigella, 234

F

Fabraea maculata, 232
Fertilization methods in strawberry, 467
Fiesta, 23
Fire blight, 82–83
Flat fruits, 444
Flower bud in quince
  damage of spring frost, 215
  electron microscopy imaging of, 214
  phenological stages from dormant, 216
Flowering in apple
  anthesis and flowering types, 55
  initiation and development of flower
    buds, 54
  pollen viability and germination, 55
  pollination and fertilization, 55–56
  pollinizer, 56
  timing, 54–55
Flowering in pear, 145
  anthesis and type, 146
  initiation and development of buds,
    145
  pollen viability and germination, 146

pollination and fertilization, 146–147
pollination period, 146–147
pollinizer, 147–148
time, 145–146
Foliar fertilization, 468
Forage, 498
Forelle, 122
Freedom apple, 22
Frigo plants, 464
Fruit set in apple
  fruit maturation, 58
  growth curve, 57
  growth, development, and ripening,
    56–60
  harvest, 66
  maturity indices, 63–65
  mechanized harvesting, 67
  respiratory rate at different
    temperatures, 58
  retention and fruit drop, 60–61
  thinning, 61–62
  yield, 62–63
Fuji, 20
Functional food of Mulberry, 498

G

Gala, 18
Gloeosporium perennans, 87
Gloster 69, 23
Gnomoniopsis castaneae, 554
Golden Delicious, 18
Grading and packing standards for
  plum, 322
Granny Smith, 19
Grapholita molesta, 89, 170, 234
Green Gage, 303–304
Greensleeves, 22
Gymnosporangium clavipes, 232
Gymnosporangium sp., 81

H

Hayward, 427
Himekami, 20

I

Idared, 21
Imperatrice, 304
Insect key pests in sweet cherry, 303–395
Insect pests of sweet cherry, 387–389
Internal breakdown/chilling injury, 292
Intraspecific hybridization in plum, 308–309
  fertility, 305
  flowering time, 305
  intersterility, 304, 308
  pollination, 308
  sterility, 308
Iron deficiency in peach, 275
Irrigation for
  apple, 40–43
  cherry tree, 367–368
  chestnut, 545–546
  kiwi plants, 433
  mulberry plants, 616–517
  peach, 271–272
  pear, 132–134
  quince, 206–208
  strawberry, 464–466
Isfahan, 194
Italian Prune, 308
Iwakami, 20

J

Japanese cultivars
  characteristics of, 307
Japanese or oriental plum, 304
Jester, 22–23
Jonagold, 19–20
Jonathan apple, 19
June drop, 60, 151
Jupiter apple, 23

K

Kalecik, 194
Kaunching, 195
Kent apple, 22

King blossom, 16
Kiwifruit
  area and production, 418–421
  botany and taxonomy, 424–426
  breeding and crop improvement, 429–430
  composition and uses of, 422–423
  export and import of, 422
  export value of, 422
  flowering, 436–437
  fruit set, 436–437
  growth, development, and ripening, 437–439
  harvesting and yield, 439
  intercropping, 435–436
  layout and planting, 432–433
  leading producing countries, 420
  marketing and trade, 421
  new varieties of, 428–429
  origin and distribution of, 423–424
  overview, 417–418
  pistillate cultivars, 426–427
  postharvest handling and storage, 439–440
  processing and value addition, 440–441
  propagation and rootstock, 431–432
  root knot nematode, 443–444
  soil and climate requirement for, 430–431
  staminate cultivars, 427
  state-wise production in India, 421
  training and pruning, 434–435
  trend in area and production, 419
  varieties of, 426
  year-wise production of, 419, 420
Kiwifruit diseases
  armillaria root rot, 441–442
  bacterial blossom blight, 442–443
  bacterial leaf spot, 442–443
  leaf infection, 442
  leaf spots, 442
  root rot, 441
  storage rot, 442

Index

L

Laspeyresia pomonella, 327
Limon, 194
Lombard, 304
Lord Lombourne, 23

M

Magnesium deficiency in peach, 274
Magnesium in
  peach, 274
Majda, 24–26
Malus domestica, 11
Malus pumila, 11
Manganese deficiency in peach, 275
Matua, 427
Max Red Bartlett, 121–122
McIntosh, 21
Mechanized harvesting, 67
Mechanized pruning in
  apple, 50–51
  pear, 141
Meech's Prolific, 195
Meloidogyne incognita, 311, 327
Meloidogyne javanica, 311
Meran, 23
Mites, 327
Mollies Delicious, 22
Monilia fructigena, 87
Monilinia fructicola, 287
Monilinia fructigena, 232
Monilinia laxa, 287
Monty, 427
Morus spp. *See* Mulberry
Mulberry
  area and production, 493
  biotech-assisted breeding, 514
  classification of, 503–504
  distribution of, 500
  export and import amount of silk raw, 495
  export and import amount of silkworm cocoons, 495
  flowering, 518–520
  forcing, 522–523
  fruit composition, 496–497
  fruiting habits of, 501
  Indian varieties, 506
  intercropping and interculture, 518
  layout and planting, 616
  leaves of, 497
  major cultivated mulberry varieties in Japan, 505
  marketing and trade, 494–495
  morphology of flowers, 502
  Morus alba route map, 492
  mutation breeding, 513–514
  net cultivation, 523
  origin and distribution, 498–499
  overview, 491–493
  phytochemicals in, 497
  plant botanical, 499–502
  production area, 494
  soil and climate required for, 514–515
  Taiwan varieties, 509–510
  taxonomy of, 502–504
  traditional breeding, 511–513
  training and pruning, 518
  traits of major cultivated mulberry varieties in China, 505
  Turkey varieties, 507
  uses of, 497–498
  varieties, 504
  varieties in the United States, 508
  yield and traits of berry in Korea, 508
Mulberry beetle, 526
Mulberry fruits, 498, 518–520
  fruit ripening index, 522
  growth, development, and ripening, 520–522
  packaging and transport, 523
  postharvest handling and storage, 523–524
  processing and value addition, 524–525
  yield, 522

Mulberry longhorn beetle *(Apriona germari)*, 527–528
Mulberry psyllid *(Paurocephala psylloptera)*, 527
Mulberry rust, 526
Mulberry trees diseases
  fruit disease, 525–526
  insect pest, 527–529
  leaf disease, 526
  popcorn disease, 526
Mulberry whiteflies *(Trialeurodes mori)*, 528–529
Mulberry wood, 497
Multi-axis tree driving (ME), 51

N

Naphthalene acetic acid (NAA), 60
Nectarines, 262–267
  newly released varieties, 263–267
Nematode, 327
Neofusicoccum australe, 165
Neofusicoccum luteum, 165
Neonectria ditissima, 80
Neonectria galligena, 80, 165
Neyshabour, 194
Nijisseiki, 123
Nitrogen level in peach, 273
Nondestructive index, 65
Nonpathogenic defects in Kiwifruit, 444
Nutrient management
  in apple, 43–47
  in cherries, 368–369
  in chestnut, 546
  in kiwifruit, 433–434
  in mulberry, 517
  in peach, 273–275
  in plum, 315–316
  in strawberry, 466–468

O

Orchard diseases
  apple mosaic virus, 83
  apple scab, 78–79

Apple Union Necrosis and Decline (AUND), 83–84
bacterial diseases, 82
canker disease, 79–80
die-back, 79–80
European canker, 80
fire blight, 82–83
flat apple virus, 84–85
fungal diseases, 78
latent virus diseases, 84
pests, 87–89
physiological disorders, 89–90
phytoplasma diseases, 85
postharvest diseases, 85–87
powdery mildew, 79
replant disease, 82
root, crown and collar rot, 81–82
rust disease, 81

P

Packham's Triumph, 122–123
Palmette system, 138
Pammene fasciana, 549
Panonychus ulmi, 327
Peach. *See also* Peach diseases
  area and production, 248–249
  botany, 251–252
  breeding and crop improvement, 268
  characteristics of important cultivars, 253–256
  climate, 268–269
  composition, 249–250
  flowering in, 279–280
  intercropping and interculture, 278–279
  internal breakdown/chilling injury, 292
  layout and planting, 270–271
  marketing and trade, 249
  newly released varieties, 257–262
  nutrient management, 272–275
  nutritive value of, 250–251
  origin and distribution, 251
  overview, 248
  pit splitting, 291–292

Index   565

propagation, 269–270
protected/high-tech cultivation,
   282–284
rootstock, 269–270
skin burning/ staining/inking/
   discoloration, 292
soils, 268–269
taxonomical details, 252
training and pruning, 275–276
uses of, 249–250
varieties and cultivars, 252–262
Peach diseases
   bacterial canker and gummosis,
      289–290
   bacterial leaf spot, 290
   brown rot, 287
   crown gall, 290
   crown rot, 288–289
   minor diseases, 289
   peach black aphid, 291
   peach leaf curl aphid, 291
   phytophthora root rot, 288–289
   plum pox, 290–291
   powdery mildew, 288
   scab, 288
   shot hole, 287–288
Peach fruit
   growth, development, and ripening,
      280
   harvesting and yield, 282
   packaging and transport, 284
   postharvest handling and storage,
      284–286
   processing and value addition,
      286–287
   retention and fruit drop, 281–282
Pear
   annual evolution of harvested area,
      109
   area and production, 109–111
   botanical description, 118–120
   breeding and crop improvement,
      124–126

canopy management, 140
climate change, impact of, 127–128
composition and uses, 114–117
distribution of production, 110
evolution of production, 112
harvesting and yield, 153–157
importer countries, 114
inflorescence, 119
intercropping with cover crops,
   142–144
irrigation, 132–134
layout and planting, 130–132
leading producing and exporter
   countries, 111, 113
marketing and trade, 112–114
micropropagation and tissue culture,
   130–131
nutrient management in, 134–137
nutrient standards for leaves, 137
nutritional composition of, 115–116
origin and distribution, 117–118
overview, 108–109
propagation and rootstock, 128
pruning, types of, 139–140
soil for, 126–127
taxonomy, 118
training type, 138–139
varieties and cultivars, 120
vegetal species, characteristics of, 143
world average values for, 110
Pear diseases, 164–166
   brown spot, 165
   canker disease, 165
   fire blight, 164
   pear scab, 164
   pests, 169–170
   physiological disorders, 168–169
   postharvest diseases, 166–168
   powdery mildew, 166
Pear fruit, 148
   controlled atmosphere, 161
   growth, development, and ripening,
      148–151

566

Index

low O$_2$ storage, 161–162
1-methylcyclopropene, use of, 162
modified atmosphere storage, 161
packaging and transport, 157–158
postharvest handling, 158–160
processing and value addition,
162–164
regular air storage, 160
retention and drop, 151–152
strategies for fruit retention, 152
thinning, 152–153
Pear fruit growth curve, 149
Pear Orchards
annual export of macronutrients, 136
Pear tree rootstocks
clonal rootstocks, 129
interstem trees, 130
scion-rootstock incompatibility,
129–130
seedling rootstocks, 128–129
stock-scion relationship, 129
Penicillium crustosum, 85–86
Penicillium expansum, 85–86, 232
Penicillium solitum, 85–86
Penicillium sp., 86–87
Penicillium verrocosum, 85–86
Pests in apple orchards
codling moth, 87–88
light brown apple moth, 87–89
oriental fruit moth, 89
San José scale, 89
Phosphorus deficiency in peach,
273–274
Phymatotrichopsis omnivora, 82
Physiological disorders in apple, 89–90
bitter pit, 90
cork spot, 90
jonathan spot, 90
russeting, 90–91
sunburn, 91
superficial scald, 91
Phytonemus pallidus, 479–480
Phytophthora cactorum, 81, 166, 232

Phytophthora sp, 288–289
Pinova, 23
Pit splitting, 291–292
Plum brandy, 325
Plum fruit moth, 327
Plum Pox virus, 290–291
Plums
botany, 301–303
climate and soil, 309–310
crop and quality regulation, 320
flowering, 318–320
fruit growth, and development,
318–320
genetic engineering, 309
harvesting and yield, 321–322
health benefits, 300–301
layout and planting, 314–315
marketing and trade, 299–300
maturity, 321–322
mutations, 309
nutritional composition, 300–301
origin and distribution, 301
overview, 297–298
pests, 326–327
processing and value addition,
323–325
production level and areas, 298–299
propagation and rootstock, 310–311
storage, 323
taxonomy, 301–303
training and pruning, 317–318
weed control and interculture
operations, 316–317
world production estimates, 298–299
Plum species, 302
Plums wine, 324
Podoshaera pannosa, 288
Podosphaera leuchotrica, 166
Podosphaera leucotricha, 79, 232
Pollination of strawberry flower, 469
Pome, 16–17
Portugal, 195
Postbloom thinners, 62

Postharvest diseases in apple, 85
  Alternaria rot, 87
  black rot, 87
  blue mold, 85–86
  brown rot, 87
  bull's eye rot, 87
  Cladosporium rot, 87
  gray mold, 86
  mold core, 86–87
  white rot, 87
Potassium deficiency in peach, 274
Potebniamyces pyri, 165
Powdery mildew, 79
Pratylenchus vulnus, 311
Precision agriculture, 69
Propagation and rootstock for apple,
  34–35
  interstem trees, 37
  micropropagation and tissue culture,
    38–39
  rootstock types, 36–37
  scion-rootstock incompatibility, 37–38
  stock-scion relationship, 35–36
Propagation and rootstock in mulberry
  air layer, 515
  cutting with dormant and woody
    twigs, 515
  cutting with new and green shoots,
    515–516
  tissue culture, 516
Propagation methods for strawberry,
  460–462
  crown division, 461
  micropropagation, 461
  phases of micropropagation, 461–462
  rooting trays, use of, 461
Protected/high-tech cultivation, 67
  fruit protectants, 67–68
  high-tech cultivation, 68–69
  precision crop management and
    protection, 69
Prune juice, 323–324
Prunes, 303

Prunus americana, 313
Prunus americana Marsh, 304
Prunus amygdatus, 314
Prunus avium. See Sweet cherry
Prunus besseyi Bailey, 304
Prunus canescens, 356
Prunus domestica, 312, 313. *See*
  European plum
Prunus hortulana Bailey, 304
Prunus incisa, 357
Prunus insititia, 304, 313
Prunus kurilensis, 357
Prunus maritima, 313
Prunus munsoniana Wight & Hedrick,
  304
Prunus nipponica, 357
Prunus persica. See Peach
Prunus salicina, 303
Prunus subcordata Benth, 305
Prunus tomentosa, 313
Pseudomonas morseprunosum, 289
Pseudomonas syringae, 165, 289
Pterochlorus persicae, 291
Pyrus communis, 129
Pyrus pyrifolia, 123
Pyrus ussuriensis, 123

Q

Quadraspidiotus perniciosus, 89, 170,
  234, 326–327
Quince. *See also* Quince fruit
  area and production, 185–186
  botany, 191–193
  breeding and crop improvement,
    197–201
  climate for, 201–202
  composition and uses, 187–189
  cultivars, 194–196
  flowering and fruit set, 213–220
  flowering of, 210
  forms of, 196
  in imperial courts and ceremonies, 186
  intercropping, 213

interculture, 213
iron chlorosis, 208
layout and planting, 204–206
macro- and micronutrients concentrations in, 208
major production areas of, 191
marketing and trade, 186–187
mean food values for minerals and vitamins, 187
nutrient management in, 208–209
origin and distribution, 189–191
overview, 184–185
propagation of, 202–204
protected/high-tech cultivation, 226
rootstock, 202–204
soil, 201–202
stages of fruit growth and maturation, 219
taxonomy, 191–193
training and pruning, 210–212
varieties, 194–196
westward and eastward spread of, 190
Quince disease, 230
affect parts of, 230–231
collar rot, 232
fire blight, 231
pests, 234
physiological disorders, 234–235
viral, 232–233
Quince fruit, 213–220
growth, development, and ripening, 220–223
harvesting and yield, 224–226
packaging and transport, 226–228
postharvest handling and storage, 228–229
processing and value addition, 229–230
retention and fruit drop, 223–224

R

Readily Available Water (RAW), 42
Red Anjou, 121

Red Chief, 19
Red Delicious, 18–19
Reine Claude, 303–304
Replant disease, 82
Roasted chestnut, 549
Rootstocks in plums
American Plum, 313
marianna, 312
myrobalan, 311
Prunus amygdatus, 314
Prunus domestica, 312
Prunus insititia selection, 313
Romanian rootstocks, 313–314
Rootstock types for apple
dwarf, 36
seedling, 37
semidwarf, 36
semivigorous, 37
Rubinette, 24

S

San Jose scale, 326–327
Sericulture, 497
Shamrock, 21
Shams, 194
Share of national apple production, 12
Shinseiki, 123
Skeena, 354
Smyrna, 194
Soil management techniques for apple, 51–53
Spartan, 21
Staccato, 354
Stanley, 308
Stem borer, 327
Stemphylium vesicarium, 165
Strawberry
absorption of nutrients by, 467
area and production, 450–451
biggest producers of, 450, 451
botany and taxonomy, 454
breeding and crop improvement, 456–458

Index

climate rquired for, 459–460
composition and uses, 452–453
construction of beds, 463
cultivars, 456
diseases in, 479
extractions of macronutrients, 466
flower and fruit, 455–456
flowering and fruit set, 468–471
under greenhouses, 477
leaves of, 455
marketing and trade, 452
mulching, 475–476
mulch or plastic mulch, 463–464
nutrition facts, 453
origin and distribution, 454
overview, 449–450
pest, 479–480
postharvest handling and storage,
   477–479
propagation methods, 460–462
root system, 454–455
selection and preparation of land,
   462–463
soil required for, 458–459
sowing, 464
species with ploidy and agronomic
   characteristics, 457–458
stem (crown), 455
stolon, 455
tunneling, 476–477
Strawberry fruit, 468–471
abnormal fruits, 469
characteristics and quality, 471–473
harvested at different ripening stages,
   473
harvesting and yield, 474–475
parthenocarpic fruit development,
   471
stages of development, 470
Suntan apple, 22
Sweet cherry
anthocyanin and phenolic contents of,
   344–346

area and production, 334–337
breeding and crop improvement,
   334–358
composition and uses, 338–348
cover crops between tree rows, 373
diseases in, 385–395
exporting and importing countries, 338
flowering, 373–376
intercropping and interculture,
   372–373
layout and planting, 365–367
marketing and trade, 337–338
nutrient concentration in, 369
origin and distribution, 348–349
overview, 334
pH, total soluble solids and titratable
   acidity of, 340–342
production (tons) and area, 336
propagation and rootstock, 361–365
soil and climate requirement for,
   359–361
sugars and organic acids contents in,
   342–343
top countries for production, 335
training and pruning, 369–372
varieties and cultivars, 351–354
Sweet cherry fruit set, 373–376
growth, development, and ripening,
   376–377
harvesting and yield, 379
packaging and transport, 381–382
postharvest handling and storage,
   382–384
processing and value addition, 384–385
protected/high-tech cultivation,
   379–381
retention and fruit drop, 377–378
ripe fruit, 377
stages of fruit growth, 377
Sweet cherry production regions
botany and taxonomy, 349–351
list of descriptors and scales for, 350
tree habit in Portugal and Spain, 351

570   Index

in Turkey, 349
in the USA, 349
Sweetheart, 354

T

Tetranychus urticae, 480
Tetranychus urticae, 327
Tomuri, 427
Top Red, 19
Torsh, 194
Torymus sinensis, 551
Training system for peach
  central leader system, 276–277
  modified leader system/palmette
    system, 277
  open center system, 276
  pillar system, 276
  pruning, 277–278
  Y system/tatura trellis, 276

V

Valsa ceratosperma, 165
Venturiacarpophila, 288
Venturia inaequalis, 78–79
Venturia pirina, 164
Vermouth, 324–325
Verticillium dahlia, 289

W

Wilsonomyces carpophilus, 287–288

X

Xiphinema americanum, 83–84, 84–85

Y

Yellow egg, 304

Z

Zinc deficiency in peach, 274